AIR CONDITIONING PRINCIPLES AND SYSTEMS

A view of the concentrating and tracking solar collectors for the 100,000 square foot corporate headquarters of Honeywell, Inc., in Minneapolis. The collectors serve a solar heating and cooling system that provides over 50% of the building's yearly heating requirements, more than 80% of the cooling, and all of the hot water. (Honeywell, Inc.)

AIR CONDITIONING PRINCIPLES AND SYSTEMS

THIRD EDITION

EDWARD G. PITA

Environmental Control Technology
New York City Technical College
The City University of New York

PRENTICE HALL
Upper Saddle River, New Jersey   Columbus, Ohio

Library of Congress Cataloging-in-Publication Data
Pita, Edward G.
   Air conditioning principles and systems /
  Edward G. Pita.—3rd ed.
      p.   cm.
   Includes bibliographical references and index.
   ISBN 0-13-505306-4 (hc : alk. paper)
   1. Air conditioning.   2. Buildings—Energy conservation.
  I. Title.
  TH7687.P446   1998   97-20014
  697.9′3--dc21   CIP

Editor: Ed Francis
Production Coordination: TSI Graphics
Design Coordinator: Karrie M. Converse
Cover Designer: Rod Harris
Production Manager: Laura Messerly
Marketing Manager: Danny Hoyt

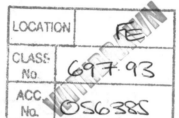
This book was set in Times Roman by TSI Graphics and was printed
and bound by R.R. Donnelley & Sons Company. The cover was printed
by Phoenix Color Corp.

© 1998, 1989, 1981 by Prentice-Hall, Inc.
Simon & Schuster/A Viacom Company
Upper Saddle River, New Jersey 07458

Printed in the United States of America

10 9 8 7 6 5 4 3

ISBN 0-13-505306-4

Prentice-Hall International (UK) Limited, *London*
Prentice-Hall of Australia Pty. Limited, *Sydney*
Prentice-Hall Canada Inc., *Toronto*
Prentice-Hall Hispanoamericana, S.A., *Mexico*
Prentice-Hall of India Private Limited, *New Delhi*
Prentice-Hall of Japan, Inc., *Tokyo*
Simon & Schuster Asia Pte. Ltd., *Singapore*
Editora Prentice Hall do Brasil, Ltda., *Rio de Janeiro*

PREFACE

This third edition of *Air Conditioning Principles and Systems* has been significantly revised. Reflecting recent developments and concerns in the industry, substantial material has been added on indoor air quality, air pollution from combustion, and the new environmental requirements on refrigerants. Consistent with the overall philosophy of this text, the practical approach to these important issues will enable the reader to effectively address them in the workplace.

With the advent of the personal computer, much software has been written to greatly simplify certain HVAC tasks. Applicable software is referred to where relevant, and a list of some sources is presented in a Software Bibliography.

In addition to new material, many chapters have been considerably revised or amplified to enhance the learning process.

This book is a fundamental text in heating, ventilation, and air conditioning (HVAC). It fills the need for a text that presents the fundamental principles and systems in a manner that is technically accurate, yet of practical use in the real working world. Today's reality, which mandates time and cost effectiveness in HVAC work, dictates this practical approach. Students in air conditioning

and refrigeration courses in college and technical institute programs, and consulting engineers, contractors, operating engineers, and service technicians will find this text useful in their studies or as a reference.

The book is designed for a two-semester course. Supplemental work may be assigned if the instructor wishes to expand on the suggested projects.

The text begins by developing the fundamental principles of air conditioning and follows this discussion by describing equipment and systems. The text emphasizes the application of theory to both the design of new systems and the troubleshooting of existing ones. This approach is enhanced by many illustrative examples and problems dealing with real situations.

An underlying theme throughout the book is energy utilization and conservation. Energy codes and standards are described, and each topic is examined from an energy conservation viewpoint, an approach that is essential for all future work in the air conditioning field. A chapter is devoted to solar heating and cooling.

Following an overview of the scope of air conditioning, the text reviews physical principles. Heating and cooling load calculations are explained in a thorough yet understandable manner. The latest methods (now required by most states) are used. Load calculation forms are furnished to aid the student. The subject of psychrometrics is presented in considerable detail, recognizing that it is at the heart of understanding air conditioning processes.

Air conditioning and refrigeration equipment and systems are covered thoroughly. Equipment construction and selection are described. Included in the discussion are reheat, dual duct, multizone, hydronic, and variable air volume systems. The presentation of refrigeration includes an explanation of absorption systems and heat pumps.

Instrumentation and balancing and the fundamentals of automatic controls are covered in separate chapters. Of special importance is the chapter devoted to energy utilization and conservation in design, installation, and operation of air conditioning systems.

Two example projects in the design of a heating and cooling system are worked out in detail. Similar projects are suggested as hands-on learning experiences. These should be of value to those who are interested in installation, operation, and service as well as design, because they require the student to analyze how the system functions.

The author sincerely hopes that this presentation, based on his more than 50 years of experience in the field working for manufacturers, as a consulting engineer, and as an educator, will contribute to your knowledge and success in the HVAC industry.

ABOUT THE AUTHOR

Edward G. Pita is Professor Emeritus and Adjunct Professor in the Environmental Control Technology Department at New York City Technical College of the City University of New York. He received a B.S. degree from Purdue University, an M.S. degree from Columbia University, and a Ph.D. degree from the University of Maryland, all in mechanical engineering. He is a member of the American Society of Heating, Refrigerating and Air-Conditioning Engineers (ASHRAE) and is a registered professional engineer.

In addition to his career as an educator, Dr. Pita was chief mechanical engineer for a large consulting engineering firm responsible for HVAC projects for the United Nations, the State City of the Vatican, the U.S. Capitol, and many other governmental and private clients.

He has also worked in applications and systems engineering for the Carrier Corporation and the Worthington Corporation.

CONTENTS

An Air Conditioning Fable xv

1 THE SCOPE AND USES OF AIR CONDITIONING 1

1.1 Scope of Air Conditioning, 2
1.2 Components of Air Conditioning Systems, 3
1.3 All-Water (Hydronic) Air Conditioning Systems, 4
1.4 All-Air Air Conditioning Systems, 5
1.5 Human Comfort, 7
1.6 Comfort Standards, 8
1.7 The HVAC System as Part of the Building Construction Field, 10
1.8 Designing the HVAC System, 10
1.9 Installing the HVAC System, 11
1.10 Operation, Maintenance, and Service of the HVAC System, 12
1.11 Employment in the HVAC Industry, 12
1.12 Description of Job Responsibilities, 13
1.13 Energy Conservation and Computers, 14

Review Questions, 15
Problems, 15

2 PHYSICAL PRINCIPLES 16

2.1 Units, 17
2.2 Conversion of Units, 17
2.3 U.S. and SI Units, 18
2.4 Mass, Force, Weight, Density, and Specific Volume, 18
2.5 Accuracy of Data, 20
2.6 Pressure, 20
2.7 Pressure of a Liquid Column, 22
2.8 Work, Power, and Energy, 25
2.9 Heat and Temperature, 26
2.10 Enthalpy, 27
2.11 The Energy Equation (First Law of Thermodynamics), 28
2.12 Liquids, Vapors, and Change of State, 29
2.13 Saturated Property Tables, 35
2.14 Refrigeration, 35
2.15 Calculation of Sensible and Latent Heat Changes, 36
2.16 Latent Heats of Fusion and Sublimation, 39
2.17 The Ideal (Perfect) Gas Laws, 39

2.18   Energy Utilization (Second Law of Thermodynamics), 40
       Review Questions, 41
       Problems, 42

3   **HEATING LOADS** 44

3.1   The Heating Load, 44
3.2   Heat Transfer, 45
3.3   Rate of Heat Transfer, 46
3.4   Overall Thermal Resistance, 49
3.5   Overall Heat Transfer Coefficient (U), 49
3.6   Heat Transfer Losses: Basement Walls and Floors, 51
3.7   Heat Transfer Losses: Floor on Ground and Floor over Crawl Space, 52
3.8   Infiltration and Ventilation Heat Loss, 54
3.9   Design Conditions, 57
3.10   Room Heat Loss and Room Heating Load, 58
3.11   The Building Net Heating Load, 59
3.12   System Heat Losses, 60
3.13   Summary of Heating Load Calculation Procedures, 61
3.14   Energy Conservation, 64
       Review Questions, 64
       Problems, 64

4   **FURNACES AND BOILERS** 69

4.1   Warm Air Furnaces, 69
4.2   Furnace Controls, 72
4.3   Heating Boilers, 73
4.4   Boiler Controls, 77
4.5   Boiler and Furnace Draft, 78
4.6   Fuels and Combustion, 80
4.7   Gas and Oil Burners, 86
4.8   Flame Safety Controls, 90
4.9   Boiler Applications, 92
4.10   Boiler Rating and Selection, 92
4.11   Boiler Installation, 96
4.12   Energy Use and Efficiency in Boilers and Furnaces, 96

4.13   Energy Conservation, 98
       Review Questions, 98
       Problems, 99

5   **HYDRONIC PIPING SYSTEMS AND TERMINAL UNITS** 100

5.1   Piping Arrangements, 100
5.2   Series Loop, 100
5.3   One-Pipe Main, 102
5.4   Two-Pipe Direct Return, 102
5.5   Two-Pipe Reverse Return, 103
5.6   Combination Arrangements, 104
5.7   Three-Pipe System, 104
5.8   Four-Pipe System, 105
5.9   Hydronic Terminal Units, 105
5.10   Radiators, 106
5.11   Convectors, 106
5.12   Baseboard, 107
5.13   Fin-Tube, 107
5.14   Radiant Panels, 108
5.15   Unit Heaters, 108
5.16   Fan-Coil Units, 109
5.17   Induction Units, 110
5.18   System Water Temperatures and Flow Rates, 111
5.19   Selection of Terminal Units, 112
5.20   System Design Procedure, 113
       Review Questions, 116
       Problems, 116

6   **COOLING LOAD CALCULATIONS** 118

6.1   The Cooling Load, 118
6.2   Cooling Load Calculation Procedures, 118
6.3   Room Heat Gains, 120
6.4   Conduction Through Exterior Structure, 121
6.5   Conduction Through Interior Structure, 128
6.6   Solar Radiation Through Glass, 128
6.7   Design Conditions, 135
6.8   Lighting, 135
6.9   People, 137
6.10   Equipment and Appliances, 138

6.11   Infiltration, 138
6.12   Room Cooling Load, 142
6.13   Room Peak Cooling Load, 142
6.14   Building Peak Cooling Load, 143
6.15   Cooling Coil Load, 144
6.16   Ventilation, 144
6.17. Heat Gain to Ducts, 145
6.18   Fan and Pump Heat, 146
6.19   Duct Air Leakage, 147
6.20   Supply Air Conditions, 147
6.21   Summary of Commercial Cooling
       Load Calculation
       Procedures, 147

RESIDENTIAL COOLING LOADS 150

6.22   Cooling Load from Heat Gain
       Through Structure, 150
6.23   Cooling Load from Heat Gain
       Through Windows, 151
6.24   People and Appliances, 152
6.25   Infiltration and Ventilation, 152
6.26   Room, Building, and Air
       Conditioning Equipment Loads, 154
6.27   Summary of Residential Cooling
       Load Calculation Procedures, 156
6.28   Energy Conservation, 158
       Problems, 158

7   PSYCHROMETRICS 162
7.1   Properties of Air, 162
7.2   Determining Air Properties, 163
7.3   The Psychrometric Chart, 166
7.4   Locating the Air Condition on the
       Chart, 166
7.5   Condensation on Surfaces, 170

AIR CONDITIONING PROCESSES 171

7.6   Process Lines on the Psychrometric
       Chart, 171
7.7   Sensible Heat Change Process
       Calculations (Sensible Heating and
       Cooling), 172
7.8   Latent Heat Change Process
       Calculations (Humidifying and
       Dehumidifying), 175

7.9   Combined Sensible and Latent
       Process Calculations, 177
7.10   The Evaporative Cooling Process and
       the Wet Bulb Temperature, 179
7.11   The Air Mixing Process, 180

PSYCHROMETRIC ANALYSIS OF THE
AIR CONDITIONING SYSTEM 182

7.12   Determining Supply Air
       Conditions, 182
7.13   Sensible Heat Ratio, 183
7.14   The RSHR or Condition Line, 184
7.15   Coil Process Line, 186
7.16   The Complete Psychrometric
       Analysis, 187
7.17   The Contact Factor and Bypass
       Factor, 189
7.18   The Effective Surface
       Temperature, 189
7.19   Reheat, 191
7.20   Part Load Operation and Control, 192
7.21   Fan Heat Gains, 193
       Problems, 193

8   FLUID FLOW IN PIPING AND DUCTS 197
8.1   The Continuity Equation, 197
8.2   The Flow Energy Equation, 199
8.3   Pressure Loss in Closed and Open
       Systems, 201
8.4   Total, Static, and Velocity
       Pressure, 202
8.5   Conversion of Velocity Pressure to
       Static Pressure (Static Regain), 204
8.6   Pressure Loss from Friction in Piping
       and Ducts, 205
8.7   Friction Loss from Water Flow in
       Pipes, 206
8.8   Pressure Loss in Pipe Fittings, 210
8.9   Piping System Pressure Drop, 211
8.10   System Pipe Sizing, 214
8.11   Friction Loss from Airflow in
       Ducts, 216
8.12   Aspect Ratio, 218
8.13   Pressure Loss in Duct Fittings, 219
8.14   Pressure Loss at Fan Inlet and
       Outlet, 230

8.15 Duct System Pressure Loss, 231
8.16 Duct Design Methods, 233
Problems, 236

9 PIPING, VALVES, DUCTS, AND INSULATION 241

9.1 Piping Materials and Specifications, 241
9.2 Fittings and Joining Methods for Steel Pipe, 244
9.3 Fittings and Joining Methods for Copper Tubing, 245
9.4 Valves, 245
9.5 Pressure Regulating and Relief Valves, 246
9.6 Valve Construction, 247
9.7 Valve Selection, 249
9.8 Pipe Expansion and Anchoring, 249
9.9 Vibration, 250
9.10 Pipe Insulation, 252
9.11 The Piping Installation, 253
9.12 Duct Construction, 253
9.13 Duct Insulation, 254
Review Questions, 255

10 FANS AND AIR DISTRIBUTION DEVICES 256

10.1 Fan Types, 256
10.2 Fan Performance Characteristics, 257
10.3 Fan Selection, 258
10.4 Fan Ratings, 259
10.5 System Characteristics, 263
10.6 Fan–System Interaction, 264
10.7 System Effect, 265
10.8 Selection of Optimum Fan Conditions, 265
10.9 Fan Laws, 266
10.10 Construction and Arrangement, 267
10.11 Installation, 268
10.12 Energy Conservation, 269

AIR DISTRIBUTION DEVICES 269

10.13 Room Air Distribution, 270
10.14 Air Patterns, 270
10.15 Location, 271

10.16 Types of Air Supply Devices, 272
10.17 Applications, 274
10.18 Selection, 275
10.19 Accessories and Duct Connections, 279
10.20 Return Air Devices, 280
10.21 Sound, 280
10.22 Sound Control, 281
Review Questions, 283
Problems, 283

11 CENTRIFUGAL PUMPS, EXPANSION TANKS, AND VENTING 285

11.1 Types of Pumps, 285
11.2 Principles of Operation, 285
11.3 Pump Characteristics, 286
11.4 Pump Selection, 289
11.5 System Characteristics, 291
11.6 System Characteristics and Pump Characteristics, 291
11.7 Pump Similarity Laws, 293
11.8 Pump Construction, 293
11.9 Net Positive Suction Head, 297
11.10 The Expansion Tank, 297
11.11 System Pressure Control, 298
11.12 Compression Tank Size, 300
11.13 Air Control and Venting, 301
11.14 Energy Conservation, 302
Review Questions, 302
Problems, 303

12 AIR CONDITIONING SYSTEMS AND EQUIPMENT 304

12.1 System Classifications, 304
12.2 Zones and Systems, 305
12.3 Single Zone System, 305
12.4 Reheat System, 307
12.5 Multizone System, 308
12.6 Dual Duct System, 309
12.7 Variable Air Volume (VAV) System, 311
12.8 All-Water Systems, 313
12.9 Air-Water Systems, 313
12.10 Unitary versus Central Systems, 314
12.11 Room Units, 314

12.12   Unitary Air Conditioners, 315
12.13   Rooftop Units, 316
12.14   Air Handling Units, 316
12.15   Cooling and Heating Coils, 317
12.16   Coil Selection, 318
12.17   Air Cleaning Devices (Filters), 319
12.18   Methods of Dust Removal, 319
12.19   Methods of Testing Filters, 320
12.20   Types of Air Cleaners, 321
12.21   Selection of Air Cleaners, 322
12.22   Indoor Air Quality, 323
12.23   Energy Requirements of Different Types of Air Conditioning Systems, 324
12.24   Energy Conservation, 328
        Review Questions, 328
        Problems, 328

13   REFRIGERATION SYSTEMS AND EQUIPMENT 330

VAPOR COMPRESSION REFRIGERATION SYSTEM 331

13.1   Principles, 331
13.2   Equipment, 332
13.3   Evaporators, 332
13.4   Types of Compressors, 333
13.5   Reciprocating Compressor, 333
13.6   Rotary Compressor, 334
13.7   Screw (Helical Rotary) Compressor, 334
13.8   Scroll Compressor, 335
13.9   Centrifugal Compressor, 335
13.10   Capacity Control of Compressors, 335
13.11   Prime Movers, 336
13.12   Condensers, 336
13.13   Flow Control Devices, 337
13.14   Safety Controls, 338
13.15   Packaged Refrigeration Equipment, 339
13.16   Selection, 340
13.17   Energy Efficiency, 341
13.18   Installation of Refrigeration Chillers, 346
13.19   Cooling Towers, 346

ABSORPTION REFRIGERATION SYSTEM 348

13.20   Principles, 348
13.21   Construction and Performance, 350
13.22   Special Applications, 351
13.23   Capacity Control, 351
13.24   Crystallization, 352
13.25   Installation, 352

THE HEAT PUMP 352

13.26   Principles, 352
13.27   Energy Efficiency, 354
13.28   Selection of Heat Pumps— The Balance Point, 355
13.29   Solar Energy–Heat Pump Application, 358
13.30   Refrigerants, 358
13.31   Ozone Depletion, 359
13.32   Refrigerant Venting and Reuse, 360
13.33   Global Warming Potential, 361
13.34   Water Treatment, 361
13.35   Energy Conservation in Refrigeration, 361
        Review Questions, 361
        Problems, 362

14   AUTOMATIC CONTROLS 363

14.1   Understanding Automatic Controls, 363
14.2   Purposes of Controls, 364
14.3   The Control System, 364
14.4   Closed-Loop (Feedback) and Open-Loop Control Systems, 366
14.5   Energy Sources, 367
14.6   Component Control Diagram, 367
14.7   Types of Control Action, 368
14.8   Controllers, 371
14.9   Controlled Devices, 374
14.10   Choice of Control Systems, 375
14.11   Control from Space Temperature, 376
14.12   Control from Outdoor Air, 377
14.13   Control from Heating/Cooling Medium, 379

14.14 Humidity Control, 380
14.15 Complete Control Systems, 380
Review Questions, 383
Problems, 383

15 ENERGY UTILIZATION AND CONSERVATION 385

15.1 Energy Standards and Codes, 386
15.2 Sources of Energy, 389
15.3 Principles of Energy Utilization, 390
15.4 Measuring Energy Utilization in Power-Producing Equipment (Efficiency), 391
15.5 Measuring Energy Conservation in Cooling Equipment—The COP and EER, 393
15.6 Measuring Energy Conservation in the Heat Pump, 395
15.7 Measuring Energy Conservation in Heating Equipment, 395
15.8 Measuring Energy Conservation in Pumps and Fans, 396
15.9 Measuring Energy Use in Existing Building HVAC Systems, 397
15.10 Measuring Energy Use in New Building HVAC Systems, 397
15.11 The Degree Day Method, 398
15.12 Other Energy Measuring Methods, 400
15.13 Air-to-Air Heat Recovery, 401
15.14 Refrigeration Cycle Heat Recovery, 403
15.15 Thermal Storage, 404
15.16 Light Heat Recovery, 405
15.17 Total Energy Systems, 405
15.18 Energy Conservation Methods, 406
15.19 Building Construction, 407
15.20 Design Criteria, 407
15.21 System Design, 408
15.22 Controls, 408
15.23 Installation, 409
15.24 Operation and Maintenance, 409
15.25 Computers in HVAC Systems, 410
Problems, 411

16 INSTRUMENTATION, TESTING, AND BALANCING 416

16.1 Definitions, 417
16.2 Instrumentation, 417
16.3 Temperature, 417
16.4 Pressure, 419
16.5 Velocity, 420
16.6 Flow Rates, 422
16.7 Heat Flow, 424
16.8 Humidity, 424
16.9 Equipment Speed, 425
16.10 Electrical Energy, 425
16.11 Testing and Balancing, 425
16.12 Preparation for Air System Balancing, 425
16.13 The Air System Balancing Process, 427
16.14 Preparation for Water System Balancing, 427
16.15 The Water System Balancing Process, 428
16.16 Energy Conservation, 429
16.17 Sound Measurement, 429
Review Questions, 429
Problems, 429

17 PLANNING AND DESIGNING THE HVAC SYSTEM 431

17.1 Procedures for Designing a Hydronic System, 431
17.2 Calculating the Heating Load, 433
17.3 Type and Location of Terminal Units, 436
17.4 Piping System Arrangement, 436
17.5 Flow Rates and Temperatures, 436
17.6 Selection of Terminal Units, 438
17.7 Pipe Sizing, 439
17.8 Piping or Duct Layout, 439
17.9 Pump Selection, 440
17.10 Boiler Selection, 440
17.11 Compression Tank, 442
17.12 Accessories, 442
17.13 Controls, 443
17.14 Plans and Specifications, 443
17.15 Energy Use and Conservation, 444

17.16   Procedures for Designing an All-Air System, 444
17.17   Calculating the Cooling Load, 444
17.18   Type of System, 449
17.19   Equipment and Duct Locations, 449
17.20   Duct Sizes, 449
17.21   Air Distribution Devices, 451
17.22   Equipment, 451
17.23   Accessories, 452
17.24   Automatic Control System, 452
17.25   Plans and Specifications, 453
17.26   Energy Conservation, 454
        Problems, 454

18   SOLAR HEATING AND COOLING SYSTEMS 455

18.1   Solar Collectors, 455
18.2   Storage and Distribution Systems, 457
18.3   Types of Solar Heating Systems, 458
18.4   Solar Cooling Systems, 459
18.5   Solar Radiation Energy, 460
18.6   Insolation Tables, 461
18.7   Clearness Factor, 462
18.8   Orientation and Tilt Angles, 467
18.9   Sunshine Hours, 468
18.10   Collector Performance, 468
18.11   Sizing the Collector, 471
18.12   Economic Analysis, 472
18.13   Storage System Sizing, 473
18.14   Approximate System Design Data, 476
18.15   Passive Solar Heating Systems, 477
        Problems, 477

Bibliography 481

Appendix 483

Table A.1   Abbreviations and Symbols, 483
Table A.2   Unit Equivalents (Conversion Factors), 485
Table A.3   Properties of Saturated Steam and Saturated Water, 486
Table A.4   Thermal Resistance R of Building and Insulating Materials, 487
Table A.5   Thermal Resistance R of Surface Air Films and Air Spaces, 490
Table A.6   Typical Building Roof and Wall Construction Cross-Sections and Overall Heat Transfer Coefficients, 491
Table A.7   Overall Heat Transfer Coefficient U for Building Construction Components, 494
Table A.8   Overall Heat Transfer Coefficient U for Glass, 496
Table A.9   Outdoor Design Conditions, 497
Figure A.1 Room Heating Load Calculations Form, 502
Figure A.2 Building Heating Load Calculations Form, 503
Figure A.3 Commercial Cooling Load Calculations Form, 504
Figure A.4 Residential Cooling Load Calculations Form, 505
Figure A.5 Psychrometric Chart, U.S. Units, 506
Figure A.6 Psychrometric Chart, SI Units, 507

Index 509

AN AIR CONDITIONING FABLE

It was a typical record-breaking July heat wave and the humidity felt like a Turkish bath. Suddenly the air conditioning system in the gigantic Acme Towers office building stopped operating. Within minutes, temperatures in the offices reached 95 F. The building did not have operable windows that could be opened to relieve the oppressive heat. Computers broke down, employees started to leave, and tenants threatened lawsuits for damages.

The building operating staff became frantic. No one knew what to do. Finally one person said, "Listen, there's a fellow named Joe Schlepper who knows an awful lot about air conditioning and refrigeration, so why don't we call him?" In desperation, the chief engineer agreed.

A few minutes later, Joe Schlepper entered the building machine room, walked around, and looked at the complex installation capable of delivering 8000 tons of refrigeration, muttered "hmm," took out a small hammer and tapped a valve. Immediately the whole plant started functioning and soon conditions in the building were comfortable again.

The building manager thanked Joe and asked him what the bill was. The answer was "$2005." "What!" the manager exclaimed. "$2005 for tapping a valve?"

"The bill for tapping the valve is $5," Joe answered, "the $2000 is for knowing which valve to tap."

CHAPTER 1

The Scope and Uses of Air Conditioning

For prehistoric people, open fires were the primary means of warming their dwellings; shade and cool water were probably their only relief from heat. No significant improvements in humankind's condition were made for millions of years. The fireplaces in the castles of medieval Europe were hardly an improvement—they only heated the area immediately around them. Paintings from those times show that the kings and queens wore furs and gloves indoors in winter!

There were a few exceptions to this lack of progress. The ancient Romans had remarkably good radiant heating in some buildings, which was achieved by warming air and then circulating it in hollow floors or walls. In the dry climate of the Middle East, people hung wet mats in front of open doorways and achieved a crude form of evaporative air cooling. In Europe, Leonardo da Vinci designed a large evaporative cooler (Figure 1.1 on page 2).

The development of effective heating, ventilating, and air conditioning (HVAC), however, was begun scarcely 100 years ago. Central heating systems were developed in the nineteenth century, and summer air conditioning using mechanical refrigeration has grown into a major industry only in the last 60 years. Yet by 1995, HVAC systems in the United States had reached a total installed value of about $35 billion yearly, with approximately $15 billion in equipment sales.

A typical person in modern society may spend up to 90% of each day indoors. It is not surprising, therefore, that providing a healthy, comfortable indoor environment has become a major factor in our economy.

OBJECTIVES

A study of this chapter will enable you to:

1. List the environmental conditions that an air conditioning system may control.
2. Describe where air conditioning is used.
3. Sketch the arrangement of the main components of an all-air air conditioning system.
4. Sketch the arrangement of the main components of a hydronic heating and cooling system.
5. Describe the internal environmental conditions that provide adequate human comfort.

Figure 1.1
Ventilator and cooling unit invented by Leonardo da Vinci in the fifteenth century. This air conditioning unit was for the boudoir of Beatrice d'Este, wife of da Vinci's patron, the Duke of Milan. The great wheel, a full story high, stood outside the palace wall and was turned by water power—sometimes assisted by slaves. Valves opened and closed automatically, drawing air into the drum, where it was washed and forced out through the hollow shaft and piped into the room. (Courtesy: IBM Corporation.)

6. Describe the business structure of the HVAC industry, including job opportunities.
7. Describe the organization of the building design team and the construction team.

1.1 SCOPE OF AIR CONDITIONING

To the average person, *air conditioning* simply means "the cooling of air." For our purposes, this definition is neither sufficiently useful nor accurate, so we will use the following definition instead:

> *Air conditioning is the process of treating air in an internal environment to establish and maintain required standards of temperature, humidity, cleanliness, and motion.*

Let us investigate how each of these conditions is controlled:

1. *Temperature.* Air temperature is controlled by heating or cooling* the air.
2. *Humidity.* Air humidity, the water vapor content of the air, is controlled by adding or removing water vapor from the air (*humidification* or *dehumidification*).
3. *Cleanliness.* Air cleanliness, or air *quality*, is controlled by either *filtration*, the removal of undesirable contaminants using filters or other devices, or by *ventilation*, the introduction of outside air into the space which dilutes the concentration of contaminants. Often both filtration and ventilation are used in an installation.
4. *Motion.* Air motion refers to air velocity and to where the air is distributed. It is controlled by appropriate air distributing equipment.

Sound control can be considered an auxiliary function of an air conditioning system, even though the system itself may be the cause of the problem. The air conditioning equipment may produce excessive noise, requiring additional sound attenuating (reducing) devices as part of the equipment.

The definition of *air conditioning* given here is not meant to imply that every HVAC system regulates all of the conditions described. A **hot water** or **steam heating** system, consisting of a boiler, piping, and radiation devices (and perhaps a pump) only controls air temperature, and only during the heating season. These types of systems are common in many individual homes (residences), apartment houses, and industrial buildings.

A **warm air** system, consisting of a furnace, ducts, and air outlet registers, also controls air temperature in winter only. However, by the addition of a humidifier in the ducts, it may also control humidity in winter. Warm air systems are popular in residences.

Some residences have combination air heating and air cooling equipment that provides control of

*"Cooling" technically means the *removal* of heat, in contrast to heating, the *addition* of heat.

temperature and humidity in both winter and summer. Some degree of control of air quality and motion is provided in air-type heating and cooling systems.

Air conditioning systems used for newer commercial and institutional buildings and luxury apartment houses usually provide year-round control of most or all of the air conditions described. For this reason, it is becoming increasingly popular to call complete HVAC systems *environmental control systems.*

Applications

Most air conditioning systems are used for either *human comfort* or for *process control.* From life experiences and feelings, we already know that air conditioning enhances our comfort. Certain ranges of air temperature, humidity, cleanliness, and motion are comfortable; others are not.

Air conditioning is also used to provide conditions that some processes require. For example, textile, printing, and photographic processing facilities, as well as computer rooms and medical facilities, require certain air temperatures and humidity for successful operation.

1.2 COMPONENTS OF AIR CONDITIONING SYSTEMS

Heat always travels from a warmer to a cooler area (see Section 2.9). In winter, there is a continual heat loss from within a building to the outdoors. If the air in the building is to be maintained at a comfortable temperature, heat must be continually supplied to the air in the rooms. The equipment that furnishes the heat required is called a *heating system.*

In summer, heat continually enters the building from the outside. In order to maintain the room air at a comfortable temperature, this excess heat must be continually removed from the room. The equipment that removes this heat is called a *cooling system.*

An air conditioning system may provide heating, cooling, or both. Its size and complexity may range from a single *space heater* or *window unit*

Figure 1.2
View of Lower Manhattan skyline with the World Trade Center Twin Towers, which have 49,000 tons of refrigeration, enough to air condition a city of 100,000 people. (Courtesy: The Port Authority of New York and New Jersey.)

for a small room to a huge system for a building complex, such as the World Trade Center (Figure 1.2), yet the basic principles are the same. Most heating and cooling systems have at a minimum the following basic components:

1. A *heating source* that adds heat to a fluid (air, water, or steam).
2. A *cooling source* that removes heat from a fluid (air or water).
3. A *distribution system* (a network of ducts or piping) to carry the fluid to the rooms to be heated or cooled.
4. Equipment (fans or pumps) for moving the air or water.
5. Devices (e.g., radiation) for transferring heat between the fluid and the room.

We will start with a brief introduction to the function and arrangement of these major components. These and other components including automatic controls, safety devices, valves, dampers, insulation, and sound and vibration reduction devices will be discussed in more detail in later chapters of the book.

Air conditioning systems that use water as the heating or cooling fluid are called **all-water** or **hydronic systems**; those that use air are called **all-air systems**. A system which uses both air and water is called a **combination** or **air-and-water system**.

1.3   ALL-WATER (HYDRONIC) AIR CONDITIONING SYSTEMS

A typical hydronic heating system is shown in Figure 1.3. Water is heated at the *heat source* (1), usually a hot water boiler. The heated water is circulated by a *pump* (2) and travels to each room through *piping* (3) and enters a *terminal unit* (4). The room air is heated by bringing it into contact

with the terminal unit. Since the water loses some of its heat to the rooms, it must return to the heat source to be reheated.

If steam is used in a heating system, the components work in the same manner, with the exception that a pump is not necessary to move the steam; the pressure of the steam accomplishes this. However, when the steam cools at the terminal unit, it condenses into water and may require a condensate pump to return the water to the boiler.

A hydronic cooling system (Figure 1.4) functions in a similar manner to a hydronic heating system. Water is cooled in refrigeration equipment called a *water chiller* (1). The chilled water is circulated by a *pump* (2) and travels to each room through *piping* (3) and enters a *terminal unit*

Figure 1.3
Arrangement of basic components of a (hydronic) hot water heating system.

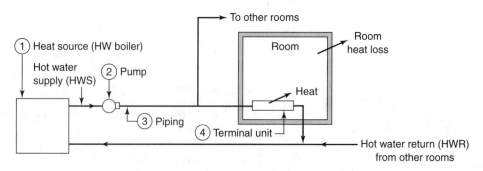

Figure 1.4
Arrangement of basic components of a (hydronic) chilled water cooling system.

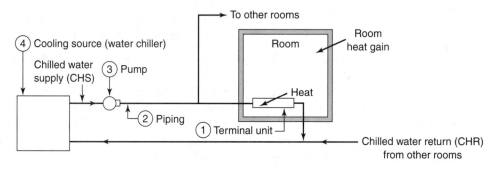

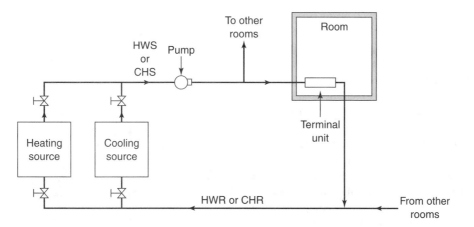

Figure 1.5
Arrangement of basic components of a hydronic heating and cooling system.

(4). The warmer room air loses its heat to the cold water in the terminal unit. Since the water is now warmed, it must return to the water chiller to be recooled.

As the reader may have guessed, hydronic systems are popular for HVAC systems that require both heating and cooling. This is because it is possible to use the same piping system for both by connecting a hot water boiler and water chiller in parallel (Figure 1.5), using each when needed.

1.4   ALL-AIR AIR CONDITIONING SYSTEMS

All-air systems use air to heat or cool rooms. They may also have the added capability of controlling humidity and furnishing outdoor ventilation, which hydronic systems cannot do.

A typical all-air heating and cooling system is shown in Figure 1.6. Air is heated at the heat source (1), such as a furnace. (It may also be a coil

Figure 1.6
Arrangement of basic components of an all-air heating and cooling system (many other arrangements are possible).

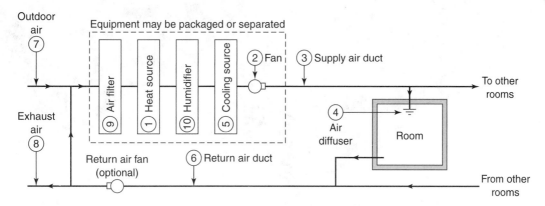

circulating hot water, or steam, heated by a remote boiler.) The heated air is circulated by a *fan* (2) and travels to each room through *supply air ducts* (3). The supply air enters the room through outlets called *air diffusers* or *registers* (4) that are designed to provide proper air distribution in the room. When the warmed supply air enters the room, the room is heated.

A *humidifier* (10) may also be included to maintain a comfortable room humidity in winter.

In summer, air is cooled when it flows over a cooling source (5), usually a coil of tubing containing a fluid cooled by refrigeration equipment (see Chapter 13). When the cooled supply air enters the room, the room is cooled.

Because a room's size is fixed, the same volume of air that enters the room must also exit. This is usually accomplished with *return air ducts* (6). The air is then heated or cooled again, and recirculated. An *outdoor air intake* duct (7) may be provided for introducing fresh outdoor air for increased air quality. Similarly, the same volume of air must be *exhausted* (8). Provisions may be made for cleaning the air with air filters (9), and for humidifying the air (10).

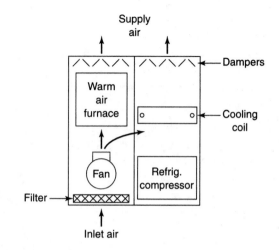

Figure 1.7
Arrangement of components of all-year air conditioning equipment for a private residence (refrigeration condenser separate).

An example of packaged all-air system equipment is shown in Figure 1.7. This arrangement is convenient for residential and light commercial air conditioning.

Figure 1.8
Rooftop-type unitary air conditioning equipment. (Courtesy: McQuay Group, McQuay-Perfex, Inc.)

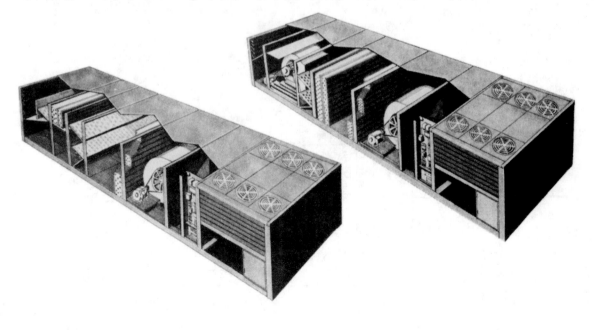

Combination Systems

It is frequently desirable to combine water and air systems. For example, a hydronic system in a central plant might generate hot or chilled water, which is then circulated to heating or cooling coils in large all-air systems in other parts of the building or even to a number of buildings.

Unitary and Central Air Conditioning Systems

A *unitary* or *package* air conditioning system uses equipment where all or most of the basic components have been assembled in the factory (e.g., room air conditioner). An example of all-air unitary equipment mounted on a roof (a "rooftop" unit), such as those used in supermarkets, is shown in Figure 1.8.

A *central* or *built-up* air conditioning system uses equipment centrally located in mechanical equipment rooms. Each piece of equipment is installed separately and connected on the job, rather than manufactured as a package. Figure 1.9 shows a portion of the equipment of a central system. This subject will be discussed in more detail in Chapter 12.

1.5   HUMAN COMFORT

Since the purpose of most air conditioning systems is to provide a comfortable indoor environment, the system designer and operator should understand the factors that affect comfort.

Body Heat Loss

The human body creates heat when it metabolizes (oxidizes) food. This body heat is continually lost to its cooler surroundings. The factor that determines whether one feels hot or cold is the *rate of body heat loss.* When the rate of heat loss is within certain limits, a comfortable feeling ensues. If the rate of heat loss is too great, cold is felt; if the rate is too low, one feels hot.

The processes by which the body loses heat to the surroundings are: *convection, radiation,* and *evaporation.*

Figure 1.9
Mechanical equipment room of a large central station air conditioning system, showing absorption refrigeration machines. (Courtesy: Syska & Hennessy, Inc., Engineers.)

In *convection,* the air immediately around the body receives heat from the body. The warmed air continually moves away, by rising naturally through the cooler air around it, or by being blown away, and is replaced by more air which in turn receives heat.

In *radiation,* body heat is transmitted through space directly to nearby objects (e.g., walls) which are at a lower temperature than the body; this is why it can be uncomfortable to sit near a window or wall in cold weather, even in a warm room. However, heating sources that are warmer than the body can radiate heat toward the body, creating a feeling of warmth even at a low surrounding air temperature; this is why one feels warm in front of a fire even on a cold day. Some restaurants now have glass-enclosed sidewalk cafés with radiant heating panels that keep the customers comfortable in winter even though the café temperature is only about 50 F (10 C).

The body is also cooled by *evaporation:* water on the skin (perspiration), which has absorbed heat from the body, evaporates into the surrounding air, taking the heat with it.

The rate of body heat loss is affected by five conditions:

1. Air temperature
2. Air humidity
3. Air motion
4. Temperature of surrounding objects
5. Clothing

The system designer and operator can control comfort primarily by adjusting three of these conditions: temperature, humidity, and air motion. How are they adjusted to improve comfort?

The indoor air temperature may be raised to decrease body heat loss (winter) or lowered to increase body heat loss (summer) by convection.

Humidity may be raised to decrease body heat loss (winter) and lowered to increase body heat loss (summer) by evaporation.

Air motion may be raised to increase body heat loss (summer) and lowered to decrease body heat loss (winter) by convection.

Occupants of the buildings, of course, have some personal control over their own comfort. For instance, they can control the amount of clothing that they wear, they can use local fans to increase convection and evaporative heat loss, and they can even stay away from cold walls and windows to keep warmer in winter.

Indoor Air Quality

Another factor, *air quality,* refers to the degree of purity of the air. The level of air quality affects both comfort and health. Air quality is worsened by the presence of contaminants such as tobacco smoke and dust particles, biological microorganisms, and toxic gases. Cleaning devices such as filters may be used to remove particles. Adsorbent chemicals may be used to remove unwanted gases. Indoor air contaminants can also be diluted in concentration by introducing substantial quantities of outdoor air into the building. This procedure is called *ventilation.*

The subject of indoor air quality (IAQ) has become of major concern and importance in recent years. Evidence has grown that there are many possible indoor air contaminants which can and have caused serious health effects on occupants. The phrases *sick building syndrome* and *building-related illnesses* have been coined to refer to these effects. Intensive research and amelioration efforts are being carried out in this branch of HVAC work. Indoor air quality will be discussed in Chapter 12.

1.6   COMFORT STANDARDS

Studies of the conditions that affect human comfort have led to the development of recommended indoor air conditions for comfort, published in ASHRAE[*] Standard 55-1992, *Thermal Environmental Conditions for Human Occupancy.* Some of the results of these studies are shown in Figure 1.10.

The shaded regions in Figure 1.10 are called the comfort zones. They show the regions of air temperature and relative humidity where at least 80% of the occupants will find the environment comfortable. Note that there are separate zones for winter and summer, with a slight overlap.

The use of Figure 1.10 is valid only for the following conditions:

1. The comfort zones apply only to sedentary or slightly active persons.
2. The comfort zones apply only to summer clothing of light slacks and a short sleeve shirt, or equivalent (0.5 clo);[**] and winter clothing of heavy slacks, long sleeve shirt, and sweater or jacket, or equivalent (0.9 clo).
3. The comfort zones apply to air motion in the occupied zone not exceeding 30 feet per minute (FPM) in winter and 50 FPM in summer.

---

[*] "ASHRAE" stands for the American Society of Heating, Refrigerating and Air-Conditioning Engineers. A list of reference sources used in this text can be found in the Bibliography.

[**] The *clo* is a numerical unit representing a clothing ensemble's thermal insulation.

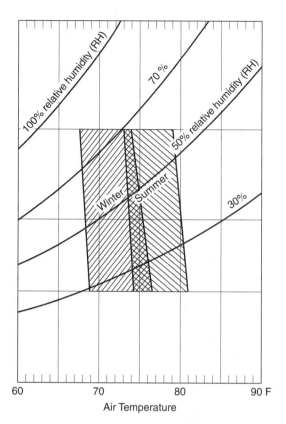

Figure 1.10
Comfort zones of indoor air temperature and relative humidity. These zones apply to persons clothed in typical summer or winter clothing engaged in sedentary activity. (**Adapted with permission from the** *1993 ASHRAE Handbook—Fundamentals.*)

4. The comfort zones apply only under certain conditions of thermal radiation between the occupant and the surroundings. For example, an individual receiving direct solar radiation through a window in summer might feel uncomfortably warm even though the room air temperature and humidity are within the comfort zone.

Although these restrictions may seem to reduce the usefulness of Figure 1.10, this is not so. First, the situations specified are very common (a typical office environment). Furthermore, for changes in conditions, the comfort zones can be adjusted to reflect these changes. The procedures for making these corrections can be found in the ASHRAE Standard.

Applications

In order to use Figure 1.10 to find whether a specific set of conditions is comfortable or not, it is necessary to know the room air temperature and humidity. The air temperature is technically called the *dry bulb temperature* (DB). The humidity is often expressed as the *percent relative humidity* (% RH). See Chapter 7 for a complete definition of these terms.

Example 1.1 _____
The conditions in an office building in the summer are 77 F DB and 50% RH. The occupants are lightly clothed. Air movement in the rooms is about 30 FPM. There is negligible radiation of heat from the surroundings to the occupants. Would this be a comfortable indoor condition?

Solution
From Figure 1.10, the condition noted (the intersection of 77 F DB and 50% RH) is within the summer comfort zone, therefore most of the occupants would feel comfortable.

Air Quality Standards

As mentioned previously, satisfactory indoor air quality is maintained by cleaning the air and by introducing outside air (ventilation). Recommended ventilation requirements are discussed in Chapter 6.

Indoor Design Conditions for Energy Conservation

The comfort zones shown in Figure 1.10 leave a wide range of choices for the air conditioning system designer and operator. In recent years, in an effort to conserve energy, more specific conditions have been recommended (Table 1.1 on page 10). The temperatures listed are at the low end of the comfort zone in winter and at the high end of the comfort zone in summer. These recommendations may not be a matter of choice: most states now mandate energy conserving design conditions. For

TABLE 1.1   RECOMMENDED ENERGY CONSERVING DESIGN CONDITIONS FOR HUMAN COMFORT

	Air Temperature (DB) F	Relative Humidity (RH) %	Maximum Air Velocity* FPM	Clothing Insulation clo
Winter	68–72	25–30	30	0.9
Summer	76–78	50–55	50	0.5

*At occupant level.

example, the New York State Energy Code requires a maximum winter indoor design temperature of 72 F and a minimum summer indoor design temperature of 78 F. (Exceptions may be granted for special situations, on application.) California Energy Standards require indoor design values of 70 F in winter and 78 F in summer.

When buildings are unoccupied on nights, holidays, and weekends, it is common practice to lower indoor air temperatures in winter ("set-back") and raise them in summer ("set-up") either manually or automatically with the control system. These and other energy-saving strategies will be discussed in appropriate places throughout the text.

The values recommended in Table 1.1 apply to general applications such as offices, residences, and public buildings, but there are exceptions. Lower indoor temperatures in winter might be used in department stores when customers are heavily clothed. Higher indoor temperatures in winter may be desirable for small children, senior citizens, and the ill. Other special applications might have different design conditions.

1.7   THE HVAC SYSTEM AS PART OF THE BUILDING CONSTRUCTION FIELD

The student who intends to work in the HVAC industry should have some understanding of how the industry is organized and how it relates to the building construction field, of which it is a part.

The development of an HVAC system for a building consists of a number of steps. These are:

1. Design
2. Installation
3. Operation and regular maintenance
4. Service

We will outline who is responsible for each step, what their tasks are, and how the HVAC system relates to other building systems.

The student is strongly advised, if possible, to locate a proposed building and follow the HVAC system development through planning, installation, and operation. No textbook can substitute for this valuable learning experience. Become a "sidewalk superintendent" for the construction of an urban building, suburban mall, or an industrial or commercial park. Take notes. Ask questions of your instructors. Watch out—the dynamism and excitement can be addictive!

1.8   DESIGNING THE HVAC SYSTEM

The design of a large building project is an extremely complex task. It may take months or even years and involve scores of people. The design of a private residence is much simpler and may involve as few as one or two people.

The design of an HVAC system for large projects is the responsibility of the *mechanical consulting engineers.* The electrical, structural, and plumbing systems are designed by consulting engineers specializing in their respective fields. Consulting engineers may also carry out other duties such as *cost estimating* and *field supervision* of construction. Each of these tasks is performed in cooperation with the *architects,* who carry out the overall building planning and design. An organizational flowchart of this arrangement is shown in Figure 1.11.

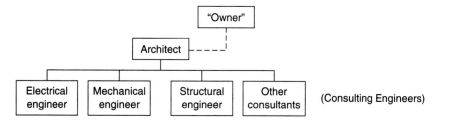

Figure 1.11
Organizational flowchart to a building planning and design team.

Coordination of the work between the architects and engineers is an important and difficult task. This includes checking that the equipment and materials to be installed do not physically interfere with each other. An error in coordination can have disastrous results.

The design of an HVAC system involves determining the type of system to use, calculations of heating and cooling loads (requirements), calculations of piping and duct sizes, selection of the type and size of equipment, and planning the locations of each piece of equipment in the building. This information is shown on the building HVAC *plans* and *specifications,* which serve as instructions on how to install the system. The plans are drawings of the system. The specifications are written descriptions of materials, equipment, and so forth.

1.9 INSTALLING THE HVAC SYSTEM

The overall construction of a building is the responsibility of the *general contractor.* The general contractor is awarded a *contract* by the *owner,* which may be a real estate company, public agency, school system, or other prospective builder. The general contractor may hire *subcontractors* (mechanical, electrical, and so forth) to install each of the building's systems. The subcontractors must coordinate their work to avoid any physical interference. The *mechanical* or *HVAC contractor* is responsible for installing the HVAC system. Figure 1.12 shows a typical organizational flowchart.

The mechanical contractor takes the mechanical consulting engineer's drawings (called *contract* or *engineering drawings*) and then prepares *shop drawings* from these. Shop drawings are larger scale, more detailed drawings of the HVAC system which will be necessary for the workers. The mechanical contractor hires these people, who include pipefitters, sheet metal workers, insulation workers, and other skilled *building trade* workers.

The mechanical contractor also purchases all necessary HVAC equipment and materials. To do this, their employees first carry out a *take-off,* that

Figure 1.12
Organizational flowchart of a building construction team.

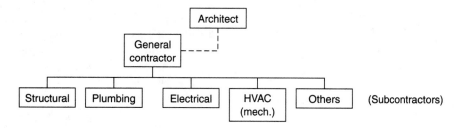

is, they list all the equipment and materials shown on the drawings and specifications. This can be a very involved task. Overhead and labor costs must also be determined.

When the installation is complete, the mechanical contractor tests, adjusts, and balances (TAB) the HVAC system (see Chapter 16). The mechanical consulting engineer may check the installation as it proceeds and may also check the TAB work.

The Design-Build (Fast Tracking) Approach

In contrast to the procedures described, there are companies that handle all of the design and construction functions as a package: architecture, consulting engineering, and contracting. This is called the *design-build* approach. Proponents of this approach claim that construction can start and continue as plans are developed for each stage—it is not necessary to wait for engineering plans and contractor drawings, there is no delay for contractor competitive bidding, better coordination is achieved, and determination of liability is easier, since one organization is responsible for everything. The proponents claim that all of these factors result in lower costs, faster construction, and better quality.

However, proponents of construction projects using an independent architect, consulting engineers, and contractors claim that costs are kept down by competitive bidding, and quality is better because the architects and engineers have independent control over the performance of the contractors.

Both of the two approaches are in common use.

1.10   OPERATION, MAINTENANCE, AND SERVICE OF THE HVAC SYSTEM

When the HVAC installation is complete and after start-up and TAB, the building *operating engineering* staff takes over. Their function is to operate the system, maintaining comfortable conditions in the building while trying to minimize energy consumption, and to keep the system in proper working order.

Regular inspection and maintenance of the system is also part of the operating engineer's duties. Some routine servicing may be performed by the building operating staff, but when more complicated work is required, a *mechanical service contractor* is called in. Using instrumentation, this contractor measures the conditions and compares them with the HVAC system plans and specifications. This troubleshooting procedure leads to the cause of the problem. Proper procedures, such as repairing or replacing equipment, or adjusting its performance, are then carried out.

1.11   EMPLOYMENT IN THE HVAC INDUSTRY

It is helpful for students intending to work in some part of the HVAC industry, and for those already working who wish to advance themselves, to know what types of positions are available and what knowledge, responsibilities, and training are required.

One important basic fact is that a fundamental knowledge of air conditioning principles is required, regardless of whether one is employed in design, installation, operation, or service, in order to succeed in the HVAC field today.

A description of the types of employers and their work, followed by a list of job titles and responsibilities, will aid the student in planning his or her career.

A *mechanical consulting engineer* is a company that designs the heating, ventilating, air conditioning, and plumbing systems for buildings. They estimate costs, perform technical calculations, prepare drawings and specifications, and supervise installations. Positions include: *Project Manager, Designer, Drafter, Inspector, Computer Programmer,* and *Energy Specialist.*

A *mechanical contractor* is a company that installs the system. This includes cost estimates, preparation of drawings, and supervision of installation. Positions include: *Sales Engineer, Inside Representative, Estimator, Contract Manager, Drafter, Purchasing Agent, Field Supervisor, Shop Technician,* and *Field Service Technician.*

A *service* company repairs and maintains the HVAC system. It is often a branch of a mechanical contractor. Positions include: *Sales Engineer, Inside Representative, Estimator, Service Manager, Field Service Engineer and Technician,* and *Shop Technician.*

A *manufacturer* is a company that makes HVAC equipment. This involves production, research and development, marketing, and sales. Positions include: *Sales Engineer, Inside Representative, Sales Manager, Application Engineer, Drafter, Purchasing Agent, Production Supervisor, Field Service Engineer and Technician, Shop Technician, Service Manager, Research and Development Engineer,* and *Technician.*

A *manufacturer's representative* is a company that sells HVAC equipment manufactured by another company. Their work involves sales and technical advice. Positions include *Sales Engineer, Inside Representative, Estimator, Sales Manager,* and *Application Engineer.*

A *building owner* may be a real estate company; a business corporation; city, state, or federal government; public authority; school system; and others. Positions within operations include: *Chief Engineer, Watch Engineer, Computer Operator,* and *Mechanic.* In addition to HVAC operating personnel, large property owners may have a permanent staff which checks and supervises the work of consulting engineers and contractors that do work for them. Positions include: *Application Engineer, Designer, Drafter, Estimator, Field Supervisor.*

1.12 DESCRIPTION OF JOB RESPONSIBILITIES

Project Manager. Supervises design of project for consulting engineer. Instructs designers and drafters, checks calculations and plans, coordinates with other consultants and the architect.

Designer. Performs calculations, selects equipment, plans layout of system and specifications, supervises drafters.

Drafter. Prepares drawings with supervision. May assist in design work.

Inspector. Inspects the system installation during construction to check conformity with plans and specifications. Checks performance after testing and balancing.

Computer Programmer. Processes work to be handled by computer, such as load calculations and energy studies.

Energy Specialist. Prepares energy use analyses and conservation studies, studying varied alternatives.

Sales Engineer. Sells equipment and installation and service contracts. Furnishes technical advice to customers.

Inside Representative. Processes sales and orders by phone and correspondence. Furnishes product information and prices.

Estimator. Uses plans and specifications to determine quantity of materials, labor, and equipment for project. Prepares costs from this data.

Contract Manager. Supervises contract. Responsible for costs, time schedule, and installation.

Purchasing Agent. Orders and purchases materials and equipment. Checks technical characteristics, follows up delivery time.

Field Supervisor. Supervises installation technically. Checks comformity with drawings and specifications and resolves problems of conflict.

Service Manager. Supervises the service installation and contract.

Field Service Engineer and Technician. Determines solutions to problems (troubleshooting), obtains materials and equipment, and directs service work.

Shop Technician. Responsible for assembly or fabrication done in shop (e.g., sheet metal duct parts) or shop service and repair.

Sales Manager. Supervises the sales and marketing of a line of products for a manufacturer.

Application Engineer. Assists consulting engineer or contractor to provide technical information and aid in selection of proper equipment.

Production Supervisor. Responsible for fabrication of equipment in factory. Supervises technical work of employees, inspection, and quality.

Research and Development Engineer and Technician. Plan, develop, and test new types of equipment.

Chief Operating Engineer. Supervises the operation and maintenance of the building system. Determines method of operation for comfort and energy conservation, plans maintenance routines, and directs work of operating personnel.

Watch Engineer. Responsible for operation and maintenance of the building system under supervision.

Computer Operator. Responsible for operations of computerized building systems.

These job descriptions do not mean that a separate individual always performs each task. Often, particularly in small companies, one person may be responsible for a number of jobs if there is not enough work to employ people in each category.

The amount of education and training required for each job varies both with the type of responsibility and how complex it is. For most of the categories, at least a technical institute or community college program in air conditioning and refrigeration is required, preferably equivalent to two years of study. A bachelor of engineering degree is needed for some of the categories and for improving opportunities for advancement. Further on-the-job training is also extremely valuable.

Licensing

Operating licenses are required by local laws for those responsible for operating many categories of refrigeration equipment, boilers, and incinerators. A *professional engineering (PE) license* is required for those responsible for preparing the engineering design drawings. A combination of education, experience, and examinations (usually supervised by

each state) determine the granting of these licenses. National certification standards have been developed for HVAC technicians. National certification licensing is now required by the U.S. Environmental Protection Agency (EPA) for refrigerant handling; that is, the recovery, recycling, and reclaiming of refrigerants (see Chapter 13).

1.13 ENERGY CONSERVATION AND COMPUTERS

Energy conservation and the use of computers have become such important aspects of the air conditioning industry that they merit a special emphasis at the beginning of our studies.

The effort to conserve energy and reduce costs has revolutionized the design and operation of air conditioning systems and equipment. Perhaps the biggest change has been in the use of computers. Computer usage in the industry has spread to such an extent that many, if not most, large air conditioning installations are now designed with the aid of computers. The selection of the equipment is done by computer software. Furthermore, virtually all large and many medium-sized new installations are operated through computers.

Computerized inventory, parts ordering, and accounting are now standard practices for most mechanical contractors. Through application of computer graphics and networks, the office or field engineer or technician can visually observe equipment parts, their arrangement in equipment, and installation procedures. Undoubtedly interactive computer graphics and text and voice communication will result in still more efficient design, installation, and service of HVAC systems in the future.

The *Internet* has also become an important tool in the HVAC industry. Many equipment manufacturers have sites on the *World Wide Web* which offer technical data, selection procedures, and specifications for their products (*electronic catalogs*). With a personal computer, modem, and appropriate software, the HVAC practitioner can access information using a point-and-click interface. To answer individual questions, manufacturers may also offer technical support via *electronic mail*.

Students who wish to be successful in the air conditioning field should consciously approach each task as to how it can be solved in an energy-efficient manner; students should also gain some understanding of computer application and programming (see Chapter 15 for a further discussion). Emphasis on energy conservation and computer applications are placed at the appropriate places within the text.

Review Questions

1. What are the two primary situations in which air conditioning is needed?

2. List the four conditions that an air conditioning system may be required to control.

3. What two methods may be used to improve air quality?

4. List the four major components of any air conditioning system. Sketch a diagram that shows their arrangement.

5. What are the major components of a hydronic heating system and a hydronic cooling system?

6. Sketch a typical all-air air conditioning system and name each component.

7. What are the indoor environmental conditions that affect human comfort?

8. Using a sketch, describe the building design team's organization and responsibilities.

9. Using a sketch, describe the building construction team's organization and responsibilities.

10. What do the terms *design-build* and *fast tracking* mean?

Problems

1.1 Sketch an environmental control system that provides heating and ventilating, but not cooling or humidity control. Label all components.

1.2 As the operating engineer of an HVAC system in a large office building, you have been instructed to raise the summer thermostat setting from 76 F to 80 F to conserve energy. Prepare a list of suggestions you might give to the building's occupants on how to minimize their decrease in comfort.

1.3 In a department store, which should be the more comfortable summer condition: 80 F DB and 40% RH or 78 F DB and 70% RH? Explain.

1.4 The conditions in an office are 70 F DB and 40% RH. Would the occupants be comfortable in winter? Would they be comfortable in summer? Explain.

1.5 The conditions in an office in summer are 75 F DB and 50% RH. Should the conditions remain as they are or should they be changed? Explain. What changes should be made, if any? Explain.

1.6 The conditions in an office in winter are 77 F DB and 10% RH. Are these conditions acceptable? Explain. What changes should be made, if any? Explain.

1.7 Select two HVAC careers that interest you. List the subjects discussed in this chapter that you think are important to learn in training for these positions. When you have completed the book, prepare a new list and compare it with this one.

CHAPTER 2
Physical Principles

The HVAC practitioner often encounters problems that cannot be solved without a knowledge of applied physics. In this chapter, the physical principles that are useful in understanding air conditioning will be explained. (One subject in applied physics, fluid flow, will be discussed in Chapter 8.) This presentation of applied physics is not intended to substitute for a course in physics; a background in that subject will be helpful as a preparation for this book.

Generally the definitions and concepts accepted today in physics will be used here, however, in some instances terms that are in practice in the HVAC industry will be used, even when they differ slightly in meaning. This approach will enable the student to communicate and work with others in the air conditioning field.

OBJECTIVES

A study of this chapter will enable you to:

1. Identify units and convert from one set of units to another.
2. Calculate density, specific volume, and specific gravity.
3. Express the relationship between pressure and head, and between absolute, gage, and vacuum pressure.
4. Distinguish between energy and power, and between stored energy and energy in transfer.
5. Explain the difference between temperature, heat, and enthalpy and show the relationship between temperature scales.
6. Describe and use the energy equation.
7. Identify the changes that occur when a substance changes between its liquid and vapor states.
8. Use the saturated property tables for water and the sensible and latent heat equations.
9. Make some general conclusions regarding energy conservation in HVAC.

16

2.1 UNITS

Concepts such as length, area, volume, temperature, pressure, density, mass, velocity, momentum, and time are called *physical characteristics*. Physical characteristics are measured by standard quantities called **units**. For instance, the foot (ft) is one of the standard units used to measure length.

For each physical characteristic, there are many different units. These units have fixed numerical relationships to each other called *equivalents* or *conversion factors*. Examples of equivalents are:

Characteristic	Unit Equivalents (Conversion Factors)
Length	1 ft = 12 inches (in.)
	= 0.30 meters (m)
Volume	1 ft^3 = 7.48 gallons (gal)
Time	1 minute (min) = 60 seconds (sec)
Mass	2.2 pounds (lb) = 1 kilogram (kg)

Table A.2 in the Appendix lists some useful unit equivalents. Table A.1 lists abbreviations and symbols used in this book.

2.2 CONVERSION OF UNITS

The equivalence between any two units can also be written as a ratio, by dividing both sides of the equality by either term. For instance, from Table A.2, the relation between area expressed in ft^2 and in.^{2} is 1 ft^2 = 144 in.^{2}. Dividing both sides by 144 in.^{2} gives

$$\frac{1 \text{ ft}^2}{144 \text{ in.}^2} = \frac{144 \text{ in.}^2}{144 \text{ in.}^2} = 1$$

Or, dividing by 1 ft^2 gives

$$\frac{1 \text{ ft}^2}{1 \text{ ft}^2} = \frac{144 \text{ in.}^2}{1 \text{ ft}^2} = 1$$

This shows that multiplying by the ratio of equivalent units is the same as multiplying by 1. This enables us to change units, as is now explained.

This ratio arrangement is used when it is desired to change a quantity expressed in one unit into a different unit. The procedure is carried out in the following manner:

1. Arrange the equivalency (conversion factor) between the units as a ratio, choosing that ratio that will give the results in the desired units, by canceling units that are the same in the numerator and denominator (units can be multiplied and divided in the same way as numbers).
2. Multiply the original quantity by the ratio. The result will be the correct value in the new units.

The following example illustrates the procedure for converting units.

Example 2.1

Some solar heating collector panels measuring 28 in. by 33 in. require insulation. The insulation is to be ordered in square feet (ft^2). How much insulation would you order for each panel?

Solution

The area of the insulation for each panel is

$$\text{Area} = 28 \text{ in.} \times 33 \text{ in.} = 924 \text{ in.}^2$$

The area is not in the units needed, however. The equivalent between the known and required units is 1 ft^2 = 144 in.^{2} (Table A.2). Here this is arranged as a ratio, multiplied by the original quantity, and units cancelled:

$$\text{Area} = 924 \text{ in.}^2 \times \frac{1 \text{ ft}^2}{144 \text{ in.}^2} = 6.42 \text{ ft}^2$$

This is the amount of insulation required for each panel, expressed in ft^2.

An important point to note in this example is that there are always two possible ratios that can be used in converting units. In that case, it was either

$$\frac{1 \text{ ft}^2}{144 \text{ in.}^2} \quad \text{or} \quad \frac{144 \text{ in.}^2}{1 \text{ ft}^2}$$

Only one can be correct. Suppose the other ratio had been used. This results would be

$$\text{Area} = 924 \text{ in.}^2 \times \frac{144 \text{ in.}^2}{1 \text{ ft}^2} = 133{,}000 \, \frac{\text{in.}^4}{\text{ft}^2}$$

We know that this is incorrect, because the units resulting are not ft². Imagine what your boss would have said if you had ordered 133,000 ft² of insulation, when he saw it! The student should adopt the habit of always writing out the unit names when doing computations.

The procedure for changing units is the same when more than one unit is to be changed, as seen in the following example.

Example 2.2

A U.S. manufacturer ships some air filters to Venezuela, with the note "Warning—maximum air velocity 600 ft/min." The contractor installing the filters wishes to inform the operating engineer what the maximum velocity is in meters per second (m/sec). What information should be given?

Solution

We must use the equivalency between feet and meters and that between minutes and seconds. Arrange the ratios in the form that will give the correct units in the result, multiply and divide values and units:

$$\text{Velocity} = 600 \; \frac{\cancel{ft}}{\cancel{min}} \times \frac{1 \; \cancel{min}}{60 \; sec} \times \frac{0.30 \; m}{1 \; \cancel{ft}}$$

$$= 3.0 \; \frac{m}{sec}$$

Combined Conversion Factors

In Example 2.2, the problem involved converting velocity from units of ft/min to m/sec. It is convenient to use combined conversion factors such as this for calculations that are frequently repeated. The following example shows how one is developed.

Example 2.3

Find the equivalence for the flow rate of water measured in units of lb/hr and gal/min (GPM), at 60 F.

Solution

From Table 2.1, the density of water is 62.4 lb/ft³ at 60 F. From Table A.2, 7.48 gal = 1 ft³. Using these values for water,

$$1 \; \frac{gal}{min} = \left(1 \; \frac{\cancel{gal}}{\cancel{min}} \times 60 \; \frac{\cancel{min}}{hr} \times \frac{1 \; \cancel{ft^3}}{7.48 \; \cancel{gal}} \times 62.4 \; \frac{lb}{\cancel{ft^3}} \; \text{water} \right) = 500 \; \frac{lb}{hr} \; \text{water}$$

That is, 1 GPM = 500 lb/hr (for water only, at about 60 F).

2.3 U.S. AND SI UNITS

There are two systems of units used in the HVAC industry. One is called the *inch-pound, U.S.,* or *English* system; the other is called the *SI* or international system. SI units are part of a broader system of units called the metric system. The inch-pound (I-P) system of units is generally used in the United States, whereas SI units are used in most other countries. In this book, U.S. units will be emphasized. However, SI units will be introduced in two ways: (1) in some examples and tables, units will be converted between U.S. and SI units; and (2) some examples and problems will be done completely in SI units. In this way, those students who wish to become familiar with SI units may do so.

Only certain units of the metric system are standard in the SI system. The SI system of units uses only one unit of measurement for each physical characteristic. For instance, the standard SI unit of length is the meter, not the centimeter or kilometer. However, occasionally we may use metric units that are not standard SI units, because this is common practice in the HVAC industry in countries using the SI system. Table A.2 includes conversion factors for both U.S. and SI units.

2.4 MASS, FORCE, WEIGHT, DENSITY, AND SPECIFIC VOLUME

The **mass** (*m*) of an object or body is the quantity of matter it contains. The U.S. unit of mass is the pound mass. The SI unit is the kilogram (kg).

A **force** is the push or pull that one body may exert on another. The U.S. unit of force is the pound force. The SI unit is the Newton (N).

The **weight** *(w)* of a body is the force exerted on it by the gravitational pull of the earth. That is, weight is a force, not mass. Unfortunately the word *weight* is often used for mass of a body. The confusion also occurs because the word *pound* is used for both mass and force in U.S. units. However, the numerical value in pounds (lb) for the mass and weight of an object is the same on earth, therefore no error should occur in calculations. In any case the nature of a problem indicates whether mass or weight is being considered.

Density and Specific Volume

Density *(d)* is the mass per unit of volume of a substance. **Specific volume** *(v)* is the reciprocal of density. That is

$$d = \frac{m}{\text{volume}} \tag{2.1}$$

$$v = \frac{\text{volume}}{m} \tag{2.2}$$

Weight density is the weight per unit volume of a substance. Although weight density and (mass) density are not the same, they are often used as such, as both are measured in lb/ft^3 in U.S. units. Density varies with temperature and pressure. Densities and other properties for some substances are shown in Table 2.1.

Example 2.4

A contractor is going to install a cooling tower on a roof. He must inform the structural engineer how much extra weight to allow for the water in the tower basin when designing the roof. The tower basin is 15 ft by 10 ft in plan and filled with water to a depth of 1.5 ft.

Solution

The weight of water in the tank is found from Equation 2.1, after finding the volume of water. The density of water is shown in Table 2.1.

$$\text{Volume} = 15 \text{ ft} \times 10 \text{ ft} \times 1.5 \text{ ft} = 225 \text{ ft}^3$$

Solving Equation 2.1 for *m*,

$$m = d \times \text{volume} = 62.4 \; \frac{\text{lb}}{\text{ft}^3} \times 225 \text{ ft}^3$$

$$= 14,000 \text{ lb}$$

Specific Gravity

The *specific gravity (s.g.)* of a substance is defined as the ratio of its weight to the weight of an equal volume of water, at 39 F. The density of water at 39 F is 62.4 lb/ft^3, so the specific gravity is

$$\text{s.g.} = \frac{d}{d_w} = \frac{d}{62.4} \tag{2.3}$$

where

d = density of substance, lb/ft^3
d_w = density of water at 39 F, 62.4 lb/ft^3

TABLE 2.1   PHYSICAL PROPERTIES OF SUBSTANCES

Substance	Density, lb/ft^3	Specific Heat, BTU/lb-F	Note
Water	62.4	1.0	At 32–60 F
Ice	57.2	0.50	
Steam	(Table A.3)	0.45	Average for water vapor in air
Air	0.075	0.24	At 70 F and 14.7 psia
Mercury	849.0		At 32 F

The value of specific gravity will change slightly with temperature, but for most calculations the values from Equation 2.3 are satisfactory.

Example 2.5
A fuel oil has a density of 58.5 lb/ft^3. What is its specific gravity?

Solution
Using Equation 2.3,

$$\text{s.g.} = \frac{d}{62.4} = \frac{58.5}{62.4} = 0.94$$

2.5   ACCURACY OF DATA

In reporting results of measurements of calculations of data, decisions must be made as to the number of *significant figures* or places of accuracy to use in numerical values. This procedure is called *rounding off*.

For example, suppose the results of some calculations produced a value of 18,342 CFM for the required air supply rate to a building. This number is said to have five significant figures, because the value of the fifth digit from the left is known. The number might be used to select a fan, and then to balance the system to obtain this flow rate. However, neither manufacturer's fan ratings or testing instruments can produce that accurate a value. Equipment and instrument ratings are often only accurate to within 2–5% of listed values, therefore there is no point in calculating or measuring data to an excess number of significant figures. Data in HVAC work are usually rounded off (i.e., the number of significant figures are reduced) to three or four places, and sometimes even two. If the above value is rounded off to three places, it would be reported as 18,300 CFM. Until students become familiar with good practice in rounding off values, they should use the examples of this book as a guide.

2.6   PRESSURE

Pressure (p) is defined as force (F) exerted per unit area (A). Expressed as an equation, this is

$$p = \frac{\text{force}}{\text{area}} = \frac{F}{A} \tag{2.4}$$

If force is measured in pounds (lb) and area in square feet (ft^2), the units of pressure will be

$$p = \frac{F}{A} = \frac{\text{lb}}{\text{ft}^2}$$

If force is measured in pounds and area in square inches (in.^{2}), units of pressure will be lb/in.^{2}.

The abbreviations *psf* for lb/ft^2 and *psi* for lb/in.^{2} are commonly used.

Example 2.6
A hot water storage tank used in a solar heating system contains 3000 lb of water. The tank is 2 ft long by 3 ft wide. What is the pressure exerted on the bottom of the tank, in lb/ft^2?

Solution
A sketch of the tank is shown in Figure 2.1. Equation 2.4 will be used to find the pressure. The pressure is being exerted on an area 2 ft × 3 ft = 6 ft^2. The force acting on the bottom is the total weight of water.

$$p = \frac{F}{A} = \frac{3000 \text{ lb}}{6 \text{ ft}^2} = 500 \text{ lb/ft}^2$$

The relation between force and pressure is illustrated in Figure 2.2. A force of 3000 lb is distributed over the 2 ft × 3 ft area. The pressure is the force on each of the six 1 ft × 1 ft areas, 500 lb/ft^2.

Figure 2.1
Sketch for Example 2.6.

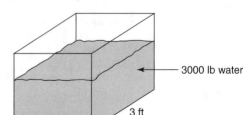

3000 lb water

3 ft

2 ft

Total force = 3000 lb
Pressure = force on each square foot = 500 lb

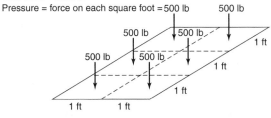

Figure 2.2
Relation between force and pressure.

Pressures of liquids and gases are of great importance in HVAC work. Some examples are the steam pressure in a boiler, the air pressure developed by a fan, the water pressure exerted on a valve, and the pressure exerted by the atmosphere.

Absolute, Gage, and Vacuum Pressure

A space that is completely evacuated of any gas or liquid (a complete vacuum) has zero pressure, because there is nothing to exert a pressure. The pressure exerted by a fluid above the zero pressure is called its *absolute pressure* (p_{abs}). This is illustrated in Figure 2.3.

The atmospheric air above the earth exerts a pressure (p_{atm}) because of its weight. The pressure it exerts at sea level has been measured and found to be approximately 14.7 lb/in.^{2} absolute (psia). (This pressure decreases at higher elevations be-

cause there is less weight of air above. For example, the atmospheric pressure in Denver, Colorado is about 12.23 psia).

Pressure measuring instruments usually measure the difference between the pressure of a fluid and the pressure of the atmosphere. The pressure measured above atmospheric pressure is called *gage pressure* (p_g). The relation between absolute, atmospheric, and gage pressures, shown in Figure 2.3, is

$$p_{abs} = p_{atm} + p_g \tag{2.5}$$

Using gage pressure is convenient because most pressure measuring instruments are calibrated to read 0 when they are subject to atmospheric pressure. Figure 2.4 (*a*) on page 22 shows the dial face of a typical *compression gage*. (Pressure gages and similar instruments will be discussed in Chapter 16.)

Example 2.7

The pressure gage connected to the discharge of a cooling tower water pump in the Trailblazers Bus Terminal in San Francisco reads 18 psi. What is the absolute water pressure at the pump discharge?

Solution
The pressure gage reads gage pressure, 18 psig (above atmospheric). San Francisco is at sea level, so the atmospheric pressure is approximately 14.7 psia. Using Equation 2.5,

$$p_{abs} = p_g + p_{atm} = 18 \text{ psi} + 14.7 \text{ psi} = 32.7 \text{ psia}$$

Figure 2.3
Relations of absolute, gage, and vacuum pressures.

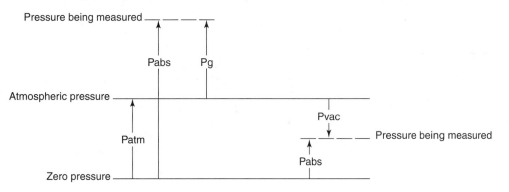

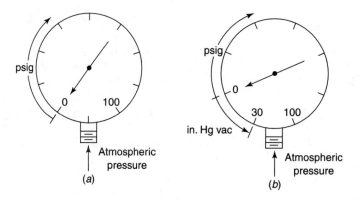

Figure 2.4

Pressure gages (a) Compression gage reads gage pressure only. (b) Compound gage reads gage and vacuum pressure.

If a fluid exerts a pressure below atmospheric pressure, it is called a "partial" vacuum; its pressure value reading down from atmospheric pressure is called *vacuum pressure* (p_{vac}). The relation among absolute, atmospheric, and vacuum pressures, shown in Figure 2.3, is

$$p_{abs} = p_{atm} - p_{vac} \qquad (2.6)$$

Some gages are constructed to read both vacuum and gage pressure. This type is called a *compound gage* and is shown in Figure 2.4(b).

Example 2.8 _____

The gages on the suction gas and discharge gas lines of a compressor read 5 psiv (lb/in.^{2} vac) and 60 psig, respectively. How much is the pressure increased by the compressor?

Solution

Referring to Figure 2.5, the pressure increase is

$$\text{pressure increase} = 60 + 5$$
$$= 65 \text{ psi}$$

2.7   PRESSURE OF A LIQUID COLUMN

A liquid exerts a pressure because of its weight, and the weight depends on the height of the column of liquid. The relation between the pressure exerted and the height, as shown in Figure 2.6, is

$$p = d \times H \qquad (2.7)$$

where

Figure 2.5

Sketch for Example 2.8.

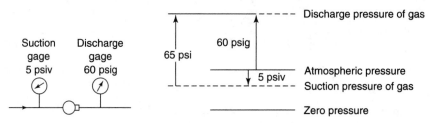

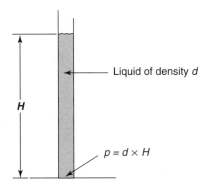

Figure 2.6
Pressure exerted by a liquid column. Pressure may be expressed as "head" (height of liquid).

p = pressure exerted by a liquid, lb/ft^2
d = density of liquid, lb/ft^3
H = height of liquid, ft

Other units can be used in the equation, but these are often convenient.

Example 2.9 _____
A 300-ft vertical pipe in a high-rise building is filled with chilled water. What is the pressure in lb/in.^{2} gage (psig) that a valve in the bottom of the line will have to withstand?

Solution
The density of water is approximately 62.4 lb/ft^3 (Table 2.1). Using Equation 2.7,

$$p = d \times H$$

$$p = 62.4 \ \frac{\text{lb}}{\text{ft}^3} \times 300 \ \text{ft} = 18,720 \ \frac{\text{lb}}{\text{ft}^2} \times \frac{1 \ \text{ft}^2}{144 \ \text{in.}^2}$$

$$= 130 \ \text{psig}$$

The relation between pressure and height of a liquid is used by pressure measuring instruments that have a column of liquid. These are called *manometers,* an example of which is shown in Figure 2.7. In Figure 2.7(*a*), the pressure exerted on both legs of the manometer (atmospheric pressure) is the same, so the liquid is at the same level. In Figure 2.7(*b*), the pressure in the duct is above atmospheric. In Figure 2.7(*c*), the pressure in the duct is below atmospheric (vacuum pressure), so the liquid is higher in the leg connected to the duct.

Example 2.10 _____
A service technician wishes to measure the pressure of air in a duct. He connects one leg of a water manometer to the duct and the other leg is exposed to the atmosphere. The difference in height of the water columns is 8 in. w.g. (inches of water gage) as shown in Figure 2.8 on page 24. What is the air pressure in the duct in psig?

Figure 2.7
Manometer reading pressures above and below atmospheric pressure. (*a*) Equal pressure on both legs. (*b*) Pressure in duct above atmospheric (gage pressure). (*c*) Pressure in duct below atmospheric (vacuum pressure).

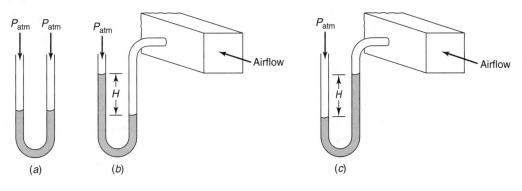

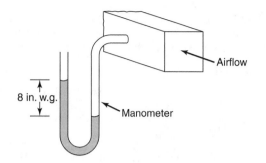

Figure 2.8
Sketch for Example 2.10.

Solution
The difference in height is related to the pressure by Equation 2.7. Changing the units of H first,

$$H = 8 \text{ in. w.g.} \times \frac{1 \text{ ft}}{12 \text{ in.}} = 0.667 \text{ ft w.g.}$$

$$p = d \times H$$

$$= 62.4 \ \frac{\text{lb}}{\text{ft}^3} \times 0.667 \text{ ft}$$

$$= 41.6 \ \frac{\text{lb}}{\text{ft}^2} \times \frac{1 \text{ ft}^2}{144 \text{ in.}^2} = 0.29 \text{ psig}$$

The air pressure in the duct is 0.29 psi above atmospheric pressure.

Water manometers are often used for measuring relatively small pressures, particularly when testing and balancing air systems. They are not convenient for high pressures because a very high liquid column would be needed. Manometers using mercury, a liquid with a much higher density than water, are often used for measuring higher pressures.

The *barometer* (Figure 2.9) is a special manometer used for measuring atmospheric air pressure. Mercury (Hg) is the liquid used. The tube is evacuated of all gas so that no atmospheric pressure acts on the top of the mercury column. Because atmospheric pressure acts on the bottom of the mercury, the height to which the mercury column is lifted represents atmospheric pressure.

Example 2.11
How high would the mercury column in a barometer be, in both in. Hg and mm Hg, at a location where atmospheric pressure is 14.7 psi and the temperature 32 F?

Solution
Using Equation 2.7 with proper units, noting the density of mercury (Table 2.1) is $d = 849 \text{ lb/ft}^3$ at 32 F,

Changing units,

$$p_{atm} = 14.7 \ \frac{\text{lb}}{\text{in.}^2} \times \frac{144 \text{ in.}^2}{1 \text{ ft}^2} = 2116.8 \ \frac{\text{lb}}{\text{ft}^2}$$

Using Equation 2.7,

$$H = \frac{p}{d} = \frac{2116.8 \text{ lb/ft}^2}{849 \text{ lb/ft}^3} = 2.49 \text{ ft} \times \frac{12 \text{ in.}}{1 \text{ ft}}$$

$$= 29.92 \text{ in. Hg}$$

$$= 29.92 \text{ in. Hg} \times \frac{25.4 \text{ mm}}{1 \text{ in.}} = 760 \text{ mm Hg}$$

Head

Is is often convenient to express pressure in units of head. *Head* is the equivalent of liquid column height (H) expressed in Equation 2.7. In Example 2.11, instead of stating that the pressure of the atmosphere was 14.7 psi, it could have been stated that it was 29.92 in. Hg or 760 mm Hg. In Example

Figure 2.9
Mercury barometer.

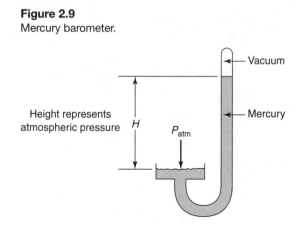

2.10, the air pressure in the duct could also have been stated both ways, $p = 0.29$ psig $= 8$ in. w.g.

That is, there does not actually have to be a column of liquid to express any pressure in head units. Equation 2.7 can be used to convert to or from units of pressure expressed as head. Some of the equivalents for pressure expressed as head, obtained from that equation, are listed in Table A.2.

Example 2.12

A contractor requires a pump that will have a discharge pressure of 42 psi. He looks in a manufacturer's catalog to find a suitable pump, but finds that the pump ratings are listed as "head, feet of water." What pump head should he specify in his purchase order?

Solution
Using the conversion factor equality (Table A.2) of

$$2.3 \text{ ft w.} = 1 \text{ psi}$$

$$H = 42 \text{ psi} \times \frac{2.3 \text{ ft w.}}{1 \text{ psi}} = 97 \text{ ft w.}$$

2.8 WORK, POWER, AND ENERGY

Work is the effect created by a force when it moves a body. It is expressed by the following equation:

$$\text{Work} = \text{force} \times \text{distance} \tag{2.8}$$

Example 2.13

A cooling tower weighing 6000 lb is hoisted from the street level to the roof of the Gusher Oil Co. building, 300 ft high. How much work is done in lifting it?

Solution
The force required is equal to the weight of the tower. Using Equation 2.8,

$$\text{Work} = 6000 \text{ lb} \times 300 \text{ ft} = 1,800,000 \text{ ft-lb}$$

Power is the time rate of doing work. It is expressed by the equation

$$\text{Power} = \frac{\text{work}}{\text{time}} \tag{2.9}$$

Power is usually of more direct importance than work in industrial applications; the capacity of equipment is based on its power output or power consumption. If work is expressed in ft-lb, some units of power that would result are ft-lb/min and ft-lb/sec. More convenient units for power are the horsepower (HP) and kilowatt (KW), because the numbers resulting are not as large.

Example 2.14

If the cooling tower in Example 2.13 is lifted by a crane in 4 minutes, what is the minimum power required?

Solution
Using Equation 2.9,

$$\text{Power} = \frac{1,800,000 \text{ ft-lb}}{4 \text{ min}} = 450,000 \text{ ft-lb/min}$$

From Table A.2, 1 HP $= 33,000$ ft-lb/min. Converting to HP,

$$450,000 \text{ ft-lb/min} \times \frac{1 \text{ HP}}{33,000 \text{ ft-lb/min}} = 13.6 \text{ HP}$$

The actual size of the engine or motor selected to hoist the cooling tower would be greater than 13.6 HP, due to friction and other losses, and to allow some excess capacity as a safety reserve.

Although it is a somewhat abstract concept, **energy** is sometimes defined as the ability to do work. For example, we use the stored chemical energy in a fuel by burning it to create combustion gases at high pressure that drive the pistons of an engine and thus do work. Work is therefore one of the forms of energy.

Energy can exist in a number of forms. They can be grouped into those forms of energy that are *stored* in bodies or those forms of energy in *transfer* or *flow* between bodies. Work is one of the

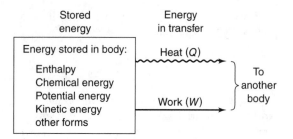

Figure 2.10
Comparison of stored energy and energy in transfer.

forms of energy in transfer between bodies. That is, one body does work on another when it moves it.

Energy can be stored in matter in many forms. Figure 2.10 is a diagram showing some types of *stored energy* and *energy in transfer*. At this time we will turn our attention to a form of energy in transfer or motion called heat. Some of the forms of stored energy will discussed in Section 2.10.

2.9 HEAT AND TEMPERATURE

Heat has been described as a form of energy in transfer.

Heat is the form of energy that transfers from one body to another due to a temperature difference.

Figure 2.11 graphically describes this definition. In Figure 2.11(*a*) heat (*Q*) flows from the high tem-

perature body, hot water, in the heating unit to the lower temperature body, the air in the room. Figure 2.11(*b*) shows that heat will flow from the higher temperature body, room air, to the lower temperature body, the air in the refrigerator interior, due to the temperature difference.

Note that heat can only flow naturally from a higher to a lower temperature—"downhill," so to speak, as seen in Figure 2.12. Of course if there is no temperature difference, there is no heat flow. The most common unit used for heat in the United States is the BTU (British Thermal Unit). The BTU is defined as *the quantity of heat required to raise the temperature of one pound of water one degree Fahrenheit (F) at 59 F.*

Temperature is a measure of the thermal activity in a body. This activity depends on the velocity of the molecules and other particles of which all matter is composed. It is not practical to measure temperature by measuring the velocity of molecules, however, so this definition is not of great importance in our work. Temperature is usually measured with thermometers. The most commonly used type relies on the fact that most liquids expand and contract when their temperature is raised or lowered. By creating an arbitrary scale of numbers, a temperature scale and units are developed. Other types of thermometers used in HVAC work will be discussed in Chapter 16.

The unit scale most often used for measuring temperature in the United States is the *degree*

Figure 2.11
Examples of heat flow. (*a*) Heat flows from heating unit at higher temperature to room air at lower temperature. (*b*) Heat flows from room air at higher temperature to refrigerator air at lower temperature.

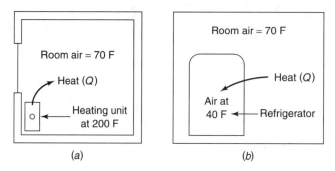

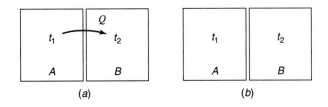

Figure 2.12
Heat can flow only from a higher to a lower temperature. (a) If t_1 is greater than t_2, heat flows from A to B. (b) If $t_1 = t_2$, no heat follows.

Fahrenheit (F), in which the boiling point of water is 212 F and the freezing point of water is 32 F at atmospheric pressure. In the SI system of units the *degree Celsius* (C) is used, in which the boiling point of water is 100 C and the freezing point is 0 C at atmospheric pressure. The relationship between these two units is therefore

$$F = 1.8\,C + 32 \qquad (2.10a)$$

$$C = \frac{F - 32}{1.8} \qquad (2.10b)$$

Example 2.15 _____

A room is supposed to be at a temperature of 78 F in an air conditioned building. The building maintenance engineer checks the temperature with a thermometer that has a Celsius scale. What should be the reading on the thermometer?

Solution
Using Equation 2.10b,

$$C = \frac{F - 32}{1.8} = \frac{78 - 32}{1.8} = 25.6\ C$$

There are also two *absolute temperature* scales. These take the value 0 for the lowest temperature that can exist. They are called the *Rankine* (R) and *Kelvin* (K) temperature scales. The Rankine is used in the U.S. system, with the difference in size between each degree the same as Fahrenheit. The Kelvin is used in the SI system with the difference between each degree equal to Celsius. The relationships are

$$R = F + 460 \qquad (2.10c)$$

$$K = C + 273 \qquad (2.10d)$$

The relations among temperature scales are shown graphically in Figure 2.13.

2.10 ENTHALPY

We have noted previously that energy can be classified into **energy in transfer** between bodies (heat and work) or **stored energy** in bodies. There are a number of types of stored energy, some of which we will briefly discuss here. (We will not always define these terms rigorously, when it will not serve our purposes.)

Chemical energy is a form of stored energy in a body that is released from a body by combustion. When a fuel is burned, its stored chemical energy is released as heat.

Kinetic energy is the stored energy in a body due to its motion, or velocity.

Figure 2.13
Relations among temperature scales.

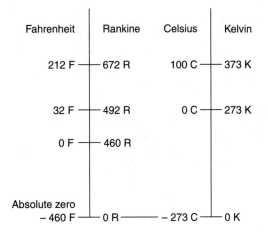

Potential energy is the stored energy a body has due to its position, or elevation.

There is a property a body has that is a combination of its energy due to temperature, pressure, and volume; it is called enthalpy.

Enthalpy *is a property of a body that measures its heat content.*

Specific enthalpy (*h*) is the enthalpy per unit mass of a substance. It is expressed in BTU/lb in U.S. units.

Although this definition of enthalpy is used extensively in the HVAC industry, it is scientifically imprecise. Its exact definition is best defined by a mathematical equation. For our purposes, the terms *heat content* and *enthalpy* are considered to have the same meaning and to be a property of a body. *Heat,* however, as defined in Section 2.9, means a form of (heat) energy in transfer or flow, not a property of a body. For this reason, it is preferable to use the word enthalpy, not heat content, so that heat is not used with two different meanings.

It should also be understood that temperature and enthalpy (heat content) are not the same thing. The temperature of a body is a measure of its thermal level or thermal intensity, but by itself does not determine how much thermal energy it has. The amount of thermal energy of a body depends not only on its temperature but also on its mass and specific heat. The enthalpy of a body, however, is a property that does reflect its amount of thermal energy.

For example, consider a thimbleful of molten steel at 2500 F as compared to a very large tank of hot water at 200 F. The hot water has a higher total enthalpy; it has more thermal energy available for space heating, despite its much lower temperature.

2.11 THE ENERGY EQUATION (FIRST LAW OF THERMODYNAMICS)

The subject we have been examining, called *thermodynamics,* is the branch of physics that deals with heat and work. The *First Law of Thermody-*

namics is a principle that may be stated in various ways, for instance, "energy can neither be created nor destroyed," or "there is conservation of energy in nature." This principle is used extensively in the HVAC industry, especially when stated as an energy balance:

The change in total energy in a system equals the energy added to the system minus the energy removed from the system.

The word *system* refers to any closed body or group of bodies for which the flow of energy in or out can be determined. It could be the air in a room (Figure 2.14), a boiler, a whole building, or a complete air conditioning system.

This energy balance can be expressed as an equation, called the *Energy Equation:*

$$E_{ch} = E_{in} - E_{out} \qquad (2.11)$$

where

E_{ch} = change in stored energy in the system
E_{in} = energy added to (entering) the system
E_{out} = energy removed from (leaving) the system

Example 2.16 illustrates the use of the energy equation.

Example 2.16

A hot water heating convector in Mr. Jones office is supplying 4000 BTU/hr of heat. Heat is being transferred from the room air to the outdoors at the rate of 6500 BTU/hr. What will happen in the

Figure 2.14
Sketch for Example 2.16.

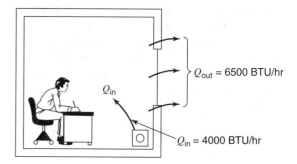

Q_{out} = 6500 BTU/hr

Q_{in}

Q_{in} = 4000 BTU/hr

room? What size electric heater should Mr. Jones temporarily use to solve the emergency?

Solution
We apply the Energy Equation 2.11. Figure 2.14 shows the energy (heat) added and removed:

$$E_{ch} = E_{in} - E_{out}$$
$$= 4000 \, \frac{BTU}{hr} - 6500 \, \frac{BTU}{hr}$$
$$= -2500 \, \frac{BTU}{hr}$$

The negative sign means the room air energy is decreasing. This loss in enthalpy (heat content) will cause the room air temperature to drop, making it uncomfortable. A solution is to install an electric heater that will make up the heat loss which the convector does not supply, 2500 BTU/hr. There will then be no net loss of heat from the room, and the temperature will not drop.

The capacity of an electric heater is normally expressed in watts (W) or kilowatts (KW) rather then BTU/hr. The heater should therefore have the following capacity.

From Table A.2,

$$3410 \, BTU/hr = 1000 \, W$$
$$2500 \, BTU/hr \times \frac{1000 \, W}{3410 \, BTU/hr} = 733 \, W$$

The nearest size larger heater manufactured would probably be 750 W.

Example 2.16 illustrates the sign convention that will be used in the Energy Equation: An energy decrease in the system is negative, an energy increase is positive.

The example also shows that any units used for energy are interchangeable, regardless of the form of energy, whether heat, work, or enthalpy. Example 2.17 will illustrate this, as well as a further application of the Energy Equation.

Example 2.17
A business equipment room has 1000 watts of lighting and some small motors with a total output of 10 HP. All of the energy in the lighting and from the motors is converted into heat. What is the increase in enthalpy of the room air from these sources?

Solution
The energy added to the room air will increase its enthalpy. Applying the Energy Equation 2.11 and converting all units to BTU/hr,

$$E_{ch} = E_{in} - E_{out}$$
$$E_{ch} = 1000 \, W \times \frac{3.41 \, BTU/hr}{1 \, W}$$
$$+ 10 \, HP \times \frac{2545 \, BTU/hr}{1 \, HP} - 0$$
$$= 28,860 \, BTU/hr$$

2.12 LIQUIDS, VAPORS, AND CHANGE OF STATE

Substances can exist in three different *states* (also called *phases*)—solid, liquid, or vapor (gas). The state that a substance is actually in depends on its temperature and pressure. The meaning of this for liquids and vapors is best understood by describing the experiment (which the student could carry out at home to check the results) shown in Figure 2.15 on page 30.

Figure 2.15(*a*) shows a pot of water at room temperature. Being open, it is subject to atmospheric pressure, 14.7 psia at sea level.

At (*b*) heat (*Q*) is being added to the water, and it is noted that its temperature continually rises as heat is added.

At some point in time (*c*), however, it is noted that the temperature stops rising (at 212 F).

Even though more heat is added after that (*d*), the temperature does not increase for a while. What is observed now, however, is that the liquid water will gradually change into its gas or vapor state (steam). This process is called *boiling* or *vaporization*. As heat is added, no further temperature increase occurs as long as some liquid remains.

At (*e*), all the water is evaporated.

If more heat is added, it will be noted that the

temperature (of the steam) will begin to rise again, above 212 F, as seen in (*f*). (This part of the experiment would be difficult to carry out, because the steam will escape into the room.)

The whole series of processes just described could also be carried out in reverse. Removal of heat (cooling) from the steam in Figure 2.15(*f*) lowers its temperature. When the cooling continues to (*e*), the temperature no longer drops, but the gas begins to condense to a liquid (*d*). After all of the steam is condensed (*c*), further removal of heat will result in a temperature drop of the liquid, (*b*) and (*a*).

A useful summary of all of this information is shown in Figure 2.16, called the *temperature-enthalpy diagram*. When heat is added to the water

between 32 F and 212 F, both its enthalpy and temperature increase.

However, if more heat is added at 212 F, note that although its enthalpy continues increasing, the temperature remains constant. What does happen is that the water gradually boils until it all vaporizes to steam, still at 212 F, assuming enough heat is added.

Once all the liquid is evaporated, if more heat is added, then and only then, does the temperature start to increase again. (The enthalpy continues to increase as before.) The temperature and enthalpy increase of the steam will then continue if further heat is provided.

Figure 2.16 also shows the temperature-

Figure 2.15
Experiment showing change of state of water at atmospheric pressure (14.7 psia). (*a*) Initial condition (subcooled liquid). (*b*) Heat added, temperature increases (subcooled liquid). (*c*) Heat added, liquid reaches boiling point (saturated liquid). (*d*) Heat added, liquid changing to vapor, no temperature increase. (*e*) Heat added, all liquid vaporized (saturated vapor). (*f*) Heat added, temperature of vapor increases (superheated vapor). *Note:* Subcooled liquid is liquid below its boiling point. Saturated liquid and saturated vapor are the liquid and vapor at the boiling (condensing) point. Superheated vapor is vapor above the boiling point.

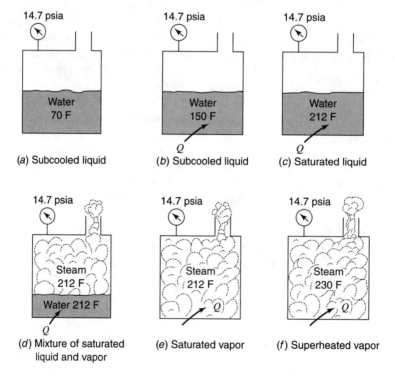

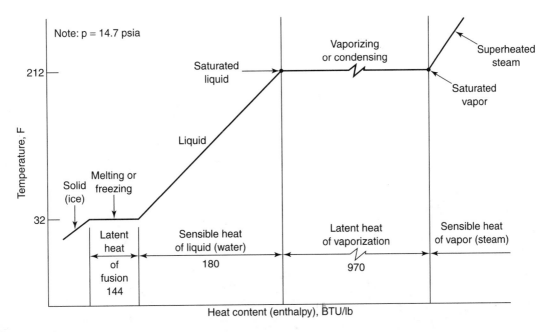

Figure 2.16
Temperature-enthalpy (heat content) change of water at 14.7 psia surrounding pressure.

enthalpy changes that occur between the liquid and solid state, which will be discussed later.

The conditions shown in Figure 2.16 are correct for water only when the surrounding pressure is 14.7 psia (atmospheric pressure at sea level). We will now examine what changes occur at different pressures.

Dependence of Boiling Temperature on Pressure

In the experiment just described the surrounding pressure was 14.7 psia. Let us conduct the same experiment where the surrounding pressure is at a higher value, say 24.9 psia. Figure 2.17 on page 32 represents the same heating process, or cooling if done in reverse.

When the water reaches 212 F (*c*) and more heat is added, *it does not boil*, but the temperature continues to rise. When the temperature reaches 240 F (*d*), however, the boiling process begins and the temperature remains constant until the liquid has completely evaporated.

This shows that the temperature at which the water boils changes with pressure. For water, the boiling point is 240 F at 24.9 psia. This means that water cannot be made to boil at a temperature below 240 F if the pressure is 24.9 psia.

If the same experiment were carried out with the surrounding pressure at 6 psia, we would find that when heat was added, the boiling process would begin at 170 F. These facts show that *the boiling/condensing temperature of water depends on its pressure.*

Figure 2.18 on page 33 shows a line representing these temperature-pressure values for water; it is called the *boiling point* curve or *saturation vapor pressure* curve. Water can exist at its boiling/condensing temperature and pressure only on this line. To the left of the line it can exist only as a liquid and to the right only as a vapor. Along the line it can exist either as liquid, vapor, or as mixture of the liquid-vapor.

Example 2.18 _____
Will water exist in the liquid state, or as steam, if its temperature is 225 F and its pressure is 25 psia?

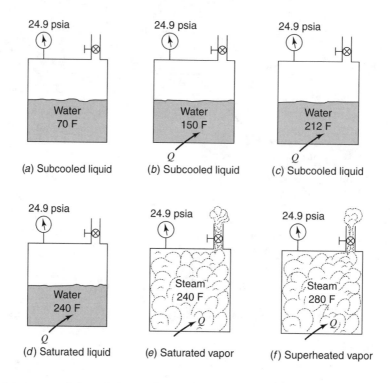

Figure 2.17
Experiment showing change of state of water at 24.9 psia. (a) Initial condition (subcooled liquid). (b) Heat added, temperature increases (subcooled liquid). (c) Heat added, temperature increases (subcooled liquid). Note that the water is not boiling at 212 F. (d) Heat added, liquid reaches boiling point, 240 F. (e) Heat added, all liquid vaporized (saturated vapor). (f) Heat added, temperature of vapor increases (superheated vapor).

Solution
Locating the pressure-temperature (*p-t*) condition on Figure 2.18, it is found to be in the liquid region. The water is in a liquid state.

This same dependence of boiling/condensing temperature on pressure holds for all fluids, except the *p-t* values are different. For example, at 14.7 psia ammonia boils at −28 F, alcohol at 170 F, and copper at 4250 F.

Note that the higher the pressure on the water, the higher the boiling temperature, and the lower the pressure, the lower the temperature at which it will boil. This same relation holds for other substances.

Let us examine what causes this change in boiling point temperature.

The Molecular (Kinetic) Theory of Liquids and Gases

The process of boiling and the dependence of boiling point temperature on surrounding pressure can be explained by referring to the molecular (kinetic) theory of liquids and gases. All matter is composed of particles called molecules. The molecules in a substance are constantly in motion. They are also attracted to each other by forces. The closer the molecules are to each other, the greater the attractive forces.

When a substance is in the liquid state, the molecules are closer together than when it is in its gaseous state, and therefore the attractive forces are greater. Also molecules in the gaseous state move more rapidly than molecules in the liquid state, and therefore they have more energy. This is why heat is required to boil a liquid. The heat energy is required to overcome the attractive forces holding the molecules relatively close together, so that they move further apart and change state to a gas.

The temperature of a substance is a measure of the average velocity of its molecules. The higher the average velocity, the higher the temperature. However, not all molecules move at the average velocity—some are moving faster, some slower.

Figure 2.19 on page 34 shows an open vessel of water at 70 F, surrounded by air at 14.7 psia. The water is therefore in a liquid state. The average velocity of the molecules is not great enough for them to escape rapidly. However, a small fraction of molecules have velocities well above the average. If some of these molecules are near the surface, they will escape. That is, there will be very slow evaporation from the surface. This leaves the remaining molecules at a slower average velocity,

Figure 2.18
Boiling point pressure-temperature curve for water, also called the saturation vapor pressure curve.

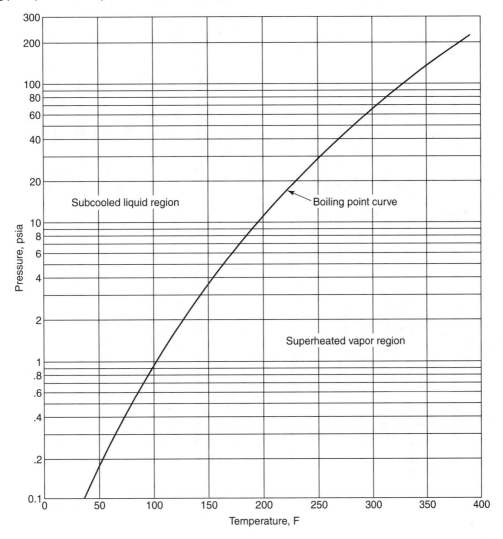

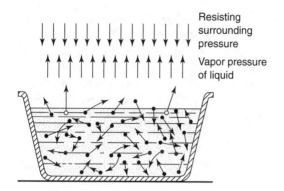

Figure 2.19
Slow evaporation of liquid. Escape of some molecules through surface causes a vapor pressure.

and therefore at a lower temperature. A slight cooling effect of the liquid has occurred as a result of the evaporation. We have all noticed this effect when alcohol is rubbed on the skin. It gradually evaporates and cools itself and the skin.

The molecules escaping from the surface of a liquid create vapor. The pressure exerted by the vapor at the surface of the liquid is called the *vapor pressure*. If the pressure exerted by a surrounding gas is above the vapor pressure, then the liquid cannot rapidly evaporate (boil). However, if the temperature of the liquid is increased enough, the molecular velocity increases to a point at which the molecules break the bonds holding them together as a liquid and the liquid boils. The vapor pressure of the liquid has been increased to a value greater than the surrounding resisting pressure.

If the resisting pressure is higher, the temperature of the liquid must be increased further to reach the boiling point. That is, a higher temperature increases the molecular velocity enough to cause boiling.

While the boiling process is occurring, the heat applied is breaking the molecular bonds that hold the molecules close together. It is not increasing the velocity of the molecules. That is why the temperature does not increase during boiling.

It is also of importance to note what happens if the pressure exerted by a gas above a liquid is reduced to a value below the vapor pressure exerted by the liquid. In this case, the liquid will suddenly boil, because the surrounding pressure is now less than the vapor pressure exerted by the liquid. The energy of the molecules is now great enough to overcome the reduced resistance, and they escape rapidly. That is, the liquid boils. This cools the remaining liquid, because energy is removed. *Boiling has been achieved by a lowering of pressure.* This process is essential in refrigeration, as will be seen in Section 2.14 and in Chapter 14.

Saturated, Subcooled, and Superheated Conditions

The pressure and temperature condition at which boiling occurs is called the saturated condition, and the boiling point is technically known as the saturation temperature and saturation pressure. As seen from the experimental description, the substance can exist as a liquid, vapor, or mixture of liquid and vapor at the saturated condition. At saturation, the liquid is called saturated liquid and the vapor is called a saturated vapor.

> **Saturated vapor** *is vapor at the boiling temperature, and* **saturated liquid** *is liquid at the boiling temperature.*
>
> *When the temperature of the vapor is above its saturation temperature (boiling point), it is called a* **superheated vapor.**
>
> *When the temperature of the liquid is below its saturation temperature, it is called a* **subcooled liquid**.

Figure 2.18, a typical boiling point curve, illustrates this. Note that a substance can exist as a subcooled liquid or superheated vapor at many temperatures for a given pressure, but it can exist as a saturated liquid or vapor at only one temperature for a given pressure.

Sensible and Latent Heat

When heat added to or removed from a substance results in a temperature change, but no change in state, the process is called a *sensible heat* change. When heat added to or removed from a substance results in a change in state, then the enthalpy change in the substance is called a *latent heat* change. The enthalpy increase as it changes from a liquid to a vapor is called the *latent heat of vaporization.* The opposite effect, the enthalpy decrease as it changes from a vapor to a liquid is called the *latent heat of condensation.* It is equal to the latent heat of vaporization.

2.13 SATURATED PROPERTY TABLES

For various substances, saturation temperatures, corresponding pressures, and other properties at saturation conditions may be found in tables designed for that purpose.

Table A.3 is a saturated property table for water. It is commonly called the "Saturated Steam Table." Note that both parts of the table have the same information. If the temperature is known, use the part of the table that lists temperatures first. If the pressure is known, use the part of the table that lists pressures first. If the known value is between two listed values, interpolate to obtain the correct value. Examples 2.19–2.21 illustrate various uses of saturated property tables.

Example 2.19 _____
At what temperature will water boil at a pressure of 10 psia?

Solution
From Table A.3, we read the saturation temperature (boiling point) at 10 psia to be about 193 F.

Example 2.20 _____
Use the steam tables to determine if water is in a liquid or gas state at 300 F and 150 psia.

Solution
Using Table A.3, the saturation (boiling) temperature at 150 psia is about 358 F. The actual temperature is less, therefore the water will be in a liquid state (subcooled liquid).

Example 2.21 _____
The operating engineer of a hot water heating system reads the temperature and pressure at the pump suction to be 200 F and 10 psia. Should the engineer be concerned?

Solution
From Table A.3, the saturation temperature at 10 psia is about 193 F. Because the actual temperature is higher (200 F), the water will exist as steam, not as liquid. Therefore the operator should be very concerned, because there is steam, not water, in the pump.

2.14 REFRIGERATION

It has been stated that the boiling point of a liquid depends on the surrounding pressure. This was explained by considering that all matter consists of particles (molecules) which are attracted to each other, but also have a considerable velocity energy. The pressure surrounding a liquid inhibits the escape of the molecules. If the liquid temperature increases, however, the molecules increase in velocity, and at some temperature (the boiling point) they will escape rapidly—the liquid will vaporize. If the pressure is increased, the molecules will have to reach a higher velocity—a higher temperature—to escape.

On the other hand, if the surrounding pressure is lowered enough (to the saturation point), the molecules will have enough energy to escape at a lower temperature. This is how refrigeration can be accomplished. A liquid is used that boils at a low temperature for the reduced pressure that can be achieved. The surrounding pressure is reduced below the saturation pressure, and the liquid suddenly boils. As also noted, liquids absorb heat when they boil (latent heat of vaporization). This heat absorbed from the surroundings at the low temperature is refrigeration.

Even boiling water can be used to achieve refrigeration, if the pressure can be lowered enough, as seen in Example 2.22.

Example 2.22 _____
The boiling of water is to be used to accomplish refrigeration at 50 F. To what value should the surrounding pressure be lowered?

Solution
From Table A.3, the saturation pressure of water at 50 F is 0.178 psia. If the surrounding pressure is reduced below this value, the water will boil. This requires (latent) heat. Heat will flow to the water from any surrounding *body at a temperature higher than 50 F, thus cooling that body.*

Boiling of water at a very low temperature to achieve refrigeration is accomplished in refrigeration equipment called absorption units (see Chapter 13).

2.15 CALCULATION OF SENSIBLE AND LATENT HEAT CHANGES

The processes that occur in HVAC systems usually involve the addition or removal of heat from air or water. The procedures for calculating the amount of heat involved will be explained in this section.

Specific Heat
The *specific heat* (*c*) of a substance is defined as the amount of heat in BTUs required to change the temperature of 1 lb of the substance by 1 F, in U.S. units.
   The specific heat of liquid water is 1 BTU/lb-F at 60 F. Values of specific heat for some other substances are shown in Table 2.1.

Sensible Heat Equation
A sensible heat change was described as a process where the temperature of a substance changes when heat is added to or removed from it, but there is no change in state of the substance. This change is described quantitatively by the *sensible heat equation*:

$$Q_s = m \times c \times TC = m \times c(t_2 - t_1) \qquad (2.12)$$

where

Q_s = rate of sensible heat added to or removed from substance, BTU/hr
m = weight rate flow of substance, lb/hr
c = specific heat of substance, BTU/lb-F
$TC = t_2 - t_1$ = temperature change of substance, F

The sensible heat equation can be used to calculate the heat added to or removed for most HVAC processes where there is a temperature change.
   For air and water, the specific heat changes slightly with temperature. However, except for processes with large temperature changes, it can be assumed constant, using the values shown in Table 2.1. For other conditions, appropriate values of specific heats can be found in handbooks.
   Examples 2.23–2.25 illustrate uses of the sensible heat equation.

Example 2.23
There are 5000 GPM of chilled water being circulated from the refrigeration plant to the air conditioning systems of the buildings at the Interplanetary Spaceport. The water is cooled from 55 F to 43 F (Figure 2.20).
   What is the cooling capacity of the refrigeration chiller in BTU/hr, tons of refrigeration, and KW?

Solution
The "capacity" of the refrigeration chiller means the amount of heat it is removing from the water.
   First, change the units for the water flow rate from GPM to lb/hr (Table A.2).

$$m = 5000 \text{ GPM} \times \frac{500 \text{ lb/hr}}{1 \text{ GPM}} \text{ (for water)}$$

$$= 2,500,000 \text{ lb/hr}$$

Using Equation 2.12 to find the heat removed (refrigeration capacity),

$$Q_s = m \times c \times TC$$

$$= 2,500,000 \ \frac{\text{lb}}{\text{hr}} \times 1 \ \frac{\text{BTU}}{\text{lb-F}} \times (43 - 55) \text{ F}$$

$$= -30,000,000 \text{ BTU/hr}$$

Figure 2.20
Sketch for Example 2.23.

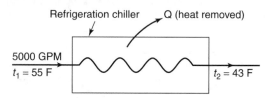

Note: the negative sign resulting for Q_s means that heat is removed; that is, the water is cooled, not heated. However, it is common practice in the HVAC industry to drop the negative sign for heat removed, so the refrigeration chiller capacity is reported as

$$Q_s = 30,000,000 \text{ BTU/hr}$$

Converting units to tons of refrigeration, the refrigeration capacity is

$$30,000,000 \text{ BTU/hr} \times \frac{1 \text{ ton}}{12,000 \text{ BTU/hr}}$$

$$= 2500 \text{ tons}$$

The capacity in KW is

$$30,000,000 \text{ BTU/hr} \times \frac{1 \text{ KW}}{3410 \text{ BTU/hr}}$$

$$= 8800 \text{ KW}$$

Example 2.24

The fuel oil preheater for a boiler has become damaged. The oil must be heated to 180 F in order to flow readily. A spare preheater with a capacity of 100,000 BTU/hr is available. The boiler requires 10 GPM of oil. The oil is at 60 F in a storage tank. The density of the oil is 8.0 lb/gal and its specific heat is 0.5 BTU/lb-F. Is the spare preheater big enough?

Solution

First, convert GPM of oil to lb/hr, and then use Equation 2.12 to find the temperature to which the oil would be heated by the spare preheater:

$$m = 10 \ \frac{\text{gal}}{\text{min}} \times \frac{60 \text{ min}}{1 \text{ hr}} \times \frac{8.0 \text{ lb}}{\text{gal}}$$

$$= 4800 \text{ lb/hr}$$

Rearranging and using Equation 2.12,

$$TC = \frac{Q}{m \times c} = \frac{100,000 \text{ BTU/hr}}{4800 \text{ lb/h} \times 0.5 \text{ BTU/lb-F}}$$

$$TC = t_2 - t_1$$

$$= 42 \text{ F}$$

$$t_2 = 42 + t_1 = 42 + 60 = 102 \text{ F}$$

The preheater will heat the fuel oil to only 102 F. It will not do the job.

Example 2.25

An electric booster heater in an air conditioning duct (Figure 2.21) has a capacity of 2 KW. The mechanical contractor is balancing the system and wants to find out how much air is flowing in the duct. The contractor measures the temperature before and after the heater as 80 F and 100 F. How much air is flowing, expressed in ft³/min (CFM)?

Solution

A. Convert the heater capacity to BTU/hr.

B. Use Equation 2.12, solving for the mass flow rate of air. From Table 2.1, the specific heat of air $c = 0.24$ BTU/lb-F.

C. Convert the flow rate units from lb/hr to ft³/min (CFM), using the density of air from Table 2.1.

Step A.

$$Q_s = 2 \text{ KW} \times \frac{3410 \text{ BTU/hr}}{1 \text{ KW}} = 6820 \text{ BTU/hr}$$

Step B.

$$m = \frac{Q_s}{c \times TC} = \frac{6820 \text{ BTU/hr}}{0.24 \text{ BTU/lb-F} \times 20F}$$

$$= 1420 \text{ lb/hr}$$

Step C.

$$\text{CFM} = 1420 \ \frac{\text{lb}}{\text{hr}} \times \frac{1 \text{ hr}}{60 \text{ min}} \times \frac{1 \text{ ft}^3}{0.075 \text{ lb}}$$

$$= 316 \text{ ft}^3/\text{min}$$

Figure 2.21
Sketch for Example 2.25.

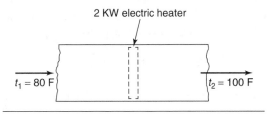

2 KW electric heater

$t_1 = 80 \text{ F}$ $t_2 = 100 \text{ F}$

The Enthalpy Equation

The heat added or removed in HVAC processes can also be found from another equation called the *enthalpy equation* (Equation 2.13), using enthalpy change instead of temperature change.

$$Q = m(h_2 - h_1) \qquad (2.13)$$

where

Q = rate of heat added or removed from substance, BTU/hr
m = weight flow rate of substance, lb/hr
$h_2 - h_1$ = specific enthalpy change of substance, BTU/lb

This equation can be used instead of Equation 2.12 for the sensible heat change process, if the enthalpy is known. The property tables list enthalpy data. Example 2.26 compares the two methods.

Example 2.26

A hot water boiler heats 10,000 lb/hr of water from 180 F to 220 F, at 30 psia. How much heat is added to the water?

Solution
The enthalpy of liquid water (h_f) is listed in Table A.3. Using Equation 2.13 at each temperature,

$$Q = m(h_{f2} - h_{f1})$$
$$= 10,000 \text{ lb/hr } (188.1 - 147.9)$$
$$= 402,000 \text{ BTU/hr}$$

The result using Equation 2.12 is

$$Q = m \times c \times TC = 10,000 \ \frac{\text{lb}}{\text{hr}} \times 1 \ \frac{\text{BTU}}{\text{lb-F}} \times 40 \text{ F}$$

$$= 400,000 \text{ BTU/hr}$$

The two equations give almost identical results. Either one is acceptable.

Note: As was done in Example 2.26, the enthalpy of the subcooled liquid is always looked up in the table at its temperature, not its pressure.

Latent Heat Equation

The change in enthalpy that occurs when a substance evaporates or condenses is determined from the *latent heat equation,* found by applying the Enthalpy Equation 2.13 to the change in state:

$$Q = m(h_g - h_f) = m \times h_{fg} \qquad (2.14)$$

where

Q = heat added to or removed from substance, BTU/hr
m = weight flow rate, lb/hr
h_f = enthalpy of saturated liquid, BTU/lb
h_g = enthalpy of saturated vapor, BTU/lb
h_{fg} = latent heat of vaporization, BTU/lb

The latent heat of vaporization for water is shown in Table A.3. Note that it changes with temperature.

When a heating or cooling process involves both a sensible and a latent heat change to the substance, the results can be found by simply adding the two effects together.

Example 2.27

A steam boiler generates 20,000 lb/hr of saturated steam at 20 psia. The water enters the boiler at 180 F. How much heat is required?

Solution
The enthalpy increase of the water is the sum of the sensible and latent heat change. The sensible change is to the boiling point. From Table A.3, the boiling temperature is 228 F at 20 psia. The change in sensible heat content of the liquid is

$$h_{f2} - h_{f1} = 196.2 - 147.9 = 48.3 \text{ BTU/lb}$$

The change in the latent heat content at 20 psia is

$$h_{fg} = 960.1 \text{ BTU/lb}$$

The total enthalpy change per pound is the sensible plus latent heat:

$$48.3 + 960.1 = 1008.4 \text{ BTU/lb}$$

The total heat required is

$$Q = 20,000 \ \frac{\text{lb}}{\text{hr}} \times 1008.4 \ \frac{\text{BTU}}{\text{lb}}$$

$$= 20,168,000 \text{ BTU/hr}$$

Equation 2.13 can be used to solve this example in one step instead of two, by using the initial and final enthalpy values of the whole process:

$$Q = m(h_2 - h_1)$$
$$= 20,000 \, (1156.3 - 147.9)$$
$$= 20,168,000 \text{ BTU/hr}$$

Note that Examples 2.26 and 2.27 illustrate that the Enthalpy Equation 2.13 can be used for both sensible and latent processes involving water if there are property tables available that list enthalpy values.

In Chapter 7, we will use this equation for air conditioning processes.

2.16 LATENT HEATS OF FUSION AND SUBLIMATION

The change of state of a substance from liquid to gas involves gaining the latent heat of vaporization. A substance in a solid state will increase in temperature when heat is added to it (sensible heat) but when a certain temperature is reached, its temperature will no longer increase when more heat is added, and the substance will begin to change state to a liquid—it will melt. If the reverse process is carried out, removal of heat from a liquid, its temperature will drop but eventually it will freeze into a solid. The heat accompanying the melting or freezing process is called the *latent heat of fusion*. For water the latent heat of fusion is 144 BTU/lb.

At very low pressure and temperature it is possible to change some substances directly from the solid to the gas state. This process is called *sublimation*. It is used in the procedure called *freeze-drying*, to prepare dried foods by first freezing them and then evaporating the ice in the food directly to vapor, at a very low pressure.

2.17 THE IDEAL (PERFECT) GAS LAWS

Under certain conditions, the pressure, volume, and temperature of gases are related by an equation called the *perfect* or *ideal gas law*. Air at the temperatures and pressures in air conditioning work follow this equation. The perfect gas equation can be expressed

$$pV = mRT \qquad (2.15)$$

where

p = pressure, lb/ft^2 absolute
V = volume, ft^3
m = weight of gas, lb
R = a gas constant
T = absolute temperature, degrees R

We will use the equation in this form in Chapter 7. By rearranging the terms in the equation for two different conditions of the gas, 1 and 2, the following equation results:

$$\frac{p_2 V_2}{T_2} = \frac{p_1 V_1}{T_1} \qquad (2.16)$$

The gas law is useful in finding changes in p, V, and T for changed conditions. If only two of these three variables change, the equation simplifies. If the temperature is constant,

$$\frac{p_2}{p_1} = \frac{V_1}{V_2} \qquad (2.17)$$

If the volume is constant,

$$\frac{p_2}{p_1} = \frac{T_2}{T_1} \qquad (2.18)$$

If the pressure is constant,

$$\frac{V_2}{V_1} = \frac{T_2}{T_1} \qquad (2.19)$$

Example 2.28 _____

Compressed air required for operating the pneumatic controls in an air conditioning system is stored in a 10 ft^3 tank at 150 psig. The air is used in the controls at 15 psig. What volume of air is available for the controls?

Solution
Using Equation 2.17, with absolute pressures ($p_{abs} = p_g + p_{atm}$), assuming the temperature remains constant,

$$V_1 = \frac{P_2}{P_1} \ V_2 = \frac{164.7 \text{ psia}}{29.7 \text{ psia}} \times 10 \text{ ft}^3 = 55.5 \text{ ft}^3$$

Example 2.29 _____

A technician testing and balancing a system measures 5000 ft^3/min (CFM) of air *entering* a heating coil at 40 F, and leaving at 120 F. However, the engineering specifications call for measuring the volume flow rate *leaving* the coil. What is the airflow leaving the coil?

Solution

Equation 2.19 can be used, with volume flow rate instead of volume, because the same unit of time is involved. Temperatures must be in absolute units $R = (F + 460)$:

$$\text{CFM}_2 = \frac{T_2}{T_1} \ \text{CFM}_1 = \frac{580}{500} \times 5000 = 5800 \text{ CFM}$$

The technician would now check the system design specifications. If it called for 5800 CFM, he knows that the proper amount of air is flowing, even though he measured 5000 CFM.

_____

2.18   ENERGY UTILIZATION (SECOND LAW OF THERMODYNAMICS)

We have seen how the First Law of Thermodynamics can be used, in the form of the Energy Equation, to solve problems in HVAC work, and we will use it again. Basically, it tells us how much energy is used for a given task (the power of a pump, the capacity of a refrigeration machine, and similar information). However, it tells us nothing about the answers to such questions as "Can I use a smaller pump, fan, or refrigeration machine?" or "How do I reduce the energy consumption of an HVAC system?"

An understanding and application of the *Second Law of Thermodynamics* will enable us to investigate problems of more efficient energy utilization. Energy conservation has become of great necessity

and concern. Unfortunately, efforts in this area have sometimes been haphazard, partly due to a lack of understanding of the Second Law.

The Second Law may be expressed as an equation but it is not simple to use in energy utilization analysis. Therefore, what we will do here is state some conclusions derived from the Second Law. Throughout the book we will suggest energy conservation steps, many based on these conclusions. In Chapter 15, these ideas will be gathered together and additional ones will be discussed. The reader may wish to treat the whole subject of energy conservation at that time, or to consider each aspect as it is brought up. Some of the conclusions that can be drawn from the Second Law are:

1. Whenever heat energy is used to do work, it is never all available for a useful purpose. Some must be lost, and unavailable for the job to be performed. For instance, if we are using an engine to drive a refrigeration compressor, only part of the energy in the fuel can be used; the rest will be wasted.

2. The maximum amount of energy that can be made available in a power-producing device such as an engine or turbine can be calculated. That is, we can determine the best efficiency possible, and compare it with an actual installation.

3. The minimum amount of energy required to produce a given amount of refrigeration can be calculated, and this can be compared with the actual system.

4. There are a number of physical effects that are called *irreversible* which cause a loss of available energy. These effects cannot be avoided, but should be reduced to a minimum. Included among them are:

   A. *The temperature difference for heat transfer.* Greater temperature differences cause greater losses, therefore temperature differences between fluids should be kept as small as practical, as for example, in evaporators and condensors.

   B. *Friction.* Friction causes loss of useful energy, and therefore should be minimized.

For example, regular cleaning of condenser water piping will reduce the roughness of the wall. Fluid friction will be reduced, and less energy will be lost in pumping power.

C. *Rapid expansion.* An example of possible wasted energy that results from this is the generation of high pressure steam and then expanding it in a "flash tank" to a low pressure before using it for heating.

D. *Mixing.* Mixing fluids of different temperatures can result in a loss of useful energy. Mixing processes are common in HVAC systems, but should be minimized or even avoided when they cause a loss of available energy. Dual duct systems and three pipe systems are two types of air conditioning systems using mixing that can result in energy waste. They will be discussed in Chapter 12.

Any process that occurs without any of these effects is called a *reversible* process. Although a reversible process is an ideal case that is impossible to achieve, we always try to minimize irreversible effects in the interests of energy conservation.

Entropy is a physical property of substances related to energy utilization and conservation. It is defined as the ratio of the heat added to a substance to the temperature at which it is added. However, this definition is not useful here. It is important to understand that entropy is a measure of the energy that is *not available* to do work.

For any process that requires work, such as driving a refrigeration compressor, the least amount of work is required if the entropy of the fluid does not change. This is called a constant entropy, or *isentropic* process. In a constant entropy process, no heat is added to or removed from the substance (*adiabatic* process) and there are no irreversible effects (e.g., friction).

A constant entropy process is an ideal reversible process that can never really occur. However, studying it gives us a goal to aim for. In any real process where work is required, the entropy increases, and we try to minimize this increase.

Practical applications of the First and Second Laws are discussed in much greater detail in later chapters, especially Chapters 3, 4, 6, 12, 13, and 15.

Example 2.30

A mechanical contractor has a choice of using copper tubing or steel piping of the same diameter in a chilled water system. Which would be the best choice to minimize energy consumption?

Solution
Copper tubing has a smoother surface and therefore less friction. Less energy will be used in the pump, according to the Second Law.

Review Questions

1. What is a unit? What problems may arise when using units?

2. What is a conversion factor?

3. What are the advantages of the SI system of units?

4. Explain what is meant by *rounding* off.

5. Define mass, force, weight, and pressure.

6. Define density, specific volume, and specific gravity.

7. With the aid of a sketch, explain gage pressure and absolute pressure.

8. What is a compound gage?

9. What is meant by stored energy and energy in transfer? Name types of energy in each category and give an example of each.

10. What is head?

11. Define heat, temperature, and enthalpy.

12. What is the difference between work and power?

13. State the energy balance as a sentence and as an equation.

14. What are the three common states in which matter may exist?

15. Define the saturated, superheated, and subcooled conditions.

16. Explain what is meant by a sensible heat change and a latent heat change.

17. List four conditions that should be sought in HVAC systems to minimize energy use, as suggested by the Second Law of Thermodynamics.

Problems

2.1 List the physical characteristics measured by each of the following units: lb/in.^{2}, HP, GPM, in. Hg, m/sec, ft^2, KW, BTU, kg/m^3, and ft^3/lb.

2.2 List the standard SI unit and a typical U.S. unit for each of the following physical characteristics: power, pressure, velocity, mass, flow rate, energy, specific volume, and density.

2.3 Change the following quantities to the new units specified:

A. 120 lb/in.^{2} to ft w.

B. 83.2 ft^3/sec to gal/min (GPM)

C. 76,500 BTU/hr to tons of refrigeration

D. 18.2 in. Hg to lb/in.^{2}

E. 0.91 HP to BTU/min

2.4 Change the following quantities from the U.S. units to the SI units specified:

A. 12.6 ft^2 to m^2

B. 629 ft^3/min (CFM) to m^3/sec

C. 347,000 BTU/hr to KW

D. 62.4 lb/ft^3 to kg/m^3

2.5 Find the area in ft^2 of a window that is 4 ft 3 in. wide by 6 ft 6 in. high.

2.6 A contractor wants to lift 100 sections of steel pipe each 20 ft long. The pipe, made in Germany, is stamped "100 kg/m." How many pounds must the crane be capable of lifting?

2.7 Round off the following numbers to three significant figures:

A. 793,242

B. 2.685

C. 542

D. 1.9

E. 0.8319

2.8 Change the following quantities to the units specified:

A. 276 gal water to lb

B. 2760 lb/hr water to GPM

C. 41,800 ft^3 air at 70 F and 14.7 psia to lb

2.9 A hot water storage tank for a solar-energy system measures 18 ft long × 9 ft wide. It is filled to depth of 6 ft. What is the weight of water in the tank? What is the water pressure on the bottom of the tank, in lb/in.^{2}?

2.10 What is the density in lb/ft^3 of a fuel oil with a s.g. = 0.93?

2.11 The pressure gage on a boiler in Boston reads 28.7 psi. What is the absolute pressure in psi? What would the absolute pressure be if the boiler were in Denver?

2.12 The absolute pressure in the suction line to a compressor is 12.2 in. Hg. What pressure would a vacuum gage read at sea level, in in. Hg?

2.13 A vat 25 ft high is filled with Big Brew Beer that has a s.g. = 0.91. What is the pressure in psi on a valve 3 ft above the bottom of the tank?

2.14 The discharge pressure of a pump is 32.6 psig. What is the pressure in ft w?

2.15 The air pressure in a tank is 3.7 ft w.g. What would be the reading on a Hg manometer attached to the tank in inches, and in mm?

2.16 Change the following temperatures to the new units specified:

A. 88 F to C

B. 630 F to R

C. −10 C to F

D. 280 K to C

E. 31 C to R

2.17 A room receives 1200 BTU/hr of heat from solar radiation, loses 1450 BTU/hr through heat transfer to the outdoors, and has 2.2 KW of appliances operating. What is the net heat gain or loss to the room?

2.18 An electric heater is to be used to heat an enclosed porch that is losing 7900 BTU/hr to the outdoors. What size heater should be used?

2.19 Determine the state of water at the following conditions:

A. 230 F and 18 psig

B. 180 F and 5 psia

C. 20 psia and 400 F

D. 0.1 psia and 50 F

2.20 A hot water boiler heats 6400 lb/hr of water from 180 F to 220 F. How many BTU/hr of heat are required? Solve by the sensible heat equation and by using Table A.3.

2.21 How many tons of refrigeration are required to cool 46 GPM of milk from 80 F to 50 F? The milk has a specific heat of 0.9 BTU/lb-F and a density of 8.1 lb/gal.

2.22 A 2.5 KW electric heater in a duct is heating 1300 lb/hr of air entering at 40 F. To what temperature is the air heated?

2.23 Water enters a steam boiler at 160 F and leaves as saturated steam at 30 psig, at a flow rate of 5300 lb/hr. How much heat is the boiler supplying?

2.24 If 520 ft^3 of air at atmospheric pressure at sea level is to be compressed and stored in a tank at 75 psig, what is the required volume of the tank?

2.25 An air conditioning unit takes in 15,700 CFM of outdoor air at 10 F and heats it to 120 F. How many CFM of air are leaving the unit?

2.26 What is the boiling point (saturation) temperature of water at pressures of 7.5 psia and 67.0 psia?

2.27 Is water liquid or vapor at 270 F and 50 psia?

2.28 A 24 ft high pipe filled with water extends from a condenser on the top floor of a building to a cooling tower on the floor above. What is the pressure exerted on the condenser in psi?

2.29 A barometer reads 705 mm Hg. What is the atmospheric pressure expressed in psi and in in. Hg?

2.30 A refrigeration unit has a cooling capacity of 327,000 BTU/hr. Express this capacity in tons of refrigeration and in KW.

2.31 A water chiller with a capacity of 150 tons of refrigeration cools 320 GPM of water entering the chiller at 52 F. At what temperature does the water leave the chiller?

CHAPTER 3
Heating Loads

In this chapter, we will discuss methods for determining the amount of heat required to keep the spaces in a building comfortable in winter. The methods presented here will be those which are believed to be the most accurate and the most energy efficient.

OBJECTIVE

After studying this chapter, you will be able to:

1. Find R- and U-values for building components.
2. Select appropriate indoor and outdoor design conditions.
3. Calculate room and building heat transfer losses.
4. Calculate room and building infiltration and ventilation losses.
5. Determine room and building heating loads.

3.1   THE HEATING LOAD

From our own experiences, we know that if the heating system in a building stops functioning in winter, the indoor air temperature soon drops. This temperature decrease occurs for two reasons: *heat transfer* from the warm inside air to the cold outside air through walls, windows, and other parts of the building envelope, and leakage of cold air through openings in the building (*infiltration*).

To counteract these heat losses, heat must be continually added to the interior of the building in order to maintain a desired air temperature. This can be shown by applying the Energy Equation (Chapter 2) to the air in a room or building. The energy added to the room air (E_{in}) is the heat supplied by the heating system (Q_{in}). The heat removed (E_{out}) is the heat loss (Q_{out}). The change in stored energy (E_{ch}) is the change in the room air enthalpy (H_{ch}). Substituting into Equation 2.1,

$$E_{ch} = E_{in} - E_{out}$$

we obtain

$$H_{ch} = Q_{in} - Q_{out} \tag{3.1}$$

where

H_{ch} = change in room air enthalpy
Q_{in} = heat supplied by heating system
Q_{out} = heat losses from room air to outside

The room air temperature depends on its enthalpy. If the air enthalpy decreases, its temperature decreases. Since we want the room air to remain at a constant elevated temperature, the enthalpy must also remain constant. That is, the enthalpy does not change, H_{ch} = 0. Substituting into Equation 3.1, we obtain

$$0 = Q_{in} - Q_{out}$$

or

$$Q_{in} = Q_{out} \qquad (3.2)$$

Figure 3.1 illustrates the heat flow into and out of a room. Equation 3.2 tells us that if the room air enthalpy (and therefore temperature) is to be maintained at a constant desired value, *the heat supplied by the heating system must equal the heat losses from the room.* This is a valuable conclusion, because the heat losses from a building can be easily calculated; then this information can be used to determine the required capacity of the heating equipment.

The amount of heat that must be supplied to keep the building or room air at the desired temperature is called the *heating load.* The heating load must be determined because it is used in the selection of the heating equipment, piping and duct sizing, and in energy utilization studies.

Accurately determining the heating load is a fundamental step in planning a heating system. In the past, many inaccurate methods have been used to find heating loads. This can result in unsatisfactory indoor air conditions and increased energy costs. States and agencies have established codes that now require accurate heating and cooling load calculation methods. (Cooling load calculation methods will be discussed in Chapter 6.)

The heating load requirements for buildings result from two types of heat losses: *heat transfer* losses and *infiltration/ventilation* losses. An explanation of these will now be presented.

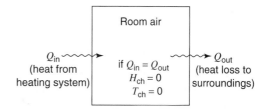

Figure 3.1
Heat exchanges between room air and surroundings. If heat furnished to room (Q_{in}) equals heat lost from room (Q_{out}), room air temperature remains constant (T_{ch} = 0).

3.2 HEAT TRANSFER

Because building heat losses are partially a consequence of heat transfer, it is necessary to understand some basic features of this process.

It was noted previously that heat is transferred only when there is a temperature difference between two locations, and that the heat always travels from the location of higher temperature to the location of lower temperature. There are three different ways that heat transfer can occur: *conduction, convection,* and *radiation.*

Conduction is the form of heat transfer through a body that occurs without any movement of the body; it is a result of molecular or electron action.

Conduction is most familiar in heat transfer through solids—for example, when the metal body of a pot is heated on a stove, the heat flows through the handle and then to your hand. Another example of conduction is heat transfer through a building wall or roof. Conduction heat transfer can also occur through liquids and gases; however, an additional form of heat transfer is more usual in fluids (convection).

Convection is the form of heat transfer that results from gross movement of liquids or gases.

A familiar example of convection is the air in a room heated by a unit such as a hot water convector.

Heat is transferred to the air adjacent to the metal surface, increasing its temperature. This warmed air then moves vertically upward because it is now less dense (lighter) than the surrounding cooler air. So air continually moves throughout the space (Figure 3.2). This form of convection is called *natural convection* because the fluid moves by natural gravity forces created by density differences. The less dense part of the fluid rises and the more dense (heavier) fluid drops. The rate of fluid motion created by natural convection effects is generally quite low, and therefore the resulting rate of heat transfer is relatively small. The rate of fluid motion and therefore the rate of heat transfer can be increased by using a fan for gases or a pump for liquids. This is called *forced convection.*

> *Thermal **radiation** is the form of heat transfer that occurs between two separated bodies as a result of a means called electromagnetic radiation, sometimes called wave motion.*

As with all forms of heat transfer, one body must be at a higher temperature than the other. Heat transfers between the two bodies even if there is a vacuum (an absence of all matter) between them. When there is a gas between the bodies, heat still transfers by radiation but usually at a lesser rate. However, the presence of an opaque solid object between the bodies will block radiation. Familiar examples of radiation are the heat our body

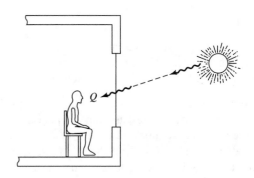

Figure 3.3
Heat transfer by radiation from the sun to objects in a room.

receives when standing in front of a fire, and the heat received from the sun (Figure 3.3).

When radiation is received by a solid surface, some is *absorbed,* heating the material, and some is *reflected.* The proportion absorbed depends on the color and the roughness of the surface. Dark, rough surfaces absorb more radiant heat than lighter-colored smooth surfaces.

Most of the radiation received passes through transparent materials like clear glass. Color tinted glass, however, called *heat absorbing* glass, can prevent the transmission of a good part of the solar radiation.

3.3 RATE OF HEAT TRANSFER

The rate at which heat is conducted through any material depends on three factors:

1. The temperature difference across which the heat flows
2. The area of the surface through which heat is flowing
3. The *thermal resistance (R)* of the material to heat transfer

This can be expressed by the following equation

$$Q = \frac{1}{R} \times A \times \text{TD} \qquad (3.3)$$

Figure 3.2
Heat transfer by natural convection from a terminal unit (hot water convector) to room air.

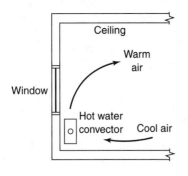

where

Q = heat transfer rate, BTU/hr
R = thermal resistance of material,
 hr-ft²-F/BTU
A = surface area through which heat flows, ft²
TD = $t_H - t_L$ = temperature difference across
 which heat flows, from higher temperature
 t_H to lower temperature t_L, F

Thermal Resistance

The *thermal resistance R* of a material is its ability to resist the flow of heat through it. Equation 3.3 allows us to understand how the thermal resistance affects building heat energy losses or gains. Since R is in the denominator, high R-values mean low heat transfer (Q), and low R-values mean high heat transfer. Materials with high R-values will transfer heat at a low rate, that is, they are good thermal insulators. Building construction materials with a high R-value are desirable because they reduce heat losses. On the other hand, using a material with a low R-value (metal) for equipment such as a boiler is desirable because it helps to increase the rate of heat transfer from the combustion gases to the water.

The thermal resistances of various building materials are listed in Table A.4. The resistance is often expressed by a symbol, for example "R-6." This means $R = 6$.

Example 3.1 _____
A 110 ft long by 20 ft high wall is made of 4 in. common brick. The temperature on the inside surface of the wall is 65 F and on the outside surface the temperature is 25 F. What is the rate of heat transfer through the wall?

Solution
Figure 3.4 illustrates the conditions. From Table A.4,

$$R = 0.20 \text{ hr-ft}^2\text{-F/BTU per in.} \times 4 \text{ in.}$$
$$= 0.80 \text{ hr-ft}^2\text{-F/BTU}$$
Area of wall $A = 110 \text{ ft} \times 20 \text{ ft} = 2200 \text{ ft}^2$
$$\text{TD} = t_H - t_L = 65 - 25 = 40 \text{ F}$$

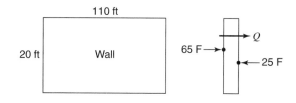

Figure 3.4
Sketch for Example 3.1.

Using Equation 3.3,

$Q = 1/R \times A \times \text{TD}$
 $= 1/0.80 \text{ hr-ft}^2\text{-F/BTU} \times 2200 \text{ ft}^2 \times 40 \text{ F}$
 $= 110,000 \text{ BTU/hr}$

Thermal Resistance of Surface Air Films

There is a very thin film of air on each side of a solid building element such as a wall or roof. These films also have a thermal resistance, just as solid materials do. The resistance of an air film depends on the spatial orientation of the surface (vertical, horizontal, or on a slope), and on the air velocity near the surface.

Table A.5 lists thermal resistances of these air films. For winter conditions (heating loads), it is assumed that the air velocity outdoors is 15 MPH. For the indoor surface of any building element, still air is assumed. Example 3.2 illustrates the use of Table A.5.

Example 3.2 _____
A wall of a supermarket measures 80 ft by 18 ft. The temperature of the air in the store is 70 F and the inside surface of the wall is 60 F. What is the heat loss through the wall?

Solution
The resistance is the inside air film on a vertical surface (Figure 3.5). From Table A.5, $R = 0.68$.

$A = 80 \times 18 = 1440 \text{ ft}^2$, TD $= 70 - 60 = 10 \text{ F}$

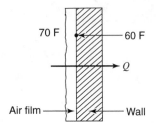

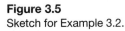

Figure 3.5
Sketch for Example 3.2.

Using Equation 3.3,

$$Q = 1/R \times A \times TD$$
$$= 1/0.68 \times 1440 \times 10$$
$$= 21,200 \text{ BTU/hr}$$

Conductance and Conductivity

Besides thermal resistance, *conductance* and *conductivity* are terms which are used to describe a material's ability to transfer heat.

The *thermal conductance (C)* of a material is the reciprocal of its resistance:

$$C = 1/R \qquad (3.4)$$

C is measured in units of BTU/hr-ft^2-F in the U.S. system. The thermal conductance may be thought of as the ability of a material to transfer heat, the opposite meaning of resistance.

The thermal conductance of the air film adjacent to a surface is often called the *film coefficient.*

The *thermal conductivity (k)* of a material is defined as its conductance per unit of thickness. The units used for conductivity are usually BTU/hr-ft^2-F per inch of thickness. Its relationship to conductance is

$$C = \frac{k}{L} \qquad (3.5)$$

where

C = conductance, BTU/hr-ft^2-F
k = conductivity, BTU/hr-ft^2-F per in. of thickness
L = thickness of material, in.

Example 3.3

A roof has 4 in. of glass fiber insulation with a thermal conductivity k = 0.24 BTU/hr-ft^2-F per in. thickness. What is the thermal resistance of the insulation?

Solution
From Equation 3.5, the conductance is

$$C = \frac{k}{L} = \frac{0.24}{4} = 0.06 \; \frac{\text{BTU}}{\text{hr-ft}^2\text{-F}}$$

Using Equation 3.4, solving for R,

$$R = \frac{1}{C} = \frac{1}{0.06} = 16.7 \; \frac{\text{hr-ft}^2\text{-F}}{\text{BTU}}$$

It is not necessary to memorize the definitions just described. The concept of thermal resistance is the important one to understand, as it is being used in all new building standards and codes. Remember that for a material with high thermal resistance, heat transfer will be low, and for a material with a low resistance, heat transfer will be high.

Example 3.4

Compare the thermal resistance R of an 8 in. thick concrete block (with sand and gravel aggregate) to that of 1 in. thick insulating board made of glass fiber.

Solution
Both R-values are found from Table A.4.

For the concrete block, $R = 1.11$
For the insulating board, $R = 4.0$

Note from Example 3.4 that the insulation has about four times the thermal resistance of the concrete block, although it is only 1/8th as thick. That is, it is 32 times more effective per inch of thickness as a thermal insulator!

In the next section, we will learn how to determine the overall thermal resistance of a building component.

3.4 OVERALL THERMAL RESISTANCE

The heat transfer through the walls, roof, floor, and other elements of a building is through the air film on one side, through the solid materials, and then through the air film on the other side.

These elements are usually made up of layers of different materials. The *overall (total) thermal resistance* of the combination can be found very simply by adding the individual thermal resistances as follows:

$$R_0 = R_1 + R_2 + R_3 + \text{etc.} \qquad (3.6)$$

where

R_0 = overall (total) thermal resistance
R_1, R_2, etc. = individual thermal resistance of each component, including air films

Once the overall resistance R_0 is known, Equation 3.3 can be used to find the heat transfer, as illustrated in Example 3.5.

Example 3.5

The exterior wall of a building is constructed of 8 in. sand and gravel aggregate concrete (not oven dried), *R*-5 insulation, and ½ in. gypsum board. The wall is 72 ft long by 16 ft high. The indoor and outdoor temperatures are 70 F and −10 F. What is the heat transfer through the wall?

Figure 3.6
Sketch for Example 3.5.

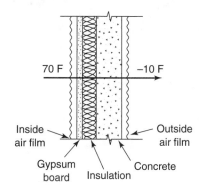

Solution
A section through the wall is illustrated in Figure 3.6. All of the resistances are found in Tables A.4 and A.5. The thermal resistances of the air films on the inside and outside surfaces must also be considered.

The overall thermal resistance R_0 is found by adding the individual resistances (See Equation 3.6).

Wall Item	*R*
Inside air film	0.68
Gypsum board	0.45
Insulation	5.0
Concrete ($R = 0.08/\text{in.} \times 8$ in.)	0.64
Outside air film	0.17
Overall resistance $R_0 =$	6.94

The wall area is

$$A = 72 \text{ ft} \times 16 \text{ ft} = 1152 \text{ ft}^2$$

The temperature difference is

$$\text{TD} = 70 - (-10) = 80 \text{ F}$$

Using Equation 3.3 and the overall resistance R_0, the heat transfer loss is

$$Q = \frac{1}{R_0} \times A \times \text{TD}$$

$$= \frac{1}{6.94} \times 1152 \times 80$$

$$= 13{,}280 \text{ BTU/hr}$$

3.5 OVERALL HEAT TRANSFER COEFFICIENT (*U*)

For each application, the designer can calculate the overall thermal resistance for each part of the building structure through which heat flows, but fortunately, these calculations have already been made for many different combinations of building materials. However, many tables do not list the results as overall resistance, but as overall conductance, called the *overall heat transfer coefficient*

(U), BTU/hr-ft^2-F. The relationship between R_0 and U is

$$U = \frac{1}{R_0} \qquad (3.7)$$

In terms of U, the heat transfer equation then becomes

$$Q = U \times A \times TD \qquad (3.8)$$

where

Q = heat transfer rate, BTU/hr
U = overall heat transfer coefficient, BTU/hr-ft^2-F
A = surface area through which heat flows, ft^2
TD = temperature difference, F

Overall U-values for some combinations of building components are listed in Tables A.6, A.7, and A.8. The following comments should help the student use these tables correctly.

1. Tables A.6 and A.7 list U-values for roofs, walls, floors, partitions, and doors. Either table may be used, depending on which one matches the construction assembly in the case encountered.
2. Table A.6 also shows graphically the sections through each building components. These should be examined so that the student learns how the construction assembly is actually arranged. Table A.6 is also valuable because many of the assemblies listed meet U and R_0 values required in energy codes.
3. The U-values in Table A.8 are for glass windows and glass doors. Note that there is a slight difference in U-values between winter and summer. This is because the R-values of the outside air film coefficients used in finding the U-values are based on a 15 MPH wind in winter and 7.5 MPH wind in summer. The summer U-values are used in cooling load calculations (see Chapter 6). The U-values also include the effect of the window or door frame (also called sash, with windows).

When U-value tables do not include the actual construction, it should be calculated by adding the individual resistances, as explained previously.

U-values and Energy Standards

State energy codes and standards attempt to limit the amount of energy used by HVAC systems. One way of doing this is to prescribe maximum allowable U-values (and minimum R_0 values). Table 3.1 is a simplified example of this type of regulation.

TABLE 3.1

Component	Min. R_0 (ft^2-hr-F/BTU)	Max U (BTU/hr-ft^2-F)
Wall	18	0.06
Roof	20	0.05
Glass	1.7	0.60

Note: This table is adapted from various state energy standards.

The values in Table 3.1 are not those from an actual state, but are similar to those prescribed in some states for a somewhat cold winter climate, one having a degree day (DD) value of about 4000–6000.

The degree day is a number that reflects the length and severity of a heating season. Degree day values for localities are shown in Table A.9.

For an actual building, of course, the designers would refer to the applicable real energy code or standard.

The degree day concept and a more detailed explanation of energy standards are discussed in Chapter 15.

Examples 3.6, 3.7, and 3.8 show how to use Equation 3.8 and the U-value tables to calculate heat transfer through building components.

Example 3.6

Find the U-value for the wall described in Example 3.5, using Table A.7. Compare it with the value that would be found from Example 3.5, using Equations 3.6 and 3.7.

Solution
From Table A.7, $U = 0.14$. From Example 3.5, $R_0 = 6.94$. Using Equation 3.7,

$$U = \frac{1}{R_0} = \frac{1}{6.94} = 0.14$$

The results agree.

Example 3.7

A building 120 ft long by 40 ft wide has a flat roof constructed of 8 in. lightweight aggregate concrete, with a finished ceiling. The inside temperature is 65 F and the outdoor temperature is 5 F. What is the heat transfer loss through the roof?

Solution
From Table A.7, U = 0.09 BTU/hr-ft²-F. Using Equation 3.8,

$$Q = U \times A \times TD$$

$$= 0.09 \frac{BTU}{hr\text{-}ft^2\text{-}F} \times 4800 \text{ ft}^2 \times 60 \text{ F}$$

$$= 25,900 \text{ BTU/hr}$$

Example 3.8

A frame wall of the bedroom of a house has the following specifications:

Wall: *12 ft by 8 ft, wood siding, wood sheathing, 2 in. of insulation with* R-7 *value, and inside finish*

Window: *3 ft by 4 ft 6 in., single glass, aluminum frame*

The room temperature is 68 F and the outdoor temperature is 2 F. What is the heat transfer loss through the wall and window combined?

Solution
Figure 3.7 illustrates the wall. There is heat transfer through the opaque part of the wall and through the window.

The U-values from Tables A.7 and A.8 and the areas are:

$$\text{Window } U = 1.10 \text{ BTU/hr-ft}^2\text{-F}$$
$$A = 3 \times 4.5 = 13.5 \text{ ft}^2$$
$$\text{Wall } U = 0.09 \text{ BTU/hr-ft}^2\text{-F}$$
$$\text{Gross } A = 12 \times 8 = 96 \text{ ft}^2$$
$$\text{Net } A = 96 - 13.5 = 82.5 \text{ ft}^2$$

The heat transfer is:

Wall

$$Q = 0.09 \times 82.5 \times 66 = 490 \text{ BTU/hr}$$

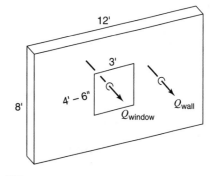

Figure 3.7
Sketch for Example 3.8.

Window

$$Q = 1.10 \times 13.5 \times 66 = \underline{980} \text{ BTU/hr}$$
$$\text{total } Q = 1470 \text{ BTU/hr}$$

3.6 HEAT TRANSFER LOSSES: BASEMENT WALLS AND FLOORS

Equation 3.8 is also used to calculate the heat losses through basement walls and floors. However, the fact that the basement floor and part or all of the wall is underground (below grade) complicates the use of the equation.

If any part of the basement is above ground the U-values from Table A.6 or Table A.7 are used. The inside and outside design air temperatures are also used to find the TD (see Section 3.9).

However, for the part of the structure which is below grade, the U-values and TD used in Equation 3.8 will be different. This is due to the effect of the surrounding ground on the thermal resistance and the heat flow path. Table 3.2 on page 52 lists recommended U-values for below grade basement walls and floors.

To find the TD for the below grade wall or floor, the outside winter design temperature is taken to be the deep ground temperature value. This temperature varies between about 40 F and 60 F in cold climates in the continental United States.

TABLE 3.2   OVERALL HEAT TRANSFER COEFFICIENT *U* FOR BASEMENT WALLS AND FLOORS, BELOW GRADE, BTU/HR-FT²-F

Component	*U*
Wall, uninsulated	0.16
Wall, *R*-4 insulation	0.08
Floor	0.04

Notes: Values are for a 7 ft high below grade basement. An uninsulated wall is not recommended in cold climates. Insulation is full depth of wall.

Example 3.9 illustrates a calculation of heat losses from a below grade basement.

Example 3.9

The recreation room of a basement has a floor area of 220 ft² and an insulated wall below ground of 400 ft² area. The room temperature is 70 F, and the ground temperature is 50 F. What is the heat loss from the room?

Solution

Using recommended *U*-values from Table 3.2,

$$\text{Floor } Q = 0.04 \times 220 \times 20 = 180 \text{ BTU/hr}$$
$$\text{Wall } Q = 0.08 \times 400 \times 20 = \underline{640} \text{ BTU/hr}$$
$$\text{total } Q = 820 \text{ BTU/hr}$$

Remember that if part of the basement wall is above ground and part is below, the heat transfer losses from each part should be calculated separately, using the appropriate *U, A,* and TD values. The separate heat losses should then be added to find the total wall heat loss.

Basement Inside Temperature

A number of possible conditions may exist in a basement.

1. *Basement heated (with terminal units).* The basement heat losses should be calculated using the basement inside design temperature. If the basement is partitioned, a separate calculation for each area is required.
2. *Unheated basement.* The temperature in an unheated basement (with no heat sources) will

be between the design inside and design outside temperatures. No heat loss calculation from the basement should be made.

3. *Basement heated by heat source equipment.* This situation exists when the furnace and ducts or boiler and piping are located in the basement. The following guidelines are recommended:
   A. For any area that the heat source is in or substantial hot ducts or piping pass through, assume the inside temperature is equal to that of the rest of the building, and calculate the resulting heat transfer loss.
   B. For other partitioned off areas, assume an unheated basement (as in item 2).

In items 2 and 3b, there will still be a heat loss from the floor above which should be added to the heat losses from those rooms. If the specific basement conditions are not known, an estimated unheated basement temperature of 50 F should be used. (A temperature below this should not be permitted anyway, because of the possibility of freezing water in piping.)

3.7   HEAT TRANSFER LOSSES: FLOOR ON GROUND AND FLOOR OVER CRAWL SPACE

Special calculations also apply for the heat transfer which occurs through a concrete floor slab on grade and through a floor with a crawl space below.

Floor Over Crawl Space

If the crawl space is vented during the heating season (to prevent moisture condensation), the crawl space air temperature will equal the outside air design temperature. The heat loss through the floor area should be calculated using Equation 3.8. In cold climates, the floor should be insulated; an overall *R*-20 value is a typical requirement in state energy codes.

If the crawl space is used for the warm air heating ductwork, the heat loss calculation is more complicated. Consult the *ASHRAE Handbook* or local energy codes.

Floor Slab on Grade

When a floor is on the ground, the heat loss is greatest near the outside edges (perimeter) of the building and is proportional to the length of these edges, rather than the area of the floor (Figure 3.8). This is the only case where heat transfer is not calculated using Equation 3.8. The following equation is used instead:

$$Q = E \times L \times TD \qquad (3.9)$$

where

Q = heat transfer loss through floor on grade, BTU/hr

E = edge heat loss coefficient, BTU/hr-F per ft of edge length

L = total length of outside (exposed) edges of floor, ft

TD = design temperature difference between inside and outside air, F

Table 3.3 lists values of E for various wall constructions, both with and without edge insulation. Insulation is usually required by energy codes. The last entry applies to a heated floor slab, which is recommended in more severe winter climates.

TABLE 3.3 EDGE HEAT LOSS COEFFICIENT, E, FOR FLOOR SLAB ON GRADE (BTU/hr-F per ft of edge)

Wall Construction	Edge Insulation	Degree Days		
		3000	5400	7400
8 in. block with face brick	None	0.62	0.68	0.72
	R-5	0.48	0.50	0.56
4 in. block with face brick	None	0.80	0.84	0.93
	R-5	0.47	0.49	0.54
Metal stud with stucco	None	1.15	1.20	1.34
	R-5	0.51	0.53	0.58
Poured concrete, heated floor	None	1.84	2.12	2.73
	R-5	0.64	0.72	0.90

Adapted from *ASHRAE 1993 Handbook—Fundamentals.*

Example 3.10

A 60 ft by 30 ft (in plan) garage built with a concrete floor slab on grade is maintained at 65 F. The outdoor air temperature is 4 F. Degree days are 5400. The floor edge has R-5 insulation. The walls are 8 in. block with face brick. What is the heat loss through the floor?

Solution
First find TD, E, and L.
 TD = 65 − 4 = 61 F
 From Table 3.3, E = 0.50 for 5400 degree days
 Edge length L = 2(60 + 20) = 160 ft
 Using Equations 3.9,

$$Q = E \times L \times TD$$
$$= 0.50 \times 160 \times 61$$
$$= 4880 \text{ BTU/hr}$$

The edge loss method (Equation 3.9) is recommended for buildings with small floor slab areas. For buildings with large floor slab areas, the edge loss method is recommended only for perimeter rooms. For the interior areas, Equation 3.8 should be used, with the U and TD values for basements.

Figure 3.8
Heat loss through floor of building without basement. (Actual insulation arrangement may differ.)

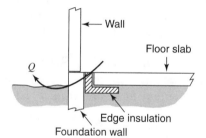

3.8 INFILTRATION AND VENTILATION HEAT LOSS

In addition to the heat required to offset heat transfer losses in winter, heat is also required to offset the effects from any cold outdoor air that may enter a building. The two means by which cold air may enter the building are called *infiltration* and *ventilation*. The resulting amounts of heat required are called the *infiltration heating load* and the *ventilation heating load*.

Sensible Heat Loss Effect of Infiltration Air

Infiltration occurs when outdoor air enters through building openings, due to wind pressure. The openings of most concern to us are cracks around window sashes and door edges, and open doors. Infiltration air entering a space in winter would lower the room air temperature. Therefore, heat must be furnished to the room to maintain its design air temperature.

The amount of heat required to offset the sensible heat loss from infiltrating air can be determined from the sensible heat equation (Section 2.15).

$$Q_s = m \times c \times TC \qquad (2.12)$$

where

Q_s = heat required to warm cold outdoor air to room temperature, BTU/hr
m = weight flow rate of outdoor air infiltration, lb/hr
c = specific heat of air, BTU/lb-F
TC = temperature change between indoor and outdoor air, F

The weight flow rate of air (m) in Equation 2.12 is expressed in lb/hr. However, air flow rates in HVAC work are usually measured in *ft³/min* (CFM). If the units are converted, using the appropriate specific heat of air, the sensible heat equation is

$$Q_s = 1.1 \times CFM \times TC \qquad (3.10)$$

where

Q_s = sensible heat loss from infiltration or ventilation air, BTU/hr
CFM = air infiltration (or ventilation) flow rate, ft³/min
TC = temperature change between indoor and outdoor air, F

Latent Heat Loss Effect of Infiltration Air

Since infiltration air is often less humid than the room air, the room air humidity may fall to an unacceptable level for comfort. If the room air humidity is to be maintained, water vapor must be added. The addition of this moisture requires heat (latent heat of vaporization of water). This is expressed by the following equation:

$$Q_l = 0.68 \times CFM \times (W_i' - W_o') \qquad (3.11)$$

where

Q_l = latent heat required for infiltration or ventilation air, BTU/hr
CFM = air infiltration or ventilation rate, ft³/min
W_i', W_o' = higher (indoor) and lower (outdoor) humidity ratio in grains water/lb dry air (gr w/lb d.a.)

To sum up, Equations 3.10 and 3.11 are used to find the room air sensible and latent heat losses resulting from infiltration air. The sensible heat loss should always be calculated. If the lower room air humidity resulting from infiltration is acceptable, then the latent heat loss effect may be neglected.

For the interested student, Equations 3.10 and 3.11 are derived in Chapter 7. The humidity ratios (W') for Equation 3.11 can be read from the psychometric chart, also introduced in Chapter 7.

Finding the Infiltration Rate

There are two methods used to estimate the CFM of infiltration air: the *crack method* and the *air change method*.

Crack Method The crack method assumes that a reasonably accurate estimate of the rate of air infiltration per foot of crack opening can be measured or established. Energy codes list maximum permissible infiltration rates for new construction or renovation upgrading. Table 3.4 lists typical allowable infiltration rates, based on a 25 MPH wind.

TABLE 3.4 TYPICAL ALLOWABLE DESIGN AIR INFILTRATION RATES THROUGH EXTERIOR WINDOWS AND DOORS

Component	Infiltration Rate
Windows	0.37 CFM per ft of sash crack
Residential doors	0.5 CFM per ft^2 of door area
Nonresidential doors	1.00 CFM per ft^2 of door area

Note: This table is adapted from various state energy standards.

The crack lengths and areas are determined from architectural plans or field measurements. Example 3.11 illustrates use of the crack method.

Example 3.11 _____
The windows in a house are to be replaced to meet local infiltration energy standards. The windows are 3 ft W x 4 ft H, double-hung type. Indoor and outdoor design temperatures are 70 F and 10 F. What will be the sensible heat loss due to infiltration?

Solution

1. From the Table 3.4, the new allowed infiltration rate is 0.37 CFM per ft of sash crack.
2. The total crack length $L = 3(3) + 2(4) = 17$ ft. (Note the allowance for the crack at the middle rail of a double-hung window, as shown in Figure 3.9.)
3. The total infiltration rate for the window is

$$\text{CFM} = 0.37 \text{ CFM/ft} \times 17 \text{ ft} = 6.3 \text{ CFM}$$

4. Using Equation 3.10, the infiltration sensible heat loss is

$$Q_s = 1.1 \times \text{CFM} \times \text{TC}$$
$$= 1.1 \times 6.3 \times (70 - 10) = 415 \text{ BTU/hr}$$

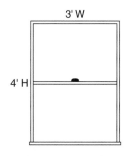

Figure 3.9
Sketch for Example 3.11.

The quality of installation and the maintenance of windows and doors greatly affect the resultant crack infiltration. Poorly fitted windows may have up to five times the sash leakage shown in Table 3.4.

Corner Room Infiltration When the infiltration rate is calculated for a room with two adjacent exposed walls (a corner room) with door or window openings on both sides, we assume that infiltration air comes through cracks on one side only, since the wind can only come from one direction at any given time. The wind changes direction, of course, but the infiltration effects cannot be additive, since they occur at different times.

If the wind comes obliquely (toward the corner), the projected crack lengths for each side are less; the overall effect is the same as if the wind came directly from one side only (using its actual crack lengths). If there are different types or sizes of openings on each side, the side that has the *greater* CFM should be used for the calculation.

The procedure for combining the infiltration rates of individual rooms on different walls, in order to find the total building infiltration rate, will be discussed in Section 3.11.

Door Usage For buildings that have frequent door usage (e.g., department stores), the infiltration that results from door opening should be included. The rate of door usage (number of people per minute) is first determined, with the advice of the architect or owner. Some average infiltration rates are shown in Table 3.5.

TABLE 3.5 INFILTRATION RATES FOR FREQUENT DOOR USAGE

Type	ft³ per Person
Swinging door, no vestibule	900
Swinging door, vestibule	550
Revolving door	60

For doors that are left indefinitely open, special means may be used to try to offset infiltration. *Unit heaters,* which can blow warm air directly at the opening, and *air curtains,* which direct a vertical warm air barrier across the opening, are two such methods. However, it is difficult to determine the effect on the building heating load of these methods.

Additional air infiltration may occur through a porous wall. If the walls have significant porosity, sealant coatings or other coverings may be applied to them.

In high-rise buildings, a thermal stack effect may increase infiltration through existing cracks. This occurs when the warmer inside air, which rises through the building and exits out through cracks on upper stories, is replaced by colder outside air entering through cracks on lower floors. Consult the *ASHRAE Handbook* for more information.

Much publicity has been given to reducing infiltration in existing buildings by use of weatherstripping and the sealing of cracks around frames, sillplates, wall penetrations, and other openings. There are many excellent publications available from governmental agencies and utility companies on this subject.

Air Change Method This procedure for finding the infiltration rate is based on the number of air changes per hour (ACH) in a room caused by the infiltration.

One air change is defined as being equal to the room air volume.

Determination of the expected number of air changes is based on experience and testing. Suggested values range from 0.5 ACH to 1.5 ACH for buildings ranging from "tight" to "loose" construction.

Using the definition of an air change, Equation 3.12 can be used to find the air infiltration rate in CFM.

$$\text{CFM} = \text{ACH} \times \frac{V}{60} \qquad (3.12)$$

where

CFM = air infiltration rate to room, CFM
ACH = number of air changes per hour for room
   V = room volume, ft³

Example 3.12 should help to clarify the meaning and use of the air change method.

Example 3.12
A 20 ft × 10 ft × 8 ft high room in a house has 0.7 air changes per hour due to infiltration. Find the infiltration rate in CFM.

Solution

1. The room volume is

   $$V = 20 \times 10 \times 8 = 1600 \text{ ft}^3$$

2. Using Equation 3.12,

   $$\text{CFM} = \text{ACH} \times \frac{V}{60} = 0.7 \times \frac{1600}{60} = 18.7 \text{ CFM}$$

Crack Method versus Air Change Method

The obvious question arises as to which of these methods should be used. There is no unqualified answer, but the following suggestions may be helpful:

1. The air change method is used primarily in residential construction heating load estimates, but there is no reason why the crack method cannot be used if reliable data are available.
2. The crack method is generally used in nonresidential construction. Reliable data from window manufacturers and quality control of installation and maintenance may provide good estimates using this method.

   It is often difficult to estimate leakage rates in older buildings because the condition of the windows is not known; but since heating load calculations for existing buildings are usually being made when upgrading for energy conservation, this should not be a problem.

Ventilation (Outside Air) Load

Some outside air is usually brought into nonresidential buildings through the mechanical ventilation equipment (air handling units) in order to maintain the indoor air quality. The outside ventilation air will be an additional part of the building heating load, since the entering air is at the outdoor temperature and humidity.

Equations 3.10 and 3.11 are also used to find the ventilation heating load. However, the ventilation air is heated (and humidified, if required) in the air conditioning equipment, *before* it enters the room. Therefore, it is part of the total building heating load, but *not* part of the individual room heating loads. The procedures for determining the appropriate quantity of outside ventilation air are explained in Chapter 6.

Mechanical ventilation systems for large buildings are usually designed and installed so that fans create a slightly positive air pressure in the building. This will reduce or even prevent infiltration.

When it is felt that the building is relatively tight and pressurized, no allowance for infiltration is made; only the outside air ventilation load is included.

A separate word of caution on pressurizing buildings: it is not uncommon to find that overpressurization results in doors that require great force to open or close.

Some nonresidential buildings have fixed windows (no openable part). In this case, of course, crack infiltration is limited to exterior doors only.

Air distributing systems in residences often use recirculated air only; in this case there is no ventilation load component. Infiltration generally provides adequate fresh air. However, in modern "tight" residences, there is concern that there may be inadequate natural infiltration, resulting in long-term health problems from indoor air pollutants. Outside ventilation air should be introduced in such cases.

Example 3.13 illustrates the calculation of ventilation loads.

Example 3.13

A building with sealed windows is maintained at 72 F, with an outdoor temperature of −5 F. The mechanical ventilation system introduces 5000 CFM of outside air. What is the additional sensible heating requirement from this effect?

Solution

The inside humidity conditions are not specified as being fixed, so only the sensible heat of the ventilation air is calculated. Using Equation 3.10,

$$Q_s = 1.1 \times \text{CFM} \times \text{TC}$$
$$= 1.1 \times 5000 \times 77 = 423{,}500 \text{ BTU/hr}$$

3.9 DESIGN CONDITIONS

The values of the indoor and outdoor air temperature and humidity that are used in heating (and cooling) load calculations are called the *design conditions.*

The indoor design conditions are generally chosen within the area of the comfort zone, as described in Section 1.6 and shown in Figure 1.10. More specifically, the indoor design conditions listed in Table 1.1 are compatible both with comfort and responsible energy conservation. The governing state energy code must also be followed in choosing the design condition.

Outside winter design temperatures are based on weather records. Table A.9 lists recommended outdoor design conditions for winter for some localities in the United States and other countries.

There are two possible design temperatures listed in the columns, entitled 99% and 97.5% dry bulb (DB) temperatures. (The term dry bulb is a way of expressing air temperature, explained in Chapter 7, and need not concern us here.)

The temperature in the 99% column means that in an average winter, for the months December through February, the outdoor temperature will be above the value listed 99% of the time. That is, 1% of the time the outdoor temperature will be below the listed value. The same concept applies to the 97.5% column.

The 99% column would be used if the major concern was to minimize any uncomfortably low indoor temperatures during the coldest hours. The 97.5% column might be used to reduce the heating equipment size selected, but possibly resulting in short periods of slightly low indoor temperatures. This might not actually occur, depending on the way the heating system is operated and other factors. State energy standards often specify the permitted outdoor design temperature.

If the building is to be humidified in winter, it is assumed that the outdoor design humidity is zero.

Table A.9 also includes summer design conditions for cooling (see Chapter 6.)

Example 3.14

Plans are being prepared for the Big Bargain Department Store in Chicago, Illinois. What outdoor winter temperature should be used for designing the heating system?

Solution
From Table A.9, the temperature is 2 F (using the 97.5% level), or −3 F (using the 99% level).

Unheated Space Temperature

Unheated rooms or spaces between a heated room and the outdoors will have a temperature lower than the heated room. The heat loss from the heated room through the separating partition should be calculated.

Some designers assume that the temperature of the unheated space is halfway between indoor and outdoor design conditions. If the unheated space has a large exposed glass area, it is better to assume the space is at outdoor temperature. If an unheated space is totally surrounded by heated spaces, it can be assumed to be at indoor design conditions, so the heat transfer can be neglected.

3.10   ROOM HEAT LOSS AND ROOM HEATING LOAD

The **room heat loss** *is the sum of each of the room heat transfer losses and infiltration heat losses.*

The **room heating load** is the amount of heat that must be supplied to the room to maintain it at the indoor design temperature. In Section 3.1, we explained that *the room heating load is equal to the room heat loss.*

Example 3.15 illustrates the procedure for finding the room heating load.

Example 3.15

The office room shown in Figure 3.10 is in a one-story building in Des Moines, Iowa. The building has a heated basement. Find the design room heating load. Construction is as follows:

Wall: 6 in. concrete (120 #/cu. ft), R-8 insulation, ½ in. gypsum board finish

Window: 5 ft H x 4 ft W pivoted type, double glass, aluminum frame. Infiltration rate is 0.50 CFM/ft

Roof: flat roof, metal deck, R-8 insulation, suspended ceiling

Ceiling height: 9 ft

Solution

1. A table is arranged to organize the data.
2. Design temperatures are selected from Tables 1.10 and A.9. Indoor temperature is 72 F; outdoor temperature is −10 F (at 99% level).
3. The *U* values are found from Table A.6 and A.8.

Figure 3.10
Floor plan for Example 3.15.

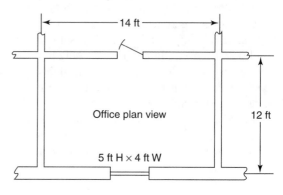

4. The heat transfer losses are found using Equation 3.8.

5. The infiltration heat loss is found using Equation 3.10.

6. The room heating load is the sum of all the losses.

	U	×	A	×	TD	=	Q
	BTU/hr-ft²-F		ft²		F		BTU/hr
Wall	0.10	×	106	×	82	=	870
Window	0.60	×	20	×	82	=	980
Roof	0.08	×	168	×	82	=	1100
Total heat transfer loss						=	2950

Infiltration heat loss = 1.1×0.50

		CFM/ft × 18 ft × 82	=	810
		Room heating load	=	3760

In calculating design heat losses, usually no credit is taken for heat gains. A solar heat gain could not be guaranteed, of course. For buildings that have steady internal gains, however, there is no reason why they should not be considered in calculating net heat losses.

3.11 THE BUILDING NET HEATING LOAD

In addition to calculating the individual room heating loads, the building heating load must also be determined.

The **building net heating load** is the amount of heat required for the building at outdoor design conditions.

The building net heating load is the sum of the building heat transfer losses, infiltration losses, and ventilation load, if any.

Building Heat Transfer Loss

This is the sum of the heat transfer losses to the outdoors through the exposed walls, windows, doors, floor, and roof of the building. (It is also the sum of the room heat transfer losses, but it is preferable to calculate it directly, measuring total building areas.)

Building Infiltration Loss

Although the building generally has more than one side with openings, it should be understood that infiltration air cannot enter through all sides at the same time. This is because, as noted earlier, the wind comes from only one direction at any given time.

Air that infiltrates on the windward side of a building leaks out through openings on the other sides. It is difficult to evaluate this, because it is affected by interior conditions such as partition arrangements.

Mechanical ventilation will reduce and often prevent significant infiltration.

For a building that is not mechanically ventilated and that has reasonably free interior passages for air movement, the following rule is suggested:

The building air infiltration CFM is equal to one-half the sum of the infiltration CFM of every opening on all sides of the building.

The following example illustrates the procedure for finding the building infiltration heat loss and the building net heating load.

Example 3.16

Find the infiltration heat loss and net heating load for the building shown in Figure 3.11. The building heat transfer loss is 170,000 BTU/hr. Infiltration rates are shown in Figure 3.11. The indoor and outdoor temperatures are 70 F and 10 F.

Figure 3.11
Sketch for Example 3.16.

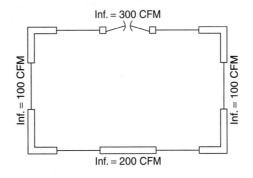

Solution

1. Find one-half the infiltration rates for all sides of the building:

   Building infiltration CFM =

   $$\frac{300 + 100 + 200 + 100}{2} = 350 \text{ CFM}$$

2. Find the infiltration heat loss (Equation 3.10):

   $$Q = 1.1 \times \text{CFM} \times TC$$
   $$= 1.1 \times 350 \times (70 - 10) = 23,100 \text{ BTU/hr}$$

3. Find the building net heating load:

Heat transfer loss	= 170,000 BTU/hr
Infiltration heat loss	= 23,100
Building net heating load	= 193,100 BTU/hr

3.12   SYSTEM HEAT LOSSES

Besides the direct room and building heat losses (heat transfer and infiltration/ventilation), there are often *system* heat losses, such as heat loss from ducts and piping.

Duct Heat Transfer Loss

In a warm air heating system, if the ducts pass through unheated spaces (e.g., attics, basements, shafts, crawl spaces), there will be heat transfer from the air in the duct to the cooler surrounding spaces.

It is suggested that 2–5% of the building sensible heat loss be added to account for duct heat loss. This range of values depends on the length of ductwork, insulation and surrounding temperatures.

A minimum of *R*-4 insulation is recommended for ductwork in all cases.

Duct Leakage

There is usually some air leakage from supply ducts at seams. This constitutes a heat loss when the air leaks into unconditioned spaces. The loss is usually insignificant in private residences, but can reach 5–10% of the load in larger buildings, depending on the quality of the sheet metal installation.

Piping Losses

In hot water or steam heating systems, there may be heat lost from the hot piping. Since piping is relatively small and always should be insulated, this loss is usually negligible. An allowance is sometimes made for heating the system itself, however, on start-up of the system. This is called the *pickup factor*. It is explained in the following discussion.

Pickup Factor or Allowance

When a building is intermittently heated, or when regular nighttime temperature setback is practiced, and if the equipment capacity just equals the building load, the heating equipment may not be able to bring the rooms up to design temperature quickly enough. In this situation, it may be desirable to allow a pickup loss in sizing the central heating equipment.

Some designers allow an extra 10% loss for intermittently heated buildings, and up to 40% for a 10 F night setback in residential equipment. Of course, this additional capacity is gained at a sacrifice to cost.

In large buildings, an equipment sizing allowance for setback is not standard practice. Many other strategies are available for bringing the space temperature up to design in sufficient time. For instance, if the weather forecast is for cold weather, the system may be started earlier. Furthermore, in large buildings multiple boilers and excess standby capacity are common.

It is normal practice in the boiler industry to specify a combined *piping and pickup factor* (*allowance*) that can be used when sizing a heating boiler. This factor combines the piping and pickup losses described earlier. The standard piping and pickup factor varies from 15–25%, depending on boiler type and size (see Chapter 4). The actual piping and pickup losses may be greater or less than these values, as suggested here; if there are doubts, they should be calculated.

Net and Gross Heating Load

The net heating load is the amount of heat needed for all the building rooms. The system heating losses (ducts, piping, pickup) are not part of the room loads; they are loads on the boiler or furnace. When they are added to the net load, the sum is called the *gross heating load*. This is the heat output that the heating equipment must furnish.

Example 3.17 illustrates this relationship.

Example 3.17 _____

A building has a building net heating load of 350,000 BTU/hr. It is estimated that the combined duct heat transfer loss and heat loss due to leakage is 10%. The space temperature will be set back at night by 10 F. What is the building gross heating load (the required furnace capacity)?

Solution
The system losses are added to the net load.

Net heating load	= 350,000 BTU/hr
Duct losses 0.10 × 350,000	= 35,000
Pickup allowance (0.40 × 350,000)	= 140,000
Furnace capacity (gross load)	= 525,000 BTU/hr

Service (Domestic) Hot Water Heating

The heat output of a boiler is sometimes used to heat water (for kitchens, baths, and so forth) as well as for space heating. This load should then be included when sizing the boiler. Procedures for determining hot water use loads may be found in the *ASHRAE Handbooks*.

3.13 SUMMARY OF HEATING LOAD CALCULATION PROCEDURES

The following step-by-step instructions summarize how to calculate heating loads. The individual room data and results are recorded on a room heating load calculations form (Figure 3.12 on page 62). Figure 3.13 on page 63 is a building heating

load calculations form. The forms are suitable for both residential and commercial estimates.

Room Heating Load

1. Select appropriate indoor and outdoor design temperatures (Tables 1.1 and A.9).
2. Obtain dimensions from architecture plans. For each room, find areas of exposed windows, walls, and so forth through which there will be heat transfer.
3. Select appropriate overall heat transfer coefficients (*U*-values) from Tables 3.1, A.6–A.8 or calculate from *R*-values if necessary (Tables A.4 and A.5).
4. Calculate the heat transfer losses through all exposed surfaces in the room (Equation 3.8). Total these to find the room heat transfer loss.
   A. For basement floors and walls below grade, use the *U*-values from Table 3.2 and outdoor ground temperature.
   B. For floor slabs on grade, and exterior rooms, use the edge loss (Equation 3.9 and Table 3.3).
5. Find the room infiltration heat loss, if any. For the crack method:
   A. Use architectural plans to find window crack lengths and door areas. Use only one wall for corner rooms.
   B. Use Table 3.4 (or the equivalent) to find the infiltration rate.
   C. Use Equation 3.10 to find the infiltration heat loss.
   Remember if the building is pressurized by the mechanical ventilation system, infiltration can usually be considered negligible.
6. Find the room heating load.
   Room heating load = room heat transfer rate + room infiltration loss

Building Heating Load

The steps for finding the building heating load are as follows.

1. Calculate the heat transfer losses through all exposed surfaces. Use the total exterior areas

Figure 3.12
Room heating load calculations form.

Room Heating Load Calculations

P. _____ of _____ PP.

Project _____   Location _____   Indoor DB _____ F
Engrs. _____   Calc. by _____   Chk. by _____   Outdoor DB _____ F

Room																			
Plan Size																			
Heat Transfer	U ×	A ×	TD =	BTU/hr		U ×	A ×	TD =	BTU/hr		U ×	A ×	TD =	BTU/hr					
Walls																			
Windows																			
Doors																			
Roof/ceiling																			
Floor																			
Partition																			
Heat Transfer Loss																			
Infiltration		(CFM)					(CFM)					(CFM)							
	$1.1 ×$	$A ×$	$B ×$	TC =		$1.1 ×$	$A ×$	$B ×$	TC =		$1.1 ×$	$A ×$	$B ×$	TC =					
Window	1.1					1.1					1.1								
Door	1.1					1.1					1.1								
Infiltration Heat Loss																			
Room Heating Load																			

Room																			
Plan Size																			
Heat Transfer	U ×	A ×	TD =	BTU/hr		U ×	A ×	TD =	BTU/hr		U ×	A ×	TD =	BTU/hr					
Walls																			
Windows																			
Doors																			
Roof/ceiling																			
Floor																			
Partition																			
Heat Transfer Loss																			
Infiltration		(CFM)					(CFM)					(CFM)							
	$1.1 ×$	$A ×$	$B ×$	TC =		$1.1 ×$	$A ×$	$B ×$	TC =		$1.1 ×$	$A ×$	$B ×$	TC =					
Window	1.1					1.1					1.1								
Door	1.1					1.1					1.1								
Infiltration Heat Loss																			
Room Heating Load																			

Infiltration CFM	Column A	Column B
Windows	CFM per ft	Crack length, ft
Doors	CFM per ft^2	Area, ft^2

of the walls, roof, and so forth. Do not use areas from each room—this leads to errors.

2. There is an infiltration heat loss if the building does not have mechanical ventilation that pressurizes the interior. To find this,

   A. Using the crack method, find the infiltration CFM for all openings in the building.

   B. Take one-half of this CFM, and using Equation 3.10, find the building infiltration heat loss.

3. If the building has mechanical ventilation that sufficiently pressures the interior, there is a ventilation heat load but no infiltration heat loss. To find this,

Figure 3.13
Building heating load calculations form.

Building Heating Load Calculations

	DB, F	W', gr/lb
Intdoor		
Outdoor		
Diff.		

Project _____
Location _____ Calc. by _____
Engineers _____ Chk. by _____

Heat Transfer	$Q =$	U	$\times$	A	$\times$	TD	$=$	BTU/hr
Roof								
Walls								
Windows								
Doors								
Floor								

Heat Transfer Subtotal	
Infiltration $Q_s = 1.1 \times$ _____ CFM $\times$ _____ TC =	
Building Net Load	
Ventilation $Q_s = 1.1 \times$ _____ CFM $\times$ _____ TC =	
$Q_L = 0.68 \times$ _____ CFM $\times$ _____ gr/lb =	
Duct Heat Loss _____ %	
Duct Heat Leakage _____ %	
Piping and Pickup Allowance _____ %	
Service HW Load	
Boiler or Furnace Gross Load	

A. Determine the CFM of outside ventilation air from Table 6.17.
B. Using Equation 3.10, find the ventilation sensible heat load.
C. If the building is to be humidified, find the ventilation latent heat load from Equation 3.11.

4. Find the building net heating load:

Building net heating load = building heat transfer loss + infiltration loss

5. Find any system losses such as duct losses and piping and pickup allowances (Section 3.11).
6. Find the service hot water load, if the boiler is to handle this.
7. Find the building gross heating load (this is the required furnace/boiler capacity):

Gross heating load = net load + ventilation loads + system losses + service hot water load

An example of heating load calculations for a building may be found in Example Project I in Chapter 17. The student is advised to read Sections 17.1 and 17.2 at this point for an account of some practical problems encountered in doing an actual estimate.

3.14 ENERGY CONSERVATION

Reducing the building heating load provides a major opportunity for energy conservation. Some ways this can be achieved are as follows:

1. Use ample insulation throughout the building. Building construction in the past has been scandalously wasteful of energy due to inadequate insulation. For instance, an overall roof-ceiling resistance of R-20 to R-30 is recommended for residential buildings in colder climates.
2. Use the 97.5% winter outdoor design temperature level where possible.
3. Use inside winter design temperatures that provide comfort but not excessive tempera-

ture. The practice of using temperatures as high as 75 F is often unnecessary. Consider using 68 to 72 F.
4. Be certain all windows and doors are weather-stripped, and use either double glass or storm windows, except in mild climates.
5. Heat losses shall be calculated using thorough, correct procedures.
6. The building architectural design (orientation, use of glass, type of materials, and so forth) should be consistent with reducing energy consumption.
7. Follow applicable energy conservation construction standards.

Review Questions

1. What is the *heating load* and what items make it up?
2. What is the *infiltration loss*?
3. What is the *ventilation load*?
4. Define resistance and conductance. What is their relationship?
5. What are the two methods for estimating infiltration?
6. How is the infiltration for a corner room found when using the air change method?
7. What is meant by the "97.5% outside design temperature"?
8. What outside temperature is used to find the heat transfer from below grade surfaces?
9. Through what part of a building is the heat transfer loss proportional to the perimeter?

Problems

3.1 A homeowner asks an energy specialist to find the heat loss from his home. On one wall, measuring 15 ft by 9 ft (without windows), the specialist measures a temperature of 66 F on the inside surface of the wall and 18 F on the outside surface. The wall has a thermal resistance of 0.35 ft^2-F-hr/BTU.

What is the rate of heat loss through the wall?

3.2   An insulating material has a thermal conductivity of $k = 0.23$ BTU/hr-ft^2-F per inch. How many inches of the material should the contractor install if energy conservation specifications call for insulation with an R-12 value?

3.3   Find the overall R value and U factor in winter for the wall with construction as shown in Figure 3.14.

3.4   The wall in Problem 3.3 is 30 ft long by 12 ft high. The indoor temperature is 70 F. What is the heat transfer loss through the wall on a day when the outdoor temperature is −5 F?

3.5   A state energy code requires a certain wall to have an overall R-15 value. What must be the R-value of insulation added to the wall in Problem 3.3?

3.6   A roof is constructed of built-up roofing on top of a metal deck, R-5.61 insulation, with a suspended ½ in. acoustical tile ceiling. Calculate the R-value of the roof in winter. Compare the result with the value from Table A.6.

3.7   Find the R-value of the roof in Problem 3.6 if 2 in. more of glass fiber insulating board is added.

3.8   Calculate the heat transfer loss through a 4 ft–6 in. wide by 3 ft high wood sash window with indoor and outdoor temperatures of 68 F and 3 F, respectively.

3.9   Calculate the heat transfer loss through a 25 ft by 30 ft roof-ceiling of a house with pitched asphalt shingle roof, vented attic, R-8 insulation, and ½ in. acoustical tile on furring ceiling. Inside and outside temperatures are 72 F and −2 F.

3.10 What is the heat transfer loss through a 40 ft by 20 ft basement floor when the room is at 65 F and the ground temperature is 50 F?

3.11 A warehouse in Cleveland, Ohio, built on grade, is 100 ft by 40 ft in plan. The wall construction is 4 in. block with brick facing. R-5 insulation is used around the edge. The inside temperature is 68 F. What is the heat loss through the floor?

3.12 Find the sensible heat loss from infiltration through a casement window with a 3 ft wide by 4 ft high operable section if the infiltration rate is 0.8 CFM/ft. The room temperature is 69 F and the outdoor temperature is −8 F.

3.13 A room 15 ft by 20 ft by 10 ft has an air infiltration rate of 1.5 air changes per hour. The room is at 72 F and the outdoor temperature is 1 F. What is the heat loss from infiltration?

3.14 A building in Milwaukee, Wisconsin has 2000 ft^2 of single-glazed vinyl frame windows. The inside temperature is 72 F. Using

Figure 3.14
Sketch for Problem 3.3.

6 in. sand and gravel aggregate (not dried) concrete

R-5 insulation

1/2 in. gypsum board

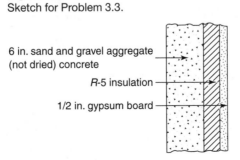

the 97.5% recommended outdoor winter design temperature, calculate the design heat transfer loss through the windows.

3.15 To save energy, it has been decided to install double glass windows on the building in Problem 3.14. What is the reduction in the design heat transfer loss?

3.16 Find the total heat loss from heat transfer and infiltration through a single-glazed 5 ft wide by 4 ft high double-hung vinyl frame window in a building in Springfield, Illinois, maintained at 71 F. The infiltration rate is 0.6 CFM/ft. Use the 97.5% design value.

3.17 A corner room in a building in Pittsburgh, Pennsylvania, has two 3 ft wide by 4 ft high casement windows on one side with an infiltration rate of 0.7 CFM/ft, and a 7 ft by 3 ft door on the other side with an infiltration rate of 1.2 CFM/ft. The room is at 68 F. Find the design infiltration heat loss from the room, using the 97.5% value.

3.18 Find the total heat loss through the exterior wall and windows of a room at 72 F, as described:

Wall: *8 in. concrete, R-5 insulation, ½ in. gypsum wallboard, 30 ft long by 15 ft high.*

Windows: *(5) 4 ft wide by 4 ft high double glass, aluminum frame, casement type. Infiltration rate 0.5 CFM/ft.*

Location: *Salt Lake City, Utah. Use 97.5% value.*

3.19 Find the design heat loss from the room shown in Figure 3.15 on an intermediate floor in an office building in Toronto, Ontario. Use recommended energy conservation design values suggested in this chapter.

Wall: *4 in. face brick, 8 in. cinder block, ½ in. furred gypsum wallboard.*

Windows: *4 ft wide by 5 ft high, double glass, double-hung, vinyl frame.*

Ceiling height: *9 ft.*

Figure 3.15
Floor plan for Problem 3.19.

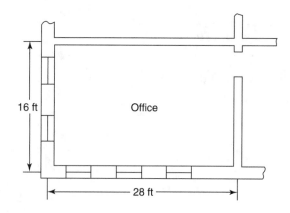

3.20 Calculate the individual room heating loads and building heating load for the house shown in Figure 3.16. Use recommended energy conservation design values.

Location: *Hartford, Connecticut.*

Walls: *wood siding, building paper, wood sheathing, R-4 insulation, ⅜ in. gypsum board interior.*

Windows: *double-hung, wood sash, single glass.*

Roof: *pitched, asphalt shingles, building paper, wood sheathing, R-8 insulation, gypsum board ceiling, attic.*

Doors: *1½ in. wood, 7 ft high.*

Ceiling height: *9 ft. No basement.*

3.21 Calculate the heating load for the building shown in Figure 3.17. The building is constructed on grade.

Figure 3.16
Building plan for Problem 3.20.

All windows 3' −6" H × 4' W
Scale 1/8 in. = 1 ft−0 in.

Figure 3.17
Building plan for Problem 3.21.

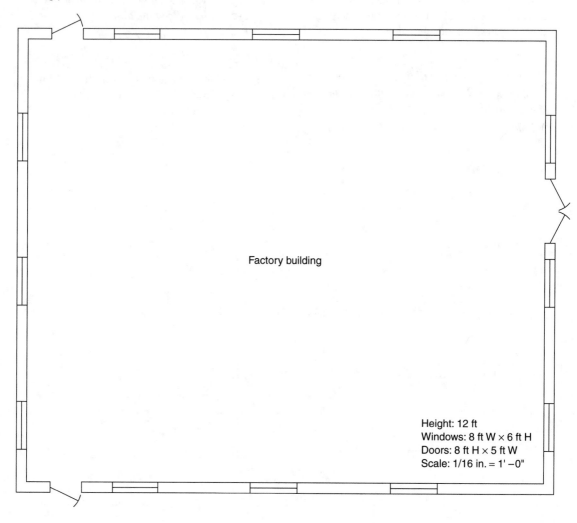

Factory building

Height: 12 ft
Windows: 8 ft W × 6 ft H
Doors: 8 ft H × 5 ft W
Scale: 1/16 in. = 1' −0"

Location: Topeka, Kansas. Factory.

Walls: 8 in. concrete block, ½ in. furred gypsum board.

Windows: double-glazed, aluminum frame.

Mechanical ventilation: 2500 CFM.

Doors: 1 in. wood.

Roof: 4 in. lightweight concrete, finished ceiling.

CHAPTER 4
Furnaces and Boilers

This chapter will examine boilers and furnaces, the most common heat source equipment used in air conditioning. The heat which is produced by this equipment is most often used for heating, but may also be used in cooling (see the discussion on absorption refrigeration in Chapter 13). The heat pump, used for both heating and cooling, will be discussed in Chapter 13. Solar heaters will be discussed in Chapter 18.

OBJECTIVES

After studying this chapter, you will be able to:

1. Describe the basic features of warm air furnaces and heating boilers.
2. Explain the functions of the basic operating and safety controls for furnaces and boilers.
3. Explain the function of flame safety controls.
4. Explain draft and how it is created and controlled.
5. Describe commonly used fossil fuels, products of combustion, pollutants, and methods of pollution control.
6. Describe the basic types of gas burners and oil burners.
7. Select a warm air furnace or heating boiler.
8. Describe the energy conservation methods that are associated with the use of furnaces and boilers.

4.1 WARM AIR FURNACES

A *warm air furnace* heats by delivering warmed air to the spaces in a building. Warm air furnaces are popular in private residences and small commercial installations, since, in very small buildings, warm air systems with ductwork are often less expensive than hydronic systems. Also, if the ductwork is already installed, by using a combined heating/cooling central unit or add-on cooling unit, summer air conditioning may be easily added at a minimal cost. A third advantage of warm air systems over hydronic systems is that when nighttime temperature setback is used, full heat can be delivered to rooms faster in the morning. The hydronic system, however, does have advantages in many applications (Chapter 5).

Components

The main parts of a warm air furnace are the heat exchanger, fuel burner, air blower (fan), controls, and insulated housing cabinet. The basic components of a warm air furnace are shown in Figure 4.1. The furnace may also have a humidifier and an air filter.

Furnaces may have coal, oil, gas, or wood burners, or electric heaters, as a heat source. Oil and gas burners are discussed in Section 4.7.

The construction of furnaces for residential or commercial use is similar, except that commercial furnaces have larger capacities, are structurally stronger, and may have more complex controls than residential furnaces.

Operation

Circulating air enters the furnace through the return air inlet (Figure 4.1). Pushed by a blower, the

Figure 4.1
Components of a warm air furnace.

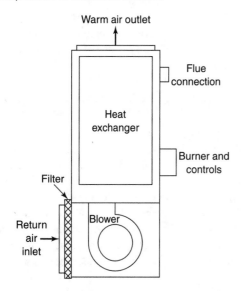

air passes over the outside of the heat exchanger, which has been warmed from the inside by hot combustion gas passing through it. The heated air exits through the warm air outlet and travels through ducts to the rooms in the building.

The hot combustion gas (also called *flue gas*) inside the heat exchanger is produced by the burning of fuel. After its heat is transferred to the circulating air, the combustion gas is exhausted through a *vent* to the outdoors. The vent may be a pipe, sheet metal stack, or masonry chimney. (Heating units in which the combustion gases are discharged to the outdoors are called *vented appliances*. Some heating units, called *unvented appliances,* discharge combustion gases directly into the room where the heater is located.)

Types of Furnaces

To fit in different spaces, furnaces are made in a variety of physical configurations. Figure 4.2 shows some arrangements of residential type furnaces. The *upflow* or *high-boy* type is suitable for full-height basement or utility room installations with overhead ductwork. The *low-boy* type may be used if there is less headroom. The *downflow* type is practical when the supply air ductwork is under a floor or grade, or in a crawl space. The *horizontal* type is suitable for an attic; when weatherized for outdoor service, it is popular for commercial rooftop installations.

Additional heating equipment that is usually grouped with furnaces includes *space heaters, wall* and *floor furnaces, duct heaters,* and *unit heaters.* Unlike warm air furnaces, these are not designed to be connected with ductwork (except for duct heaters), but instead deliver air directly into the space to be heated.

Space heaters are usually freestanding units resting on the floor. Some space heaters have blowers; others rely solely on the convected motion of the warm air. The wall furnace and floor furnace are designed to be recessed into a wall or floor. Duct heaters are mounted in a section of duct; airflow in duct heaters is created by a separate blower-housed unit. Unit heaters are generally hung from a ceiling; they may be gas or oil fired, or may use heating coils (Chapter 5). Another version is the gas-fired *radiant* heater which works by using the flame and hot combustion gases to heat an element to a very high temperature; the heat is then radiated directly from the element to solid objects in the space, rather than having warm air circulate.

Because incomplete combustion may cause toxic pollutants, the use of unvented appliances is often restricted by legal codes. If unvented appliances are to be used, they must have ample room ventilation and special safety shut-off devices.

Capacity and Performance

Manufacturers rate heating capacity in BTU/hr at the furnace outlet (bonnet). Residential type furnaces are available in capacities from about 35,000 –175,000 BTU/hr. Commercial furnaces are available up to about 1 million BTU/hr.

The system designer needs to know both the net heat available to heat the room or building and the gross furnace output at the bonnet. Allowances must be made for any duct or pickup losses (see Chapter 3).

In addition to the heating capacity, the CFM of air to be circulated and the duct system air static

Figure 4.2
Arrangements of residential warm air furnaces.

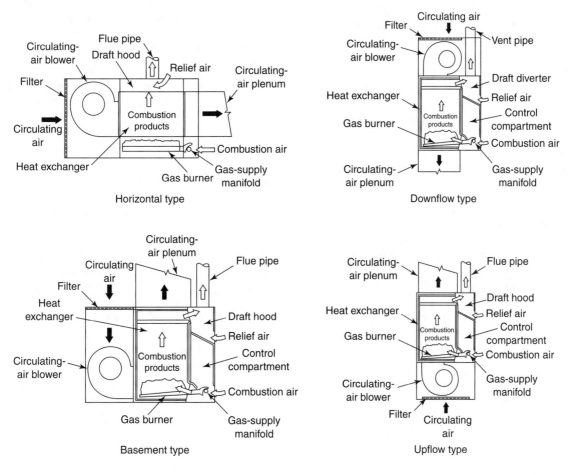

Horizontal type

Downflow type

Basement type

Upflow type

pressure loss requirements must be determined (see Chapter 8). Combination heating/cooling units generally have two-speed fans to enable them to provide more airflow in the summer.

The steady-state efficiency of warm air furnaces typically ranges from 75–80%, except for so-called high-efficiency furnaces (Section 4.12), which may have efficiencies as high as 95%.

4.2   FURNACE CONTROLS

The controls for a warm air furnace are of two types: *operating controls* and *safety* (limit) *controls.* Operating controls regulate the burner (Section 4.7) and the air circulation fan during normal operation. Safety controls (also called limit controls) stop or prevent furnace operation if safe limits are exceeded. Safety controls may sense air and combustion gas temperatures, and air and gas flow; a safety control that detects flame presence is also used (Section 4.8).

The controls used in each application depend on the equipment, type of fuel, and safety code regulations. The following sequence outlines the operation of a typical residential gas-fired furnace:

1. On a call for heat, a switch in the room thermostat closes. A pilot flame (pilot light) safety control checks for the standing (continuous) pilot flame.
2. If the flame is present, the main gas valve opens. The pilot flame ignites the main gas. The safety control will continually check for flame presence; if the flame fails later, the valve will close.

    Instead of a standing (continuous) pilot flame, the pilot gas itself may be ignited on a call for heating. When the furnace is not heating, the pilot flame is off. This arrangement, called *intermittent ignition,* is used to save energy.
3. A fan control thermostat located in the circulating air fan plenum (discharge) automatically starts the fan when the air has been warmed to a comfortable level and stops it when the air is too cool. An alternate arrange-

ment, which achieves the same result, is to use a timer to delay the fan's start until a short period after the air starts warming. Fan shutdown can also be delayed until after the valve closes, utilizing the "free" remaining heat left in the heat exchangers, thus saving energy.

Manual control of the fan is provided to allow continual air circulation in the summer.

4. A fan limit switch thermostat (high limit) will stop the fan if the air temperature becomes dangerously high (about 200 F). A simple occurrence such as blocked airflow from dirty filters can cause this. When the temperature drops, the switch may reclose to start the fan at a lower high limit (say 175 F); alternately, a manual reset switch may be used, if preferred.

The operation of a residential oil-fired furnace is similar, except that the space thermostat starts the oil burner pump motor and activates the circuit providing an ignition spark.

Commercial furnaces need more complex control arrangements. For instance, the quantity of gas-air mixture that remains in a large furnace after a shutdown is enough to be an explosion hazard; therefore, purge cycles are incorporated into the furnace control operations. In these more complex arrangements, the sequence of events is often called a *programming control sequence,* because it consists of programmed steps. A typical programming control sequence for a larger gas-fired furnace might be:

1. On a call for heating, the space thermostat closes the heating control circuit (only if all safety controls are also closed).
2. The air circulating fan starts.
3. Air circulation is tested by a pressure switch, and if airflow is proven,
4. The combustion fan starts.
5. Combustion gas flow is tested, and if proven,
6. A timed prepurge cycle (typically 30–60 seconds) exhausts combustion gases which may remain in the furnace from the last operation.
7. The main gas valve opens and the spark ignition circuit is activated.

8. Ignition is tested (with the flame safety control).
   A. If ignition does not occur (in about 4 seconds), the ignition circuit is deactivated, and steps 6 and 7 are repeated (prepurge cycle and attempted ignition).
   B. If the second ignition attempt fails, the system shuts down and "locks out" (a manual reset is necessary). (During operation, the flame safety control continually checks the flame. If the flame fails or is unstable, the system shuts down.)
9. When the thermostat is satisfied, the main gas valve closes.
10. A timed postpurge cycle exhausts combustion gases from the furnace.

4.3 HEATING BOILERS

Boilers produce hot water or steam, which is then delivered through pipes to space heating equipment. A *hot water boiler* heats water to a high temperature, but does not boil it. Since it does not actually boil water, a hot water boiler would be better named a hot water *generator.* A *steam boiler,* also called a steam generator, heats water to the boiling point to make steam. Both hot water and steam boilers, having similar features, will be discussed together; any important differences will be noted.

At this point, a few words about safety should be said. Of all the equipment used in HVAC systems, probably the boiler is where one should be most sensitive about safety to life. This is because there is a tremendous amount of energy packed into a boiler. In New York City in 1962, a heating boiler in a telephone company building exploded, killing 21 and injuring 95. This tragic incident is mentioned to stress the importance of obtaining a thorough working knowledge of boilers.

Components

The main parts of a boiler are the combustion chamber, burner, heat exchanger, controls, and enclosure.

Boilers may be classified in various ways, according to:

1. Their specific application
2. Their pressure and temperature ratings
3. Their materials of construction
4. Whether water or combustion gas is inside the tubes (watertube or firetube)
5. Whether the boiler and accessories are assembled on the job site or at the factory
6. The type of fuel used
7. How draft (airflow) is achieved

The basic features of these groupings will be explained separately. Items 6 and 7, fuels and draft, will be discussed in relation to boilers and furnaces together, since they apply to both types of equipment.

Pressure and Temperature Ratings

The *American Society of Mechanical Engineers* (ASME) has developed standards for the construction and permissible operating pressure and temperature limits for low pressure *heating boilers* which are used in the United States.

The ASME *Code for Heating Boilers* limits maximum working pressure to 15 psig for steam and 160 psig for water. Hot water temperatures are limited to 250 F.

Hot water boilers are usually manufactured for 30 psig maximum working pressure, since this is more than adequate for the vapor pressure exerted by 250 F water.

For higher temperatures and pressures, the ASME *Code for Power Boilers* applies. *Power boilers*, although so named because they are used to generate steam to be utilized in generating electric power, can be used in high temperature hot water (HTW) hydronic heating systems. Another application for high pressure boilers in HVAC systems is to use them with steam turbine-driven centrifugal refrigeration machines, because steam turbines require steam at a relatively high pressure. Low pressure heating boilers do not require the attendance of a licensed operating engineer in many locations, whereas high pressure boilers do. This may affect the choice of low or high pressure boilers for an installation. The HVAC engineer should always check that a boiler conforms to ASME codes.

Materials of Construction

Cast iron boilers have a heat exchanger constructed of hollow cast iron sections. The water flows inside these with the combustion gases outside. The sections are assembled together similar to cast iron radiators. An advantage of this construction is that when the boiler is too large to fit through the building access opening, it can be shipped in parts and assembled on site. They range from small to fairly large capacity, up to about 10 million BTU/hr. Figure 4.3 shows a small cast iron boiler complete with burner, controls, and housing.

Figure 4.3
Cutaway view of small gas-fired cast iron hot water package boiler. Note draft hood and automatic flue gas damper. **(Courtesy: Burnham Corporation—Hydronics Division.)**

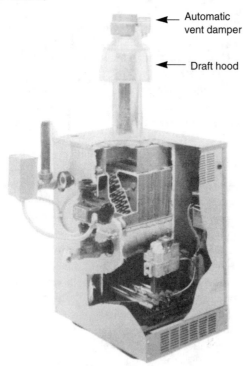

← Automatic vent damper

← Draft hood

Steel boilers have a heat exchanger constructed of steel tubes arranged in a bundle, as seen in Figures 4.4 and 4.5.

Boilers that have copper tube heat exchangers are also available.

Firetube or Watertube

Steel boilers can also be classified as either *firetube* or *watertube.* In firetube boilers (Figure 4.4), the combustion gases flow inside the tubes and the water circulates outside. In watertube boilers, the water flows inside the tubes and the combustion gases outside. Firetube boilers are less expensive than watertube boilers but are less durable. Firetube boilers range from small capacities to about 20 million BTU/hr. Watertube heating boilers range from medium size to about 100 million BTU/hr.

The differences among types of firetube boilers (e.g., *locomotive, horizontal return tube,* and *Scotch marine* type) are in their construction (a specialized subject, and not important for our purposes.) The Scotch marine type firetube boiler (Figures 4.4 and 4.5) is the most popular for commercial heating service because of its compactness, low cost, and reliability.

Watertube boilers are not often used in HVAC installations. Their main application is for large steam power plants or for creating process steam to be used in industry.

Figure 4.4
Steel firetube boiler arrangement.

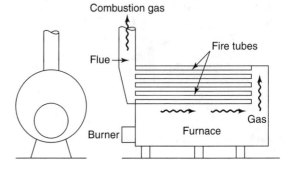

Combustion gas

Fire tubes

Flue →

Gas

Burner

Furnace

Built-Up and Package Boilers

A *built-up* boiler is a boiler whose components are assembled at the job site; this includes the combustion chamber, heat exchanger, burner, and accessories. A *package boiler* is completely assembled and tested in the factory. This procedure reduces cost, increases reliability by ensuring that components are properly matched, and decreases the contractor's field work. The small cast iron residential type boiler shown in Figure 4.3, and the Scotch marine type boiler shown in Figure 4.5, are both package boilers.

Boiler Accessories

Certain accessories are needed for the proper operation, maintenance, and safety of boilers. Some accessories are optional; others are required by codes or by law.

Figure 4.5
Package firetube boiler. (Courtesy: Cleaver-Brooks, Inc.)

Accessories Required for a Steam Boiler

A *low water cut-off* (Figure 4.6) senses water level in a steam boiler; it will stop burner operation if the water level falls below a safe level.

A *water column* with a *gauge glass* (Figure 4.7), when mounted on the side of a steam boiler, allows the operator to see the water level.

A pressure gauge and thermometer, mounted on or near the boiler outlet, aids the operating engineer in checking performance.

Accessories Required for a Hot Water Boiler

A *dip tube,* which is a piece of pipe from the boiler outlet extending down below the water line, prevents air that may be trapped in the top of the boiler from getting into the water supply line.

An *expansion tank* provides space for the increased volume of water when it is heated (Chapter 11).

A *flow check valve* closes when the pump stops. Without this valve, hot water would circulate by natural convection, heating rooms even when no heat is called for.

A *make-up water connection* allows for filling the system and replenishing water losses.

A *pressure reducing valve* (PRV) prevents excess pressure from being exerted on the boiler from the water make-up source.

Air control devices may be required in the water circuit; these devices divert air in the system to the expansion tank.

A typical piping arrangement with accessories for a hot water heating boiler is shown in Figure 4.8.

Accessories Required for Both Steam and Hot Water Boilers

A *safety relief valve* (Figure 4.9) which opens if boiler pressure is excessive. The ASME Code specifies the type of valve which is acceptable for a particular application. This valve must be connected separately at the boiler. In the New York explosion incident mentioned earlier, one of the claims in the investigation was that the safety relief valve did not open and relieve the excessive pressure which had developed.

Figure 4.6
Low water cut-off. (Courtesy: McDonnell & Miller ITT.)

Figure 4.7
Water column and gage glass.

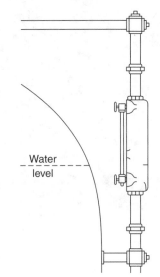
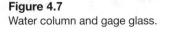

Water level

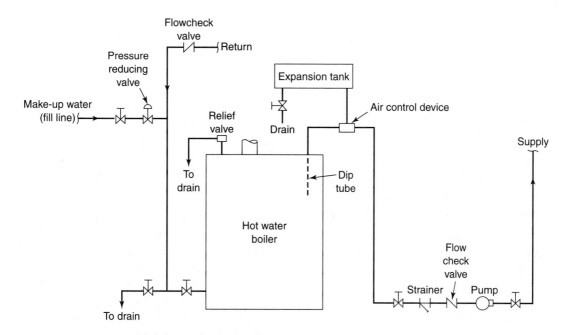

Note: Unions or equivalent for service not shown.

Figure 4.8
Typical piping arrangement and accessories for a hot water boiler. (In small systems, the pump may be located in return line.)

A combustion gas connection, called a *vent* on small boilers and a *breeching* on larger boilers, through which combustion gas travels from the boiler to the chimney or flue.

Figure 4.9
Safety relief valve. (Courtesy: TACO, Inc., Cranston, RI.)

Thermal insulation, which when applied around the boiler, reduces heat loss. It may be field installed or applied in the factory.

A *preheater*, if heavy fuel oil is to be used; it heats the oil to a temperature at which it will flow easily.

4.4 BOILER CONTROLS

Boiler controls, like furnace controls, are of two types, operating controls and safety (limit) controls.

Operating Controls

Operating controls regulate the burner (Section 4.7) during normal operation.

In smaller units, a room thermostat starts and stops the burner in response to space conditions.

In larger units, a controller sensing a condition in the boiler regulates the burner operation: in steam boilers a pressure controller is used; in hot water boilers, a temperature controller ("aquastat") is used.

When controls sensing conditions in the boiler are used, secondary operating controls are often also used. For instance, room thermostats may be used to control the flow of hot water to the terminal units in each room.

Safety Controls

As discussed earlier, safety (limit) controls prevent or stop an unsafe condition from occurring. Conditions that boiler safety controls may check for are

1. High steam pressure (steam boilers)
2. High hot water temperature (hot water boilers)
3. High or low fuel oil/fuel gas pressure
4. High or low fuel oil temperature
5. Low water level
6. Flame failure

Programming Control Sequence

Commercial boiler controls, like furnace controls (Section 4.2), operate automatically in a specific programmed sequence. A typical programming control sequence for a small commercial gas-fired boiler might be:

1. *Prepurging.* On a call for heat, a fan starts, purging any combustion gases that might remain in the boiler from the last operation.
2. After an automatically timed period (a few seconds), the pilot gas valve opens, an ignition transformer is energized, and the pilot flame is lit. If the pilot flame does not light, the flame safety control shuts the pilot flame gas valve.
3. When the pilot flame lights, the main gas valve opens. If the main flame does not light, the flame safety control shuts down the burner. During operation, the flame safety control continually checks the flame. If the flame fails or is unstable, the system shuts down.
4. When the thermostat is satisfied, the main fuel valve closes.
5. *Postpurging.* On shutdown, the fan continues to run for a short time after the burner stops to purge the remaining combustion gases.

4.5   BOILER AND FURNACE DRAFT

Since boilers and furnaces need a constant supply of fresh combustion air, a pressure differential, called *draft,* must be created to force air and gas through the equipment and chimney. The pressure differential must be great enough to allow the flowing air and gas to overcome the resistance from friction in the combustion chamber, heat exchanger, and flue.

The term *draft* is also used in an associated manner to refer to the air or gas flow itself caused by the pressure differential. It is important to keep in mind both meanings of draft when dealing with combustion problems: *draft is both the pressure to move the air and gas, and the flow itself.*

Draft can be created either naturally or mechanically. *Natural draft* results from the difference in densities between the hot gas in the combustion chamber and the surrounding cool air. Since the heated gas is lighter than the cool air outside, it rises through the chimney, drawing cool air into the boiler through the openings at the bottom (the chimney flue effect).

A small negative pressure (relative to the atmospheric pressure) exists in the furnace in a natural draft unit. Under normal circumstances, this pressure prevents leakage of potentially toxic combustion gas into the equipment room.

When operating correctly, the overfire (combustion chamber) draft pressure reading in a typical residential furnace will be about −0.02 in. w.g. The manufacturer will provide precise values. Note that this is an extremely small pressure. The pressure in the flue outlet will be slightly more negative, perhaps −0.04 in. w.g., to allow for the pressure drop through the furnace heat exchanger.

Natural draft provides enough combustion air for simpler equipment, but as the equipment's size and complexity increases, the resistance to airflow also increases, resulting in a need for more draft. Natural draft can be increased by using a taller chimney, but this approach has physical limits, plus problems with aesthetics, legal restrictions, and cost.

When the natural draft would be insufficient, mechanical devices must be used. The terms *powered combustion, power burners,* and *mechanical draft* all refer to the use of fans to develop sufficient pressure to move the combustion gases through the boiler/furnace and flue. Either an *induced draft* fan or a *forced draft* fan may be used, or both. An induced draft fan, located at the equipment's outlet, pulls combustion air through the equipment and discharges it into the stack. A forced draft fan, located at the equipment's air inlet, blows air through the furnace. Since the forced draft fan creates positive pressure in the boiler/furnace, care must be taken in the equipment's design and maintenance to prevent combustion gas leakage into the room.

Generally, if mechanical draft is to be used, a fan with adequate pressure is used to create both furnace and stack draft, and only a short stack is needed. However, a draft fan may be used with a tall chimney, so that the two together develop the needed draft. In this arrangement, the fan creates only the furnace draft, and the chimney flue effect handles the required stack draft.

In addition to being able to develop more pressure than natural draft, mechanical draft fans provide closer control of draft (airflow quantity) than natural draft does, as will be seen in the following discussion.

Draft Control

Draft (airflow) should remain constant for a given fuel firing rate. Too low a draft supplies insufficient air so that combustion is incomplete, thereby wasting fuel; excess draft results in too much air, reducing efficiency because the excess air is being heated and then thrown away.

Changes in temperature and outside air conditions (e.g., wind) can cause changes in the draft through the equipment, adversely affecting the equipment's performance. A sudden downdraft (gas flow down the chimney) due to outside disturbances can even blow out the flame, creating a dangerous situation. Because of these potential problems, a means of maintaining constant draft is needed.

For smaller units, adequate draft control can usually be achieved using a *draft hood* (also called a *draft diverter*) or *barometric damper.*

A draft hood (Figures 4.3 and 4.10) is used on vented gas-fired equipment. A momentary increase in updraft will draw more surrounding air from the room into the flue via the hood. The greater volume of air in the flue increases the resistance to airflow; this reduces the chimney draft to its previous level. A momentary decrease in updraft will be canceled by an opposite reaction in the flue. If a downdraft occurs, the draft hood diverts the air into the surrounding space rather than into the combustion chamber, where it could blow out the pilot or main flame or cause poor combustion. Note that the draft hood is a safety device as well as a means of maintaining approximately constant draft.

The barometric damper, also called a barometric draft regulator (Figure 4.11 on page 80), is used to control draft in oil-fired equipment and some power gas-fired units. (A power unit has a combustion air fan in the burner.) As stack updraft increases or

Figure 4.10
Draft hood.

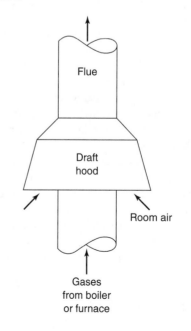

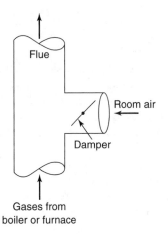

Figure 4.11
Barometric damper.

decreases, the damper will open or close, canceling the momentary change in draft in the same way that a draft hood does.

Because the draft hood and barometric damper are located in the vent stack, they control stack draft better than controlling furnace draft. Consequently, the furnace draft may still vary too much to maintain an efficient air/fuel ratio. This may be acceptable with small equipment, but the inefficiency is too costly for larger units. In larger equipment, an *automatically controlled outlet damper* is often used instead of a barometric damper. This device is regulated by a pressure sensor in the furnace and adjusts automatically to maintain the proper furnace draft.

Furnaces or boilers that have power burners (i.e., with burner draft fans) may use one of the draft regulating devices discussed, or may rely only on the burner fan to control draft.

Newer higher efficiency residential units often use a mechanical draft combustion air fan and an *automatic vent damper* (Figure 4.3). (The vent damper can also be installed in existing systems.) The fan is needed to overcome the greater resistance of the larger heat exchanger. The fan also improves draft control. The vent damper in the flue is not modulated, but instead closes automatically to prevent airflow through the stack when the system is shut down. This may reduce energy losses; for

example, if the boiler/furnace is in a heated space, a closed damper prevents the loss of warm air from the building. However, a new problem arises. When the vent damper closes, the chimney cools down for long periods and water vapor may condense on start-up, eventually causing corrosion; special chimney or vent materials may be required to prevent this occurrence.

4.6   FUELS AND COMBUSTION

Combustion *is the rapid chemical combination of the combustible substances in a fuel with oxygen (in the air).*

In the process, stored energy is released as thermal energy (heat) in the products of combustion.

The three major fossil fuels used in boilers and furnaces are gas, oil, and coal. The possible combustible substances in these fuels are carbon (C), hydrogen (H_2), and sulfur (S). These may exist in the fuel in their *element* form (C, H_2, S) or as *compounds* of those elements. For example, the hydrocarbon methane (CH_4) is a combustible compound often present in fuel. Fossil fuels may also contain small amounts of noncombustible substances. Not all fuels contain all of the combustible elements or compounds.

The products of combustion are mostly gases at a high temperature. This is the thermal energy that the boiler or furnace attempts to capture (as much of as is practical). These gases are used to heat water or air, or to generate steam.

When combustion is *complete,* the possible products are carbon dioxide (CO_2), water vapor (H_2O), and sulfur dioxide (SO_2). Combustion may be incomplete, producing a different product. For instance, *incomplete* combustion of carbon produces carbon monoxide (CO), a highly toxic gas.

Gas and oil have largely replaced coal as the fossil fuels for space heating boilers and furnaces. They are easier to handle and generally have less pollutant products. Coal is still extensively used in many large power plant boilers because of its lower cost.

Gas

Natural gas is the most commonly used gaseous fuel. Its greatest convenience is that it is delivered directly from the gas wells through pipelines to the consumer. This eliminates deliveries and storage needs, as required with oil and coal. It also contains virtually no pollutants, unlike both oil and coal.

Natural gas is composed of a number of hydrocarbons, primarily methane (CH_4) and lesser amounts of ethane (C_2H_6). The exact composition varies, depending on the source. The products of complete combustion are carbon dioxide and water vapor.

The amount of heat released by complete combustion of natural gas (called the *heating value*) is about 1000 BTU/ft^3. Heating values for various fuels are shown in Table 4.1.

Liquified petroleum gases (LPG) is the name given to both of the hydrocarbon gaseous fuels *propane* and *butane,* because they are liquified and bottled for use. They are convenient where piped natural gas is not available.

Biogas is a fuel gas (largely methane) that results from decomposing garbage. There are already a number of installations tapping biogas from large garbage landfills in the United States.

Oil

Fuel oils are available in different grades (Numbers 1, 2, 4, 5, and 6). The lower number grades are lighter in density, have a lower viscosity, and slightly lower heating values. Use of No. 1 oil (kerosene) is limited to small space heaters. No. 2 oil is generally used in residential and small commercial furnaces and boilers. No. 6 and No. 5 oils require preheating before they become fluid enough to be used. This limits their usage to larger installations with auxiliary heating equipment. Their advantage is that they cost less than No. 2 oil.

Since oil must be stored in tanks, it is less convenient than natural gas. In smaller quantities, oil may be stored indoors; for larger installations, oil storage tanks are buried underground or placed outdoors.

Fuel oil is composed largely of hydrocarbons and a small amount of sulfur.

Fuel Choice

The system designer should consider availability, cost, convenience, and pollution effects of the various fuels. Wood-fired units are growing in use, particularly in those areas where wood is abundant. Waste-fired boilers, which use garbage, save depletable fuel resources and aid in garbage disposal problems. A few cities are already successfully using large waste-fired boilers.

For residences, the popular choices are No. 2 heating oil and natural gas. The cost of each varies with location, market conditions, and legislated price controls. In some cases, gas is cheaper, in others, oil costs less.

Combustion

The chemical reaction in which a fuel combines with oxygen in the air and releases heat is called

TABLE 4.1   FUEL COMBUSTION DATA

Fuel	Theoretical Air/Fuel Ratio	Percent CO_2 in Combustion Gas Quantity of Air Supplied			Heating Value
		Theoretical	20% Excess	40% Excess	
Natural gas	9.6 ft^3/ft^3	12.1	9.9	8.4	1000 BTU/ft^3
No. 2 fuel oil	1410 ft^3/gal	15.0	12.3	10.5	140,000 BTU/gal
No. 6 fuel oil	1520 ft^3/gal	16.5	13.6	11.6	153,000 BTU/gal
Bituminous coal	940 ft^3/lb	18.2	15.1	12.9	13,000 BTU/lb

Notes: Air/fuel ratios based on air density of 0.075 lb/ft^3, are by volume. Values are approximate, as the composition of fuels varies.

combustion, or more commonly, burning. The amount of air required for complete combustion is called the *theoretical air* quantity and the resulting ratio of air to fuel is called the *theoretical air/fuel ratio.*

Because it is not practical or economical to construct equipment to mix air and fuel perfectly, if a boiler or furnace is furnished with the exact theoretical air quantity, combustion will not be complete. The result is unburned fuel and a waste of energy. Furthermore, incomplete combustion produces carbon monoxide (CO), a highly toxic pollutant.

To prevent this problem, *excess air* (air above the theoretical quantity) is always furnished. However, the efficiency of a boiler/furnace is maximized by using the minimum excess air needed for complete combustion. If too much excess air is used, efficiency is unnecessarily reduced, because the excess air is being heated and then thrown away. In practice, many installations are operated with huge oversupplies of excess air, resulting in a tremendous energy waste.

The minimum amount of excess air actually needed for complete combustion to occur depends on the type of fuel and the construction of the heating device and controls, and may vary from 5–50% above 100% theoretical air. Generally, larger units need less excess air.

Manufacturers furnish data on recommended air/fuel ratios for their equipment. Table 4.1 and Figure 4.12 show theoretical air/fuel ratios, CO_2 content in the combustion gases for different excess air quantities, and fuel heating values.

Figure 4.12
Effect of excess air on CO_2 percentage in flue gas. **(Courtesy: Dunham Bush/Iron Fireman.)**

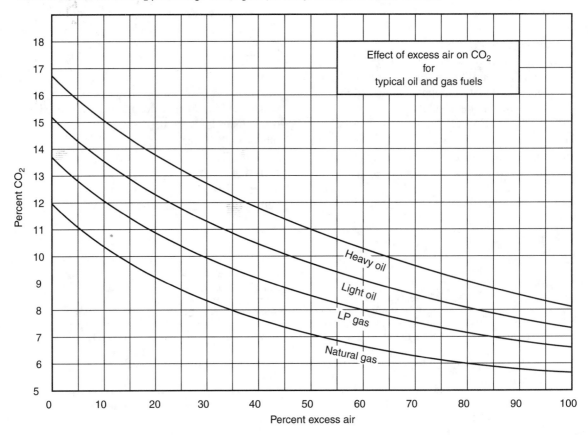

Examples 4.1 and 4.2 illustrate how the heating specialist can use measurement of percentage of CO_2 in the stack gas to determine the necessary amount of combustion air and the percent excess air used. After such a test, the technician may adjust the air/fuel ratio if the air is excessive, in accordance with the manufacturer's data.

Example 4.1
A boiler firing 15 gal/hr of No. 2 oil requires 20% excess air for complete combustion. How many CFM of combustion air should be supplied?

Solution
Using the air/fuel ratio data from Table 4.1, and noting that the actual air quantity should be 1.2 times the theoretical (20% excess).

$$CFM = 1410 \ \frac{ft^3}{gal} \times 15 \ \frac{gal}{hr} \times \frac{1 \ hr}{60 \ min} \times 1.2$$

$$= 423 \ CFM$$

Example 4.2
A technician measures 8% CO_2 in the combustion gas of a natural gas-fired boiler that requires 15% excess air for complete combustion. Is the air/fuel ratio satisfactory?

Solution
From Figure 4.12, 45% excess air is being used. The answer is a loud NO! A great deal of excess hot gas is going up the stack.

Measuring Combustion Efficiency

The effectiveness of the boiler or furnace in utilizing the heat released from combustion of the fuel can be determined quite easily in the field, after taking two measurements in the stack (flue gas):

1. *Percent CO_2.* The percent of CO_2 of the total gases in the stack is an indication of the amount of excess air used (Figure 4.12). The higher the CO_2, the less excess air is used, and therefore the higher the efficiency.
2. *Flue gas temperature.* The lower the flue gas temperature, the greater the amount of heat

that is being absorbed in the equipment heat exchanger.

Table 4.2 on page 84 shows the effect of both flue gas percent CO_2 and temperature on the percent of heat input from the fuel that is lost to the flue gas.

The *combustion efficiency* of a fuel burning heating unit is defined as

Combustion efficiency =

$$\frac{heat \ input - flue \ gas \ loss}{heat \ input} \times 100\% \quad (4.1)$$

Using CO_2 flue gas analysis equipment (see Chapter 16) and a stack gas thermometer, the combustion efficiency of a boiler or furnace can be determined from Table 4.2 (note that the table uses the difference between flue gas and boiler room temperatures).

Later in this chapter, we will discuss what efficiencies to expect from heating equipment.

Measuring Efficiency

The *thermal efficiency* of a boiler or furnace is defined as the ratio of useful heat output to heat input, expressed as a percentage (see Section 4.10). The heat input is the amount of heat released from complete combustion of the fuel. Two measurements that can be used to determine thermal efficiency are *stack (flue) gas temperature* and *percent CO_2*.

The stack gas temperature should be relatively low, because this indicates that the heat exchanger has transferred more heat to the working fluid, and less heat is wasted in the stack gas. Stack gas temperatures typically range from 300–600 F. Stack gas temperatures below about 300 F have traditionally not been desired, because of possible water vapor condensation and the resulting corrosion of the chimney or equipment. In some newer equipment, however, lower stack gas temperatures are being used to save energy (see Section 4.12).

Percent CO_2 in the stack gas should be relatively high, because this indicates that combustion is complete. The amount of CO_2 in the combustion gas can be measured with flue gas analysis equipment (Chapter 16).

TABLE 4.2   EFFECT OF FLUE GAS TEMPERATURE AND CO_2 ON HEAT LOSS (%) TO FLUE GAS

	% CO_2	Difference Between Flue Gas and Room Temperature in Degrees F										
		300	350	400	450	500	550	600	650	700	750	800
NATURAL GAS	4.0	25.1	27.7	30.4	33.1	35.8	38.3	40.9	43.5	46.2	48.8	
	4.5	23.6	25.9	28.3	30.7	33.0	35.4	37.8	40.1	42.6	44.8	47.2
	5.0	22.2	24.4	26.8	28.7	30.9	33.0	35.7	37.3	39.7	41.8	43.8
	5.5	21.2	23.4	25.2	27.3	29.2	31.3	33.2	35.3	37.3	39.2	41.0
Fuel Analysis	6.0	20.4	22.3	24.1	25.8	27.8	29.6	31.5	33.3	35.2	36.8	38.8
1120 BTU/cu ft	6.5	19.8	21.4	23.2	24.8	26.5	28.3	30.0	31.7	33.5	34.6	36.8
% by Volume	7.0	19.1	20.7	22.3	23.9	25.5	27.1	28.8	30.4	32.0	33.8	35.3
CH_4 79.9	7.5	18.5	20.0	21.5	23.0	24.6	26.1	27.7	29.1	30.8	32.2	33.8
C_2H_6 17.3	8.0	18.0	19.5	20.9	22.4	23.8	25.2	26.7	28.1	29.5	31.0	32.4
CO_2 0.3	8.5	17.6	19.0	20.4	21.7	23.1	24.5	25.8	27.1	28.6	29.9	31.3
N_2 2.5	9.0	17.2	18.5	19.9	20.1	22.5	23.8	25.2	26.4	27.8	29.0	30.3
	9.5	16.9	18.1	19.5	20.7	21.9	23.1	24.4	25.6	26.9	28.2	29.4
	10.0	16.6	17.8	19.0	20.2	21.4	22.6	23.8	25.0	26.2	27.4	28.6
	11.0	16.1	17.1	18.4	19.4	20.5	21.6	22.8	23.9	25.0	26.2	27.2
NO. 2 FUEL OIL	5.0	22.7	25.4	28.2	30.9	33.8	36.3	39.2	42.0	44.7	47.4	50.1
	5.5	21.3	23.8	26.3	28.9	31.4	34.0	36.4	39.0	41.7	44.0	46.5
	6.0	20.4	22.5	24.9	27.2	29.5	31.9	34.2	36.4	38.9	41.0	43.5
	6.5	19.3	21.3	23.6	25.7	27.8	30.1	32.3	34.4	36.5	38.7	40.8
Fuel Analysis	7.0	18.4	20.5	22.4	24.5	26.5	28.5	30.5	32.6	34.6	36.5	38.6
BTU 19,750/lb	7.5	17.7	19.6	21.3	23.4	25.2	27.1	29.0	30.9	32.9	34.8	36.5
% by Weight	8.0	17.1	18.9	20.7	22.4	24.2	26.0	27.8	29.6	31.5	33.2	35.0
C 86.1	8.5	16.5	18.2	20.0	21.6	23.3	24.9	26.7	28.4	30.1	31.8	33.5
H 13.6	9.0	16.0	17.6	19.3	20.9	22.4	24.1	25.7	27.3	28.9	30.5	32.1
O 0.2	9.5	15.7	17.1	18.6	20.2	21.7	23.2	24.8	26.3	27.9	29.4	31.0
N 0.1	10.0	15.2	16.6	18.1	19.6	21.0	22.5	24.0	25.4	27.0	28.3	29.9
	11.0	14.5	15.8	17.2	18.5	20.0	21.2	22.6	23.9	25.3	26.7	28.0
	12.0	13.9	15.1	16.4	17.6	18.9	20.2	21.4	22.6	24.0	25.2	26.5
	13.0	13.4	14.5	15.8	16.9	18.1	19.3	20.5	21.5	22.8	24.0	25.2
	14.0	13.0	14.0	15.3	16.3	17.4	18.5	19.7	20.8	21.8	22.9	24.1
NO. 6 FUEL OIL	5.0	22.8	25.9	29.0	32.0	35.3	38.1	41.0	44.3	47.5	50.1	53.6
	5.5	21.3	24.0	26.9	29.7	32.5	35.2	37.9	40.7	43.8	46.1	49.1
	6.0	20.0	22.5	25.2	27.7	30.3	32.9	35.3	37.9	40.5	43.0	45.8
	6.5	18.9	21.3	23.7	26.0	28.5	30.9	33.4	35.6	38.0	40.2	42.8
Fuel Analysis	7.0	17.9	20.1	22.4	24.6	26.8	28.8	31.2	33.4	35.8	37.9	40.1
BTU 18,150/lb	7.5	17.3	19.2	21.4	23.3	25.4	27.5	29.6	31.6	33.9	35.9	37.9
% by Weight	8.0	16.4	18.4	20.4	22.3	24.2	26.2	28.2	30.2	32.1	34.1	36.0
C 89.36	8.5	15.7	17.6	19.6	21.3	23.3	25.1	26.8	28.6	30.8	32.6	34.2
H 9.30	9.0	15.3	17.0	18.8	20.5	22.3	24.0	25.7	27.5	29.4	31.1	32.9
S 0.90	9.5	14.7	16.3	18.1	19.7	21.4	23.1	24.7	26.4	28.1	29.8	31.2
N 0.20	10.0	14.4	15.8	17.5	19.0	20.6	22.2	23.8	25.4	27.0	28.7	30.1
O 0.19	11.0	13.5	15.0	16.5	17.9	19.4	20.9	22.3	23.8	25.2	26.8	28.1
Ash 0.05	12.0	12.8	14.1	15.6	16.9	18.3	19.6	21.0	22.4	23.8	25.0	26.4
	13.0	12.3	13.5	14.8	16.0	17.3	18.6	19.8	21.1	22.4	23.8	24.9
	14.0	11.8	13.0	14.3	15.4	16.6	17.7	18.8	20.1	21.2	22.5	23.7

(*Courtesy:* Dunham Bush/Iron Fireman)

Using measurements of the stack gas temperature and percent CO_2 in the flue gas, the heating specialist can determine the thermal efficiency of the equipment from Table 4.2. After the results are compared to the manufacturer's data, adjustments may be performed to improve the unit's efficiency. This testing should be performed regularly. (*Note:* Table 4.2 uses the difference between the temperatures of the stack gas and of the boiler room.)

Example 4.3

In carrying out an energy study, readings on a boiler burning No. 2 oil are taken. The flue gas analyzer reads 10% CO_2. The stack gas temperature is 570 F and the boiler room temperature is 70 F. What is the thermal efficiency of the boiler?

Solution

The difference between the stack gas and room temperatures is $570 - 70 = 500$ F. Using Table 4.2, at 500 F and 10% CO_2, for No. 2 oil, the heat loss to the flue gas is 21.0%. The thermal efficiency is $100.0 - 21.0 = 79.0\%$.

Example 4.4

A soot blower is used to clean off the heating surfaces of the boiler in Example 4.3. The stack gas temperature now reads 470 F. What is the thermal efficiency?

Solution

The difference between the stack gas and room temperatures is now $470 - 70 = 400$ F. From Table 4.2, the heat loss is 18.1%. The thermal efficiency is $100.0 - 18.1 = 81.9\%$. A significant fuel savings (about 3%) has been accomplished.

Combustion and Air Pollution

The combustion of fossil fuels can unfortunately produce air pollutants. Some of the pollutants contribute to respiratory illnesses such as bronchitis, emphysema and lung cancer. They can also result in damage to forests, agricultural crops, and the quality of lakes.

The HVAC specialist needs to be aware of what these pollutants are and how they can be controlled. The pollutants include smoke, ash, soot, sulfur dioxide (SO_2), sulfur trioxide (SO_3), carbon monoxide (CO), and nitrogen oxides (NO_x). Fuel oil is generally more of a problem than natural gas.

Smoke is very small particles from the combustion process formed by

1. Insufficient oxygen (that is, less than the theoretical air)
2. Poor mixing of fuel and air (even with sufficient air)
3. Premature chilling of a partially burned mixture
4. Burning with too much air

Proper adjustment of the air-fuel ratio and maintenance of burners and combustion controls are necessary to prevent smoke. Smoke is easily measured by the *Ringlemann Chart.* This is a card with four sections numbered from 1 to 4, ranging from light to dark, representing the opacity or density of smoke. Pollution codes limit the density number. For instance, the New York City Air Pollution Code sets the following smoke limits.

A. Smoke as dark or darker than #2 density shall not be allowed at all.
B. Smoke darker than #1 but less than #2 shall not be given off for more than 2 minutes in any one-hour period.

Ash consists of particles of noncombustible solids produced after combustion. Although it is formed primarily from coal combustion, it may result from fuel oil combustion. When present, it can be removed by filters or similar means.

Soot is carbon-ash particles, larger in size than smoke. Control methods are the same as for smoke and ash.

Sulfur dioxide (SO_2) results from the combustion of sulfur present in fuel oil and coal. Fuel oil with very low sulfur content is required in many urban areas. For instance, No. 2 fuel oil used in New York City cannot contain more than 0.2% sulfur. Another method of control is to remove the SO_2 gas in the stack with appropriate devices.

Sulfur trioxide (SO_3) can be removed by neutralizing it with an additive compound.

Nitrous oxides result from high flame temperatures. They react with other substances in the atmosphere to form *smog,* which has serious respiratory effects. Nitrous oxide control methods include using natural gas instead of oil and maintaining low excess air and low flame temperatures.

Carbon monoxide (CO) is an extremely toxic gas resulting from incomplete combustion of carbon or hydrocarbons, due to insufficient excess air. Proper maintenance and adjustment of burners and draft will prevent its formation.

4.7   GAS AND OIL BURNERS

The *fuel burner* is a device for delivering the fuel and part or all of the combustion air to the furnace/boiler. It also helps to mix the fuel and air, and when fuel oil is used, breaks up the liquid oil into a spray of small droplets.

Gas Burners

Atmospheric gas burners and *power gas burners* differ in how the air and fuel gas are delivered to the combustion chamber. In the atmospheric burner (Figure 4.13), the flow of fuel through a Venturi (a nozzle shaped tube) draws part of the combustion air (called *primary air*) through an opening, into the gas stream; the air/gas mixture then goes into the combustion chamber. The amount of primary air may be varied by using adjustable shutters or dampers at the opening. The remainder of the combustion air (called *secondary air*) is drawn by natural draft directly into the combustion chamber around the burner head ports (openings). (This arrangement is also called a *premix type burner,* since some of the air and gas mix before entering the burner ports.) In some burners, mixing of the fuel gas and primary air is enhanced by vanes or other devices.

Various burner head arrangements are available, each designed to match the furnace characteristics and size. *Inshot* burners have ports located to deliver the gas horizontally; *upshot* burners (Figure 4.13) deliver the gas vertically upward. The burner may have one or more ports in a pipe; multiple pipes can be arranged in parallel. Another version has a narrow slot (ribbon). A ring burner has the

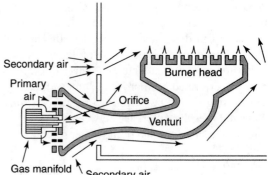

Figure 4.13
Atmospheric gas burner. Burner head is upshot type.

ports arranged around a ring-shaped pipe (like a kitchen stove).

Power gas burners use fans to deliver the air. The fan creates turbulence to promote air and gas mixing. The fan may be designed for complete forced draft (to overcome furnace and stack draft loss) or only to overcome furnace draft loss, relying on natural induced draft for the stack.

Gas Burner Ignition

Fuel ignition in a gas burner may be achieved by a *standing pilot,* an *intermittent pilot,* or *direct spark ignition.* A standing pilot is a continuously burning small gas flame. When the main gas fuel enters the combustion chamber, the pilot flame ignites it. Since the pilot flame remains burning when the burner is off, there is a perpetual waste of heat up the stack.

To save energy, intermittent pilot ignition may be used. The pilot is lit by a spark, only on a call for heating. The pilot flame then lights the main fuel. The pilot and main gas are both shut off when the heating requirement is satisfied. Intermittent pilot ignition is also useful on rooftop equipment, where a standing pilot may be blown out by wind.

Some gas fuel equipment uses direct spark ignition. There is no pilot. On a call for heating, the fuel gas is fed into the combustion chamber and a spark ignites the gas directly. Descriptions of typical ignition procedures were given in Sections 4.2 and 4.4.

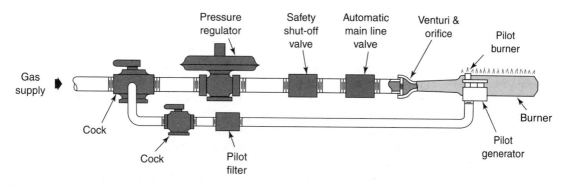

Figure 4.14
Gas manifold valves and burner arrangements. (Courtesy: North American Heating & Air Conditioning Wholesalers Association, Home Study Institute Division.)

Many states have banned standing pilots on new gas-fired equipment as an energy conservation measure. Although not usually legally required, existing equipment may often be easily converted to intermittent pilot ignition.

Typical gas manifold connections to a burner with a pilot flame (Figure 4.14) include a manual shut-off valve or cock; a gas pressure regulator; a pilot safety control; the electrically controlled main valve; and the pilot piping, burner, and sensor. On newer residential equipment, a single *combination valve* (Figure 4.15) serves the functions of gas cock, pressure regulator, pilot safety control, and main gas valve.

Oil Burners

An oil burner mixes fuel oil and combustion air and delivers the mixture to the combustion chamber.

Except for one type, the *vaporizing pot burner,* all oil burners have the additional function of

Figure 4.15
Combination gas valve. (Courtesy: North American Heating & Air Conditioning Wholesalers Association, Home Study Institute Division.)

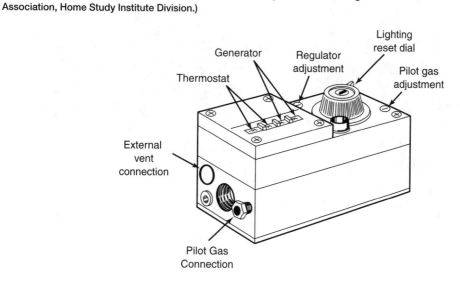

mechanically assisting in vaporizing the fuel oil. Vaporizing the oil is necessary since oil will not burn in its liquid state.

The vaporizing pot burner is basically a bowl filled with fuel oil. The oil at the surface vaporizes naturally due to its vapor pressure, aided by the turbulence and heat from the combustion gas. Because the vaporization is slow and difficult to control, it is used mainly in space heaters burning No. 1 oil, which vaporizes more rapidly than heavier oils.

The other types of oil burners help vaporize oil by breaking it up into very small droplets. This process, called *atomizing,* increases the oil's surface area, causing it to vaporize faster. Each burner type has an oil pump, a combustion air fan, and an ignition system. They differ considerably from each other, both in the means of atomizing the oil and introducing it into the combustion chamber.

Steam atomizing or *air atomizing* burners use steam or air under pressure to create fuel and air mixing, turbulence, and atomizing. The *horizontal rotary cup burner* (Figure 4.16) has a rotating cup which throws the oil into the air stream, causing atomization; these types of burners are used with larger commercial equipment and are suitable for both heavy and light oils.

The *gun burner,* or *mechanical pressure atomizing burner* (Figure 4.17), atomizes oil by pumping it

Figure 4.16
Rotary oil cup burner.

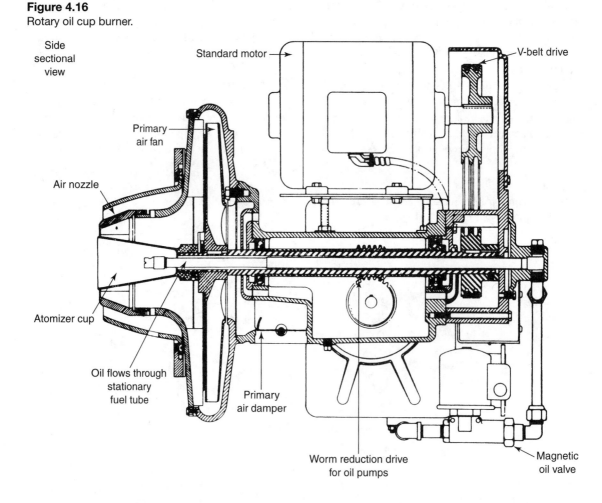

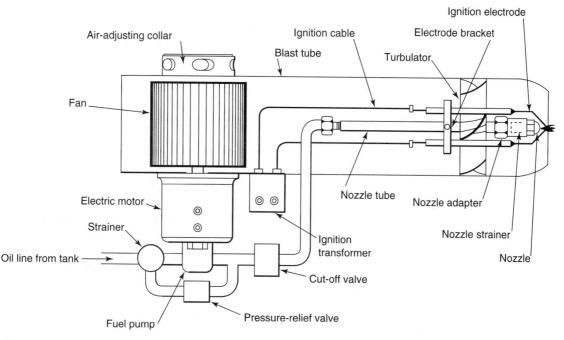

Figure 4.17
High pressure atomizing ("gun type") burner. (**Courtesy: Air Conditioning Contractors of America (ACCA).**)

under high pressure through a small diameter nozzle. A combustion air fan introduces air through a tube surrounding the nozzle. Deflector vanes at the burner outlet (head) cause mixing and proper distribution of the oil-air mixture. The gun burner, which usually uses No. 2 oil, is used in both residential and smaller commercial oil-fired boilers and furnaces.

Retention head gun burners have a head (outlet) that is designed to improve mixing and combustion, resulting in higher combustion efficiency.

Steam or air atomizing burners and rotary cup burners have a relatively high *turn-down ratio,* which is the ratio of maximum to minimum fuel oil flow that the burner can handle. A high turn-down ratio is desirable because the boiler/furnace can operate at a low capacity when necessary, thereby saving fuel.

Gun burners have a relatively low turn-down ratio, which limits their ability to operate at part capacity. Small gun burners may operate only in an on-off mode.

Combination Burners

A combination burner can burn both oil and gas. Essentially it has the components of both burner types in the burner housing. It is useful when there is the possibility of a shortage of one of the two fuels, or where prices may change so that the relative costs of the fuels reverse.

Oil Burner Ignition

Ignition of the oil-air mixture in gun burners is done with a high voltage electric spark. Ignition of rotary cup and steam or air atomizing burners is usually done with a gas pilot flame, or a spark ignited pilot oil burner.

Gas and Oil Burner Firing Rate Control

Burner capacity (firing rate) may be controlled by a space thermostat, which is used with warm air furnaces and residential boilers; or by a steam pressure or water temperature controller, which is often used with commercial boilers. Other controls are available, too. The methods of burner capacity control are: *On/Off, High/Low/Off,* and *modulation.*

An On/Off control simply starts and stops the burner; this limits the boiler/furnace to full capacity operation only. A High/Low/Off (or High/Medium/Low/Off) control provides more flexibility in operating capacity, allowing the furnace/boiler to be more closely matched to load variations. Although considerable heat is wasted in starting up the heating equipment and when it is in the off cycle, use of these control methods is common in smaller equipment because of their low cost and simplicity.

To achieve maximum efficiency in larger units, full modulation of burner capacity, by control of fuel and airflow rates over the entire turn-down range, is standard. Fuel valves and air dampers are automatically modulated by the temperature or pressure controllers. A common method for doing this is to interlock the fuel valve and air damper after the initial adjustments have been made to provide the most efficient air-fuel ratio.

However, the air-fuel ratio may change even for fixed relative positions of fuel valve and damper due to changes in temperature, humidity, fuel characteristics, and equipment conditions. To correct for this, a microprocessor-based combustion control system can be used that continually measures the percent oxygen in the combustion gases. When deviations from the correct amount occur, the programmed controller takes action to adjust the air-fuel ratio. The result is improved efficiency under all operating conditions.

4.8   FLAME SAFETY CONTROLS

A safety control that deserves special consideration is the *flame safety control,* because it ensures safe burner operation. This control shuts off the fuel supply if the fuel does not ignite, or if the flame fails during operation. If fuel were to continue to enter the furnace and not be burned, a serious explosion hazard would quickly arise.

The flame safety control consists of a flame sensing element and a means of relaying its signal to start or stop fuel flow. The sensor can detect one of three possible effects of the flame: temperature (heat), flame electrical conductivity, or radiation (either visible light or infrared or ultraviolet radiation).

Gas-fired residential furnaces and boilers with a standing pilot use a heat sensing *thermocouple* for flame detection. The thermocouple consists of two wires of different metals that at a high temperature create a very small voltage. The thermocouple is placed in the pilot flame. If the flame fails, the control circuit will be deenergized and the gas valve will close.

Because it reacts slowly, the thermocouple is not a satisfactory flame safety control for larger heating equipment nor for equipment that uses intermittent pilot or direct ignition. Since it may take the main gas valve 30–40 seconds to close, enough gas will collect in the combustion chamber so that when the spark ignites the mixture on a call for heat, and explosion may occur.

The *flame rod* is a suitable flame safety control for gas ignition systems, since it closes the gas valve quickly (1–3 seconds). It consists of two electrodes placed at the flame location. Since a flame conducts electricity, the flame's presence completes a circuit opening the gas valve. Flame failure will open the circuit causing the valve to close. The flame rod is not used in oil-fired systems because the flame temperature is too high for the electrodes.

Oil-fired residential units use either a *photo cell* or a *stack switch* for flame detection. The photo cell (Figure 4.18) is a light-sensitive device whose conductivity increases in the presence of strong light radiation. It is pointed at the flame location and allows the burner to operate only when the cell conducts. (The photo cell is not used with a gas flame because the intensity of light from a gas flame is too low.) The photo cell for small boilers and furnaces

Figure 4.18
Photo-cell-type flame safety control sensor ("cad cell" element). **(Courtesy: Air Conditioning Contractors of America (ACCA).)**

is often made of cadmium sulfide, and is then given the name "cad cell."

The stack switch (Figure 4.19) is a heat sensitive device placed in the stack to sense gas temperature. The stack switch is often a bimetal type thermostat. The bimetal closes a circuit on high temperature, permitting burner operation. The photo cell has replaced the stack switch as flame safety control in newer equipment both because of its faster response and its direct sensing of flame presence.

Flame safety controls for larger commercial equipment must react more quickly because fuel enters at a greater rate, reaching an explosive concentration sooner. Radiation sensing safety controls are

Figure 4.19
Stack switch flame safety control. **(Courtesy: Air Conditioning Contractors of America (ACCA).)**

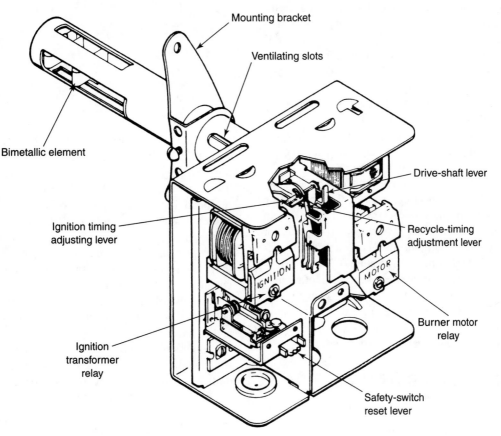

ideal, since they react very fast. The three types of radiation flame sensors used in commercial equipment are: photo, infrared, and ultraviolet cells. In each case, the type of radiation from the flame changes an electrical property of the detector, which is then used in a relay to start or stop fuel flow.

4.9   BOILER APPLICATIONS

Hot water heating boilers generate hot water that is used directly in hydronic heating systems. Steam heating boilers generate steam that may be used directly in steam heating systems, or the steam may be used to heat the hot water with a heat exchanger called a *converter.* In one circuit steam from the boiler flows and in the other the water to be heated. An obvious question arises as to why a hot water boiler should not be used, considering the additional expense and complication of the converter. The answer is that low pressure hot water boilers usually are designed for a maximum pressure of 30 psig. This corresponds to a head of $2.3 \times 30 = 69$ ft water. If a hot water boiler were installed in a basement or lower floor with more than 69 ft of height of water piping above it, it would be subject to an unsafe pressure. Very high buildings may be split into zones to prevent excess pressure on equipment. Figure 4.20 shows this arrangement. Another solution is to put the boiler on the roof.

The hot water or steam generated in boilers may be used for space heating, space cooling, or heating of service (domestic) hot water. It may seem strange that hot water or steam can be used for cooling, but absorption refrigeration machines require heat to produce refrigeration (Chapter 13).

The boiler used for space heating is often also used to heat service hot water. It is possible to obtain small boilers with a service hot water heating coil furnished internally. With large boilers, a separate heat exchanger is usually specified.

4.10   BOILER RATING AND SELECTION

Manufacturers present rating data in tables, from which the proper boiler can be selected for a given

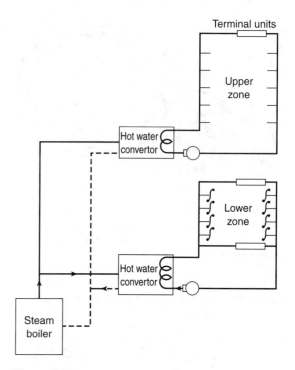

Figure 4.20
Arrangement of steam boiler and hot water convertors in hydronic heating system for high-rise building.

application. Boilers are rated by their heat output in BTU/hr. It is also desirable to specify the temperature and flow rate of steam or hot water required. Two other units beside the BTU/hr have been used in specifying boiler capacity, boiler horsepower, and equivalent direct radiation (EDR). The use of these units is disappearing. They are confusing and sometimes misleading, and it is recommended that they be avoided. If necessary, their conversion equalities can be found in tables.

It is possible to produce increased capacity output from a boiler by firing it at very high fuel rates and by other variations in operating procedures. It is also possible to produce increased output by sacrificing other characteristics, such as increasing the draft loss. These practices may shorten the life of the boiler. For these and further reasons, standards have been adopted on procedures for testing and rating boilers. These standards may also recommend characteristics such as required amount of heating surface, furnace volume, fuel firing rates, and draft loss. The *Hydronics Institute,* an independent industry organization, has established recommended standards for hot water boilers, called I-B-R ratings. It is suggested that the HVAC engineer check if a boiler has been tested and rated in accordance with I-B-R standards before selection. This is usually stated in the manufacturer's catalog. The *American Gas Association* (AGA) also recommends standards for gas-fired boilers.

Piping and Pickup Loss

The heat output of a boiler is used to deliver the building heating load. However, the actual output capacity of the boiler must be greater than the building heating load, because of two factors:

1. There is a constant loss of heat through hot piping to surrounding areas, some of which is not useful heating (piping in unheated areas). Insulation will reduce, but not eliminate, this loss. This is called the *piping loss.*

2. There is an additional heat loss when starting up a cold system. Before the boiler can deliver heat to the building, all the piping, water, and equipment of the heating system itself must be heated. This is called the *pickup loss, pickup factor,* or *pickup allowance.*

Boiler Gross and Net Output

The boiler *gross output* is the actual heat output of the boiler at its nozzle (exit). The boiler *net output* is the gross output less the piping and pickup losses. That is

Gross output =

$$net \ output + piping \ loss + pickup \ loss \quad (4.2)$$

The boiler net output can be considered as equal to the building heating load for a hot water or steam heating system (plus service hot water load on the boiler, if any). The piping and pickup losses are not the same for every building and often are not easy to determine accurately. They depend both on the building and heating system configuration, building operating procedures, as well as other factors.

Experience has led to standard allowances for the piping and pickup losses that are often adequate. For hot water boilers, the I-B-R standard allowance is 15% of the net output for the combined piping and pickup losses. (The standard allowance for steam boilers is different. Consult manufacturers' data for this information.)

The 15% allowance is recommended for commercial buildings that are continually heated and that do not have night setback of temperatures. For buildings that are intermittently heated (e.g., a house of worship), an additional 10% pickup loss is recommended by some authorities.

In residential (and some small commercial) applications, if nighttime temperature setback is practiced, a large pickup allowance is needed if the building is to be brought up to a comfortable temperature within a sufficient time in the morning. For instance, with 10 F night setback and one hour required pickup time, a 40% piping and pickup allowance is recommended for sizing the boiler.

The values of pickup losses suggested for intermittent heating and night setback also apply to furnaces.

The piping and pickup allowance is not usually necessary when sizing boilers for larger commercial installations. There are two reasons for this. First, standby (reserve) boiler capacity is usually provided by using two or more boilers. Typically, the excess capacity is from 25–100% of design load; this is used to cover breakdown or maintenance of one boiler. Under normal conditions, the excess capacity can be used to cover large pickup requirements. Second, the capable operating engineer knows how to operate the heating system to ensure that, in the morning, temperatures are brought to a comfortable level before the building is occupied. Techniques using computer managed automatic control systems aid in this.

Example 4.5 _____

A building has a net space heating load of 370,000 BTU/hr and a service hot water load of 32,000 BTU/hr. The building is heated intermittently. A piping allowance of 15% and pickup of 10% are required. What should the gross output be?

Solution
From the definitions given in Equation 4.2,

Boiler net output	$= 370,000 + 32,000$
	$= 402,000$ BTU/hr
Piping allowance	$= 0.15 \times 402,000$
	$= 60,000$ BTU/hr
Pickup allowance	$= 0.10 \times 402,000$
	$= 40,000$ BTU/hr
Required boiler gross output	$= 502,000$ BTU/hr

Setback should not be so excessive and rapid that it could cause thermal shock in a boiler. This is a situation where extreme water temperature fluctuations cause stress damage to the boiler.

Steady-state Efficiency

The gross output of a boiler or furnace is less than the heat input due to unavoidable losses. This may be expressed in an efficiency term, which is given different names, such as *operating efficiency, thermal efficiency, overall efficiency,* or *steady-state efficiency.* (We will use the term steady-state.) The steady-state efficiency is defined as

Steady-state efficiency =

$$\frac{\text{gross heat output}}{\text{heat input}} \times 100\% \quad (4.3)$$

The losses that occur in the heating equipment are the flue gas losses and heat lost from the hot surface of the unit to the surrounding space.

The steady-state efficiency is slightly less than the combustion efficiency defined in Section 4.6, because the combustion efficiency includes only the flue gas losses. The heat loss from the jacket or casing of the heating unit is quite small compared to the flue gas losses, but nevertheless the unit should be well insulated. Occasionally the heat transferred to the surrounding space is useful, but this is unusual.

The combustion efficiency and steady-state efficiency terms serve two different purposes. The combustion efficiency is used in field testing of the heating unit to see if it is operating satisfactorily.

The steady-state efficiency is used in the rating and selection of the boiler or furnace for a given application.

Ratings (and other data) for a group of small cast iron, gas-fired hot water heating boilers are shown in Figure 4.21. This information can be used in selecting a boiler and determining its steady-state efficiency. The column titled D.O.E. CAPACITY is the gross output of the boiler.

Example 4.6

Select a gas-fired hot water boiler for the Moneybags Mansion. The heating load is 220,000 BTU/hr.

Solution
The net output (rating) of the boiler must be at least equal to the building heating load. From the ratings in Figure 4.21, a Model GG-325 is the smallest boiler that will do the job, with a net I-B-R rating of 226,100 BTU/hr. Note that the gross output is 260,000 BTU/hr, which includes a 15% piping and pickup allowance.

Example 4.7

For a Model GG-200H boiler, determine the steady-state efficiency and ft³/hr (CFH) of natural gas consumed at full load.

Solution
Using Equation 4.3,

Steady-state Efficiency =

$$\frac{\text{gross heat output}}{\text{heat input}} \times 100\%$$

$$= \frac{167,000}{200,000} \times 100 = 83.5\%$$

The heating value of natural gas is about 1000 BTU/ft³ (Table 4.1), therefore the amount of gas required is

$$\text{CFH of gas} = 200,000 \ \frac{\text{BTU}}{\text{hr}} \times \frac{1 \ \text{ft}^3}{1000 \ \text{BTU}}$$

$$= 200 \ \text{CFH}$$

SPECIFICATIONS: GG SERIES Hot Water Model

ratings

NATURAL AND L.P. PROPANE GAS RATINGS

MODEL NUMBER†	RATINGS FOR WATER		NET I = B = R RATING WATER (Btuh)	WATER (Sq. Ft.)	MODEL NUMBER†	RATINGS FOR WATER		NET I = B = R RATING WATER (Btuh)	WATER (Sq. Ft.)
	A.G.A. INPUT (Btuh)	D.O.E. CAPACITY (Btuh)				A.G.A. INPUT (Btuh)	D.O.E. CAPACITY (Btuh)		
GG-75H	75,000	64,000	55,700	371.3	GG-250H	250,000	209,000	181,700	1211.3
GG-100H	100,000	83,000	72,200	481.3	GG-275H	275,000	228,000	198,300	1322.0
GG-125H	125,000	103,000	89,600	597.3	GG-300	300,000	240,000**	208,700	1390
GG-150H	150,000	125,000	108,700	742.6	GG-325	325,000	260,000**	226,100	1510
GG-175H	175,000	145,000	126,100	840.6	GG-350	350,000	280,000**	243,500	1620
GG-200H	200,000	167,000	145,200	968.0	GG-375	375,000	300,000**	260,900	1740
GG-225H	225,000	186,000	161,700	1078.0					

* Net ratings are based on a piping and pick-up allowance of 1.15 (hot water). Slant/Fin should be consulted before selecting a boiler for installation having unusual piping and pick-up requirements. Ratings must be reduced by 4% at 2,000 feet elevation and an additional 4% for every additional 1,000 feet elevation over 2,000 feet.

† Add suffix "S" for standard water boiler, "P" for packaged water boiler. Add suffix "E" for intermittent pilot ignition system (available only with 24 volt gas valve). Type of gas: After model number, specify gas by name "Natural," or "Propane."

** A.G.A. gross output rating (Btuh)

NOTE: All boilers under 300,000 Btu input are tested and rated for capacity under the U.S. Dept. of Energy (D.O.E.) Test Procedure for boilers.

dimensions

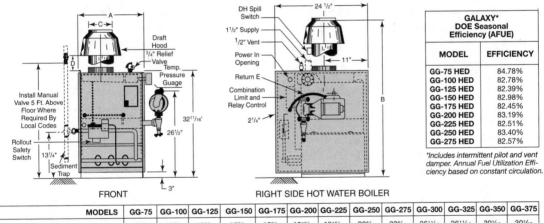

	GALAXY* DOE Seasonal Efficiency (AFUE)	
MODEL	**EFFICIENCY**	
GG-75 HED	84.78%	
GG-100 HED	82.78%	
GG-125 HED	82.39%	
GG-150 HED	82.98%	
GG-175 HED	82.45%	
GG-200 HED	83.19%	
GG-225 HED	82.51%	
GG-250 HED	83.40%	
GG-275 HED	82.57%	

*Includes intermittent pilot and vent damper. Annual Fuel Utilization Efficiency based on constant circulation.

FRONT   RIGHT SIDE HOT WATER BOILER

	MODELS	GG-75	GG-100	GG-125	GG-150	GG-175	GG-200	GG-225	GG-250	GG-275	GG-300	GG-325	GG-350	GG-375
A	JACKET WIDTH	13³⁄₁₆	13³⁄₁₆	13³⁄₁₆	16⁹⁄₁₆	16⁹⁄₁₆	19¹⁵⁄₁₆	19¹⁵⁄₁₆	23⁵⁄₁₆	23⁵⁄₁₆	26¹¹⁄₁₆	26¹¹⁄₁₆	30¹⁄₁₆	30¹⁄₁₆
B	DRAFT HOOD HEIGHT	46⅜	53¼	53¼	53¼	53¼	57½	57½	59⅝	59⅝	59⅝	59⅝	66⅛	66⅛
C	FLUE COLLAR DIAMETER	5	6	6	6	6	7	7	8	8	8	8	9	9
D	JACKET TOP TO DRAFT HOOD	6¾	13	13	13	13	16	16	17	17	17	17	22½	22½
E	CIRCULATOR RETURN FLANGE	1¼	1¼	1¼	1¼	1¼	1¼	1¼	1½	1½	1½	1½	1½	1⅓
F	GAS CONNECTION NAT./PROP.	½/½	½/½	½/½	½/½	½/½	½/½	½/½	½/½	¾/½	¾/½	¾/½	¾/½	¾/¾

Crates for all models are 30" wide, 38" high, depth 30" (GG-75 thru GG-225), 38" (GG-250 thru GG-375).

equipment

BASIC WATER BOILER (SUFFIX S) includes pre-assembled heat exchanger with built-in air eliminator, base; flue collector; gas burners, gas orifices and manifold assembly; combination gas valve including manual shut-off, pressure regulator, pilot adj. and automatic pilot-thermocouple safety; hi limit control; altitude, pressure and temperature gauge; pressure relief valve (ASME); draft hood; draft hood spill switch; rollout safety switch; pre-assembled insulated semi-extended jacket (extended as shown); automatic vent damper (except GG-300 thru GG-375 and GXH-300).

PACKAGED WATER BOILER (SUFFIX P) includes all equipment listed for model S, plus 2-way combination control (hi limit and circulator relay), instead of hi limit control; circulator; drain cock.

OPTIONAL EQUIPMENT: Room thermostat. Millivolt (self energized) controls,* combination gas valve, combination limit controls and millivolt thermostat. Air package consisting of diaphragm expansion tank, fill and pressure reducing valve, and automatic air vent. Combustible floor kit. Intermittent pilot ignition system.

*For GG-300 thru GG-375 and GXH-300

Figure 4.21
Capacity ratings for a group of small cast iron gas-fired hot water boilers. (Courtesy: The Slant/Fin Corporation.)

4.11   BOILER INSTALLATION

Each manufacturer furnishes specific instructions for the installation of the boiler when it is shipped to the job. We will not attempt to repeat these detailed instructions, but will instead list some procedures that are generally useful.

1. Allow ample size openings and passages into the boiler room for the boiler. The architect must be informed of the dimensions needed so that he or she can provide them. If an existing building that requires a new boiler does not have adequate openings for a tubular boiler, a sectional cast iron boiler may be the solution.
2. On high-rise buildings, consider a penthouse location for a gas-fired boiler. This eliminates the need for a flue running the whole height of the building.
3. Provide ample space on all sides of the boiler for maintenance. Allow adequate distance in front of the boiler for tube cleaning and removal.
4. Locate the boiler as close to the flue as possible. Install the breeching to the flue without offsets.
5. Provide sufficient openings to the outdoors for both combustion air and ventilation air. Fixed grilles in walls or doors are one method. This is an extremely important point. If the openings are not adequate, the boiler may be starved of sufficient air for combustion, resulting in the production of toxic carbon monoxide.
6. Follow fire and safety codes.

4.12   ENERGY USE AND EFFICIENCY IN BOILERS AND FURNACES

Until the advent of "high efficiency" units, the best residential and small commercial boilers and furnaces could achieve an overall steady-state efficiency (when well maintained) of about 70–80% by recovering enough heat from the combustion gases to reduce flue gas temperatures to about 400–500 F and using about 50–60% excess air.

More recently, higher efficiency equipment has been made available. These units have a larger or a secondary heat exchanger. This provides more heat transfer surface and a longer path for the hot combustion gases, resulting in the utilization of more of the heat released and a corresponding lower flue gas temperature.

One group, medium-high efficiency boilers and furnaces, reduce the combustion gas temperature to about 300 F, resulting in an operating efficiency of about 85%. Another group, very high efficiency units, reduce the stack gas temperature to about 110 F, with an operating efficiency of about 90–95%. At this low temperature, the water vapor in the flue gas condenses. The very high efficiency results from the additional sensible heat recovered, and from the heat of condensation of the water vapor given up.

The heat exchangers that handle the lower temperature gases in the very high efficiency units are made of stainless steel or other corrosion-resistant materials because of the moisture present. For the same reasons, stack vents and drains must be made of plastic pipe or other noncorrosive materials. Drainage of water collected is important both in the design and installation of these units.

Some high efficiency units have a *sealed combustion* system. Combustion air is drawn directly from outdoors to the combustion chamber through a sealed pipe, instead of being drawn from the equipment room. The combustion gases are vented directly to the outdoors through a plastic pipe, sometimes through a side wall instead of a chimney. This is called *direct venting*. No draft hood is necessary.

Because of the high resistance of gas flow caused by the greater, more tortuous heat exchanger surface, high efficiency units usually are furnished with combustion air fans. Natural draft would not be adequate.

In all of the furnaces and boilers discussed so far, fuel combustion takes place continuously. There is another high efficiency type of unit (Figure 4.22) that uses *pulse combustion,* an intermittent form of burning. Initially, a small charge of fuel and air are introduced into the combustion chamber and

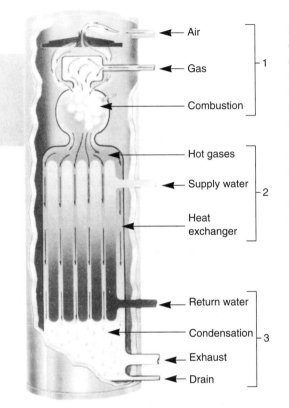

Figure 4.22
Pulse combustion type high efficiency hot water
boiler. (Courtesy: Hydrotherm, Inc.)

ignited. This creates a pressure pulse that drives the combustion gases out. Another small charge then enters and is ignited from the residual heat. One possible concern with pulse type equipment is that the pulsating noise created may be more disturbing than the continual noise of a steady burning unit.

The features discussed until now concerning high efficiency boilers and furnaces improve the steady-state efficiency. Also of major concern is the system efficiency over the full heating season; that is, we want to minimize the annual fuel use. The heating equipment does not operate in a steady-state condition continuously. There are heat losses associated with the equipment when it is not operating, some of which we will now discuss, including how to minimize these losses.

In conventional flue and equipment arrangements, when the boiler or furnace is in the off cycle, the natural draft effect will cause warm equipment room air to continuously vent up the stack. This air will be replaced naturally by cold outside air entering the room, just as in operating conditions. This represents a considerable infiltration heat loss into the building.

To prevent this loss, an *automatic vent damper* (Figure 4.3) may be installed in the flue. This damper is closed when the unit is not operating. When the boiler or furnace starts up, the damper opens. Automatic vent dampers can be retrofitted into existing systems, as well as installed with new ones. They are often required by local codes. Sealed combustion type units would not use a vent damper, since they draw air directly from outdoors.

Another factor reducing system efficiency is a standing pilot flame. This is continual energy loss when the unit is in the off cycle. Intermittent ignition systems solve this problem.

Large boilers have combustion control systems that enable them to use much less excess air than do residential boilers. Thus, higher steady-state efficiency is achieved in this manner. Some large units are equipped with a flue gas heat exchanger that extracts some of the waste heat, increasing system efficiency further. This heat exchanger might be used to preheat combustion air (this is called an *economizer*), or to heat service hot water.

Intermittent ignition and automatic vent dampers are common practice with large systems.

The heat losses associated with the actual working conditions of the boiler or furnace such as the vent stack losses on shutdown and standing pilot need to be accounted for in determining the annual energy efficiency of the unit. This is approximated by the Annual Fuel Utilization Efficiency (AFUE), which can be defined as

$$AFUE = \frac{annual\ heat\ output}{annual\ heat\ input} \times 100\% \quad (4.4)$$

The actual AFUE can really only be truly measured by taking continual measurements of heat output and heat input over the year, an almost impossible task. It is approximated by tests specified by the U.S. Department of Energy. The AFUE of a conventional boiler or furnace with natural draft, standing pilot, and no vent damper can be as low as 50–60%. With energy-saving improvements noted here, the AFUE can reach 80–95%. The AFUE is listed for the boilers shown in Figure 4.21.

4.13   ENERGY CONSERVATION

Some methods of conserving energy with boilers and furnaces are as follows:

1. Adjust the air-fuel ratio so that excess air is the minimum recommended for the equipment.
2. Clean all heat transfer surfaces regularly (boilertubes, heat exchangers).
3. Do not use unnecessarily oversized boilers or furnaces.
4. For larger projects install multiple boilers. At part loads boilers will then operate closer to full capacity, where efficiency is higher.
5. Consider the use of heat exchange devices to use some of the waste heat in the hot flue gases.
6. Use proper boiler water treatment methods. Consult a specialist for advice.
7. Clean burner nozzles regularly.
8. Consider installation of a solar heating system for domestic hot water.
9. Install an automatic vent damper in the flue. This device closes when the combustion unit is not operating, thereby reducing extra infiltration air that would go up the stack.
10. Install a flame retention type of oil burner in residential equipment. This type of burner uses less excess air and results in better heat transfer than other types.
11. Use intermittent ignition, rather than a standing pilot flame. (Care must be taken that this or automatic vent dampers do not result in water vapor condensation, which might cause corrosion.)
12. Consider the use of high efficiency boilers or furnaces.
13. Use temperature setback when feasible.

Review Questions

1. List the major components of a warm air furnace.
2. List four physical arrangements of warm air furnaces and where they would be located in a residence.
3. Name four types of warm air heating devices.
4. Describe the basic operating and safety controls for a warm air furnace.
5. Describe a typical programming control sequence for the operation of a warm air furnace.
6. List the major components of a hot water boiler and a steam boiler. List and explain the purpose of common boiler accessories.
7. Sketch a typical hot water boiler piping arrangement, listing essential components. What are piping loss and pickup loss? What are gross and net boiler output?
8. What are the pressure and temperature ratings for low pressure boilers?
9. Explain the difference between a firetube and watertube boiler.
10. Describe the basic operating and safety controls for hot water and steam boilers.
11. Describe a typical programming control sequence for the operation of a boiler.
12. What are the two meanings of draft?
13. Describe the different methods of achieving draft.
14. How is draft usually controlled in small gas- and oil-fired equipment? How is draft controlled in larger boilers?
15. List the major fuels used in heating plants and their relative advantages and disadvantages. List the by-products of the combustion of these fuels. What pollutants may result from incomplete combustion of these fuels?
16. Explain the terms theoretical air and excess air.
17. Describe the basic types of gas and oil burners.

18. Describe the methods of burner firing rate control.

19. List the types of flame safety controls. How do these controls work?

20. What problem may arise if stack gas temperature is too low?

21. What is the meaning of AFUE?

22. List five possible ways of increasing heating equipment efficiency.

Problems

4.1 A residence has a net heating load of 120,000 BTU/hr. Select a natural gas-fired hot water boiler, assuming a standard piping and pickup allowance. Determine the full load efficiency of the boiler.

4.2 If the boiler in Problem 4.1 is operating at full load efficiency, how much gas would be required at full load?

4.3 A building has a net heating load of 175,000 BTU/hr. The piping heat loss is 25,000 BTU/hr and the pickup loss is 30,000 BTU/hr. Select a natural gas-fired hot water boiler for this application.

4.4 A hot water boiler has a design heat input of 800,000 BTU/hr and a full load efficiency of 78%. It is to be used in a building with a piping and pickup loss of 100,000 BTU/hr. What is the maximum heating load the boiler can handle?

4.5 A boiler is using 1.3 GPM of No. 2 fuel oil. It has an efficiency of 72% and the piping and pickup loss is 22%. What is gross output and net output of the boiler?

4.6 A furnace burning natural gas is designed to operate with 30% excess air. What percentage reading of CO_2 will indicate proper operation? If the difference between the flue gas temperature and room temperature is 400 F, what would be the furnace efficiency at design conditions (neglecting other losses)?

4.7 A boiler uses No. 2 fuel oil. Using a combustion gas analyzer, a technician measures 12% CO_2 in the stack gases. What is the percent excess air? Flue gas temperature reads 520 F. Room temperature is 70 F. What is the boiler efficiency (neglecting other losses)?

4.8 For the boiler referred to in Problem 4.7, a stack heat exchanger is installed, reducing flue gas temperature to 370 F. What is the approximate boiler efficiency?

CHAPTER 5
Hydronic Piping Systems and Terminal Units

The piping that is used to circulate hot or chilled water for air conditioning is called a *hydronic piping system*. The terminal units are the heat exchangers that transfer the heat between the water and the spaces to be heated or cooled. In this chapter, we will examine types of hydronic piping arrangements and terminal units.

OBJECTIVES

After studying this chapter, you will be able to:

1. Identify the types of hydronic piping system arrangements and describe their features.
2. Identify the types of hydronic terminal units and describe their features.
3. Select baseboard radiation.
4. Lay out a hydronic system and determine its water temperatures and flow rates.

5.1 PIPING ARRANGEMENTS

The connections between the piping and the terminal units may be made in any of these four basic ways:

1. Series loop
2. One-pipe main
3. Two-pipe direct return
4. Two-pipe reverse return

5.2 SERIES LOOP

A diagram of a *series loop* arrangement is shown in Figure 5.1. It is so named because all of the units are in a series, and one loop is formed.

Note that the entire water supply flows through each terminal unit and then returns to the generator and pump. Because all of the water flows through each unit, and the units cannot be isolated from each other, the series loop has several disadvantages:

1. The maintenance or repair of any terminal unit requires shutdown of the entire system.
2. Separate capacity control of each unit by changing its water flow rate or temperature is

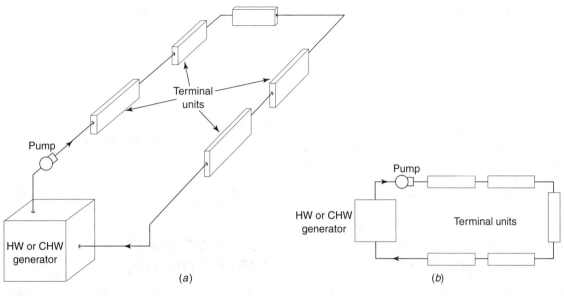

Figure 5.1
Series loop piping system. (*a*) Isometric. (*b*) Schematic.

not possible. (Control is possible by use of air dampers, however.)

3. The number of units is limited. Since in heating systems the water temperature continually decreases as it gives up heat in each unit in series, the water temperature in later units may be too low for adequate heating.

These disadvantages can be partially remedied by arranging the piping in two or more split series loops, as shown in Figure 5.2. This creates two or more zones which can be controlled separately.

The series loop arrangement is simple and inexpensive. It is limited to small, low-budget applications such as residences.

Figure 5.2
Split series loop piping system. (*a*) Isometric. (*b*) Schematic.

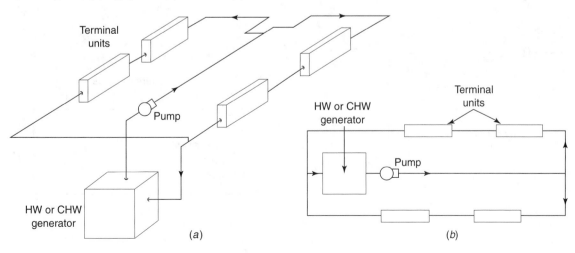

5.3   ONE-PIPE MAIN

A diagram of a *one-pipe main* arrangement is shown in Figure 5.3. As in the series loop, there is one main pipe through which the water flows, but instead of being in series with this main, each terminal unit is connected by a supply and a return branch pipe to the main. By locating valves in the branch lines, each unit can be separately controlled and serviced. As in the series loop, if there are too many units, the water going to the later units may be too cool to heat the rooms adequately.

Flowing water seeks the path of least resistance. Consequently, water circulating in the main tends to flow through the straight run of the tee fitting at each supply branch, thus starving the terminal unit. To overcome this problem, special *diverting tees* (Figure 5.4) are used at each supply branch take-off, directing some of the water to the branch. Additionally, if the terminal unit is below the main, a

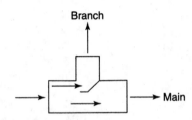
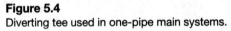

Figure 5.4
Diverting tee used in one-pipe main systems.

special tee is also needed at the return branch to prevent backflow.

5.4   TWO-PIPE DIRECT RETURN

To get the water temperature supplied to each terminal unit to be equal, the two-pipe (also called

Figure 5.3
One-pipe main piping system. (*a*) Isometric. (*b*) Schematic.

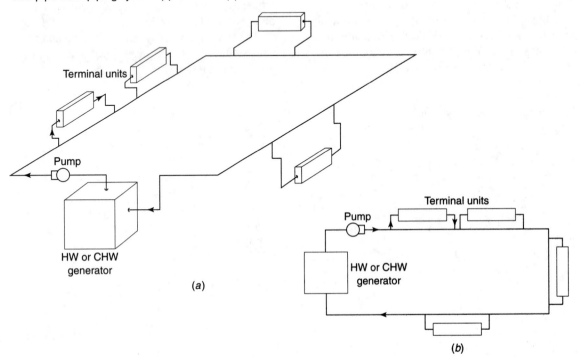

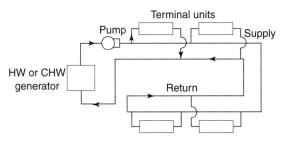

Figure 5.5
Two-pipe direct return system.

parallel piping) arrangement is used. Figure 5.5 shows a *two-pipe direct return* system. There are two mains, one for supply water and one for return. Each terminal unit is fed by an individual supply branch; a return branch carries the water back to the return main. In this manner, all units receive water directly from the source.

The total system flow rate (GPM) is split up among the terminal units, according to the design.

Although its cost is higher than one-pipe main and series loop arrangements, the two-pipe system allows each terminal unit to be separately controlled and serviced, and because the supply water temperature to each unit is the same, it can be used

on any size installation. All larger systems use two-pipe arrangements.

The two-pipe arrangement in Figure 5.5 is called *direct return* because the return main is routed to bring the water back to the source by the shortest path. However, this creates a problem. Note in Figure 5.5 that the path the water takes from the pump to the first units and back is shorter than the units further away. Since flowing water prefers the path with the least resistance, there will be too much water going to the units nearest the pump and too little to the units furthest from the pump. To overcome this problem, balancing valves can be installed in every branch, but the balancing process is difficult and requires considerable expense. The problem is largely solved with a *reverse return*.

5.5 TWO-PIPE REVERSE RETURN

The balancing problem in the direct return arrangement would be overcome if the circuit length out to each terminal unit and back was made approximately the same. This is accomplished by piping the return main in a *reverse return* arrangement, as shown in Figure 5.6. Note that the path length for

Figure 5.6
Two-pipe reverse return system. (*a*) Isometric—two-pipe reverse return to a number of buildings. (*b*) Schematic.

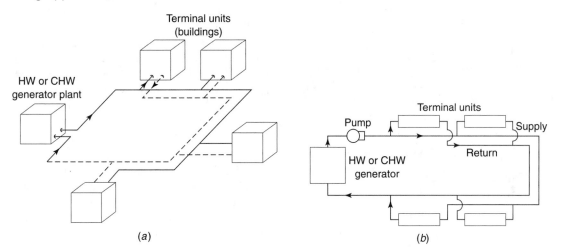

(*a*)   (*b*)

the water is about the same regardless of which unit it passes through. With this arrangement, it is a relatively simple process to balance the flow rates.

The relative costs of the direct return and reverse return piping arrangements depend on the building shape and location of terminal units. In some cases, the costs are not significantly different; in others, the reverse return piping may be more expensive.

It may seem from the prior discussion on balancing that the two-pipe reverse return system would always be chosen over the direct return. In some situations, however, it may not be difficult to balance a direct return system. These are:

1. If the terminals are all far from the pump and grouped near each other, there may be little difference between the length of each path.
2. A very high resistance in the terminal units may make fluid flow through them approximately equal.
3. It is possible to make the fluid resistance in each circuit approximately equal in a direct return system by using smaller diameter piping in the closer branches. This depends on the piping layout, but it may then cause other problems (see Chapter 9).

In each case, the planner must examine the layout before making a choice.

The two-pipe and one-pipe main arrangements can be split into two or more systems (if this is useful), as was shown with the series loop arrangement.

Two-pipe arrangements are almost always used for chilled water distribution to terminal cooling units. The water temperature to units far from the chiller would be too high for adequate cooling with series loop or one-pipe main arrangement.

5.6 COMBINATION ARRANGEMENTS

It is sometimes useful to combine some of the four basic piping arrangements, taking advantage of the best features of each. Figure 5.7 shows an example

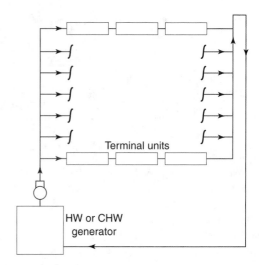

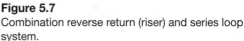

Terminal units

HW or CHW generator

Figure 5.7
Combination reverse return (riser) and series loop system.

of a combined two-pipe reverse return with a group of units on each floor in series. This might be chosen for a high-rise building where separate control of each unit on a floor is not needed, yet flow balance will be simple and costs reduced as compared with a complete reverse return.

5.7 THREE-PIPE SYSTEM

The supply main in the two-pipe arrangements can be furnished with either chilled or hot water for cooling or heating, if the system is connected to both a water chiller and hot water boiler. However, only one can be used at any given time. In modern buildings, heating is often required in some rooms and cooling in others at the same time. An instance of this might occur on a cool day with solar radiation on one side of the building only.

Simultaneous heating or cooling can be made available by use of a *three-pipe* arrangement (Figure 5.8). There are two supply mains, one circulating chilled water, the other hot water. Three-way control valves in the branch to each terminal unit will determine whether the unit receives hot or chilled water. The return main receives the water

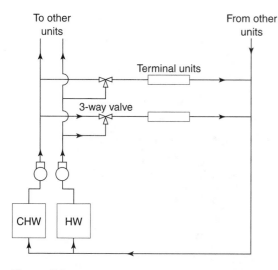

Figure 5.8
Three-pipe system.

from each unit. The connections to units can be made either direct or reverse return.

Because the return main mixes hot and chilled water, the three-pipe system can waste energy. The chilled water is warmed and the hot water is cooled in the mixing process, which results in extra heating and cooling in the boiler and chiller. This problem can be minimized by careful design, but it should be recognized. Some energy codes prohibit three-pipe systems.

5.8 FOUR-PIPE SYSTEM

The *four-pipe* system is actually two separate two-pipe systems, one for chilled water and one for hot water, and therefore no mixing occurs. This is an ideal arrangement, but of course it is expensive.

5.9 HYDRONIC TERMINAL UNITS

The *terminal units* are heat exchangers that transfer the heat between the room air and the circulating water. Generally, the type of units used for heating and cooling are different from each other. The following will be discussed here:

Heating

1. *Radiators*
2. *Convectors*
3. *Baseboard*
4. *Fin-tube*
5. *Radiant panels* (heating and cooling)
6. *Unit heaters*

Cooling

1. *Fan-coil units* (heating and cooling)
2. *Induction units* (heating and cooling)

Radiators, convectors, baseboard, and fin-tube are collectively called *radiation*. This is a misleading name because they transfer some of the heat to the room largely by natural convection. The air adjacent to the unit is warmed and rises naturally, creating a natural circulation.

All types of radiation should be located along exposed walls and, preferably, under all windows in the colder climates. In this location, heat is supplied where the heat loss is greatest, and cold downdrafts are prevented. Figure 5.9 shows good and poor locations of radiation.

Except for radiators and some convectors, the heating or cooling element of all hydronic terminal units is usually made of finned tubing. The fins increase the heat transfer. The material may be steel pipes with steel fins, or copper tube with either aluminum or copper fins.

Figure 5.9
Correct and incorrect location of radiation.

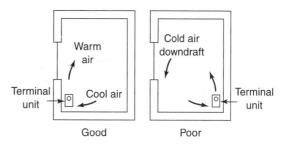

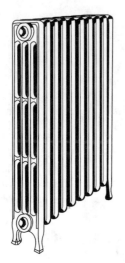

Figure 5.10
Cast iron sectional radiator.

5.10   RADIATORS

This type of radiation is constructed of hollow cast iron sections through which the hot water flows (Figure 5.10). Once very popular, radiators are not used much in new construction. The other types of radiation are often less expensive, require less space, and are more attractive looking.

This type of radiation is available in three forms:

1. Hollow sections made either of cast iron or fabricated steel sheet metal (Figure 5.10).
2. Hollow metal panels.
3. Steel tubing in various assembly arrangements.

Large cast iron sectional radiators are less commonly used in new installations because of their bulkiness, cost, and appearance.

5.11   CONVECTORS

Convectors have a finned tube or small cast iron heating element enclosed by a sheet metal cabinet (Figure 5.11). Room air enters through an opening in the bottom and leaves through an outlet grille at the top.

Convectors are available in varied arrangements to suit the architectural needs of the building. *Flush* units are mounted against the wall, whereas *re-*

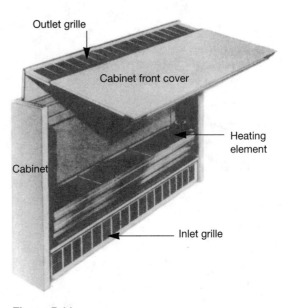

Figure 5.11
Convector (free-standing type).

cessed units are recessed into an opening provided in the wall (Figure 5.12). Recessed units have the advantage of not taking up floor space.

Free-standing units rest on the floor whereas *wall hung* units are off the floor and are supported by the wall (Figure 5.13). Wall hung units allow easier floor and carpet cleaning.

Flush-type units are available with the outlet grille on top, at the top front, or with a sloping top (Figure 5.14). The sloping top prevents placement of objects or sitting on the cabinet.

Figure 5.12
Recessed convector.

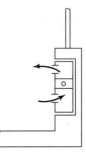

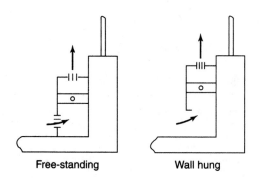

Figure 5.13
Free-standing and wall hung convector.

Convectors are used in rooms, vestibules, and stairwells. They are available in a number of standard lengths and heights.

5.12 BASEBOARD

This type of radiation is located close to the floor in front of the architectural baseboard strip. It consists of a finned tube heating element with a sheet metal cover open at the bottom and with a slotted opening in the top (Figure 5.15).

The cover is often installed along the whole length of the wall for a neater appearance, even when the heating element is not required for the whole length. Baseboard radiation is very popular in residences because it is inexpensive and unobtrusive. Tubing diameter is small, usually ½ or ¾ in. The cover and fins are thin and therefore will not withstand heavy abuse.

Figure 5.14
Outlet arrangements.

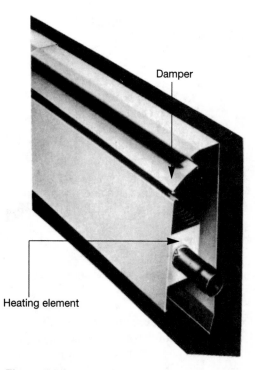

Figure 5.15
Baseboard radiation. **(Courtesy: Slant/Fin Corporation.)**

5.13 FIN-TUBE

This type of radiation is similar to baseboard radiation. The heating element is usually made of larger tubing (¾ to 2 in.) and both the element and cover are heavier and stronger than that used for baseboard radiation. (Figure 5.16 on page 108).

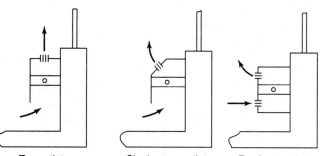

Figure 5.16
Fin-tube radiation. (Courtesy: Vulcan Radiator Company.)

Covers are available with flat or sloping tops and varied quality of appearance. The capacity of fin-tube radiation is greater per foot of length than baseboard radiation because of larger fins and larger pipe. Tubing may be stacked more than one row high to increase output. (However, capacity does not increase proportionally with number of rows.) Fin-tube radiation is widely used in commercial and industrial applications where radiation is desired along exposed walls.

The capacity of convectors and baseboard and fin-tube radiation can all be manually controlled by dampers located at the air outlet.

5.14   RADIANT PANELS

A radiant panel system has tubing installed in walls, floors, or ceiling, and extending over all or a considerable part of the surface. Both heating and cooling systems are available. Ceiling panels are used for cooling, so that the cooled air drops and circulates through the room. Because the heating or cooling source is spread out, radiant panel systems produce uniform temperatures and comfortable air motion. It is an ideal system, but it can be very expensive.

5.15   UNIT HEATERS

The unit heater differs from the previous types of terminal units in having a fan that forces the air through the unit at a greater rate than would be achieved by natural convection. The heating element is finned tubing, arranged in coils to achieve a more compact arrangement. As a result, unit heaters have a high heating capacity for a given physical size. Two kinds of unit heaters will be discussed here.

Propeller Unit Heaters

This type of unit heater is available in two versions —horizontal or vertical discharge. Each has a finned-tube coil heating element, propeller fan, motor, and casing (Figure 5.17).

The *horizontal blow* heater is usually mounted at 7–10 ft elevations. It has adjustable outlet

Figure 5.17
Horizontal and vertical propeller unit heaters. (*a*) Horizontal propeller unit heater. (*b*) Vertical down-blow propeller unit heater.

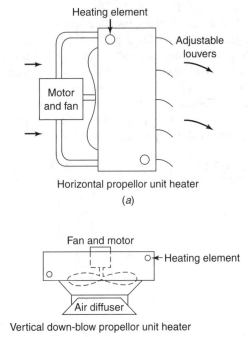

Horizontal propellor unit heater
(*a*)

Vertical down-blow propellor unit heater
(*b*)

dampers to control air direction. Air is directed toward work spaces or door openings. It is often used at loading platforms, vestibules, garage doors—wherever doors may be opened frequently and "spot" heating is needed.

The *vertical down-blow* unit heater is suitable for heating spaces with high ceilings and large floor areas. The units are mounted at high elevations. Adjustable outlet diffusers are available so that the amount of floor area heated can be varied. These units are often used in factories and warehouses.

Propeller fan unit heaters are generally limited to industrial applications or service areas of commercial buildings because they are unsightly, bulky, and noisy.

Cabinet Unit Heaters

This type has a finned-tube heating element arranged as a serpentine coil, small centrifugal fans, an air filter, and a cabinet enclosure. In outward appearance, it looks like a convector (Figure 5.18).

The cabinet unit heater is often used where a convector would be suitable, but where the required heat output is larger, as in vestibules. It can also be mounted flat against a ceiling when this is architecturally desirable, because the outlet grilles will direct the air in the desired direction (Figure 5.19).

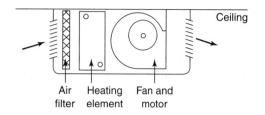

Figure 5.19
Cabinet unit heater—ceiling mounted.

Cabinet unit heaters can be used in commercial applications because they have a pleasing appearance and are relatively quiet. They are sometimes called fan-coil units, but we will use this name for a unit quite similar in construction that is used for either heating or cooling.

5.16 FAN-COIL UNITS

This type of hydronic terminal unit is suitable for both cooling and heating. It consists of a cabinet enclosing one or two serpentine-shaped finned-tube coils, small centrifugal fans with motors, and an air filter (Figure 5.20). Depending on system design, it may have one coil for heating or cooling or separate heating and cooling coils. Alternately, some units have an electric strip heater instead of a hot water heating coil. As with cabinet unit heaters, fan-coil units can be mounted in various horizontal and vertical arrangements, as required.

Fan-coil units include a drain pan under the coil to collect the condensate created from dehumidifying

Figure 5.18
Cabinet unit heater—floor mounted.

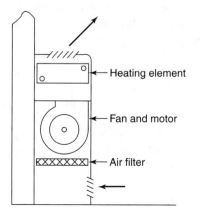

Figure 5.20
Fan-coil unit.

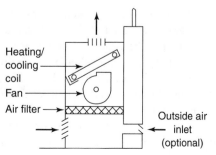

the air when operating in the cooling mode. These drains usually must be piped to a central building drain.

Some fan-coil units include an opening and damper in the rear of the cabinet to connect directly through the wall for outdoor ventilation air. There are problems associated with this. The changing wind effects can greatly affect the amount of outside air brought in. Too much air will waste energy; too little will result in poor air quality.

In addition, the filter is of minimal efficiency. (Higher filtering efficiency not only raises the cost of the filter, but also the fans, since the resistance to airflow increases with higher efficiency). The filter is basically suitable for cleaning only recirculated room air, not the often quite dirty outside air. For these reasons, bringing ventilation directly into the fan-coil unit will often lead to unnecessary operating and maintenance problems. Instead, ventilation air can be furnished from central air handling units with better filters.

Capacity variation of a fan-coil unit can be achieved through room thermostat control of either fan speed or coil water flow. Central HVAC systems using fan-coil units are very popular due to their flexibility and often competitive total system costs.

Considerable maintenance is a major aspect of a fan-coil system, because it has so many units. Coils, drain pans, and filters must be cleaned of lint and dirt regularly and often. Maintenance of the large number of motors must also be compared to the few in an HVAC system with only central air handling units.

5.17   INDUCTION UNITS

This type of terminal unit is suitable for both cooling and heating. It is used in air-water type central HVAC systems. The cabinet contains a cooling/heating coil, lint screen type filter, a connection for *primary* air, and air nozzles or jets (Figure 5.21).

The induction unit does not require a fan to circulate room air across the coil. The air is moved by an induction effect. Primary air from a central air handling unit is delivered at high pressure to the

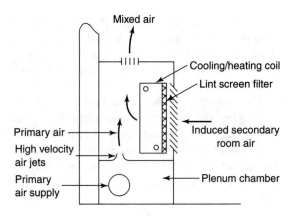

Figure 5.21
Induction unit.

plenum (chamber). This air is forced out through small nozzles at a high velocity. This induces (draws) room (secondary) air into the unit across the coil. The mixed air (primary and secondary) exits through the top grille.

In the cooling mode, chilled water is delivered to the coil. In the heating mode, hot water is delivered to the coil. The primary air can be delivered cold, neutral, or warm as needed. A more detailed explanation of the induction unit air conditioning system will be discussed in Chapter 12.

The lint screen filter is very thin and porous, with a minimal cleaning quality, about on the level of the familiar window unit filter. This is because the induction effect can only overcome a very small air resistance. Frequent cleaning of the lint screen is often required to keep the induction effect going; otherwise little room air is circulated and the heating or cooling is inadequate.

However, the absence of a unit fan as compared to the fan-coil unit means no terminal unit motors to maintain (a large project may have hundreds of units). Furthermore, the fan noise, especially if the fan starts and stops, is usually more objectionable than the very low level air noise from the induction unit. The induction unit has a drain pan. It depends on the system design and operating conditions whether or not the drain pan must be piped to a central drain.

Water heating and cooling coils that are installed in central air handling units will be discussed in Chapter 12.

5.18 SYSTEM WATER TEMPERATURES AND FLOW RATES

Hydronic heating systems are classified into temperature categories as follows:

LTW (low temperature hot water)
—temperature below 250 F.

MTW (medium temperature hot water)
—250–350 F.

HTW (high temperature hot water)
—350–450 F.

These categories are important because different types of boilers and equipment are required for each. For example, as the water temperature increases, the boiler pressure must be increased to prevent the water from evaporating. Consequently, equipment of greater strength is required to handle the increased pressure.

In theory, a high supply water temperature is preferred because the terminal units may be made smaller. A high temperature drop is also desirable because less water is required, allowing smaller pumps and piping to be used, and pump energy consumption is reduced.

However, there are reasons for limiting the water temperature. As mentioned earlier, lower water temperature/pressures do not require the heavy and expensive equipment that higher water temperatures do. In addition, the lower water temperature lessens the severity of a burn from accidental contact, a greater possibility with children, for instance.

The usual practice in designing LTW systems is to select a supply water temperature between 180–240 F and a system temperature drop between 10–40 F. For private residences, supply water temperatures of 180–210 F and a temperature drop of 20 F or less are satisfactory, since the room heating loads are small. The designer should consider supply temperatures up to 240 F and temperature drops up to 40 F for commercial applications.

In HTW systems, much greater temperature drops are often chosen (up to 100 F) to reduce pipe sizes and power use. HTW systems are used in very large projects.

Hydronic cooling systems using chilled water (CHW) do not have temperature categories. The supply temperature required in CHW systems depends on the dehumidification needed (Chapter 7) and usually ranges from 40–50 F. The system temperature rise usually ranges from 5–15 F. Here also, high temperature rises should be considered when planning in order to reduce energy consumption. Manufacturers often suggest desirable temperature ranges for their heating and cooling equipment.

The relationship between water temperature, flow rate, and heat gain or loss was shown previously by Equation 2.12:

$$Q = m \times c \times TC$$

Because the specific heat $c = 1.0$ BTU/lb-F for water, this becomes

$$Q = m \times TC \qquad (5.1)$$

where

Q = heat gain or loss of water, BTU/hr

m = flow rate of water, lb/hr

$TC = t_1 - t_2$ = temperature change of water, F

A more convenient form of the equation is to express the flow rate in GPM. Because (approximately) 1 GPM = 500 lb/hr of water, the equation becomes

$$Q = 500 \times GPM \times TC \qquad (5.2)$$

where Q, TC are as before and

$$GPM = \text{flow rate, gal/min}$$

Although the conversion factor of 1 GPM = 500 lb/hr is correct only at cold water temperatures, it may be used with insignificant error to 250 F.

Example 5.1
A hydronic heating system is to be installed in the Square Tire Company factory. The building heating load is 8 million BTU/hr. A system water supply temperature of 240 F and return temperature of 200 F is chosen. What is the required system flow rate in GPM?

Solution
The flow rate is found by using Equation 5.2:

$$GPM = \frac{Q}{500 \times TC} = \frac{8,000,000}{500 \times (240 - 200)}$$

$$= 400 \text{ GPM}$$

Example 5.2
A water chiller with a capacity of 30 tons of refrigeration cools 80 GPM of water entering at 54 F. What is the temperature of the water leaving the chiller?

Solution
Changing units of cooling capacity,

$$Q = 30 \text{ tons} \times \frac{12,000 \text{ BTU/hr}}{1 \text{ ton}}$$

$$= 360,000 \text{ BTU/hr}$$

Using Equation 5.2,

$$TC = t_1 - t_2 = \frac{Q}{500 \times GPM} = \frac{360,000}{500 \times 80} = 9 \text{ F}$$

Solving for t_2,

$$t_2 = t_1 - TC = 54 - 9 = 45 \text{ F}$$

5.19 SELECTION OF TERMINAL UNITS

The rating (capacity) of terminal units is measured and reported by manufacturers in their equipment catalogs. Standard testing procedures for measuring ratings have been established. The designer and installer should check that any unit being considered has been tested according to a standard rating procedure, such as that of the *Hydronics Institute*.

The manufacturer's catalog ratings are used to select the terminal units required. Table 5.1 shows the ratings for a typical baseboard radiation. The heating capacity is listed in BTU/hr per foot of length. From this, the required length of baseboard can be chosen. Note that the capacity depends on both the flow rate and average water temperature in the unit. The capacity also depends on the entering

TABLE 5.1   RATINGS FOR TYPICAL BASEBOARD RADIATION

Nominal Tube Size in.	Flow Rate GPM	Velocity FPS	Hot Water Ratings, BTU/hr per foot Length at Following Average Water Temperatures, F							
			170	180	190	200	210	220	230	240
	1	0.6	510	580	640	710	770	840	910	970
	2	1.2	520	590	650	730	790	860	930	990
¾	3	1.8	530	600	670	740	800	870	950	1010
	4	2.4	540	610	680	750	810	890	960	1030
	1	1.2	550	620	680	750	820	880	950	1020
	2	2.4	560	630	700	770	840	900	970	1040
½	3	3.6	570	640	710	780	850	920	990	1060
	4	4.8	580	660	720	790	870	930	1000	1080

Notes: Tubing is copper with 2⅛ × 2⁵⁄₁₆ in. aluminum fins, 55 fins per foot. Height of unit with enclosure is 8 in. Ratings are based on air entering at 65 F. For flow rates over 4 GPM, use 4 GPM ratings. Type M tubing.

air temperature. Most ratings are listed for 65 F air entering the unit. For a room maintained at 68–70 F, no correction for the ratings at 65 F is usually necessary.

Example 5.3

Select the length of ½ in. baseboard required for a room with a heat load of 12,000 BTU/hr. The unit has 2 GPM of water flowing through it at an entering temperature of 218 F.

Solution

Table 5.1 will be used to select the unit.

The average water temperature in the unit must first be determined. To find this, Equation 5.2 will be used.

Using Equation 5.2,

$$TC = \frac{Q}{500 \times GPM} = \frac{12,000}{500 \times 2} = 12 \text{ F}$$

Leaving $t = 218 - 12 = 206$ F

Average $t = \dfrac{218 + 206}{2} = 212$ F

Using Table 5.1, at a flow rate of 2 GPM and an average water temperature of 210 F (the next lowest temperature rating listed is used to be certain that the unit has adequate capacity), the rated capacity is listed as 840 BTU/hr per foot of length. The length required for a capacity of 12,000 BTU/hr is therefore

$$Length = \frac{12,000 \text{ BTU/hr}}{840 \text{ BTU/hr per ft}} = 14.3 \text{ ft (use 15 ft)}$$

The contractor would order 15 ft of the radiation, rather than deal with the fractional amount. This also provides a little extra capacity.

The manufacturer should be consulted before using flow rates greatly outside the range shown in their tables. Generally, however, a good guideline is to use flow rates between values that result in water velocities between 1 and 5 ft/sec. Velocities above 5 ft/sec in occupied areas may result in objectionable noise and velocities below 1 ft/sec may not be enough to carry dirt particles through the

TABLE 5.2 WATER VELOCITIES, FT/SEC, IN TYPE L TUBING

Tube Diameter	Flow Rate, GPM				
	1	2	4	6	8
½ in.	1.4	2.7	5.5	8.5	
¾ in.	0.6	1.3	2.6	4.0	5.5

unit. Table 5.2 lists water velocities for different flow rates.

Example 5.4

A contractor is about to install a hydronic system and notes that the engineer's specifications call for ½ in. diameter tubing with a flow rate of 7 GPM. What should the contractor do?

Solution

This flow rate results in a very high velocity, as seen in Table 5.2, and would probably be very noisy. The contractor should call the engineer and discuss possible changes in the design.

Rating tables and selection procedures are similar for other types of radiation, and will not be included here. Catalogs can be obtained from manufacturers.

5.20 SYSTEM DESIGN PROCEDURE

In the previous sections, we discussed piping arrangements, water temperatures and flow rates, and selection of terminal units. It is often difficult for the student to put all this information together in planning a system. The following procedures should be helpful:

1. Choose the types of terminal units best suited for the application (Sections 5.9–5.19).
2. Choose the type of piping arrangement best suited for the application (Sections 5.1–5.8).
3. Prepare a diagrammatic sketch of the piping system and the terminal units connected together.

4. Select a trial value of the system water temperature change (Section 5.20) and calculate the system water flow rate required to handle the building load (Equation 5.2).
5. Check to see if this is a satisfactory flow rate. Velocities through units should be within recommended values (Section 5.21). Also check manufacturer's recommendations on flow rates.
   A. For a series loop system, the flow rate through every unit is the same, of course.
   B. For a one-pipe main system, the flow rates through each unit may be arbitrarily selected within recommended values, but cannot of course be greater than the total flow rate.
   C. For a two-pipe system, the total system flow rate is distributed among all the units. The flow rate through each unit may be arbitrarily chosen within the manufacturer's recommended values, but of course the sum of the flow rates through each of the units must equal the total system flow rate.
6. If the flow rate is not satisfactory according to this check, a new trial value of the system water temperature change is taken and a new flow rate calculated. After a little experience, a designer can usually select an appropriate temperature change on the first or second trial.
7. Calculate the water temperature change through each unit, based on the required capacity of each unit (Equation 5.2).
8. Choose a suitable system supply temperature (Section 5.20). Determine the water temperature entering and leaving each unit. It is helpful to record all flow rates and temperatures on the piping sketch.
9. Select the terminal units from the manufacturer's catalog. Prepare a table showing all the information collected.

Two examples will illustrate this procedure.

Example 5.5

The S.O. Smith residence has a design heating load of 68,000 BTU/hr. A series loop hydronic heating system will be used, with ¾ in. diameter baseboard radiation. A sketch of the piping system and units is shown in Figure 5.22. Table 5.3 lists the required heating capacity of each unit.

Determine appropriate water temperatures, flow rates, and select the terminal heating units.

Solution
The system design procedure recommended will be followed. (The results are noted in Table 5.3.)

1–3. For this small house, a series loop baseboard system has been chosen.
4. Try a system temperature drop of 10 F. Using Equation 5.2 to find the resulting flow rate,

$$GPM = \frac{Q}{500 \times TC} = \frac{68,000}{500 \times 10} = 13.6 \text{ GPM}$$

5. This flow rate will result in a greatly excessive velocity (Table 5.2).
6. Try a temperature drop of 30 F.

$$GPM = \frac{68,000}{500 \times 30} = 4.5 \text{ GPM}$$

This is a satisfactory flow rate.
7. The first unit has a required capacity of 9200 BTU/hr. The temperature change is

$$TC = \frac{9200}{500 \times 4.5} = 4 \text{ F}$$

8. A supply temperature of 220 F is chosen. The return temperature is then $220 - 30 = 190$ F. The temperature entering the first unit is 220 F. The leaving temperature is $220 - 4 = 216$ F.

Figure 5.22
Sketch for Example 5.5.

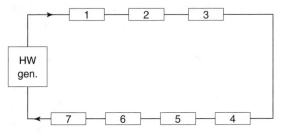

TABLE 5.3   RESULTS OF TERMINAL UNIT SELECTION FOR EXAMPLE 5.5

Unit	Capacity, BTU/hr	Flow Rate, GPM	t_{in}, F	t_{out}, F	t_{ave}, F	Length, ft
1	9,200	4.5	220	216	218	11
2	12,400	4.5	216	211	213	16
3	8,700	4.5	211	207	209	11
4	9,600	4.5	207	203	205	13
5	6,300	4.5	203	200	201	9
6	13,600	4.5	200	194	197	20
7	11,800	4.5	194	189	191	18

The same procedure is used to find the remaining temperatures.

9. The required length of baseboard is now determined. The average water temperature in the first unit is

$$\text{Average } t = \frac{220 + 216}{2} = 218 \text{ F}$$

From Table 5.1, using the rating at 4 GPM and 220 F the heat output is 890 BTU per foot of length. The length required is therefore

$$\text{Length} = \frac{9200}{890} = 10.3 \text{ ft (use 11 ft)}$$

The choice of 11 ft of radiation instead of 10.3 ft will make up for selecting it at 220 F instead of 218 F. For more accuracy, values can be interpolated between listed temperatures. The results for the other units are shown in Table 5.3. The student should check these. (The sum of the unit capacities is slightly greater than the building load, as explained in Chapter 3, due to infiltration. This results in a calculated return temperature slightly less than originally chosen which, however, will not seriously affect the accuracy of the selection of the terminal units.)

If the radiation selected above is excessively long, the problem might be resolved by raising the water supply temperature. Another solution is to use radiation that has a greater output per unit length.

Example 5.6

Determine the chilled water temperatures and flow rates for the two-pipe system shown in Figure 5.23. The terminal units are fan-coil units. Capacities are listed in Table 5.4 on page 116. The building is a small group of medical offices, with a cooling load of 220,000 BTU/hr.

Solution

The design procedures recommended will be followed. All results are listed in Table 5.4.

1–3. After a study of the building plans and use, it has been decided to use fan-coil units and a two-pipe reverse return system. The fan-coil units will fit nicely under the window in each office. The building shape is such that there may be unbalanced flow if a direct return layout is used.

4. Try a system temperature rise of 12 F. Using

Figure 5.23
Sketch for Example 5.6.

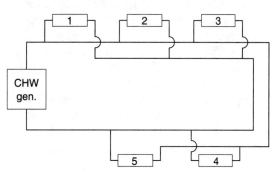

TABLE 5.4   RESULTS OF EXAMPLE 5.6

Unit	Total[a] Capacity, BTU/hr	Flow Rate, GPM	$t_{in'}$ F	$t_{out'}$ F
1	52,000	7.3	44	58
2	41,000	7.3	44	55
3	53,000	7.3	44	59
4	47,000	7.3	44	57
5	27,000	7.3	44	51

[a] In selecting cooling units, the sensible and latent capacities must also be determined. These are not shown here. Furthermore, the building load will not generally be as great as the sum of the room loads, as will be seen in Chapter 6.

Equation 5.2, the flow rate is

$$GPM = \frac{Q}{500 \times TC} = \frac{220,000}{500 \times 12}$$

$$= 36.7 \text{ GPM}$$

The flow rate is arbitrarily distributed equally among all five units, giving 36.7/5 = 7.3 GPM each.

5–6. Referring to a manufacturer's catalog of fan-coil units (not shown here) it is noted that their coil has a ¾ in. nominal diameter tubing. From Table 5.2, the velocity will be satisfactory.

7. The temperature change for the first unit is

$$TC = \frac{52,000}{500 \times 7.3} = 14 \text{ F}$$

The temperature change for the other units is found in the same manner.

8. The system supply water temperature chosen depends on the selection of the refrigeration equipment and costs. Assume a CHW supply temperature of 44 F has been found satisfactory. The temperature leaving the first unit is then 44 + 14 = 58 F. The same procedure is carried out for each unit.

9. From the information above, the terminal units can be selected from the manufacturer's catalog. We will not describe that process here. Each manufacturer has slightly different procedures, which are described in their catalogs.

In the previous example, an alternate procedure could have been used—to assume that every unit has exactly the same water temperature rise and then to calculate the required flow rate. This procedure is just as acceptable as those chosen, but remember that the flow rates found should be checked to see if they are within recommended values. The student should work out this solution as a learning exercise.

Review Questions

1. List the four basic types of hydronic system piping arrangements.

2. Sketch the arrangements for the four types of hydronic piping system arrangements.

3. List the applications and the advantages and disadvantages of the four types of hydronic piping system arrangements.

4. What application do three- and four-pipe systems have?

5. What undesirable feature does a three-pipe system have?

6. List the types of hydronic terminal units used for heating and/or for cooling. Describe a suitable application for each.

7. What are the basic parts of a unit heater?

8. List the types of unit heaters and one application for each type.

Problems

5.1   The operating engineer wants to check the capacity of a refrigeration water chiller. The instruments show 240 GPM of water flowing through the chiller, entering at 52 F and leaving at 40 F. What is the chiller's capacity in tons?

5.2   A building has a heating load of 630,000 BTU/hr. A hydronic heating system is used, supplying 40 GPM of water at 240 F. What is the return water temperature?

5.3   A fan-coil unit is to be used to cool a room with a cooling load of 12,000 BTU/hr. If the water temperature rise in the unit is 14 F, what is the flow rate in GPM?

5.4   The flow rate through a convector is 4.5 GPM. Water enters the unit at 220 F and leaves at 208 F. What is the heat output of the convector?

5.5   In Figure 5.24, terminal unit A has a heat output of 9300 BTU/hr and unit B of 8100 BTU/hr. What is the water temperature leaving unit A and unit B?

5.6   In Figure 5.25, terminal units A, B, and C have cooling capacities of 14,000 BTU/hr, 7200 BTU/hr, and 12,700 BTU/hr, respectively. Determine the water temperatures and flow rates at points (1), (2), and (3). The flow rate through each unit is 3 GPM.

5.7   What is the heating capacity of a 7 ft length of ½ in. baseboard radiation of the type listed in Table 5.1, with a flow rate of 2 GPM and an average water temperature of 200 F?

5.8   A room has a heating load of 9600 BTU/hr. Find the required length of ¾ in. baseboard radiation to heat the room, of the type shown in Table 5.1, if the baseboard is supplied with 3 GPM of water at 235 F.

5.9   Select the terminal units for the residence described in Example 5.5, using ½ in. baseboard, a supply temperature of 230 F, and a temperature drop of 35 F.

5.10   Select baseboard radiation for the residence described in Example 5.5 using a split series loop piping system and suitable temperatures and flow rates.

5.11   Find the flow rate in each section of pipe for the hydronic cooling system of Example 5.6.

5.12   Using a system temperature rise of 10 F for Example 5.6 and equal flow rates to each unit, calculate the temperature rise in each unit and the flow rate in each section of pipe.

5.13   Assuming a system temperature rise of 14 F for Example 5.6 and the same temperature rise in each unit, calculate the flow rate through each unit.

5.14   Lay out a hydronic piping system and terminal units for the house shown in Problem 3.20. (Use the type of system and terminal units as assigned by your instructor, or select your own. Use the heating loads calculated previously, or those specified by the instructor.)

5.15   Lay out a hydronic piping system and terminal units for the building shown in Problem 3.21. (Use the type of system and terminal units as assigned by your instructor, or select your own. Use the heating loads calculated previously, or those specified by the instructor.)

Figure 5.24
Sketch for Problem 5.5.

Figure 5.25
Sketch for Problem 5.6.

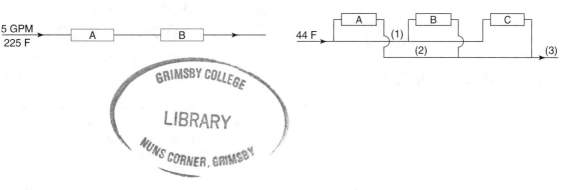

CHAPTER 6
Cooling Load Calculations

The objective of this chapter is to learn how to determine the amount of cooling required to keep the rooms in a building comfortable in summer. In Chapter 3, we learned how to calculate the winter heating requirements of a building. The procedures described for calculating cooling needs are similar but involve additional items that make the subject more complicated.

OBJECTIVES

After studying this chapter, you will be able to:

1. Calculate the heat gains to a space.
2. Select appropriate design conditions for cooling.
3. Determine peak load conditions.
4. Find required ventilation rates.
5. Perform a commercial cooling load analysis.
6. Perform a residential cooling load analysis.

6.1 THE COOLING LOAD

The air inside a building receives heat from a number of sources during the cooling season. If the temperature and humidity of the air are to be maintained at a comfortable level, this heat must be re-moved. The amount of heat that must be removed is called the *cooling load.*

The cooling load must be determined because it is the basis for selection of the proper size air conditioning equipment and distribution system. It is also used to analyze energy use and conservation.

6.2 COOLING LOAD CALCULATION PROCEDURES

In Chapter 3, we noted that the heat loss from a room at any instant was equal to the heating load at that time.

With cooling, the situation is more complex. The amount of heat that must be removed (the cooling load) is not always equal to the amount of heat received at a given time.

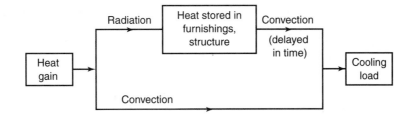

Figure 6.1
Heat flow diagram showing building heat gain, heat storage, and cooling load.

This difference is a result of the *heat storage* and *time lag* effects. Of the total amount of heat entering the building at any instant, only a portion of it heats the room air immediately; the other part (the radiation) heats the building mass—the roof, walls, floors, and furnishings. This is the heat storage effect. Only at a later time does the stored heat portion contribute to heating the room air. This is the time lag effect, as shown in Figure 6.1.

*The **room cooling load** is the rate at which heat must be removed from the room air to maintain it at the design temperature and humidity.*

The thermal storage effect and resulting time lag cause the cooling load to often be different in value from the entering heat (called the *instantaneous heat gain*). An example is shown in Figure 6.2. Note that during the time of day at which the instantaneous heat gain is the highest (the afternoon), the cooling load is less than the instantaneous heat gain. This is because some of this heat is stored in the building mass and is not heating the room air. Later in the day, the stored heat plus some of the new entering heat is released to the room air, so the cooling load becomes greater than the instantaneous heat gain.

Figure 6.2
Difference between instantaneous heat gain and cooling load as a result of heat storage effect.

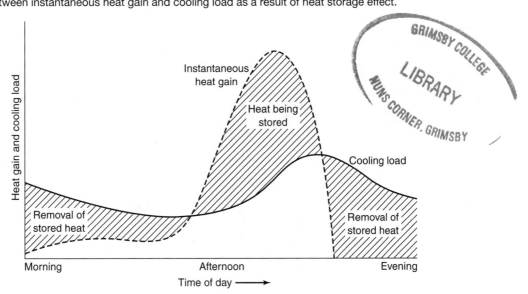

This effect is noticed in the huge southern European cathedrals built of massive, thick stone walls. Even on a sunny, very hot day the church interior remains quite cool, even though it is not air-conditioned. The entering heat doesn't reach the interior, it merely heats the walls (heat storage). By the time the heat reaches the interior (time lag), night has come. In this extreme example of time delay, the building may even have a reverse heat flow at night—heat flows out from the hot walls to the cool outdoors.

There are a few different, acceptable procedures for calculating cooling loads that take into account the phenomena we have discussed. All of them are more accurate than past methods and are often required in state energy codes and standards. These methods often lead to use of smaller equipment and sometimes in less energy use.

The cooling load calculation procedure that will be explained here is called the CLF/CLTD method. This procedure is relatively easy to understand and use. One of its valuable features is that in learning it, one may understand better the effects we have been discussing. The CLF/CLTD method can be carried out manually or by using a computer. The software bibliography in the rear of this text lists some of the available computer software for cooling load calculations.

6.3   ROOM HEAT GAINS

The heat gain components that contribute to the room cooling load consist of the following (Figure 6.3):

1. Conduction through exterior walls, roof, and glass
2. Conduction through interior partitions, ceilings, and floors
3. Solar radiation through glass
4. Lighting
5. People
6. Equipment
7. Heat from infiltration of outside air through openings

It is convenient to arrange these heat gains into two groups—those from external sources outside the room, and those internally generated. From the earlier description, it is seen that items 1 through 3 are *external heat gains,* and items 4 through 6 are *internal heat gains.* Infiltration can be considered as a separate class.

It is also convenient to arrange the heat gains into a different set of two groups: *sensible* and *latent* heat gains. Sensible heat gains result in increasing the air temperature; latent heat gains are due to addition of

Figure 6.3
Room heat gain components, Q.

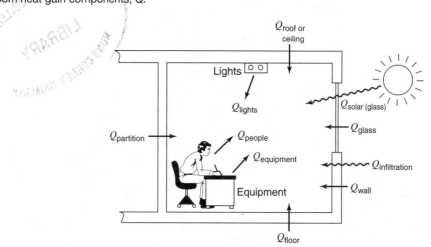

water vapor, thus increasing humidity. Items 1 through 4 are solely sensible gains. Items 5 and 7 are part sensible and part latent, and item 6 can fall in either category or both, depending on the type of equipment. As will be noted in the study of psychrometrics (Chapter 7), it is necessary to separate the sensible and latent gains because the selection of cooling equipment depends on their relative values.

6.4 CONDUCTION THROUGH EXTERIOR STRUCTURE

The cooling loads caused by conduction heat gains through the exterior roof, walls, and glass are each found from the following equation:

$$Q = U \times A \times CLTD_c \qquad (6.1)$$

where

Q = cooling load for roof, wall, or glass, BTU/hr

U = overall heat transfer coefficient for roof, wall, or glass, BTU/hr-ft^2-F

A = area of roof, wall, or glass, ft^2

$CLTD_c$ = corrected cooling load temperature difference, F

The cooling load temperature difference (CLTD) is not the actual temperature difference between the outdoor and indoor air. It is a modified value that accounts for the heat storage/time lag effects.

Tables 6.1 and 6.2 on pages 122–124 list CLTD values for some roof and wall constructions.

The CLTD values in Tables 6.1 and 6.2 are based on the following conditions:

1. Indoor temperature is 78 F DB.

2. Outdoor average temperature on the design day is 85 F DB.

3. Date is July 21st.

4. Location is 40°N latitude.

If the actual condition differs from any of the above, the CLTD must be corrected as follows:

$$CLTD_c = CLTD + LM + (78 - t_R) + (t_a - 85) \quad (6.2)$$

where

$CLTD_c$ = corrected value of CLTD, F

$CLTD$ = temperature from Table 6.1, 6.2 or 6.5

LM = correction for latitude and month, from Table 6.4 (page 126)

t_R = room temperature, F

t_a = average outside temperature on a design day, F

The temperature t_a can be found as follows:

$$t_a = t_o - (DR/2) \qquad (6.3)$$

where

t_o = outside design dry bulb temperature, F

DR = daily temperature range, F

Both t_o and DR (the daily temperature range) are found in Table A.9.

Tables 6.1 and 6.3 (page 125) include U-values for the roofs and walls described. However, it is always advisable to confirm these values by calculation from individual R-values, as described in Chapter 3.

The hours listed in Tables 6.1 and 6.2 are Solar Time. This is approximately equal to Standard Time. Add one hour for Daylight Savings Time.

The following two examples illustrate the procedure for finding the cooling load due to conduction heat gain through a roof and a wall.

Example 6.1

A 30 ft × 40 ft roof of a building in Washington, D.C., is constructed of 4 in. heavy weight concrete with 1 in. insulation and a suspended ceiling. The inside temperature is 76 F.

Find the roof cooling load at 2 PM Solar Time on July 21.

Solution

1. From Table 6.1, roof is type No. 9 with suspended ceiling. At 2 PM (14 hrs), CLTD = 29 F.

2. Find the corrected $CLTD_c$ from Equation 6.2, first finding each correction:

    A. Correct for LM (Table 6.4). D.C. is at 38°N latitude (use 40°N).

TABLE 6.1 COOLING LOAD TEMPERATURE DIFFERENCES (CLTD) FOR CALCULATING COOLING LOAD FROM FLAT ROOFS, F

Roof No	Description of Construction	Weight, lb/ft²	U-value, BTU/h·ft²·°F	Solar Time																								Hour of Maximum CLTD	Minimum CLTD	Maximum CLTD	Difference CLTD
				1	2	3	4	5	6	7	8	9	10	11	12	13	14	15	16	17	18	19	20	21	22	23	24				
				Without Suspended Ceiling																											
1	Steel sheet with 1-in. (or 2-in.) insulation	7 (8)	0.213 (0.124)	1	-2	-3	-3	-5	-3	6	19	34	49	61	71	78	79	77	70	59	45	30	18	12	8	5	3	14	-5	79	84
2	1-in. wood with 1-in. insulation	8	0.170	6	3	0	-1	-3	-3	-2	4	14	27	39	52	62	70	74	74	70	62	51	38	28	20	14	9	16	-3	74	77
3	4-in. lightweight concrete	18	0.213	9	5	2	0	-2	-3	-3	1	9	20	32	44	55	64	70	73	71	66	57	45	34	25	18	13	16	-3	73	76
4	2-in. heavyweight concrete with 1-in. (or 2-in.) insulation	29	0.206 (0.122)	12	8	5	3	0	-1	-1	3	11	20	30	41	51	59	65	66	66	62	54	45	36	29	22	17	16	-1	67	68
5	1-in. wood with 2-in. insulation	9	0.109	3	0	-3	-4	-5	-7	-6	-3	5	16	27	39	49	57	63	64	62	57	48	37	26	18	11	7	16	-7	64	71
6	6-in. lightweight concrete	24	0.158	22	17	13	9	6	3	1	1	3	7	15	23	33	43	51	58	62	64	62	57	50	42	35	28	18	1	64	63
7	2.5-in. wood with 1-in. ins.	13	0.130	29	24	20	16	13	10	7	6	6	9	13	20	27	34	42	48	53	55	56	54	49	44	39	34	19	6	56	50
8	8-in. lightweight concrete	31	0.126	35	30	26	22	18	14	11	9	7	7	9	13	19	25	33	39	46	50	53	54	53	49	45	40	20	7	54	47
9	4-in. heavyweight concrete with 1-in. (or 2-in.) insulation	52 (52)	0.200 (0.120)	25	22	18	15	12	9	8	8	10	14	20	26	33	40	46	50	53	53	52	48	43	38	34	30	18	8	53	45
10	2.5-in. wood with 2-in. ins.	13	0.093	30	26	23	19	16	13	10	9	9	13	13	17	23	29	36	41	46	49	51	50	47	43	39	35	19	8	51	43
11	Roof terrace system	75	0.106	34	31	28	25	22	19	16	14	13	13	15	18	22	26	31	36	40	44	45	46	45	43	40	37	20	13	46	33
12	6-in. heavyweight concrete with 1-in. (or 2-in.) insulation	75 (75)	0.192 (0.117)	31	28	25	22	20	17	15	14	14	16	18	22	26	31	36	40	43	45	45	44	42	40	37	34	19	14	45	31
13	4-in. wood with 1-in. (or 2-in) insulation	17 (18)	0.106 (0.078)	38	36	33	30	28	25	22	20	18	17	16	17	18	21	24	28	32	36	39	41	43	43	42	40	22	16	43	27

122

TABLE 6.1 COOLING LOAD TEMPERATURE DIFFERENCES (CLTD) FOR CALCULATING COOLING LOAD FROM FLAT ROOFS, F (Continued)

Roof No	Description of Construction	Weight, lb/ft²	U-value, BTU/h·ft²·°F	1	2	3	4	5	6	7	8	9	10	11	12	13	14	15	16	17	18	19	20	21	22	23	24	Hour of Maximum CLTD	Minimum CLTD	Maximum CLTD	Difference CLTD
															Solar Time (With Suspended Ceiling)																
1	Steel Sheet with 1-in. (or 2-in.) insulation	9 (10)	0.134 (0.092)	2	0	-2	-3	-4	-4	-1	9	23	37	50	62	71	77	78	74	67	56	42	28	18	12	8	5	15	-4	78	82
2	1-in. wood with 1-in. ins.	10	0.115	20	15	11	8	5	3	2	3	7	13	21	30	40	48	55	60	62	61	58	51	44	37	30	25	17	2	62	60
3	4-in. lightweight concrete	20	0.134	19	14	10	7	4	2	0	0	4	10	19	29	39	48	56	62	65	64	61	54	46	38	30	24	17	0	65	65
4	2-in. heavyweight concrete with 1-in. insulation	30	0.131	28	25	23	20	17	15	13	13	14	16	20	25	30	35	39	43	46	47	46	44	41	38	35	32	18	13	47	34
5	1-in. wood with 2-in. ins	10	0.083	25	20	16	13	10	7	5	5	7	12	18	25	33	41	48	53	57	57	56	52	46	40	34	29	18	5	57	52
6	6-in. lightweight concrete	26	0.109	32	28	23	19	16	13	10	8	7	8	11	16	22	29	36	42	48	52	54	54	54	47	42	37	20	7	54	47
7	2.5-in. wood with 1-in. insulation	15	0.096	34	31	29	26	23	21	18	16	15	15	16	18	21	25	30	34	38	41	43	44	44	42	40	37	21	15	44	29
8	8-in. lightweight concrete	33	0.093	39	36	33	31	29	26	23	20	18	15	14	14	15	17	20	25	29	34	38	42	45	46	44	42	21	14	46	32
9	4-in. heavyweight concrete with 1-in. (or 2-in.) ins.	53 (54)	0.128 (0.090)	30	29	27	26	24	22	21	20	20	21	22	24	27	29	32	34	36	38	38	38	37	36	34	33	19	20	38	18
10	2.5-in. wood with 2-in. ins	15	0.072	35	33	30	28	26	24	22	20	18	18	18	20	22	25	28	32	35	38	40	41	41	40	39	37	21	18	41	23
11	Roof terrace system	77	0.082	30	29	28	27	26	25	24	23	22	22	22	23	23	25	26	28	29	31	32	33	33	33	33	32	22	22	33	11
12	6-in. heavyweight concrete with 1-in. (or 2-in.) insulation	77 (77)	0.125 (0.088)	29	28	27	26	25	24	23	22	21	21	22	23	25	26	28	30	32	33	34	34	34	33	32	31	20	21	34	13
13	4-in. wood with 1-in. (or 2-in.) insulation	19 (20)	0.082 (0.064)	35	34	33	32	31	29	27	26	24	23	22	21	22	22	24	25	27	30	32	34	35	36	37	36	23	21	37	16

Reprinted with permission from the *1989 ASHRAE Handbook—Fundamentals*.

TABLE 6.2   COOLING LOAD TEMPERATURE DIFFERENCES (CLTD) FOR CALCULATING COOLING LOAD FROM SUNLIT WALLS, F

North Latitude Wall Facing

Group A Walls

Wall Facing	0100	0200	0300	0400	0500	0600	0700	0800	0900	1000	1100	1200	1300	1400	1500	1600	1700	1800	1900	2000	2100	2200	2300	2400	Hr of Maximum CLTD	Minimum CLTD	Maximum CLTD	Difference CLTD
N	14	14	14	13	13	13	12	12	11	11	10	10	10	10	10	10	11	11	12	12	13	13	14	14	2	10	14	4
NE	19	19	19	18	17	17	16	15	15	15	15	15	16	16	17	18	18	18	19	19	20	20	20	20	22	15	20	5
E	24	24	23	23	22	21	20	19	19	18	19	19	20	21	22	23	24	24	25	25	25	25	24	24	22	18	25	7
SE	24	23	23	22	21	20	20	19	18	18	18	18	19	20	21	22	23	23	24	24	24	24	24	24	22	18	24	6
S	20	20	19	19	18	18	17	16	16	15	14	14	14	14	14	15	16	17	18	19	19	20	20	20	23	14	20	6
SW	25	25	25	24	24	23	22	21	20	19	19	18	17	17	17	17	18	19	20	22	23	24	25	25	24	17	25	8
W	27	27	26	26	25	24	24	23	22	21	20	19	19	19	18	18	18	18	19	20	22	23	26	26	1	18	27	9
NW	21	21	21	20	20	19	19	18	17	16	16	15	15	14	14	14	15	15	16	17	18	19	20	21	1	14	21	7

Group B Walls

Wall Facing	0100	0200	0300	0400	0500	0600	0700	0800	0900	1000	1100	1200	1300	1400	1500	1600	1700	1800	1900	2000	2100	2200	2300	2400	Hr of Maximum CLTD	Minimum CLTD	Maximum CLTD	Difference CLTD
N	15	14	14	13	12	11	11	10	9	9	9	8	9	9	9	10	11	12	13	14	14	15	15	15	24	8	15	7
NE	19	18	17	16	15	14	13	12	12	13	14	15	16	17	18	19	19	20	20	21	21	20	20	20	20	12	21	9
E	23	22	21	20	18	17	16	15	15	15	17	19	21	22	24	25	26	26	27	27	26	26	25	24	20	15	27	12
SE	23	22	21	20	18	17	16	15	14	14	15	16	18	20	21	23	24	25	26	26	26	26	25	24	21	14	26	12
S	21	20	19	18	17	15	14	13	12	11	11	11	11	12	14	15	17	19	20	21	22	22	21	21	23	11	22	11
SW	27	26	25	24	22	21	19	18	16	15	14	14	13	13	14	15	17	20	22	25	27	28	28	28	24	13	28	15
W	29	28	27	26	24	23	21	19	18	17	16	15	14	14	15	15	17	19	22	25	27	29	29	30	24	14	30	16
NW	23	22	21	20	19	18	17	15	14	13	12	12	12	11	12	12	13	15	17	19	21	22	23	23	24	11	23	9

Group C Walls

Wall Facing	0100	0200	0300	0400	0500	0600	0700	0800	0900	1000	1100	1200	1300	1400	1500	1600	1700	1800	1900	2000	2100	2200	2300	2400	Hr of Maximum CLTD	Minimum CLTD	Maximum CLTD	Difference CLTD
N	15	14	13	12	11	10	9	8	8	7	7	8	8	9	10	12	13	14	15	16	17	17	17	16	22	7	17	10
NE	19	17	16	14	13	11	10	10	11	13	15	17	19	20	21	22	22	23	23	23	22	21	20	20	20	10	23	13
E	22	21	19	18	17	15	14	12	12	14	16	19	22	25	27	29	30	30	29	29	28	27	26	24	18	12	30	18
SE	22	21	19	18	17	15	14	12	12	13	16	19	22	24	26	28	29	29	29	29	28	27	26	24	19	12	29	17
S	21	19	18	16	15	13	12	10	9	9	9	10	11	14	17	20	22	24	25	26	25	24	22	22	20	9	26	17
SW	29	27	25	22	20	18	16	15	13	12	11	11	11	13	15	18	22	26	29	32	33	33	32	31	22	11	33	22
W	31	29	27	25	22	20	18	16	14	13	12	12	12	13	13	14	16	20	24	29	32	35	35	33	22	12	35	23
NW	25	23	21	20	18	16	14	13	11	10	10	10	10	11	12	13	15	18	22	25	27	27	26	22	22	10	27	17

Group D Walls

Wall Facing	0100	0200	0300	0400	0500	0600	0700	0800	0900	1000	1100	1200	1300	1400	1500	1600	1700	1800	1900	2000	2100	2200	2300	2400	Hr of Maximum CLTD	Minimum CLTD	Maximum CLTD	Difference CLTD
N	15	13	12	10	9	7	6	6	6	6	6	7	8	10	12	13	15	17	18	19	19	19	18	16	21	6	19	13
NE	17	15	13	11	10	8	7	8	10	14	17	20	22	23	23	24	24	25	25	24	23	22	20	18	19	7	25	18
E	19	17	15	13	11	9	8	9	12	17	22	27	30	32	33	33	32	32	31	30	28	26	24	22	16	8	33	25
SE	20	17	15	13	11	10	8	8	10	13	17	22	26	29	31	32	32	32	31	30	28	26	24	22	17	8	32	24
S	19	17	15	13	11	9	8	7	6	7	9	12	16	20	24	27	29	29	29	27	26	24	22	19	18	6	29	23
SW	28	25	22	19	16	14	12	10	9	8	8	8	10	12	16	21	27	32	36	38	38	37	34	31	21	8	38	30
W	31	27	24	21	18	15	13	11	10	9	9	9	10	11	14	18	24	30	36	40	41	40	38	34	21	9	41	32
NW	25	22	19	17	14	12	10	9	8	7	7	8	9	10	12	14	18	22	27	31	32	32	30	27	22	7	32	25

Group E Walls

Wall Facing	0100	0200	0300	0400	0500	0600	0700	0800	0900	1000	1100	1200	1300	1400	1500	1600	1700	1800	1900	2000	2100	2200	2300	2400	Hr of Maximum CLTD	Minimum CLTD	Maximum CLTD	Difference CLTD
N	12	10	8	7	5	4	3	4	5	6	7	9	11	13	15	17	19	20	21	22	20	18	16	14	20	3	22	19
NE	13	11	9	7	6	4	5	9	15	20	24	25	25	26	26	26	26	26	25	24	22	19	17	15	16	4	26	22
E	14	12	10	8	6	5	5	6	11	18	26	33	36	38	37	36	34	33	32	30	28	25	22	20	14	5	38	33
SE	15	12	10	8	7	5	5	8	12	19	25	31	35	37	37	36	34	33	31	28	26	23	20	17	15	5	37	32
S	15	12	10	8	7	5	4	3	4	5	9	13	19	24	29	32	34	33	31	29	26	23	20	17	17	3	34	31
SW	22	18	15	12	10	8	6	5	5	5	6	7	9	12	18	24	32	38	43	45	44	40	35	30	19	5	45	40
W	25	21	17	14	11	9	7	6	6	6	7	9	11	14	20	27	36	43	49	49	45	40	34	29	20	6	49	43
NW	20	17	14	11	9	7	6	5	5	5	6	8	10	13	16	20	26	32	37	38	36	32	28	24	20	5	38	33

Group F Walls

Wall Facing	0100	0200	0300	0400	0500	0600	0700	0800	0900	1000	1100	1200	1300	1400	1500	1600	1700	1800	1900	2000	2100	2200	2300	2400	Hr of Maximum CLTD	Minimum CLTD	Maximum CLTD	Difference CLTD
N	8	6	5	3	2	1	2	4	6	7	9	11	14	17	19	21	22	23	24	23	20	16	13	11	19	1	24	23
NE	9	7	5	3	2	1	5	14	23	28	30	29	28	27	27	27	26	24	22	19	16	13	11	11	11	1	30	29
E	10	7	6	4	3	2	6	17	28	38	44	45	43	39	36	34	32	30	27	24	21	17	15	12	12	2	45	43
SE	10	7	6	4	3	2	4	10	19	28	36	41	43	43	42	39	36	34	32	30	28	25	21	18	14	2	43	41
S	10	8	6	4	3	2	1	1	3	7	13	20	27	34	38	39	38	35	31	26	22	18	15	12	16	1	39	38
SW	15	11	9	6	5	3	2	2	4	5	8	11	17	26	35	44	50	53	52	45	37	28	23	18	18	2	53	51
W	17	13	10	7	5	4	3	3	4	6	8	11	14	20	28	39	49	57	60	54	43	34	27	21	19	3	60	57
NW	14	10	8	6	4	3	2	2	3	5	8	10	13	15	21	27	35	42	46	43	35	28	22	18	19	2	46	44

Group G Walls

Wall Facing	0100	0200	0300	0400	0500	0600	0700	0800	0900	1000	1100	1200	1300	1400	1500	1600	1700	1800	1900	2000	2100	2200	2300	2400	Hr of Maximum CLTD	Minimum CLTD	Maximum CLTD	Difference CLTD
N	3	2	1	0	-1	2	7	8	9	12	15	18	21	23	24	24	25	26	22	15	11	9	7	5	18	-1	26	27
NE	3	2	1	0	-1	9	27	36	39	35	30	26	26	27	27	26	25	22	18	14	11	9	7	5	9	-1	39	40
E	4	2	1	0	-1	11	31	47	54	55	50	40	33	31	30	29	27	24	19	15	12	10	8	6	10	-1	55	56
SE	4	2	1	0	-1	5	18	32	42	49	51	48	42	36	32	30	27	24	19	15	12	10	8	6	11	-1	51	52
S	4	2	1	0	-1	0	1	5	12	22	31	39	45	46	43	37	31	25	20	15	12	10	8	5	14	-1	46	47
SW	5	4	3	1	0	0	2	5	8	12	16	26	38	50	59	63	61	52	37	24	17	13	10	8	16	0	63	63
W	6	5	3	2	1	1	2	5	8	11	15	19	27	41	56	67	72	67	48	29	20	15	11	8	17	1	72	71
NW	5	3	2	1	0	0	2	5	8	11	15	18	21	27	37	47	55	55	41	25	17	13	10	7	18	0	55	55

Reprinted with permission from the *1989 ASHRAE Handbook—Fundamentals.*

TABLE 6.3   WALL CONSTRUCTION GROUP DESCRIPTION

Group No.	Description of Construction	Weight (lb/ft²)	*U*-Value (BTU/h•ft²•°F)
4-in. Face brick + (brick)			
C	Air space + 4-in. face brick	83	0.358
D	4-in. common brick	90	0.415
C	1-in. insulation or air space + 4-in. common brick	90	0.174–0.301
B	2-in. insulation + 4-in. common brick	88	0.111
B	8-in. common brick	130	0.302
A	Insulation or air space + 8-in. common brick	130	0.154–0.243
4-in. Face brick + (heavyweight concrete)			
C	Air space + 2-in. concrete	94	0.350
B	2-in. insulation + 4-in. concrete	97	0.116
A	Air space or insulation + 8-in. or more concrete	143–190	0.110–0.112
4-in. Face brick + (light or heavyweight concrete block)			
E	4-in. block	62	0.319
D	Air space or insulation + 4-in. block	62	0.153–0.246
D	8-in. block	70	0.274
C	Air space or 1-in. insulation + 6-in. or 8-in. block	73–89	0.221–0.275
B	2-in. insulation + 8-in. block	89	0.096–0.107
4-in. Face brick + (clay tile)			
D	4-in. tile	71	0.381
D	Air space + 4-in. tile	71	0.281
C	Insulation + 4-in. tile	71	0.169
C	8-in. tile	96	0.275
B	Air space or 1-in. insulation + 8-in. tile	96	0.142–0.221
A	2-in. insulation + 8-in. tile	97	0.097
Heavyweight concrete wall + (finish)			
E	4-in. concrete	63	0.585
D	4-in. concrete + 1-in. or 2-in. insulation	63	0.119–0.200
C	2-in. insulation + 4-in. concrete	63	0.119
C	8-in. concrete	109	0.490
B	8-in. concrete + 1-in. or 2-in. insulation	110	0.115–0.187
A	2-in. insulation + 8-in. concrete	110	0.115
B	12-in. concrete	156	0.421
A	12-in. concrete + insulation	156	0.113
Light and heavyweight concrete block + (finish)			
F	4-in. block + air space/insulation	29	0.161–0.263
E	2-in. insulation + 4-in. block	29–37	0.105–0.114
E	8-in. block	47–51	0.294–0.402
D	8-in. block + air space/insulation	41–57	0.149–0.173
Clay tile + (finish)			
F	4-in. tile	39	0.419
F	4-in. tile + air space	39	0.303
E	4-in. tile + 1-in. insulation	39	0.175
D	2-in. insulation + 4-in. tile	40	0.110
D	8-in. tile	63	0.296
C	8-in. tile + air space/1-in. insulation	63	0.151–0.231
B	2-in. insulation + 8-in. tile	63	0.099
Metal curtain wall			
G	With/without air space + 1- to 3-in. insulation	5–6	0.091–0.230
Frame wall			
G	1-in. to 3-in. insulation	16	0.081–0.178

Reprinted with permission from the *1989 ASHRAE Handbook—Fundamentals.*

TABLE 6.4   CLTD CORRECTION FOR LATITUDE AND MONTH APPLIED TO WALLS AND ROOFS, NORTH LATITUDES, F

Lat.	Month	N	NNE NNW	NE NW	ENE WNW	E W	ESE WSW	SE SW	SSE SSW	S	HOR
0	Dec	−3	−5	−5	−5	−2	0	3	6	9	−1
	Jan/Nov	−3	−5	−4	−4	−1	0	2	4	7	−1
	Feb/Oct	−3	−2	−2	−2	−1	−1	0	−1	0	0
	Mar/Sept	−3	0	1	−1	−1	−3	−3	−5	−8	0
	Apr/Aug	5	4	3	0	−2	−5	−6	−8	−8	−2
	May/Jul	10	7	5	0	−3	−7	−8	−9	−8	−4
	Jun	12	9	5	0	−3	−7	−9	−10	−8	−5
8	Dec	−4	−6	−6	−6	−3	0	4	8	12	−5
	Jan/Nov	−3	−5	−6	−5	−2	0	3	6	10	−4
	Feb/Oct	−3	−4	−3	−3	−1	−1	1	2	4	−1
	Mar/Sept	−3	−2	−1	−1	−1	−2	−2	−3	−4	0
	Apr/Aug	2	2	2	0	−1	−4	−5	−7	−7	−1
	May/Jul	7	5	4	0	−2	−5	−7	−9	−7	−2
	Jun	9	6	4	0	−2	−6	−8	−9	−7	−2
16	Dec	−4	−6	−8	−8	−4	−1	4	9	13	−9
	Jan/Nov	−4	−6	−7	−7	−4	−1	4	8	12	−7
	Feb/Oct	−3	−5	−5	−4	−2	0	2	5	7	−4
	Mar/Sept	−3	−3	−2	−2	−1	−1	0	0	0	−1
	Apr/Aug	−1	0	−1	−1	−1	−3	−3	−5	−6	0
	May/Jul	4	3	3	0	−1	−4	−5	−7	−7	0
	Jun	6	4	4	1	−1	−4	−6	−8	0	−7
24	Dec	−5	−7	−9	−10	−7	−3	3	9	13	−13
	Jan/Nov	−4	−6	−8	−9	−6	−3	9	3	13	−11
	Feb/Oct	−4	−5	−6	−6	−3	−1	3	7	10	−7
	Mar/Sept	−3	−4	−3	−3	−1	−1	1	2	4	−3
	Apr/Aug	−2	−1	0	−1	−1	−2	−1	−2	−3	0
	May/Jul	1	2	2	0	0	−3	−3	−5	−6	1
	Jun	3	3	3	1	0	−3	−4	−6	−6	1
32	Dec	−5	−7	−10	−11	−8	−5	2	9	12	−17
	Jan/Nov	−5	−7	−9	−11	−8	−15	−4	2	9	12
	Feb/Oct	−4	−6	−7	−8	−4	−2	4	8	11	−10
	Mar/Sept	−3	−4	−4	−4	−2	−1	3	5	7	−5
	Apr/Aug	−2	−2	−1	−2	0	−1	0	1	1	−1
	May/Jul	1	1	1	0	0	−1	−1	−3	−3	1
	Jun	1	2	2	1	0	−2	−2	−4	−4	2
40	Dec	−6	−8	−10	−13	−10	−7	0	7	10	−21
	Jan/Nov	−5	−7	−10	−12	−9	−6	1	8	11	−19
	Feb/Oct	−5	−7	−8	−9	−6	−3	3	8	12	−14
	Mar/Sept	−4	−5	−5	−6	−3	−1	4	7	10	−8
	Apr/Aug	−2	−3	−2	−2	0	0	2	3	4	−3
	May/Jul	0	0	0	0	0	0	0	0	1	1
	Jun	1	1	1	0	1	0	0	−1	−1	2
48	Dec	−6	−8	−11	−14	−13	−10	−3	2	6	−25
	Jan/Nov	−6	−8	−11	−13	−11	−8	−1	5	8	−24
	Feb/Oct	−5	−7	−10	−11	−8	−5	1	8	11	−18
	Mar/Sept	−4	−6	−6	−7	−4	−1	4	8	11	−11
	Apr/Aug	−3	−3	−3	−3	−1	0	4	6	7	−5
	May/Jul	0	−1	0	0	1	1	3	3	4	0
	Jun	1	1	2	1	2	1	2	2	3	2

Reprinted with permission from the *1989 ASHRAE Handbook—Fundamentals.*

Roof surface is horizontal (The HOR column).

For July, LM = 1 F

B. Use Equation 6.3 to find t_a, first finding t_o and DR from Table A.9.

For Washington, D.C., t_o = 91 F, DR = 18 F.

t_o = 91 − 18/2 = 82 F

C. Using Equation 6.2,

$$\text{CLTD}_c = \text{CLTD} + \text{LM} + (78 - t_R) +$$
$$(t_a - 85)$$
$$= 29 + 1 + (78 - 76) + (82 - 85)$$

$$\text{CLTD}_c = 29 \text{ F}$$

D. From Table 6.1, U = 0.128 BTU/hr-ft²-F

E. The roof area A = 30 ft × 40 ft = 1200 ft²

3. Use Equation 6.1 to find the cooling load,

$$Q = U \times A \times \text{CLTD}_c$$
$$= 0.128 \times 1200 \times 29$$
$$= 4455 \text{ BTU/hr}$$

Example 6.2

A south-facing wall of a building in Pittsburgh, Pennsylvania, has a net opaque area of 5600 ft². The wall is constructed of 4 in. face brick + 2 in. insulation + 4 in. heavy weight concrete. The inside air temperature is 77 F.

Find the wall cooling load at 4 PM Solar Time on June 21.

Solution

1. Using Table 6.3, the wall is in Group B.
2. From Table 6.2, CLTD = 15 F.
3. From Table 6.4, LM = − 1 F
4. Find t_a. From Table A.9, t_o = 88 F, DR = 19 F. Using Equation 6.2, t_a = 88 − 19/2 = 78 F (rounded off).
5. Using Equation 6.2,

$$\text{CLTD}_c = \text{CLTD} + \text{LM} + (78 - t_R) + (t_a - 85)$$
$$= 15 - 1 + (78 - 77) + (78 - 85)$$

$$\text{CLTD}_c = 8 \text{ F}$$

6. From Table 6.3, U = 0.116 BTU/hr-ft²-F
7. Using Equation 6.1 to find the cooling load,

TABLE 6.5 COOLING LOAD TEMPERATURE DIFFERENCES (CLTD) FOR CONDUCTION THROUGH GLASS

Solar time, h	CLTD °F	Solar time, h	CLTD °F
0100	1	1300	12
0200	0	1400	13
0300	−1	1500	14
0400	−2	1600	14
0500	−2	1700	13
0600	−2	1800	12
0700	−2	1900	10
0800	0	2000	8
0900	2	2100	6
1000	4	2200	4
1100	7	2300	3
1200	9	2400	2

Reprinted with permission from the *1993 ASHRAE Handbook—Fundamentals.*

$$Q = U \times A \times \text{CLTD}_c$$
$$= 0.116 \times 5600 \times 8$$
$$= 5200 \text{ BTU/hr}$$

Table 6.5 lists CLTD values for glass. Equation 6.2 is used to correct the CLTD, except that there is no latitude and month (LM) correction.

The following example illustrates the use of Table 6.5.

Example 6.3

A room has 130 ft² of single glass windows with vinyl frames. Inside air temperature is 75 F and outdoor average temperature on a design day is 88 F.

Find the cooling load due to conduction heat gain through the windows at 2 PM Daylight Savings Time.

Solution

1. From Table 6.5 (2 PM DST = 1 PM ST = 13 hrs), CLTD = 12 F.
2. Using Equation 6.2,

$$\text{CLTD}_c = \text{CLTD} + \text{LM} + (78 - t_R) + (t_a - 85)$$
$$= 12 + (78 - 75) + (88 - 85)$$

$$\text{CLTD}_c = 18 \text{ F}$$

3. From Table A.8, $U = 0.90$. Using Equation 6.1 to find the cooling load,

$$Q = U \times A \times \text{CLTD}_c$$
$$= 0.90 \times 130 \times 18$$
$$= 2110 \text{ BTU/hr}$$

Example 6.3 _____

A room has 130 ft^2 of exterior single glass with no interior shading. The inside design condition is 78 F and the outdoor daily average temperature is 88 F. Determine the cooling load from conduction heat gain through the glass at 12 noon Solar Time in summer.

Solution

From Table 6.5, CLTD = 9 F. Correcting this by Equation 6.3,

$$\text{CLTD}_c = \text{CLTD} + (78 - t_R) + (t_o - 85)$$
$$= 9 + 0 + 3 = 12 \text{ F}$$

From Table A.8, $U = 1.04$ BTU/hr-ft^2-F. Using Equation 6.1,

$$Q = U \times A \times \text{CLTD} = 1.04 \times 130 \times 12$$
$$= 1620 \text{ BTU/hr}$$

6.5 CONDUCTION THROUGH INTERIOR STRUCTURE

The heat that flows from interior unconditioned spaces to the conditioned space through partitions, floors, and ceilings can be found from Equation 3.5:

$$Q = U \times A \times \text{TD} \tag{3.5}$$

where

Q = heat gain (cooling load) through partition, floor, or ceiling, BTU/hr

U = overall heat transfer coefficient for partition, floor, or ceiling, BTU/hr-ft^2-F

A = area of partition, floor, or ceiling, ft^2

TD = temperature difference between unconditioned and conditioned space, F

If the temperature of the unconditioned space is not known, an approximation often used is to assume that it is at 5 F less than the outdoor temperature. Spaces with heat sources, such as boiler rooms, may be at a much higher temperature.

6.6 SOLAR RADIATION THROUGH GLASS

Radiant energy from the sun passes through transparent materials such as glass and becomes a heat gain to the room. Its value varies with time, orientation, shading, and storage effect. The solar cooling load can be found from the following equation:

$$Q = \text{SHGF} \times A \times \text{SC} \times \text{CLF} \tag{6.4}$$

where

Q = solar radiation cooling load for glass, BTU/hr

SHGF = maximum solar heat gain factor, BTU/hr-ft^2

A = area of glass, ft^2

SC = shading coefficient

CLF = cooling load factor for glass

The maximum **solar heat gain factor** (SHGF) is the maximum solar heat gain through single clear glass at a given month, orientation, and latitude. Values are shown in Table 6.6 for the twenty-first day of each month.

Example 6.4 _____

What is the maximum solar heat gain factor through the windows on the southwest side of a building located at 32°N latitude on September twenty-first?

Solution

From Table 6.6, SHGF = 218 BTU/hr-ft^2.

The SHGF gives maximum heat gain values only for the type of glass noted and without any shading devices.

To account for heat gains with different fenestration arrangements, the *shading coefficient* SC is introduced. Table 6.7 (page 130) lists some values of SC.

TABLE 6.6   MAXIMUM SOLAR HEAT GAIN FACTOR (SHGF) BTU/HR • FT² FOR SUNLIT GLASS, NORTH LATITUDES

20° N. Lat

	N	NNE/NNW	NE/NW	ENE/WNW	E/W	ESE/WSW	SE/SW	SSE/SSW	S	HOR
Jan.	29	29	48	138	201	243	253	233	214	232
Feb.	31	31	88	173	226	244	238	201	174	263
Mar.	34	49	132	200	237	236	206	152	115	284
Apr.	38	92	166	213	228	208	158	91	58	287
May	47	123	184	217	217	184	124	54	42	283
June	59	135	189	216	210	173	108	45	42	279
July	48	124	182	213	212	179	119	53	43	278
Aug.	40	91	162	206	220	200	152	88	57	280
Sep.	36	46	127	191	225	225	199	148	114	275
Oct.	32	32	87	167	217	236	231	196	170	258
Nov.	29	29	48	136	197	239	249	229	211	230
Dec.	27	27	35	122	187	238	254	241	226	217

24° N. Lat

	N	NNE/NNW	NE/NW	ENE/WNW	E/W	ESE/WSW	SE/SW	SSE/SSW	S	HOR
Jan.	27	27	41	128	190	240	253	241	227	214
Feb.	30	30	80	165	220	244	243	213	192	249
Mar.	34	45	124	195	234	237	214	168	137	275
Apr.	37	88	159	209	228	212	169	107	75	283
May	43	117	178	214	218	190	132	67	46	282
June	55	127	184	214	212	179	117	55	43	279
July	45	116	176	210	213	185	129	65	46	278
Aug.	38	87	156	203	220	204	162	103	72	277
Sep.	35	42	119	185	222	225	206	163	134	266
Oct.	31	31	79	159	211	237	235	207	187	244
Nov.	27	27	42	126	187	236	249	237	224	213
Dec.	26	26	29	112	180	234	247	247	237	199

28° N. Lat

	N (Shade)	NNE/NNW	NE/NW	ENE/WNW	E/W	ESE/WSW	SE/SW	SSE/SSW	S	HOR
Jan.	25	25	35	117	183	235	251	247	238	196
Feb.	29	29	72	157	213	244	246	224	207	234
Mar.	33	41	116	189	231	237	221	182	157	265
Apr.	36	84	151	205	228	216	178	124	94	278
May	40	115	172	211	219	195	144	83	58	280
June	51	125	178	211	213	184	128	68	49	278
July	41	114	170	208	215	190	140	80	57	276
Aug.	38	83	149	199	220	207	172	120	91	272
Sep.	34	38	111	179	219	226	213	177	154	256
Oct.	30	30	71	151	204	236	238	217	202	229
Nov.	26	26	35	115	181	232	247	243	235	195
Dec.	24	24	24	99	172	227	248	251	246	179

32° N. Lat

	N (Shade)	NNE/NNW	NE/NW	ENE/WNW	E/W	ESE/WSW	SE/SW	SSE/SSW	S	HOR
Jan.	24	24	29	105	175	229	249	250	246	176
Feb.	27	27	65	149	205	242	248	232	221	217
Mar.	32	37	107	183	227	237	227	195	176	252
Apr.	36	80	146	200	227	219	187	141	115	271
May	38	111	170	208	220	199	155	99	74	277
June	44	122	176	208	214	189	139	83	60	276
July	40	111	167	204	215	194	150	96	72	273
Aug.	37	79	141	195	219	210	181	136	111	265
Sep.	33	35	103	173	215	227	218	189	171	244
Oct.	28	28	63	143	195	234	239	225	215	213
Nov.	24	24	29	103	173	225	245	246	243	175
Dec.	22	22	22	84	162	218	246	252	252	158

36° N. Lat

	N (Shade)	NNE/NNW	NE/NW	ENE/WNW	E/W	ESE/WSW	SE/SW	SSE/SSW	S	HOR
Jan.	22	22	24	90	166	219	247	252	252	155
Feb.	26	26	57	139	195	239	248	239	232	199
Mar.	30	33	99	176	223	238	232	206	192	238
Apr.	35	76	144	196	225	221	196	156	135	262
May	38	107	168	204	220	204	165	116	93	272
June	47	118	175	205	215	194	150	99	77	273
July	39	107	165	201	216	199	161	113	90	268
Aug.	36	75	138	190	218	212	189	151	131	257
Sep.	31	31	95	167	210	228	223	200	187	230
Oct.	27	27	56	133	187	230	239	231	225	195
Nov.	22	22	24	87	163	215	243	248	248	154
Dec.	20	20	20	69	151	204	241	253	254	136

40° N. Lat

	N (Shade)	NNE/NNW	NE/NW	ENE/WNW	E/W	ESE/WSW	SE/SW	SSE/SSW	S	HOR
Jan.	20	20	20	74	154	205	241	252	254	133
Feb.	24	24	50	129	186	234	246	244	241	180
Mar.	29	29	93	169	218	238	236	216	206	223
Apr.	34	71	140	190	224	223	203	170	154	252
May	37	102	165	202	220	208	175	133	113	265
June	48	113	172	205	216	199	161	116	95	267
July	38	102	163	198	216	203	170	129	109	262
Aug.	35	71	135	185	216	214	196	165	149	247
Sep.	30	30	87	160	203	227	209	200	215	
Oct.	25	25	49	123	180	225	238	236	234	177
Nov.	20	20	20	73	151	201	237	248	250	132
Dec.	18	18	18	60	135	188	232	249	253	113

44° N. Lat

	N (Shade)	NNE/NNW	NE/NW	ENE/WNW	E/W	ESE/WSW	SE/SW	SSE/SSW	S	HOR
Jan.	17	17	17	64	138	189	232	248	252	109
Feb.	22	22	43	117	178	227	246	248	247	160
Mar.	27	27	87	162	211	236	238	224	218	206
Apr.	33	66	136	185	221	224	210	183	171	240
May	36	96	162	201	219	211	183	148	132	257
June	47	108	169	205	215	203	171	132	115	261
July	37	96	159	198	215	206	179	144	128	254
Aug.	34	66	132	180	214	215	202	177	165	236
Sep.	28	28	80	152	198	226	227	216	211	199
Oct.	23	23	42	111	171	217	237	240	239	157
Nov.	18	18	18	64	135	186	227	244	248	109
Dec.	15	15	15	49	115	175	217	240	246	89

48° N. Lat

	N (Shade)	NNE/NNW	NE/NW	ENE/WNW	E/W	ESE/WSW	SE/SW	SSE/SSW	S	HOR
Jan.	15	15	15	53	118	175	216	239	245	85
Feb.	20	20	36	103	168	216	242	249	250	138
Mar.	26	26	80	154	204	234	239	232	228	188
Apr.	31	61	132	180	219	225	215	194	186	226
May	35	97	158	200	218	214	192	163	150	247
June	46	110	165	204	215	206	180	148	134	252
July	37	96	156	196	214	209	187	158	146	244
Aug.	33	61	128	174	211	216	208	188	180	223
Sep.	27	27	72	144	191	223	228	223	220	182
Oct.	21	21	35	96	161	207	233	241	242	136
Nov.	15	15	15	52	115	172	212	234	240	85
Dec.	13	13	13	36	91	156	195	225	233	65

Reprinted with permission from the *1989 ASHRAE Handbook—Fundamentals*.

TABLE 6.7   SHADING COEFFICIENTS FOR GLASS WITHOUT OR WITH INTERIOR SHADING DEVICES

Type of Glazing	Nominal Thickness, in. (Each light)	Without Shading	With Interior Shading				
			Venetian Blinds		Roller Shades		
					Opaque		Translucent
			Medium	Light	Dark	Light	Light
Single glass							
Clear	¼	0.94	0.74	0.67	0.81	0.39	0.44
Heat absorbing	¼	0.69	0.57	0.53	0.45	0.30	0.36
Double glass							
Clear	¼	0.81	0.62	0.58	0.71	0.35	0.40
Heat absorbing	¼	0.55	0.39	0.36	0.40	0.22	0.30

Note: Venetian blinds are assumed set at a 45° position. Adapted with permission from the *1993 ASHRAE Handbook—Fundamentals*.

Example 6.5

What is the value of SC to be applied to the solar heat gain for ¼ in. single clear glass and medium-colored inside venetian blinds?

Solution

From Table 6.7, SC = 0.74.

The *cooling load factor* CLF accounts for the storage of part of the solar heat gain. Values of CLF to be applied to the solar load calculation are shown in Tables 6.8, 6.9, and 6.10. Note that there are separate listings for Light (L), Medium (M), and Heavy (H) construction, as described.

Table 6.8 is used without interior shading devices and with carpeting. Table 6.9 (page 132) is used without interior shading devices and no carpeting. Table 6.10 (page 133) is used with interior shading devices (in this case the carpeting has no storage effect).

Example 6.6

A building wall facing southwest has a window area of 240 ft². The glass is ¼ in. single clear glass with light-colored interior venetian blinds. The building is of medium construction, and is located at 40°N latitude. Find the solar cooling load in August at 3 PM Solar Time.

Solution

1. From Table 6.6, SHGF = 196
2. From Table 6.7, SC = 0.67
3. From Table 6.10, CLF = 0.83

4. Using Equation 6.4,

$$Q = SHGF \times A \times SC \times CLF$$
$$= 196 \times 240 \times 0.67 \times 0.83$$
$$= 26,160 \text{ BTU/hr}$$

External Shading Effect

The values for the SHGF shown in Table 6.6 are for *direct* solar radiation—when the sun shines on the glass. External shading from building projections (or other objects) may shade all or part of the glass. In these cases, only an *indirect* radiation reaches the glass from the sky and ground. The SHGF values for any shaded glass is the same as the N (north) side of the building, which also receives only indirect radiation.

In order to find the total radiation through partly shaded glass, the shaded area portion must first be found. Table 6.11 (page 134) can be used to find the shading from overhead horizontal projections. The values in the table are the vertical feet of shade for each foot of horizontal projection. The following example illustrates the use of Table 6.11.

Example 6.7

A building at 32°N latitude has a wall facing west with a 4 ft overhang, and a 5 ft wide by 6 ft high window whose top is 1 ft below the overhang. How much of the glass receives direct solar radiation at 3 PM?

TABLE 6.8 COOLING LOAD FACTORS (CLF) FOR GLASS WITHOUT INTERIOR SHADING, IN NORTH LATITUDE SPACES HAVING CARPETED FLOORS

Dir.	Room Mass	0100	0200	0300	0400	0500	0600	0700	0800	0900	1000	1100	1200	1300	1400	1500	1600	1700	1800	1900	2000	2100	2200	2300	2400
N	L	.00	.00	.00	.00	.01	.64	.73	.74	.81	.88	.95	.98	.98	.94	.88	.79	.79	.55	.31	.12	.04	.02	.01	.00
	M	.03	.02	.02	.02	.02	.64	.69	.69	.77	.84	.91	.94	.95	.91	.86	.79	.79	.56	.32	.16	.10	.07	.05	.04
	H	.10	.09	.08	.07	.07	.62	.64	.64	.71	.77	.83	.87	.88	.85	.81	.75	.76	.55	.34	.22	.17	.15	.13	.11
NE	L	.00	.00	.00	.00	.01	.51	.83	.88	.72	.47	.33	.27	.24	.23	.20	.18	.14	.09	.03	.01	.00	.00	.00	.00
	M	.01	.01	.00	.00	.01	.50	.78	.82	.67	.44	.32	.28	.26	.24	.22	.19	.15	.11	.05	.03	.02	.02	.01	.01
	H	.03	.03	.03	.02	.03	.47	.71	.72	.59	.40	.30	.27	.26	.25	.23	.20	.17	.13	.08	.06	.05	.05	.04	.04
E	L	.00	.00	.00	.00	.00	.42	.76	.91	.90	.75	.51	.30	.22	.18	.16	.13	.11	.07	.02	.01	.00	.00	.00	.00
	M	.01	.01	.01	.00	.01	.41	.72	.86	.84	.71	.48	.30	.24	.21	.18	.16	.13	.09	.04	.03	.02	.01	.01	.01
	H	.03	.03	.03	.02	.02	.39	.66	.76	.74	.63	.43	.29	.24	.22	.20	.18	.15	.12	.08	.06	.05	.05	.04	.04
SE	L	.00	.00	.00	.00	.00	.27	.58	.81	.93	.93	.81	.59	.37	.27	.21	.18	.14	.09	.03	.01	.00	.00	.00	.00
	M	.01	.01	.01	.00	.01	.26	.55	.77	.88	.87	.76	.56	.37	.29	.24	.20	.16	.11	.05	.04	.03	.02	.02	.01
	H	.04	.04	.03	.03	.03	.26	.51	.69	.78	.78	.68	.51	.35	.29	.22	.22	.19	.15	.09	.08	.07	.06	.05	.05
S	L	.00	.00	.00	.01	.01	.07	.15	.23	.39	.62	.82	.94	.93	.80	.59	.38	.26	.16	.06	.02	.01	.00	.00	.00
	M	.01	.01	.01	.01	.03	.07	.14	.22	.38	.59	.78	.88	.88	.76	.57	.38	.28	.18	.09	.06	.04	.03	.02	.02
	H	.05	.05	.04	.04	.03	.09	.15	.21	.35	.54	.70	.79	.79	.69	.52	.37	.29	.21	.13	.10	.09	.08	.07	.06
SW	L	.00	.00	.00	.00	.00	.04	.09	.13	.16	.19	.23	.39	.62	.82	.94	.94	.81	.54	.19	.07	.03	.01	.00	.00
	M	.02	.02	.01	.01	.01	.05	.09	.13	.16	.19	.22	.38	.60	.78	.89	.89	.77	.52	.20	.10	.07	.05	.04	.03
	H	.07	.06	.05	.05	.04	.07	.11	.14	.16	.18	.21	.35	.55	.71	.80	.79	.69	.48	.20	.14	.11	.10	.08	.07
W	L	.00	.00	.00	.00	.00	.03	.07	.10	.13	.14	.16	.18	.31	.55	.78	.92	.93	.73	.25	.10	.04	.01	.01	.00
	M	.02	.02	.02	.01	.01	.04	.07	.10	.13	.14	.16	.17	.30	.53	.74	.87	.88	.69	.24	.12	.07	.05	.04	.03
	H	.06	.06	.05	.04	.04	.06	.09	.11	.13	.15	.16	.17	.28	.49	.67	.78	.79	.62	.23	.14	.11	.09	.08	.07
NW	L	.00	.00	.00	.00	.00	.04	.09	.13	.17	.20	.22	.23	.24	.31	.53	.78	.92	.81	.28	.10	.04	.02	.01	.00
	M	.02	.02	.01	.01	.01	.05	.10	.13	.17	.19	.21	.22	.23	.30	.52	.75	.88	.77	.26	.12	.07	.05	.04	.03
	H	.06	.05	.05	.04	.04	.07	.11	.14	.17	.19	.20	.21	.22	.28	.48	.68	.79	.69	.23	.14	.10	.09	.08	.07
Hor.	L	.00	.00	.00	.00	.00	.08	.25	.45	.64	.80	.91	.97	.97	.91	.80	.64	.44	.23	.08	.03	.01	.00	.00	.00
	M	.02	.02	.01	.01	.01	.08	.24	.43	.60	.75	.86	.92	.92	.87	.77	.63	.45	.26	.12	.07	.05	.04	.03	.02
	H	.07	.06	.05	.05	.04	.11	.25	.41	.56	.68	.77	.83	.83	.80	.71	.59	.44	.28	.17	.13	.11	.10	.09	.08

Values for nominal 15 ft by 15 ft by 10 ft high space, with ceiling, and 50% or less glass in exposed surface at listed orientation.

L = Lightweight construction, such as 1 in. wood floor, Group G wall.

M = Mediumweight construction, such as 2 to 4 in. concrete floor, Group E wall.

H = Heavyweight construction, such as 6 to 8 in. concrete floor, Group C wall.

Reprinted with permission from the *1989 ASHRAE Handbook—Fundamentals*.

TABLE 6.9 COOLING LOAD FACTORS (CLF) FOR GLASS WITHOUT INTERIOR SHADING, IN NORTH LATITUDE SPACES HAVING UNCARPETED FLOORS

Dir.	Room Mass	Solar Time																							
		0100	0200	0300	0400	0500	0600	0700	0800	0900	1000	1100	1200	1300	1400	1500	1600	1700	1800	1900	2000	2100	2200	2300	2400
N	L	.00	.00	.00	.00	.01	.64	.73	.74	.81	.88	.95	.98	.98	.94	.88	.79	.79	.55	.31	.12	.04	.02	.01	.00
	M	.12	.09	.07	.06	.05	.33	.45	.53	.61	.69	.76	.82	.85	.86	.85	.81	.80	.70	.60	.43	.32	.24	.19	.15
	H	.24	.21	.19	.18	.16	.43	.48	.51	.56	.61	.66	.71	.73	.74	.73	.71	.71	.62	.52	.42	.36	.32	.29	.26
NE	L	.00	.00	.00	.00	.01	.51	.83	.88	.72	.47	.33	.27	.24	.23	.20	.18	.14	.09	.03	.01	.00	.00	.00	.00
	M	.03	.02	.02	.02	.02	.24	.45	.57	.58	.49	.41	.36	.32	.29	.27	.24	.21	.17	.13	.10	.07	.06	.05	.04
	H	.08	.07	.07	.06	.06	.27	.43	.49	.45	.37	.32	.29	.28	.27	.26	.24	.22	.19	.16	.14	.12	.11	.10	.09
E	L	.00	.00	.00	.00	.00	.42	.76	.91	.90	.75	.51	.30	.22	.18	.16	.13	.11	.07	.02	.01	.00	.00	.00	.00
	M	.03	.02	.02	.02	.01	.20	.41	.57	.65	.64	.55	.44	.36	.31	.26	.23	.19	.16	.12	.09	.07	.06	.04	.04
	H	.08	.08	.07	.06	.06	.24	.40	.50	.53	.50	.41	.33	.30	.28	.26	.23	.22	.19	.16	.14	.13	.11	.10	.09
SE	L	.00	.00	.00	.00	.00	.27	.58	.81	.93	.93	.81	.59	.37	.27	.21	.18	.14	.09	.03	.01	.00	.00	.00	.00
	M	.04	.03	.02	.02	.02	.13	.31	.48	.62	.69	.69	.61	.50	.41	.35	.30	.25	.20	.15	.12	.09	.07	.06	.05
	H	.10	.09	.08	.08	.07	.18	.32	.45	.53	.56	.54	.47	.39	.35	.32	.29	.26	.23	.19	.17	.15	.14	.12	.11
S	L	.00	.00	.00	.00	.00	.07	.15	.23	.39	.62	.81	.94	.93	.80	.59	.38	.26	.16	.06	.02	.01	.00	.00	.00
	M	.05	.04	.04	.03	.02	.05	.09	.14	.24	.38	.53	.65	.72	.71	.63	.52	.42	.33	.24	.18	.14	.11	.09	.07
	H	.13	.12	.10	.09	.09	.11	.14	.17	.25	.36	.47	.55	.58	.56	.49	.41	.36	.30	.25	.21	.19	.17	.16	.14
SW	L	.00	.00	.00	.00	.00	.04	.09	.13	.16	.19	.23	.39	.62	.82	.94	.94	.81	.54	.19	.07	.03	.01	.00	.00
	M	.08	.07	.05	.04	.03	.05	.07	.09	.12	.15	.17	.26	.40	.54	.66	.73	.72	.61	.43	.31	.23	.17	.13	.10
	H	.15	.14	.12	.11	.10	.11	.12	.14	.15	.17	.18	.26	.37	.48	.56	.59	.57	.47	.33	.27	.23	.21	.19	.17
W	L	.00	.00	.00	.00	.00	.04	.06	.08	.10	.12	.13	.15	.18	.24	.45	.71	.92	.93	.73	.25	.10	.04	.01	.00
	M	.08	.07	.05	.04	.04	.04	.06	.08	.10	.12	.13	.15	.21	.35	.50	.63	.71	.67	.46	.33	.24	.18	.14	.11
	H	.14	.13	.12	.11	.10	.10	.11	.12	.13	.14	.15	.16	.21	.33	.45	.54	.58	.52	.33	.26	.22	.19	.18	.16
NW	L	.00	.00	.00	.00	.00	.04	.07	.10	.13	.15	.16	.18	.19	.20	.28	.51	.78	.81	.46	.32	.13	.04	.01	.00
	M	.08	.06	.05	.04	.03	.05	.07	.10	.13	.16	.17	.19	.20	.24	.31	.36	.51	.53	.46	.32	.23	.17	.13	.10
	H	.13	.12	.11	.10	.09	.10	.12	.13	.15	.16	.17	.18	.19	.23	.33	.46	.55	.53	.33	.25	.21	.18	.16	.15
Hor.	L	.00	.00	.00	.00	.00	.13	.25	.45	.64	.80	.91	.97	.97	.91	.80	.64	.44	.23	.08	.03	.01	.00	.00	.00
	M	.07	.06	.05	.04	.03	.06	.14	.26	.40	.53	.64	.73	.78	.80	.77	.70	.59	.45	.33	.24	.19	.14	.11	.09
	H	.16	.15	.13	.12	.11	.13	.20	.29	.39	.48	.56	.61	.65	.65	.63	.57	.49	.40	.32	.28	.25	.22	.20	.18

Values for nominal 15 ft by 15 ft by 10 ft high space, with ceiling, and 50% or less glass in exposed surface at listed orientation.

L = Lightweight construction, such as 1 in. wood floor, Group G wall.

M = Mediumweight construction, such as 2 to 4 in. concrete floor, Group E wall.

H = Heavyweight construction, such as 6 to 8 in. concrete floor, Group C wall.

Reprinted with permission from the *1989 ASHRAE Handbook—Fundamentals.*

TABLE 6.10 COOLING LOAD FACTORS (CLF) FOR GLASS WITH INTERIOR SHADING, NORTH LATITUDES (ALL ROOM CONSTRUCTIONS)

Fenestration Facing	Solar Time, h																							
	0100	0200	0300	0400	0500	0600	0700	0800	0900	1000	1100	1200	1300	1400	1500	1600	1700	1800	1900	2000	2100	2200	2300	2400
N	0.08	0.07	0.06	0.06	0.07	0.73	0.66	0.65	0.73	0.80	0.86	0.89	0.89	0.86	0.82	0.75	0.78	0.91	0.24	0.18	0.15	0.13	0.11	0.10
NNE	0.03	0.03	0.02	0.02	0.03	0.64	0.77	0.62	0.42	0.37	0.37	0.37	0.36	0.35	0.32	0.28	0.23	0.17	0.08	0.07	0.06	0.05	0.04	0.04
NE	0.03	0.02	0.02	0.02	0.02	0.56	0.76	0.74	0.58	0.37	0.29	0.27	0.26	0.24	0.22	0.20	0.16	0.12	0.06	0.05	0.04	0.04	0.03	0.03
ENE	0.03	0.02	0.02	0.02	0.02	0.52	0.76	0.80	0.71	0.52	0.31	0.26	0.24	0.22	0.20	0.18	0.15	0.11	0.06	0.05	0.04	0.04	0.03	0.03
E	0.03	0.02	0.02	0.02	0.02	0.47	0.72	0.80	0.76	0.62	0.41	0.27	0.24	0.22	0.20	0.17	0.14	0.11	0.06	0.05	0.05	0.04	0.03	0.03
ESE	0.03	0.03	0.02	0.02	0.02	0.41	0.67	0.79	0.80	0.72	0.54	0.34	0.27	0.24	0.21	0.19	0.15	0.12	0.07	0.06	0.05	0.04	0.04	0.03
SE	0.03	0.03	0.02	0.02	0.02	0.30	0.57	0.74	0.81	0.79	0.68	0.49	0.33	0.28	0.25	0.22	0.18	0.13	0.08	0.07	0.06	0.05	0.04	0.04
SSE	0.04	0.03	0.03	0.02	0.02	0.12	0.31	0.54	0.72	0.81	0.81	0.71	0.54	0.38	0.32	0.27	0.22	0.16	0.09	0.08	0.07	0.06	0.05	0.04
S	0.04	0.04	0.03	0.03	0.03	0.09	0.16	0.23	0.38	0.58	0.75	0.83	0.80	0.68	0.50	0.35	0.27	0.19	0.11	0.09	0.08	0.07	0.06	0.05
SSW	0.05	0.04	0.04	0.03	0.03	0.09	0.14	0.18	0.22	0.27	0.43	0.63	0.78	0.84	0.80	0.66	0.46	0.25	0.13	0.11	0.09	0.08	0.07	0.06
SW	0.05	0.05	0.04	0.04	0.03	0.07	0.11	0.14	0.16	0.19	0.22	0.38	0.59	0.75	0.83	0.81	0.69	0.45	0.16	0.12	0.10	0.09	0.07	0.06
WSW	0.05	0.05	0.04	0.04	0.03	0.07	0.10	0.12	0.14	0.16	0.17	0.23	0.44	0.64	0.78	0.84	0.78	0.55	0.16	0.12	0.10	0.09	0.07	0.06
W	0.05	0.05	0.04	0.04	0.03	0.06	0.09	0.11	0.13	0.15	0.16	0.17	0.31	0.53	0.72	0.82	0.81	0.61	0.16	0.12	0.10	0.08	0.07	0.06
WNW	0.05	0.05	0.04	0.03	0.03	0.07	0.10	0.12	0.14	0.16	0.17	0.18	0.22	0.43	0.65	0.80	0.84	0.66	0.16	0.12	0.10	0.08	0.07	0.06
NW	0.05	0.04	0.04	0.03	0.03	0.07	0.11	0.14	0.17	0.19	0.20	0.21	0.22	0.30	0.52	0.73	0.82	0.69	0.16	0.12	0.10	0.08	0.07	0.06
NNW	0.05	0.05	0.04	0.03	0.03	0.11	0.17	0.22	0.26	0.30	0.32	0.33	0.34	0.34	0.39	0.61	0.82	0.76	0.17	0.12	0.10	0.08	0.07	0.06
HOR.	0.06	0.05	0.04	0.04	0.03	0.12	0.27	0.44	0.59	0.72	0.81	0.85	0.85	0.81	0.71	0.58	0.42	0.25	0.14	0.12	0.10	0.08	0.07	0.06

Reprinted with permission from the *1989 ASHRAE Handbook—Fundamentals*.

TABLE 6.11 SHADING FROM OVERHEAD PROJECTIONS

Latitude	24°				32°				40°				48°				56°			
St'd Time	9 am	Noon	3 pm	6 pm	9 am	Noon	3 pm	6 pm	9 am	Noon	3 pm	6 pm	9 am	Noon	3 pm	6 pm	9 am	Noon	3 pm	6 pm
(Facing) N	—	—	—	.58	—	—	—	.63	—	—	—	.83	—	—	—	1.37	—	—	—	1.61
NE	1.89	—	—	—	2.17	—	—	—	2.13	—	—	—	3.03	—	—	—	3.45	—	—	—
E	1.00	—	—	—	97	—	—	—	.89	—	—	—	.83	—	—	—	.74	—	—	—
SE	.93	4.55	—	—	1.00	3.33	—	—	.86	2.33	—	—	.73	1.67	—	—	.61	1.33	—	—
S	4.35	3.57	4.35	—	2.63	2.38	2.63	—	1.85	1.59	1.85	—	1.33	1.19	1.33	—	1.08	.93	1.08	—
SW	—	4.55	.93	—	—	3.33	1.00	—	—	2.33	.86	—	—	1.67	.73	—	—	1.33	.61	—
W	—	—	1.00	*	—	—	.97	*	—	—	.89	*	—	—	.83	*	—	—	.74	*
NW	—	—	1.89	*	—	—	2.17	*	—	—	2.13	*	—	—	3.03	*	—	—	3.45	*

Reprinted with permission from the *1985 Fundamentals, ASHRAE Handbook & Product Directory*.
Note: Values apply from April to September.
*Shading not effective.
—Completely shaded.

Solution

Figure 6.4 shows the arrangement. The vertical proportion of shade, from Table 6.11 is 0.97. The total vertical distance the shade extends down is therefore

$$L = 0.97 \times 4 = 3.9 \text{ ft}$$

The height of shade on the window is 3.9 − 1 = 2.9 ft, and the unshaded height is 6 − 2.9 = 3.1 ft. The unshaded area of window is

$$A = 3.1 \times 5 = 15.5 \text{ ft}^2$$

Figure 6.4
Sketch for Example 6.7.

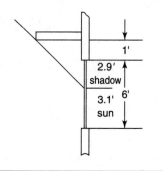

Example 6.8

A room with no carpeting and a wall facing east at 40°N latitude has a total window glass area of 80 ft². The building is of heavyweight (H) construction. The glass is ¼ in. single heat-absorbing glass with no interior shading device. At 10 AM ST in June, an adjacent building shades 30 ft² of the window. What is the solar cooling load?

Solution

Equation 6.4 will be used. The externally shaded and unshaded portions of the glass must be handled separately, however, because they receive different radiation. For the part receiving direct radiation,

$$Q = \text{SHGF} \times A \times \text{SC} \times \text{CLF}$$
$$Q = 216 \times 50 \times 0.69 \times 0.50 = 3730 \text{ BTU/hr}$$

For the part receiving only diffuse radiation, using the SHGF for north orientation,

$$Q = 48 \times 30 \times 0.69 \times 0.50 = 500 \text{ BTU/hr}$$

The total solar cooling load is

$$Q = 3730 + 500 = 4230 \text{ BTU/hr}$$

6.7 DESIGN CONDITIONS

The cooling load calculations are usually based on inside and outdoor *design conditions* of temperature and humidity. The inside conditions are those that provide satisfactory comfort. Table 1.1 lists some suggested values.

The outdoor summer design conditions are based on reasonable maximums, using weather records. Table A.6 lists these conditions for some cities. These values will normally equal or exceed 2.5% of the summertime hours. The DB (dry bulb temperature) and *coincident* WB (wet bulb temperature) occurring at the same time are listed together, and should be used as the *corresponding* design values.

The separate design WB column listed does not usually occur at the same time as the DB listed and therefore should not be used in load calculations. Previous weather data tables showed this value, which if used, would result in too large a design cooling load. The separate WB value may be needed, however, in selecting a cooling tower or for special applications. Definitions of DB and WB are given in Chapter 7.

The 2.5% values will provide a reasonable compromise between comfort and cost for most applications. For different requirements, 1% and 5% values can be found in the ASHRAE Handbook tables.

The table also lists latitudes and mean daily DB temperature ranges.

June to September can be used as months for summer outdoor design temperatures in the northern hemisphere. Occasionally, maximum cooling loads occur in other months due to solar radiation, and therefore it is necessary to know the expected design conditions at those times. Table 6.12 on page 136 lists appropriate values.

6.8 LIGHTING

The equation for determining cooling load due to heat gain from lighting is

$$Q = 3.4 \times W \times \text{BF} \times \text{CLF} \qquad (6.5)$$

where

Q = cooling load from lighting, BTU/hr

W = lighting capacity, watts

BF = ballast factor

CLF = cooling load factor for lighting

The term W is the rated capacity of the lights in use, expressed in watts. In many applications, all of the lighting is on at all times, but if it is not, the actual amount should be used. The value 3.4 converts watts to BTU/hr.

The factor BF accounts for heat losses in the ballast in fluorescent lamps, or other special losses. A typical value of BF is 1.25 for fluorescent lighting. For incandescent lighting, there is no extra loss, and BF = 1.0.

The factor CLF accounts for storage of part of the lighting heat gain. The storage effect depends on how long the lights and cooling system are operating, as well as the building construction, type of lighting fixture, and ventilation rate.

No storage effect can be allowed for any of the following conditions:

1. Cooling system operates only during occupied hours
2. Cooling system operates more than 16 hr
3. Temperature of the space is allowed to rise during nonoccupied hours (temperature swing)

These conditions cover so many possible situations that it is suggested that heat storage effects for lighting should be used with extreme caution. Building use patterns often change and may be unpredictable. Energy conservation operating techniques may also result in one of the conditions discussed earlier, even though not planned for originally.

For these reasons, the CLF tables for lighting are not presented here. For those cases where they are applicable, they may be found in the ASHRAE *Fundamentals Volume.* Otherwise use a value of CLF = 1.0.

Example 6.9 _____

A room has eight 40 W fluorescent lighting fixtures in use. The cooling system operates only during

TABLE 6.12   COOLING DESIGN DRY BULB AND MEAN COINCIDENT WET BULB

City	LAT Deg. Min.		LONG Deg. Min.		ELEV	DESIGN DB (2.5%) Jan Feb Mar Apr May Oct Nov Dec								DESIGN COINCIDENT WB (2.5%) Jan Feb Mar Apr May Oct Nov Dec							
Birmingham, Al	33	34	86	45	630	67	72	76	83	91	84	74	68	62	63	64	67	71	69	64	60
Yuma, Az	32	40	114	36	206	76	82	87	96	99	99	85	75	54	57	58	61	63	66	58	54
Little Rock, Ar	34	44	92	14	265	68	70	76	83	89	84	74	67	63	61	64	67	72	71	65	61
Arcata, Ca	40	59	124	6	217	61	60	59	59	62	66	63	59	53	54	53	54	57	58	56	55
Bishop, Ca	37	22	118	22	4112	60	66	72	81	88	85	72	64	43	45	48	52	57	55	49	45
Los Angeles, Ca	33	56	118	23	122	72	71	72	74	74	80	79	74	54	55	55	61	61	63	56	53
San Diego, Ca	32	44	117	10	37	69	69	70	73	72	79	74	71	53	56	56	60	60	64	55	54
Colorado Springs	38	49	104	42	6170	56	61	63	73	79	77	64	59	39	43	43	47	52	50	44	40
Wilmington, De	39	40	75	36	78	50	55	63	78	84	78	65	55	48	52	54	64	70	69	60	51
Jacksonville, Fl	30	25	81	39	24	75	80	83	86	93	86	81	75	66	68	68	69	73	74	69	67
Augusta, Ga	33	22	81	58	182	70	73	78	85	91	84	77	71	62	63	66	68	73	71	65	63
Boise, Id	43	34	116	13	2857	51	53	62	72	83	74	59	48	44	46	49	53	62	55	48	43
Chicago-O'Hare, Il	41	59	87	54	667	40	49	58	74	83	78	64	50	45	47	52	63	68	66	56	49
Fort Wayne, In	41	0	85	12	828	47	52	60	76	82	78	63	53	47	50	53	62	68	63	58	51
Indianapolis, In	39	44	86	16	793	53	57	64	78	82	79	66	55	52	54	57	64	68	65	59	53
Des Moines, Ia	41	32	93	39	963	42	49	59	76	84	79	63	52	39	44	49	62	69	64	55	48
Dodge City, Ks	37	46	99	58	2592	58	64	72	82	89	83	68	59	46	49	51	58	64	60	52	45
Covington, Ky	39	4	84	40	888	55	61	65	79	84	79	68	58	53	56	55	64	70	64	59	53
Louisville, Ky	38	11	85	44	488	58	63	69	82	88	81	71	61	56	57	59	66	72	67	61	57
Lake Charles, La	30	13	93	9	32	72	75	78	84	89	88	79	74	67	68	69	70	75	74	73	68
New Orleans, La	29	59	90	15	20	74	77	79	84	89	86	79	74	69	70	70	72	74	75	72	69
Portland, Me	43	39	70	19	61	41	44	49	62	77	69	58	48	33	43	44	53	64	60	55	47
Battle Creek, Mi	42	18	85	14	939	49	48	64	72	86	77	62	49	50	45	57	59	66	64	55	49
Minneapolis, Mn	44	53	93	15	838	36	42	52	73	83	76	57	43	34	39	45	58	65	62	52	43
Jackson, Ms	32	20	90	13	332	70	74	78	84	89	87	77	70	64	65	67	70	73	72	68	63
Kansas City, Mo	39	7	94	35	750	54	60	69	81	87	83	69	58	49	51	55	65	71	67	57	52
Springfield, Mo	37	14	93	23	1270	59	62	72	80	84	83	69	59	54	54	59	65	70	66	60	53
Billings, Mt	45	48	108	32	3583	51	54	62	70	80	77	59	52	41	44	47	52	59	56	47	41
North Platte, Ne	41	8	100	42	2787	52	58	64	77	83	80	64	54	42	45	49	56	63	57	49	43
Tonopah, Ne	38	4	117	8	5422	53	58	65	75	81	78	64	55	40	43	45	49	54	52	46	41
Albuquerque, NM	35	3	106	37	5314	55	63	69	79	87	79	64	56	42	45	47	51	56	55	46	42
Albany, NY	42	45	73	48	277	43	47	54	73	81	73	60	49	42	45	49	59	67	65	58	46
Greensboro, NC	36	5	79	57	891	63	64	70	83	87	81	71	63	57	56	59	65	70	68	59	58
Bismarck, ND	46	46	100	45	1660	41	42	56	70	82	77	53	43	37	39	46	53	62	57	43	38
Akron-Canton, Oh	40	55	81	26	1236	49	53	60	74	79	75	64	54	48	50	53	63	67	62	57	52
Toledo, Oh	41	36	83	48	692	44	48	59	76	84	77	64	52	42	46	53	63	69	63	58	51
Tulsa, Ok	36	11	95	54	674	63	69	75	83	88	86	73	63	57	58	60	66	73	69	61	53
Medford, Or	42	23	122	52	1329	55	59	66	75	84	78	63	53	48	50	53	58	64	60	52	49
Portland, Or	45	36	122	36	24	54	57	60	69	79	73	59	54	50	51	51	56	63	59	54	53
Pittsburgh, Pa	40	30	80	13	1151	49	53	63	78	82	77	64	55	46	50	53	63	68	64	57	51
Sioux Falls, SD	43	34	96	44	1422	41	45	57	75	84	78	58	47	37	41	48	58	65	60	49	43
Bristol, Tn	36	30	82	21	1566	59	64	69	81	86	79	70	60	54	55	58	64	70	66	61	54
Amarillo, Tx	35	14	101	46	3700	63	69	75	84	90	84	70	65	46	50	51	56	61	59	50	47
Midland, Tx	31	56	102	12	2858	72	74	81	88	94	88	77	71	53	54	55	59	65	63	56	52
Wichita Falls, Tx	33	59	98	31	1039	66	74	82	88	93	89	76	67	56	58	61	66	71	67	60	54
Cedar City, Ut	37	42	113	6	5616	53	56	63	73	81	78	63	54	41	44	46	50	55	53	46	42
Burlington, Vt	44	28	73	9	331	39	40	49	68	79	70	58	46	38	38	44	56	66	61	55	44
Blackstone, Va	37	4	77	58	438	65	64	71	83	87	82	72	65	59	56	61	67	72	70	61	60
Roanoke, Va	37	19	79	58	1174	60	63	69	82	87	81	69	62	54	52	56	63	69	66	56	55
Everett, Wa	47	54	122	17	596	50	54	61	64	72	64	54	53	47	49	49	55	61	58	52	49
Charleston, WV	38	22	81	36	989	62	64	70	83	86	80	72	64	55	55	56	63	67	66	60	55
Huntington, WV	38	25	82	27	565	65	65	72	83	87	84	71	63	58	56	60	66	70	69	60	57
Green Bay, Wi	44	29	88	8	699	36	39	48	70	77	69	58	42	36	39	44	57	65	60	54	41
Madison, Wi	43	8	89	20	866	38	44	53	73	81	74	61	44	37	42	47	60	68	62	55	43
Cheyenne, Wy	41	9	104	49	6144	51	56	58	69	76	74	60	54	37	41	41	48	52	50	43	41

Reprinted with permission from the *1979 ASHRAE Load Calculation Manual.*

occupied hours. What is the solar cooling load from the lighting?

Solution
A value of BF = 1.25 for the ballast heat will be assumed. CLF = 1.0 for the operating conditions. Using Equation 6.5,

$$Q = 3.4 \times W \times BF \times CLF$$
$$= 3.4 \times 320 \times 1.25 \times 1.0 = 1360 \text{ BTU/hr}$$

6.9 PEOPLE

The heat gain from people is composed of two parts, sensible heat and the latent heat resulting from perspiration. Some of the sensible heat may be absorbed by the heat storage effect, but not the latent heat. The equations for cooling loads from sensible and latent heat gains from people are

$$Q_s = q_s \times n \times CLF \tag{6.6}$$
$$Q_l = q_l \times n \tag{6.7}$$

where

Q_s, Q_l = sensible and latent heat gains (loads)

q_s, q_l = sensible and latent heat gains per person

n = number of people

CLF = cooling load factor for people

The rate of heat gain from people depends on their physical activity. Table 6.13 lists values for some typical activities. The rates are suitable for a

TABLE 6.13 RATES OF HEAT GAIN FROM OCCUPANTS

Degree of Activity	Typical Applications	Total Heat Adults, Male, Btu/h	Total Heat Adjusted,[d] Btu/h	Sensible Heat, Btu/h	Latent Heat, Btu/h
Seated at theater	Theater—Matinee	390	330	225	105
Seated at theater	Theater—Evening	390	350	245	105
Seated, very light work	Offices, hotels, apartments	450	400	245	155
Moderately active office work	Offices, hotels, apartments	475	450	250	200
Standing, light work; walking	Department store, retail store	550	450	250	200
Walking; standing	Drug store, bank	550	500	250	250
Sedentary work	Restaurant[e]	490	550	275	275
Light bench work	Factory	800	750	275	475
Moderate dancing	Dance hall	900	850	305	545
Walking 3 mph; light machine work	Factory	1000	1000	375	625
Bowling[f]	Bowling alley	1500	1450	580	870
Heavy work	Factory	1500	1450	580	870
Heavy machine work; lifting	Factory	1600	1600	635	965
Athletics	Gymnasium	2000	1800	710	1090

[a] Tabulated values are based on 75°F room dry-bulb temperature. For 80°F room dry-bulb, the total heat remains the same, but the sensible heat values should be decreased by approximately 20%, and the latent heat values increased accordingly.
[c] All values are rounded to nearest 5 Btu/h.
[d] *Adjusted heat gain* is based on normal percentage of men, women, and children for the application listed, with the postulate that the gain from an adult female is 85% of that for an adult male, and that the gain from a child is 75% of that for an adult male.
[e] Adjusted total heat gain for *Sedentary work, Restaurant,* includes 60 Btu/h for food per individual (30 Btu/h sensible and 30 Btu/h latent.)
[f] For *Bowling,* figure one person per alley actually bowling, and all others at sitting (400 Btu/h) or standing and walking slowly (550 Btu/h).
Reprinted with permission from the *1989 ASHRAE Handbook—Fundamentals.*

75 F DB room temperature. Values vary slightly at other temperatures, as noted.

The heat storage effect factor CLF applies to the sensible heat gain from people. If the air conditioning system is shut down at night, however, no storage should be included, and CLF = 1.0. Table 6.14 lists values of CLF for people.

Example 6.10 _____

What is the heat gain from 240 people at night in a movie theater at 75 F DB?

Solution

Equations 6.6, 6.7, and Table 6.15 (page 140) will be used. Because the air conditioning system of a theater is normally shut down overnight, no storage effect is included in calculating the cooling load.

$$Q_s = 245 \times 240 \times 1.0 = 58,800 \text{ BTU/hr}$$
$$Q_l = 105 \times 240 = 25,200 \text{ BTU/hr}$$
$$\text{Total } Q = 84,000 \text{ BTU/hr}$$

6.10 EQUIPMENT AND APPLIANCES

The heat gain from equipment may sometimes be found directly from the manufacturer or the nameplate data, with allowance for intermittent use. Some equipment produces both sensible and latent heat. Some values of heat output for typical appliances are shown in Table 6.15. CLF factors (not shown) apply if the system operates 24 hours.

Example 6.11 _____

Diane's Deli Diner has the following equipment operating in the air-conditioned area, without hoods:

  1 coffee burner (2 burners)

  1 coffee heater (1 burner)

  1 toaster (large)

What are the sensible, latent, and total heat gains, (cooling loads) from this equipment?

Solution
Using values from Table 6.15,

	Q_S	Q_L	Q_T
Coffee burner	3750	1910	5660
Coffee heater	230	110	340
Toaster	9590	8500	18,090
Total heat gains (loads)	13,570 BTU/hr	10,520 BTU/hr	24,090 BTU/hr

The heat output from motors and the equipment driven by them results from the conversion of the electrical energy to heat. The proportion of heat generated that is gained by the air-conditioned space depends on whether the motor and driven load are both in the space or only one of them is. Table 6.16 lists heat outputs for each condition.

Example 6.12 _____

A hotel with 150 rooms has a fan-coil air conditioning unit in each room, with a 0.16 HP motor. What is the heat gain (load) to the building from the units?

Solution

Both the motor and fan are in the conditioned spaces. From Table 6.16 (page 141), the heat gain (load) is

$$Q = 1160 \text{ BTU/hr} \times 150 = 174,000 \text{ BTU/hr}$$

For any lighting and equipment that operates on a periodic intermittent basis, the heat gains should be multiplied by the proportion of operating time. However, it is often not possible to guarantee predicted operations, so using such factors should be approached with caution.

6.11 INFILTRATION

Infiltration of air through cracks around windows or doors results in both a sensible and latent heat

TABLE 6.14 SENSIBLE HEAT COOLING LOAD FACTORS FOR PEOPLE

Total hours in space	Hours After Each Entry Into Space																							
	1	2	3	4	5	6	7	8	9	10	11	12	13	14	15	16	17	18	19	20	21	22	23	24
2	0.49	0.58	0.17	0.13	0.10	0.08	0.07	0.06	0.05	0.04	0.04	0.03	0.03	0.02	0.02	0.02	0.02	0.01	0.01	0.01	0.01	0.01	0.01	0.01
4	0.49	0.59	0.66	0.71	0.27	0.21	0.16	0.14	0.11	0.10	0.08	0.07	0.06	0.06	0.05	0.04	0.04	0.03	0.03	0.03	0.02	0.02	0.02	0.01
6	0.50	0.60	0.67	0.72	0.76	0.79	0.34	0.26	0.21	0.18	0.15	0.13	0.11	0.10	0.08	0.07	0.06	0.06	0.05	0.04	0.04	0.03	0.03	0.03
8	0.51	0.61	0.67	0.72	0.76	0.80	0.82	0.84	0.38	0.30	0.25	0.21	0.18	0.15	0.13	0.12	0.10	0.09	0.08	0.07	0.06	0.05	0.03	0.04
10	0.53	0.62	0.69	0.74	0.77	0.80	0.83	0.85	0.87	0.89	0.42	0.34	0.28	0.23	0.20	0.17	0.15	0.13	0.11	0.10	0.09	0.08	0.07	0.06
12	0.55	0.64	0.70	0.75	0.79	0.81	0.84	0.86	0.88	0.89	0.91	0.92	0.45	0.36	0.30	0.25	0.21	0.19	0.16	0.14	0.12	0.11	0.09	0.08
14	0.58	0.66	0.72	0.77	0.80	0.83	0.85	0.87	0.89	0.90	0.91	0.92	0.93	0.94	0.47	0.38	0.31	0.26	0.23	0.20	0.17	0.15	0.13	0.11
16	0.62	0.70	0.75	0.79	0.82	0.85	0.87	0.88	0.90	0.91	0.92	0.93	0.94	0.95	0.95	0.96	0.49	0.39	0.33	0.28	0.24	0.20	0.18	0.16
18	0.66	0.74	0.79	0.82	0.85	0.87	0.89	0.90	0.92	0.93	0.94	0.94	0.95	0.96	0.96	0.97	0.97	0.97	0.50	0.40	0.33	0.28	0.24	0.21

CLF = 1.0 for systems shut down at night and for high occupant densities such as in theaters and auditoriums.
Reprinted with permission from the 1989 ASHRAE Handbook—Fundamentals.

TABLE 6.15   HEAT GAIN FROM EQUIPMENT

Appliance	Size	Recommended Rate of Heat Gain, BTU/hr			With Hood
		Without Hood			
		Sens.	Latent	Total	Sensible
Restaurant, electric blender, per quart of capacity	1 to 4 qt	1000	520	1520	480
Coffee brewer	12 cups/2 brnrs	3750	1910	5660	1810
Coffee heater, per warming burner	1 to 2 brnrs	230	110	340	110
Display case (refrigerated), per ft³ of interior	6 to 67 ft³	62	0	62	0
Hot plate (high-speed double burner)		7810	5430	13,240	6240
Ice maker (large)	220 lb/day	9320	0	9320	0
Microwave oven (heavy-duty commercial)	0.7 ft³	8970	0	8970	0
Toaster (large pop-up)	10 slice	9590	8500	18,080	5800

Appliance	Size	Recommended Rate of Heat Gain, BTU/hr
Computer Devices		
Communication/transmission		5600–9600
Disk drives/mass storage		3400–22,400
Microcomputer/word processor	16–640 kbytes	300–1800
Minicomputer		7500–15,000
Printer (laser)	8 pages/min	1000
Printer (line, high-speed)	5000 or more pages/min	2500–13,000
Tape drives		3500–15,000
Terminal		270–600
Copiers/Typesetters		
Blue print		3900–42,700
Copiers (large)	30–67 copies/min	1700–6600
Copiers	6–30 copies/min	460–1700
Miscellaneous		
Cash register		160
Cold food/beverage		1960–3280
Coffee maker	10 cup	sensible 3580
		latent 1540
Microwave oven	1 ft³	1360
Paper shredder		680–8250
Water cooler	8 gal/hr	6000

Abridged with permission from the *1993 ASHRAE Volume—Fundamentals.*

TABLE 6.16 HEAT GAIN FROM TYPICAL ELECTRIC MOTORS

| Motor Name-plate or Rated Horse-power | Motor Type | Nom-inal rpm | Full Load Motor Effici-ency in Percent | Location of Motor and Driven Equipment with Respect to Conditioned Space or Airstream | | |
| | | | | A | B | C |
				Motor in, Driven Equip-ment in Btu/h	Motor out, Driven Equip-ment in Btu/h	Motor in, Driven Equip-ment out Btu/h
0.05	Shaded Pole	1500	35	360	130	240
0.08	Shaded Pole	1500	35	580	200	380
0.125	Shaded Pole	1500	35	900	320	590
0.16	Shaded Pole	1500	35	1160	400	760
0.25	Split Phase	1750	54	1180	640	540
0.33	Split Phase	1750	56	1500	840	660
0.50	Split Phase	1750	60	2120	1270	850
0.75	3-Phase	1750	72	2650	1900	740
1	3-Phase	1750	75	3390	2550	850
1	3-Phase	1750	77	4960	3820	1140
2	3-Phase	1750	79	6440	5090	1350
3	3-Phase	1750	81	9430	7640	1790
5	3-Phase	1750	82	15,500	12,700	2790
7.5	3-Phase	1750	84	22,700	19,100	3640
10	3-Phase	1750	85	29,900	24,500	4490
15	3-Phase	1750	86	44,400	38,200	6210
20	3-Phase	1750	87	58,500	50,900	7610
25	3-Phase	1750	88	72,300	63,600	8680
30	3-Phase	1750	89	85,700	76,300	9440
40	3-Phase	1750	89	114,000	102,000	12,600
50	3-Phase	1750	89	143,000	127,000	15,700
60	3-Phase	1750	89	172,000	153,000	18,900
75	3-Phase	1750	90	212,000	191,000	21,200
100	3-Phase	1750	90	283,000	255,000	28,300
125	3-Phase	1750	90	353,000	318,000	35,300
150	3-Phase	1750	91	420,000	382,000	37,800
200	3-Phase	1750	91	569,000	509,000	50,300
250	3-Phase	1750	91	699,000	636,000	62,900

Reprinted with permission from the *1993 ASHRAE Handbook—Fundamentals.*

gain to the rooms. Procedures and equations for calculating infiltration heat losses were explained in detail in Chapter 3. The same procedure is used for calculating infiltration heat gains.

Most summer air conditioning systems have *mechanical ventilation* using some outside air, which reduces or eliminates infiltration by creating a positive air pressure within the building.

Ventilation air is not a load on the room, but is a load on the central cooling equipment. Many modern buildings have fixed (sealed) windows and therefore have no infiltration loss, except for entrances.

6.12   ROOM COOLING LOAD

The **room cooling load** *is the sum of each of the cooling load components (roof, walls, glass, solar, people, equipment, and infiltration) in the room.*

When calculating cooling loads, a prepared form is useful. A load calculation form is shown in Figure 6.5 on page 149 and the Appendix. It can be used for individual rooms or for a small building.

The following abbreviations will be used for convenience.

TCL, SCL, LCL = component total, sensible, latent cooling loads

RTCL, RSCL, RLCL = room total, sensible, latent cooling loads

BTCL, BSCL, BLCL = building total, sensible, latent cooling loads

CTCL, SCSL, CLCL = coil total, sensible, latent cooling loads

6.13   ROOM PEAK COOLING LOAD

We have learned how to calculate the cooling loads, but not how to determine their peak (maximum) value. Because the air conditioning system must be sized to handle *peak* loads, we must know how to find them.

The external heat gain components vary in intensity with time of day and time of year because of changing solar radiation as the orientation of the sun changes and because of outdoor temperature changes. This results in a change in the total room cooling load. Sometimes it is immediately apparent by inspecting the tables at what time the peak load occurs, but often calculations are required at a few different times. Some general guidelines can be offered to simplify this task. From the CLTD, SHGF, and CLF tables we can note the following:

1. For west-facing glass, maximum load is in mid-summer in the afternoon.
2. For east-facing glass, maximum solar load is in early or mid-summer in the morning.
3. For south-facing glass, maximum solar load is in the fall or winter in early afternoon.
4. For southwest-facing glass, maximum solar load is in the fall in the afternoon.
5. For roofs, maximum load is in the summer in the afternoon or evening.
6. For walls, maximum load is in the summer in the afternoon or evening.

These generalizations can be used to localize approximate times of room peak loads. For instance, we might expect a south-facing room with a very large window area to have a peak load in early afternoon in the fall—not in the summer! If the room had a small glass area, however, the wall and glass heat conduction might dominate and the peak load time would be a summer afternoon. Once the appropriate day and time are located, a few calculations will determine the exact time and value of the peak load.

Example 6.14

A room facing east, in the Shelton Motel in St. Louis, Missouri, has a 60 ft^2 window with an aluminum frame with a thermal break. The window is ¼ in. single heat-absorbing glass. Light colored interior venetian blinds are used. The wall is a metal curtain wall with a *U*-value of 0.14. Building construction is lightweight. Find the time and value of room peak cooling load. The room is at 78 F DB.

Solution

The glass area in the room is large enough compared with the wall so that the solar load determines the peak load time.

From Table 6.6 for 40°N latitude, the peak SHGF is in April (224 BTU/hr-ft^2). However, there will be a large conduction heat loss through both glass and wall in the morning at that time of year. In August the SHGF = 216, almost as large as in April, so the total heat gain will be maximum in August.

Referring to the tables, the CLF for the glass is maximum at 8 AM and the CLTD for the wall is maximum at 10 AM. It appears as if 8, 9, or 10 AM are the possible peak times for the room cooling load. We are assuming that the room is not on the top floor, otherwise the roof load might affect the peak time.

Proceeding to check the possibilities:

at 8 AM

Solar, glass $Q = 216 \times 60 \times 0.53 \times 0.80 = 5500$
  Conduction, glass $Q = 1.01 \times 60 \times 0 = \quad\ \ 0$
  Conduction, wall $Q = 0.14 \times 40 \times 47 = \underline{\ 260}$
  Total $= 5760$ BTU/hr

at 9 AM

Solar, glass $Q = 216 \times 60 \times 0.53 \times 0.76 = 5220$
  Conduction, glass $Q = 1.01 \times 60 \times 2 = \ 120$
  Conduction, wall $Q = 0.14 \times 40 \times 54 = \underline{\ 300}$
  Total $= 5640$ BTU/hr

at 10 AM

Solar, glass $Q = 216 \times 60 \times 0.52 \times 0.62 = 4180$
  Conduction, glass $Q = 1.01 \times 60 \times 4 = \ 240$
  Conduction, wall $Q = 0.14 \times 40 \times 55 = \underline{\ 310}$
  Total $= 4730$ BTU/hr

The peak load for this room is at 8 AM in August. Even though the conduction heat gain through the glass and wall increases later in the morning, the solar gain is large enough to dominate. On the other hand, if the window were smaller, the peak time might be later.

Another point that needs comment here is the possibility of peak load in April, mentioned earlier. But the early morning outside temperature in April would result in a considerable conduction heat loss from the room, and the net gain would probably be less. If in doubt, however, the calculation should be made.

Each building must be analyzed in a similar way to determine time of room peak loads so that the proper room load is calculated.

6.14 BUILDING PEAK COOLING LOAD

*The **building cooling** load is the rate at which heat is removed from all air-conditioned rooms in the building at the time the building cooling load is at its peak value.*

If peak cooling loads for each room were added, the total would be greater than the peak cooling load required for the whole building, because these peaks do not occur at the same time. Therefore, the designer must also determine the time of year and time of day at which the building cooling load is at a peak, and then calculate it. A reasoning and investigation similar to that carried out in finding room peak loads is used. From our previous discussion and a study of the tables, the following guidelines emerge:

1. For buildings that are approximately square-shaped in plan with similar construction on all four walls, the peak load is usually in late afternoon in summer. This is because the outside temperature is highest then, and there is no differential influence of solar radiation on one side of a building.
2. For buildings with a long south or southwest exposure having large glass areas, the peak load may occur in the fall, around mid-day, because radiation is highest then. This case requires careful analysis.
3. For one-story buildings with very large roof areas, the peak load usually occurs in the afternoon in summer.

These suggestions must be verified in each case because there are so many variations in building orientation and construction. Once the peak load time is determined, the total building heat gains can be calculated.

The search for the time and value of peak room and building cooling loads is greatly simplified by using computer software programs. After the necessary data are entered, a complete time profile of loads for many hours can be developed in a few minutes.

Diversity

On some projects, the actual building peak load may be less than the calculated value because of load *diversity*. In some buildings, at the time of peak load, usage practice may be such that all of the people are not present and some of the lights and equipment are not operating. In these cases, a *diversity factor* or usage factor is sometimes estimated and applied to the calculated building peak load in order to reduce it. For example, if it is estimated that only 90% of the lighting is actually on at peak load time, the calculated lighting load would be multiplied by a factor of 0.90. Choosing proper diversity factors requires both experience and judgment about building use practices.

6.15   COOLING COIL LOAD

After the building cooling load is determined, the cooling coil load is found.

> The **cooling coil load** *is the rate at which heat must be removed by the air conditioning equipment cooling coil(s).*

The cooling coil load will be greater than the building load because there are heat gains to the air conditioning system itself. These gains may include:

1. Ventilation (outside air)
2. Heat gains to ducts
3. Heat produced by the air conditioning system fans and pumps
4. Air leakage from ducts

6.16   VENTILATION

Some outside air is generally brought into a building for health and comfort reasons. The sensible and latent heat of this air is usually greater than that of the room air, so it becomes part of the cooling load. The excess heat is usually removed in the cooling equipment, however, so it is part of the cooling coil load but not the building load.

The equations for determining the sensible and latent cooling loads from ventilation air, explained in Chapters 3 and 7, are

$$Q_s = 1.1 \times \text{CFM} \times \text{TC} \qquad (3.10)$$
$$Q_l = 0.68 \times \text{CFM} \times (W_0' - W_i') \qquad (3.11)$$

where

Q_s, Q_l = sensible and latent cooling loads from ventilation air, BTU/hr

CFM = air ventilation rate, ft^3/min

TC = temperature change between outdoor and inside air, F

W_0', W_i' = outdoor and inside humidity ratio, gr w./lb d.a.

The total heat Q_t removed from the ventilation air is $Q_t = Q_s + Q_l$.

Recommended outdoor air ventilation rates for some applications are listed in Table 6.17. This table is based on ventilation rates required in many state codes and standards.

The ventilation rates in Table 6.17 are often higher than the minimum listed in earlier standards. For instance, it requires 15 CFM per person in an office space. An earlier standard permitted a minimum of 5 CFM per person with a recommended value of 15 CFM per person.

Note that in some applications the ventilation requirements in Table 6.17 are listed as a CFM per ft^2 basis.

In order to save energy, beginning in the 1970s, many designers and operating personnel often provided only the minimum CFM required. At the same time, outside infiltration air was being reduced by improved weatherfitting in both existing and new buildings. This contributed to a deterioration in indoor air quality. The new requirements improve this situation.

There are still further changes in ventilation requirements that are being considered. The present standard allows for "moderate smoking." However,

TABLE 6.17 MINIMUM MECHANICAL VENTILATION REQUIREMENT RATES

Outdoor air shall be provided at a rate no less than the greater of either
A. 15 CFM per person, times the expected occupancy rate.
B. The applicable ventilation rate from the following list, times the conditioned floor area of the space.

Type of Use	CFM per Square Foot of Conditioned Floor Area
Auto Repair Workshops	1.50
Barber Shops	0.40
Bars, Cocktail Lounges, and Casinos	1.50
Beauty Shops	0.40
Coin-Operated Dry Cleaning	0.30
Commercial Dry Cleaning	0.45
Hotel Guest Rooms (less than 500 sq ft)	30 CFM/Guest Room
Hotel Guest Rooms (500 sq ft or greater)	0.15
Retail Stores	0.20
Smoking Lounges	1.50
All Others	0.15

Abridged from *Energy Efficiency Standards,* California Energy Commission, 1995.

smoking is now prohibited in many indoor locations. Also the values shown in Table 6.17 do not make special allowances for the amount of indoor air pollutants being generated. Undoubtedly new standards will reflect this and other information that is being found in this rapidly developing field.

Example 6.15

The Stellar Dome enclosed athletic stadium seats 40,000 people. The space design conditions are 80 F and 50% RH, and outdoor design conditions 94 F DB and 74 F WB. What is the cooling load due to ventilation?

Solution

Equations 3.10 and 3.11 will be used. Table 6.17 lists 15 CFM per person.

$$Q_s = 1.1 \times \text{CFM} \times \text{TC}$$
$$= 1.1 \times 15 \times 40,000 \times 14$$
$$= 9,240,000 \text{ BTU/hr}$$

The humidity ratios at the inside and outdoor conditions are 77.0 and 95.0 gr.w./lb d.a. (see Chapter 7).

$$Q_l = 0.68 \times \text{CFM} \times (W_0' - W_i')$$
$$= 0.68 \times 15 \times 40,000 \times (95.0 - 77.0)$$
$$= 7,344,000 \text{ BTU/hr}$$
$$Q_t = 9,240,000 + 7,344,000$$
$$= 16,584,000 \text{ BTU/hr}$$
$$\times \frac{1 \text{ ton}}{12,000 \text{ BTU/hr}} = 1382 \text{ tons}$$

If the peak load does not occur at the time of the day that the outdoor temperature is at a maximum, a correction must be made to the outdoor temperature used for calculating ventilation and infiltration loads. Table 6.18 on page 146 lists this correction.

6.17 HEAT GAIN TO DUCTS

The conditioned air flowing through ducts will gain heat from the surroundings. If the duct passes through conditioned spaces, the heat gain results in a useful cooling effect, but for the ducts passing through unconditioned spaces it is a loss of sensible heat that must be added to the BSCL. The heat

TABLE 6.18 DECREASE FROM PEAK DESIGN OUTDOOR DB TEMPERATURE, F

Daily Range, F	1	2	3	4	5	6	7	8	9	10	11	12	13	14	15	16	17	18	19	20	21	22	23	24
10	9	9	10	10	10	10	9	8	7	6	4	2	1	0	0	0	1	2	3	5	6	7	8	8
15	13	14	14	15	15	15	14	13	11	8	6	3	2	0	0	0	2	3	5	7	9	10	11	12
20	17	18	19	20	20	20	19	17	14	11	8	5	2	1	0	1	2	4	7	9	12	14	15	16
25	22	23	24	25	25	25	23	21	18	14	10	6	3	1	0	1	3	5	9	12	15	17	19	21
30	26	28	29	30	30	30	28	25	21	17	12	7	3	1	0	1	3	6	10	14	17	20	23	25
35	30	33	34	35	35	35	33	29	25	20	14	8	4	1	0	1	4	7	12	16	20	24	27	29

Reprinted with permission from the *1979 ASHRAE Load Calculation Manual.*

gain can be calculated from the heat transfer Equation 3.5:

$$Q = U \times A \times TD \qquad (3.5)$$

where

Q = duct heat gain, BTU/hr

U = overall coefficient of heat transfer, BTU/hr

A = duct surface area, ft^2

TD = temperature difference between air in duct and surrounding air, F

It is recommended that cold air ducts passing through unconditioned areas be insulated to at least an overall value of *R*-4 ($U = 0.25$).

Example 6.16

A 36 in. × 12 in. duct, 50 ft long, carrying air at 60 F, runs through a space at 90 F. The duct is insulated to an overall $U = 0.25$. What is the heat gain to the air in the duct?

Solution

The surface area of the duct is

$$A = \left(2 \times 36 \text{ in.} \times \frac{1 \text{ ft}}{12 \text{ in.}} + 2 \times 12 \text{ in.} \right.$$

$$\left. \times \frac{1 \text{ ft}}{12 \text{ in.}} \right) \times 50 \text{ ft} = 400 \text{ ft}^2$$

Using Equation 3.5,

$$Q = U \times A \times TD = 0.25 \times 400 \times (90 - 60)$$
$$= 3000 \text{ BTU/hr}$$

If there is significant heat gain to return air ducts, it should also be calculated, but it is only added to the CSCL, not the BSCL.

Although the heat gain to supply ducts in conditioned spaces is not wasted, care should be taken that it does not affect the distribution of cooling. If there is a long run of duct with a number of outlets, the heat gains in the first sections of duct might be enough so that the air temperature at the last outlets is too high. In this case, it might be useful to insulate the duct even though it is in the conditioned area (see Chapter 10).

Some designers find it reasonably accurate to add a percentage to the supply duct heat gain, rather than going through elaborate calculations. For insulated supply ducts, 1–3% of the building sensible load (BSCL) is suggested, depending on the extent of ductwork.

6.18   FAN AND PUMP HEAT

Some of the energy from the system fans and pumps is converted into heat through friction and other effects, and becomes part of the sensible heat gain that should be added to the load.

For a draw-through fan arrangement (fan downstream from the cooling coil) the heat is added to the BSCL, whereas for a blow-through arrangement (fan upstream from the coil) the heat gets added to the CSCL load. An approximate allowance for fan heat can be made as follows:

For 1 in. w.g. pressure add 2.5% to BSCL

For 2 in. w.g. pressure add 5% to BSCL

For 4 in. w.g. pressure add 10% to BSCL

The heat from the chilled water pump on small systems is generally small and may be neglected, but for large systems it may range from 1–2% of sensible load.

For central systems with remote chilled water cooling coils, the pump heat is a load on the refrigeration chiller, but not the cooling coil. This leads to a new term, the *refrigeration load.*

> The **refrigeration load (RL)** *is the load on the refrigeration equipment.*

For a direct expansion system, the refrigeration load and cooling coil load are equal. For a chilled water system, the refrigeration load is the cooling coil(s) load plus the chilled water pump heat.

6.19 DUCT AIR LEAKAGE

Duct systems will leak air at joints. Unfortunately, many systems have unnecessarily high air leakage due to sloppy installation. However, a careful job should limit duct leakage to 5% or less of the total CFM. If ducts are outside the conditioned space, the effect of leakage must be added to the BSCL and BLCL. If the air leaks into the conditioned space, then it does useful cooling, but care should be taken that it is not distributed to the wrong location.

6.20 SUPPLY AIR CONDITIONS

After the sensible and latent heat gains are calculated, the required supply air conditions (flow rate, temperature, and humidity) necessary to satisfy room conditions are determined. This subject is covered in Chapter 7.

6.21 SUMMARY OF COMMERCIAL COOLING LOAD CALCULATION PROCEDURES

The steps in determining commercial cooling loads can be summarized as follows.

1. Select indoor and outdoor design conditions from Tables 1.1 and A.9.
2. Use architectural plans to measure dimensions of all surfaces through which there will be external heat gains, for each room.
3. Calculate areas of all these surfaces.
4. Select heat transfer coefficient *U*-values for each element from appropriate tables, or calculate from individual *R*-values.
5. Determine time of day and month of peak load for each room by calculating external heat gains at times that they are expected to be a maximum. Search Tables 6.1, 6.2, 6.6, and 6.8 to find maximum values. Often calculations at a few different times will be required, but the suggestions in Section 6.13 should be helpful.
6. Calculate each room peak load, using the values for the external heat gains determined above and by calculating and adding the internal heat gains from people, lights, and equipment. The architect or building owner will furnish the data needed for the calculations.

   If there is infiltration, this must be added to the room load.
7. Find the time of building peak load using a similar search process as in item 5, and the suggestions in Section 6.14.
8. Calculate the building load at peak time, adding all external and internal gains and infiltration, if any. Add supply duct heat gain (Section 6.17), duct heat leakage (Section 6.19), and draw-through supply fan heat gain (Section 6.18), if significant.

9. Find the cooling coil and refrigeration load by adding the ventilation load (Table 6.17) to the building heat gains; adding blow-through fan, return air fan, and pump heat gains, if significant.

10. Calculate required supply air conditions (Chapter 7).

Example 6.17 will illustrate these procedures. The data and results are tabulated on a Commercial Cooling Load Calculation form (Figure 6.5), which should be carefully studied in relation to the explanations in the example.

Example 6.17

The Superb Supermarket, shown in Figure 6.6 on page 150, is located in Indianapolis, Indiana. It is a one-story building with a basement used for storage. Construction and conditions are as follows:

Roof is 4 in. h.w. concrete slab, 2 in. insulation, gypsum board ceiling, $U = 0.09$ BTU/hr-ft^2-F

Floor is 4 in. concrete slab, $U = 0.35$

Walls are 4 in. face brick, 4 in. common brick, 2 in. insulation, ½ in. gypsum wallboard, $U = 0.11$

Front window is ¼ in. single heat absorbing glass, 10 ft high, aluminum frame, not shaded

Doors are ¼ in. single clear glass, aluminum frame

Receiving door is 1½ in. steel with urethane core

Occupancy is 60 people

Construction is medium (M) weight

Lighting is 3 watts per square foot of floor area, fluorescent fixtures

Outdoor ventilation rate as per Table 6.17

Store is open from 10 AM to 8 PM (9 AM to 7 PM Standard Time)

Determine the required cooling load.

Solution

The procedures recommended previously will be followed.

1. Indoor and outdoor design conditions are 76 F DB/50% RH and 90 F DB/74 WB. Latitude is 39°N. Daily temperature range is 22 F. Inside and outdoor humidity ratios are 66 and 101 gr w./lb d.a.

2. Dimensions are shown on the plan.

3. Areas are calculated and recorded on the form.

4. U-values specified or found from tables are listed on the form.

5–7. The building has only one room. The construction and orientation indicates that the roof and West glass will determine the peak load time. Peak glass load is in both July and August, therefore either will be peak time of year. Peak CLF for the glass is at 5 PM. Peak CLTD for the roof is at 5 PM. Roof is No. 9. Therefore the peak load time is 5 PM. Conduction through the glass has not reduced appreciably, so this does not affect the results. Besides, wall loads will be higher at 5 PM than at 4 PM.

8. Individual heat gains items are calculated and recorded on the form. Students should see if they obtain the same values from the appropriate tables. The basement is assumed to be at the outside temperature. Walls are in Group B.

No storage effect for people or lighting is taken because the system is shut down when the store closes, and does not operate until shortly before the store opens.

No infiltration is included. Ventilation air is assumed to prevent any significant infiltration, because the doors are not used heavily.

The supply duct is exposed in the store, therefore heat gain and leakage are useful cooling and do not add to the load.

9. Ventilation loads are calculated and shown on the form. Return air fan gain is negligible, and there is no pump. A draw-through unit will be used, and it is assumed the fan gain is 3%.

10. The procedure for finding the supply air conditions will be explained in Chapter 7.

Another cooling load calculation example will be carried out as part of Example Project II (Chapter 17).

COMMERCIAL COOLING LOAD CALCULATIONS

Project __Superb Supermarket__ Bldg./Room __Building (peak)__

Location __Indianapolis, IN__ Engrs. __Energy Associates__ Calc. by __EP 8/7/96__ Chk. by __VL 8/10/96__

		DB F	WB F	RH %	W' gr/lb		
Design Conditions	Outdoor	90	74		101	Daily Range __22__ F	Ave. __79__ F
	Room	76		50	66	Day __July 21__	Time __5 PM (ST)__
						Lat. __39° N__	

Conduction	Dir.	Color	U	Gross	A, ft² Net	CLTD, F Table	Corr.	SCL BTU/hr
Glass	W		1.01		830	13	9	7540
	W		1.01		42	13	9	380
	E		1.01		42	13	9	380
Wall Group B	N	D	0.11		840	11	7	650
	S	D	0.11		840	17	12	1110
	E	D	0.11	1260	1176	26	22	2850
	W	D	0.11	1260	388	17	13	550
Roof/ceiling		D	0.09		5400	36	32	15,550
Floor			0.35		5400		14	26,460
Partition								
Door		E	0.39		42	27	23	380

Solar	Dir.	Sh.	SHGF	A	SC	CLF		
Glass	W	no	216	830	0.69	0.58		71,750
	W	no	216	42	0.94	0.58		4950
	E	no	216	42	0.94	0.22		1880

$RSHR = \dfrac{RSCL}{RTCL}$ $RSCL = \dfrac{RSCL}{1.1 \times (t_r - t_{sa})} =$ $CFM_{sa} =$

Lights __16,200__ W × 3.41 × __1.2__ BF × __1.0__ CLF → 66,290

Lights _____ W × 3.41 × _____ BF × _____ CLF → LCL BTU/hr

People __250__ SHG × __60__ n × __1.0__ CLF → 15,000

__200__ LHG × __60__ n → 12,000

Equipment __—__

Equipment __—__

Infiltration 1.1 × __—__ CFM × _____ TC

0.68 × __—__ CFM × _____ gr/lb

Subtotal → 215,720 | 12,000

SA duct gain __0__

SA duct leakage __0__

__0__

SA fan gain (draw through) __3%__ → 6470   Total CL BTU/hr

Room/Building Cooling Load → 222,190 | 12,000 | 234,190

SA fan gain (blow through) _____

Ventilation 1.1 × __1080__ CFM × __14__ TC → 16,630

0.68 × __1080__ CFM × __35__ gr/lb → 25,700

RA duct gain __0__

RA fan gain __0__

Cooling Coil Load → 238,820 | 37,700 | 276,520

Pump gain __—__

Refrigeration Load → 276,520

Figure 6.5

Commercial cooling load calculations form.

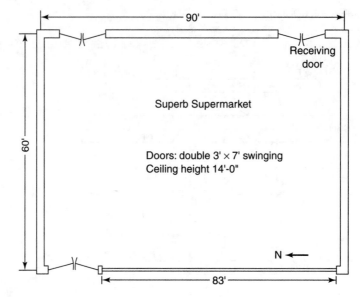

Figure 6.6
Sketch for Example 6.17.

Residential Cooling Loads

The procedures described previously are used for calculating cooling loads for commercial and industrial buildings. The procedures for determining cooling loads for residences are based on the same heat transfer principles, but are simplified somewhat. There are a number of reasons for this. Residential air conditioning equipment and controls usually do not have refined provisions for zoning, humidity control, and part load operation. Homes are often conditioned 24 hours a day. These factors all lead to a simplification of load calculations.

Only sensible loads are calculated. An allowance is made for latent loads, and lighting loads are neglected. Approximations are used for people and infiltration loads. The procedure does not require determination of peak time of load or of heat storage effect, this being included in the data.

6.22   COOLING LOAD FROM HEAT GAIN THROUGH STRUCTURE

The cooling loads from walls, roof, ceiling, and floor are each calculated by use of the following equation

$$Q = U \times A \times \text{CLTD} \qquad (6.10)$$

where

Q = sensible cooling load, BTU/hr

U = overall heat transfer coefficient, BTU/hr-ft^2-F

A = area, ft^2

CLTD = cooling load temperature difference, F

The CLTD values are listed in Table 6.19. The L (low), M (medium), and H (high) outdoor temperature ranges are listed in the footnotes to the table; these are found from Table A.6.

TABLE 6.19 CLTD VALUES FOR SINGLE-FAMILY DETACHED RESIDENCES[a]

| | Design Temperature, °F | | | | | | | | | | | |
| | 85 | | 90 | | | 95 | | | 100 | | 105 | 110 |
Daily Temp. Range[b]	L	M	L	M	H	L	M	H	M	H	M	H
All walls and doors												
North	8	3	13	8	3	18	13	8	18	13	18	23
NE and NW	14	9	19	14	9	24	19	14	24	19	24	29
East and West	18	13	23	18	13	28	23	18	28	23	28	33
SE and SW	16	11	21	16	11	26	21	16	26	21	26	31
South	11	6	16	11	6	21	16	11	21	16	21	26
Roofs and ceilings												
Attic or flat built-up	42	37	47	42	37	51	47	42	51	47	51	56
Floors and ceilings												
Under conditioned space, over unconditioned room, over crawl space	9	4	12	9	4	14	12	9	14	12	14	19
Partitions												
Inside or shaded	9	4	12	9	4	14	12	9	14	12	14	19

[a]Cooling load temperature differences (CLTDs) for single-family detached houses, duplexes, or multifamily, with both east and west exposed walls or only north and south exposed walls, °F.
[b]L denotes low daily range, less than 16 °F; M denotes medium daily range, 16 to 25 °F; and H denotes high daily range, greater than 25 °F.
Reprinted with permission from the *1993 ASHRAE Handbook—Fundamentals.*

The CLTD table is based on an indoor temperature of 75 F. For other indoor temperatures, the CLTD should be corrected by 1 F for each 1 F temperature difference from 75 F.

The CLTD values should also be interpolated between the listed outdoor temperature values.

Colors of all exposed surfaces are assumed dark.

Example 6.18

A home has a roof area of 1600 ft^2. The inside design condition is 78 F; outdoor design condition is 90 F. The outdoor daily temperature range is 20 F. The combined roof-ceiling U-factor is 0.09. Find the roof cooling load.

Solution
Equation 6.10 will be used. The CLTD will be found from Table 6.19, correcting it for the inside

design temperature of 78 F. The outdoor temperature range falls in the *M* class.

$$CLTD = 42 - (78 - 75) = 39 \text{ F}$$
$$Q = 0.09 \times 1600 \times 39 = 5620 \text{ BTU/hr}$$

6.23 COOLING LOAD FROM HEAT GAIN THROUGH WINDOWS

The sensible cooling load due to heat gains through glass (windows and doors) are found by using glass load factors (GLF). These are listed in Table 6.20. The GLF values account for both solar radiation and conduction through glass. Values should be interpolated between listed outdoor temperatures.

The glass sensible cooling load is determined from Equation 6.11:

$$Q = A \times GLF \qquad (6.11)$$

where

Q = sensible cooling load due to heat gain through glass, BTU/hr

A = area of glass, ft^2

GLF = glass load factor, BTU/hr-ft^2

Example 6.19

A residence has 80 ft^2 of regular single glass on the west side, with draperies. The outdoor design temperature is 95 F. Find the cooling load through the glass.

Solution

From Table 6.20, the CLF is 50 BTU/hr-ft^2, and the window cooling load is, using Equation 6.11,

$$Q = 50 \text{ BTU/hr-ft}^2 \times 80 \text{ ft}^2 = 4000 \text{ BTU/hr}$$

If the glass is shaded by permanent outside overhangs, the calculation is carried out differently. First, the extent of shading is determined. This can be done with the aid of Table 6.21 on page 154. The shade line factors listed in the table are multiplied by the width of the overhang to find the vertical length of shading. For that part of the glass which is shaded, the values from Table 6.20, for north facing glass are used.

No shade line factors for northwest and northeast are shown, because it is not feasible to shade those orientations with overhangs.

Example 6.20

A south wall has a 6 ft high picture window with a roof overhang as shown in Figure 6.7. The house is in Savannah, Georgia. How much of the glass area is shaded?

Solution

Savannah is at 32°N latitude. From Table 6.21, the shade line factor is 5.0. The vertical length of shade is 2 × 5.0 = 10.0 ft. The bottom of the glass is 8 ft below the overhang, so the glass is completely

shaded! Note how effective the overhang is. Look what would happen if the wall faced southwest, however. The shade extends vertically 1.6 × 2 = 3.2 ft, and barely covers one foot of glass. The orientation and design of the building can have a major effect on energy use!

6.24 PEOPLE AND APPLIANCES

The sensible heat gain per person is assumed to be an average of 225 BTU/hr. If the number of occupants is not known in advance, it can be estimated as two times the number of bedrooms. Because the maximum load usually occurs in late afternoon, it is usual to assume that the occupants are in living and dining areas for purposes of load distribution.

A sensible heat gain allowance of 1200–1600 BTU/hr is typical for kitchen appliances. If the kitchen is open to an adjacent room, 50% of this load should be assigned to that room. If large special appliances are used, their output should be individually evaluated.

6.25 INFILTRATION AND VENTILATION

Infiltration rates are listed in Table 6.22 on page 154 in air changes per hour (ACH). Three categories of construction tightness are shown, described as follows:

Tight. Well-fitted windows and doors, weatherstripping, no fireplace.

Figure 6.7
Sketch for Example 6.20.

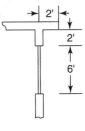

TABLE 6.20 WINDOW GLASS LOAD FACTORS (GLF) FOR SINGLE-FAMILY DETACHED RESIDENCES

Design Temperature, °F	Regular Single Glass						Regular Double Glass						Heat-Absorbing Double Glass						Clear Triple Glass		
	85	90	95	100	105	110	85	90	95	100	105	110	85	90	95	100	105	110	85	90	95
No inside shading																					
North	34	36	41	47	48	50	30	30	34	37	38	41	20	20	23	25	26	28	27	27	30
NE and NW	63	65	70	75	77	83	55	56	59	62	63	66	36	37	39	42	44	44	50	50	53
E and W	88	90	95	100	102	107	77	78	81	84	85	88	51	51	54	56	59	59	70	70	73
SE and SW[b]	79	81	86	91	92	98	69	70	73	76	77	80	45	46	49	51	54	54	62	63	65
South[b]	53	55	60	65	67	72	46	47	50	53	54	57	31	31	34	36	39	39	42	42	45
Horizontal skylight	156		161	166	167	171	137	138	140	143	144	147	90	91	93	95	96	98	124	125	127
Draperies, venetian blinds, translucent roller shades fully drawn																					
North	18	19	23	27	29	33	16	16	19	22	23	26	13	14	16	18	19	21	15	16	18
NE and NW	32	33	38	42	43	47	29	30	32	35	36	39	24	24	27	29	29	32	28	28	30
E and W	45	46	50	54	55	59	40	41	44	46	47	50	33	33	36	38	38	41	39	39	41
SE and SW[b]	40	41	46	49	51	55	36	37	39	42	43	46	29	30	32	34	35	37	35	36	38
South[b]	27	28	33	37	38	42	24	25	28	31	31	34	20	21	23	25	26	28	23	24	26
Horizontal skylight	78	79	83	86	87	90	71	71	74	76	77	79	58	59	61	63	63	65	69	69	71
Opaque roller shades fully drawn																					
North	14	15	20	23	25	29	13	14	17	19	20	23	12	12	15	17	17	20	13	13	15
NE and NW	25	26	31	34	36	40	23	24	27	30	30	33	21	22	24	26	27	29	23	23	26
E and W	34	36	40	44	45	49	32	33	36	38	39	42	29	30	32	34	35	37	32	32	35
SE and SW[b]	31	32	36	40	42	46	29	30	33	35	36	39	26	27	29	31	32	34	29	29	31
South[b]	21	22	27	30	32	36	20	20	23	26	27	30	18	19	21	23	24	26	19	20	22
Horizontal skylight	60	61	64	68	69	72	57	57	60	62	63	65	52	52	55	57	57	59	56	57	59

[a]Glass load factors (GLFs) for single-family detached houses, duplexes, or multi-family, with both east and west exposed walls or only north and south exposed walls, Btu/h • ft².

[b]Correct by +30% for latitude of 48° and by −30% for latitude of 32°. Use linear interpolation for latitude from 40 to 48° and from 40 to 32°.

To obtain GLF for other combinations of glass and/or inside shading: $GLF_a = (SC_a/SC_t)(GLF_t - U_tD_t) + U_aD_a$, where the subscripts a and t refer to the alternate and table values, respectively. SC_t and U_t are given in Table 5. $D_t = (t_a - 75)$, where $t_a = t_o - (DR/2)$; t_o is the outdoor design temperature and DR is the daily range.

Reprinted with permission from the *1993 ASHRAE Handbook—Fundamentals*.

TABLE 6.21   SHADE LINE FACTORS (SLF)

Direction Window Faces	Latitude, Degrees N						
	24	32	36	40	44	48	52
East	0.8	0.8	0.8	0.8	0.8	0.8	0.8
SE	1.8	1.6	1.4	1.3	1.1	1.0	0.9
South	9.2	5.0	3.4	2.6	2.1	1.8	1.5
SW	1.8	1.6	1.4	1.3	1.1	1.0	0.9
West	0.8	0.8	0.8	0.8	0.8	0.8	0.8

Shadow length below the overhang equals the shade line factor times the overhang width. Values are averages for the 5 h of greatest solar intensity on August 1. Reprinted with permission from the *1993 ASHRAE Handbook—Fundamentals.*

TABLE 6.22   AIR CHANGE RATES AS A FUNCTION OF OUTDOOR DESIGN TEMPERATURES

Class	Outdoor Design Temperature, °F					
	85	90	95	100	105	110
Tight	0.33	0.34	0.35	0.36	0.37	0.38
Medium	0.46	0.48	0.50	0.52	0.54	0.56
Loose	0.68	0.70	0.72	0.74	0.76	0.78

Values for 7.5 mph wind and indoor temperature of 75°F. Reprinted with permission from the *1993 ASHRAE Handbook—Fundamentals.*

Medium. Average fit windows and doors, fireplace that can be closed off.

Loose. Poorly fitted windows and doors, fireplace without shut-off.

The quantity of air infiltrating into the room is found from Equation 3.12:

$$\text{CFM} = \text{ACH} \times \frac{V}{60} \qquad (3.12)$$

where

CFM = air infiltration rate into room, CFM

ACH = number of air changes per hour (Table 6.22)

V = room volume, ft^3

The heat gain due to the infiltrating air is found from Equation 3.10:

$$Q = 1.1 \times \text{CFM} \times TC \qquad (3.10)$$

where

Q = sensible cooling load due to infiltrating air

CFM = from Equation 3.12

TC = temperature change between inside and outdoor air

If the infiltration air is expected to be less than 0.5 ACH, indoor air quality may be unsatisfactory. In this case, some outdoor air should be introduced through the air conditioning equipment, with its sensible heat contribution evaluated from Equation 3.10.

6.26   ROOM, BUILDING, AND AIR CONDITIONING EQUIPMENT LOADS

Room Sensible Cooling Load. The sensible cooling load for each room (RSCL) is found by adding up each of the room's cooling load components described.

Building Sensible Cooling Load. The building sensible cooling load (BSCL) is found by adding up the room sensible cooling loads for each room.

It still remains to find the air conditioning equipment cooling load. To do this, the duct heat gains and leakage and the latent heat gain must be accounted for.

Duct Heat Gains. Suggested values for heat gains to ducts are:

Ducts in attics: add 10% to the building sensible cooling load

Ducts in crawl space or basement: add 5% to the building sensible cooling load

Duct Leakage. An additional 5% is suggested to be added to the building sensible cooling load due to leakage of air from the ducts.

Equipment Sensible Cooling Load. This is the sum of the building sensible cooling load and the duct heat gains and leakage.

Latent Cooling Load. The latent loads are not separately calculated when using the abbreviated residential calculation procedure. Instead the building sensible load is multiplied by an approximated latent factor (LF) to obtain the building total load.

Figure 6.8 is used to find the LF value, using the outdoor design humidity ratio from the psychrometric chart (see Chapter 7).

The equipment total cooling load is then found from the following equation:

$$Q_T = Q_S \times \text{LF} \qquad (6.12)$$

where

Q_T = equipment total cooling load, BTU/hr

Q_S = equipment sensible cooling load, BTU/hr

LF = latent factor (Figure 6.8)

The air conditioning unit is then selected on the basis of the calculated equipment total cooling load. The unitary (packaged) equipment used in residential work may not have quite the sensible and latent heat proportion removal capacities desired, but it is rare that the resulting room conditions are in an uncomfortable range. If the loads are such that this is suspected, a more detailed analysis is necessary.

Figure 6.8

Effect of infiltration on latent load factor. (Reprinted with permission from the *1993 ASHRAE Handbook—Fundamentals.*)

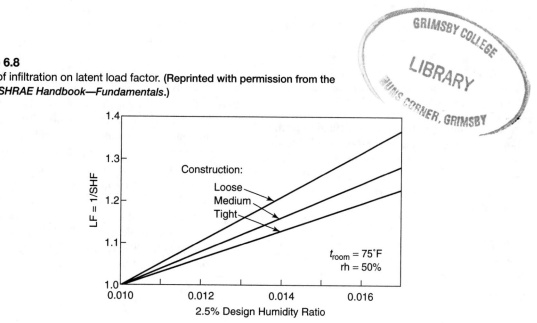

6.27 SUMMARY OF RESIDENTIAL COOLING LOAD CALCULATION PROCEDURES

The steps in determining residential cooling loads can be summarized as follows.

1. Select inside and outdoor design temperatures from Tables 1.1 and A.9.
2. Use architectural plans to measure dimensions of all surfaces through which there will be external heat gain, for each room.
3. Calculate areas of all these surfaces.
4. Select heat transfer coefficient U each element from Tables A.7 or A.8, or calculate from individual R-values.
5. Calculate heat gains through walls, roof, and floors for each room using the CLTD values from Table 6.19.
6. Calculate heat gains through glass, using Tables 6.20 and 6.21, for each room.
7. Determine occupancy and appliance load (Section 6.23).
8. Determine infiltration or ventilation load (Section 6.25).
9. Add individual loads to find sensible load for each room and building.
10. Add duct heat gains and leakage to SCL of building.
11. Multiply the building sensible cooling load by the latent factor LF (Figure 6.8). The result is the air conditioning equipment total cooling load.

A Residential Cooling Load Calculation form is shown in Figure 6.9 and in the Appendix. The following example illustrates use of the calculation procedure. The student should go through each step independently to confirm agreement with the values shown in Figure 6.9 on page 158.

Example 6.21 _____

Calculate the room and building cooling loads for the residence shown in Figure 6.10 on page 158.

Solution

The steps will be carried out as recommended in the summary. The results of each step are shown in Figure 6.9.

1. The indoor and outdoor design temperatures are 75 F and 96 F.
2–3. The dimensions are taken from the building plans and the gross and net areas of each element are calculated and recorded in Figure 6.9. Note that large closets in a room are included as part of the room. The hallway is included as part of the living room because there is no separating door.
4. Heat transfer coefficients for the materials listed are found from Tables A.7 and A.8 and recorded.
5. Select the CLTD values from Table 6.19. Outdoor temperature range is 22 F, in the M class. The wall heat gains in each room are then calculated using Equation 6.10, and recorded in Figure 6.9. Other elements are calculated in the same way.
6. From Table 6.20 for the windows on the south side, the CLF for the type of glass and shading is 28 BTU/hr-ft^2 at 95 F outdoors and 75 F indoors. The heat gains are calculated and recorded for this and all other windows.
7. For a two-bedroom house, assume an occupancy of four: two in the living room, two in the dining area at peak load times. Assume a 1200 BTU/hr kitchen appliance load.
8. Infiltration is found from Table 6.22 and Equation 3.12. (If air quality were poor from too little infiltration, some outside air would have to be mechanically introduced, adding to the load.)
9. The individual gains are added to find the RSCL for each room and the building.
10. The duct system is in the basement. Allow 5% for heat gain and 5% for leakage to add to the building sensible cooling load (BSCL).
11. Multiply the BSCL by the LF factor (Figure 6.8) to find the total load.

RESIDENTIAL COOLING   Project **A. Jones Residence**   Out. DB **96** F   In. DB **75** F   D.R. **22** F
LOAD CALCULATIONS   Location **Little Rock, AR**   Out. WB **77** F   In. RH **50** %   ACH **0.5**

Room Name		Living Room					Dining Room					Bedroom No. 1					Bedroom No.2			
Plan Size		21 × 12 + 21 × 4					9 × 10					14 × 12 + 3 × 4					10 × 11 + 3 × 4			
	D.	U	A	CLTD	BTU/hr	D.	U	A	CLTD	BTU/hr	D.	U	A	CLTD	BTU/hr	D.	U	A	CLTD	BTU/hr
Wall	S	.20	104	16	333	N	.20	58	13	151	W	.20	106	23	488	W	.20	74	23	340
	E	.20	128	23	589	E	.20	66	23	304	S	.20	82	16	262	N	.20	74	13	192
Roof/ceiling		.10	336	47	1579		.10	90	47	423		.10	180	47	846		.10	122	47	573
Floor																				
Partition																				
Door	S	.47	21	16	158															
		D.		CLF			D.		CLF			D.		CLF			D.		CLF	
		S	40	28	1120		N	14	23	322		W	14	50	700		W	14	50	700
Windows							E	14	50	700		S	14	28	392		N	14	23	322
Infiltration					443					119					238					161
People		2 × 225			450		2 × 275			550										
Appliances																				
RSCL					4672					2569					2926					2288

Room Name		Kitchen					Bathroom													
Plan Size		9 × 10					6 × 7													
	D.	U	A	CLTD	BTU/hr	D.	U	A	CLTD	BTU/hr	D.	U	A	CLTD	BTU/hr	D.	U	A	CLTD	BTU/hr
Wall	N	.20	37	13	96	N	.20	34	13	88										
Roof/ceiling		.10	90	47	423		.10	42	47	197										
Floor																				
Partition																				
Door	N	.47	21	13	128															
		D.		CLF			D.		CLF			D.		CLF			D.		CLF	
		N	14	23	322		N	14	23	322										
Windows																				
Infiltration					119					55										
People																				
Appliances					1200															
RSCL					2288					662										

Building Total
Sum RSCL = **15,405**
Duct gain **5** % = **770**
Duct leak **5** % = **770**
BSCL = **16,945**
BTCL = **1.25** × BSCL = **21,181** BTU/hr
Unit size = **2 tons**

NOTES: Ceiling ht. = 8'0". Single clear glass, blinds.
$U_w = 0.20$, $U_r = 0.10$, $U_d = 0.47$.
Windows 3.5' W × 4' H, except as noted.
Doors 3' × 7'.

Figure 6.9
Residential cooling load calculations form.

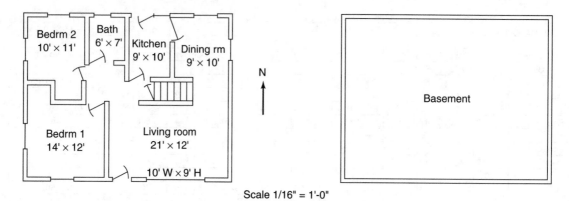

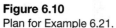

Scale 1/16" = 1'-0"

Figure 6.10
Plan for Example 6.21.

6.28 ENERGY CONSERVATION

Reducing the building cooling load provides a major opportunity for energy conservation. Some ways this can be achieved are:

1. Use high *R*-value insulation throughout the building.
2. Use the 2.5% summer outdoor design DB temperature and coincident WB temperature.
3. Use inside design DB temperatures of 78–80 F. These provide adequate comfort for most applications. Past practice of designing for 75 F or even lower is wasteful.
4. Consider use of heat-absorbing glass.
5. Provide effective interior shading devices.
6. Minimize use of glass in building unless used on the south side for receiving solar heat in the winter.
7. Consider outside construction features that provide shading of glass.
8. Orient the building so that solar radiation in summer is minimum on sides with large glass areas.
9. Avoid unnecessarily excessive lighting levels.
10. Use types of lighting that more efficiently convert electrical energy into light.
11. Above all, use proper calculation procedures that account for heat storage and time lag.

A more detailed discussion of energy conservation in HVAC system design will be presented in Chapter 15.

PROBLEMS

6.1 A building with a 120 ft × 80 ft roof, located in Cincinnati, Ohio, has a roof constructed of 1 in. wood with *R*-5.5 insulation and a suspended ceiling. The inside design condition is 78 F. Determine the net roof cooling load at

A. September 21 at noon

B. Time of peak roof heat gain

6.2 A southeast facing wall of a building located in Las Vegas, Nevada, is 90 ft × 24 ft. The wall is constructed of 8 in. concrete, *R*-5 insulation and ½ in. gypsum wallboard. The inside design condition is 77 F. Determine the cooling load through the wall at

A. June 15 at 11 AM

B. Time of peak wall heat gain

6.3 A building in Baltimore, Maryland, has 2300 ft² of exterior single glass with no interior shading. The inside design condition is 78 F.

Determine the net conduction heat gain through the glass at 2 PM in summer.

6.4 A building in Dallas, Texas, has 490 ft² of windows facing west, made of ¼ in. single clear glass, with medium color interior venetian blinds. The building is of light construction. Find the maximum net solar cooling load through the windows. In what month and hour is this?

6.5 A room in a building in New York City has a 12 ft W by 6 ft H window facing south. The building is of mediumweight construction. The window is ¼ in. single clear glass, with dark roller shades. There is a 3 ft outside projection at the top of the window. Find the solar cooling load through the window at 12 noon on July 1.

6.6 A room has four 40 W fluorescent lighting fixtures and two 200 W incandescent fixtures in use. The cooling system is shut down during unoccupied hours. What is the cooling load from the lighting?

6.7 Find the sensible and latent load from 180 people dancing in the Get Down Disco. The temperature is 78 F.

6.8 The Squidgit factory, which is air conditioned 24 hours a day, operates from 8 AM to 5 PM. It has 76 (male and female) employees doing light bench work. What is the cooling load at 1 PM? The temperature is 78 F.

6.9 The Greasy Spoon Cafe has a 20 ft² steam table, without a hood. What are the sensible and latent heat gains?

6.10 Find the peak cooling load for the general office shown in Figure 6.11, with the following conditions:

Location: Sacramento, California. Inside conditions 78 F DB, 50% RH.

Wall: U = 0.28 BTU/hr-ft²-F, Group E.

Window: 20 ft W × 6 ft H single clear glass, dark interior blinds.

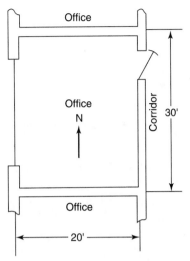

Note: lightweight construction

Figure 6.11
Plan for Problem 6.10.

Occupancy: 10 people. Lights 4 W/ft² with ballast.

Floor-to-floor height: 10 ft.

6.11 The building shown in Figure 6.12, located in Ottawa, Canada, has the following conditions:

Figure 6.12
Plan for Problem 6.11.

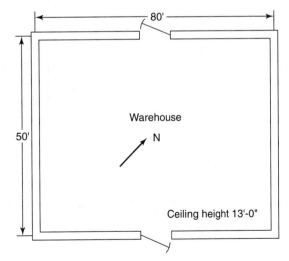

Doors: *7 ft H × 3 ft W, 1½ in. wood.*

Walls: U = 0.18 BTU/hr-ft²-F, Group C. *No windows. Light color.*

Roof: *4 in. h.w. concrete, R-5.5 insulation, finished ceiling.*

Inside design conditions: *77 F, 55% RH.*

Occupancy: *80 ft²/person. Lighting is 2.5 W/ft².*

Determine the peak cooling load and the load at 11 AM June 30.

6.12 Repeat Problem 6.12 for the building turned 45° clockwise, and 90° clockwise, and located in or near your community.

6.13 A room in a building in Memphis, Tennessee, has one exposed wall, facing east, with the following conditions:

Wall: A = 68 ft², U = 0.21 BTU/hr-ft²-F, *Group E.*

Glass: A = 130 ft², *single heat-absorbing glass, no shading.*

Lightweight construction. Room temperature = 78 F. Find the peak cooling load.

6.14 The Beetle Concert Hall in London, England, seats 2300 people. Inside design conditions are 75 F and 50% RH. Calculate the ventilation loads in summer.

Figure 6.13
Plan for Problem 6.15.

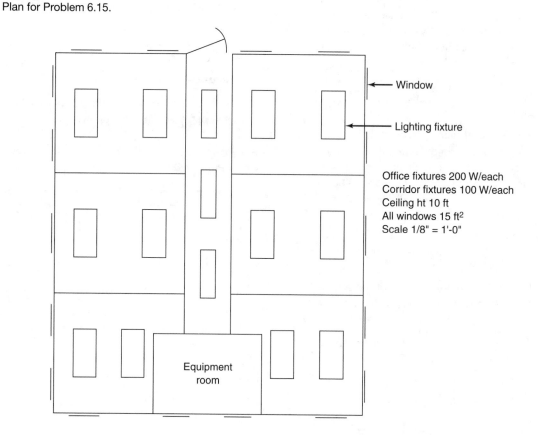

Window

Lighting fixture

Office fixtures 200 W/each
Corridor fixtures 100 W/each
Ceiling ht 10 ft
All windows 15 ft²
Scale 1/8" = 1'-0"

Equipment room

6.15 Perform a complete cooling load calculation for the one-story office building shown in Figure 6.13. Conditions are as follows:

Location: *Your town.*

Walls: U = *0.20 BTU/hr-ft^2-F. Group C.*
Glass: *Single heat-absorbing glass, light interior blinds.*

Roof: *2.5 in. wood, R-2.8 insulation.*

Orientation as assigned by instructor. Light construction. People: 1 per 60 ft^2. Make assumptions based on recommendations in the text on all other data (ventilation, duct and fan heat gains, and so on).

6.16 Perform a complete cooling load calculation for the residence described in Problem 3.20, to be located in your town.

CHAPTER 7
Psychrometrics

The atmospheric air that surrounds us is a mixture of **dry air** and **water vapor**, called **moist air**. Because this gas mixture is conditioned in environmental control systems, it is necessary to understand how it behaves. **Psychrometrics** is the name given to the study of air-water vapor mixtures. Hereafter, as conventionally done, we will use the word *air* to refer to the air-water vapor mixture that is the atmosphere.

In this chapter, we will first learn how to determine the physical properties of air, and then examine how air is processed in air conditioning systems. Because the water vapor content in air can change, these processes can be complex and their understanding may require a special effort by the student. Some comprehension of psychrometrics is an absolute necessity, however, in order to become a competent air conditioning practitioner. A psychrometric analysis is required in selecting the proper air conditioning equipment for a job, and in troubleshooting systems that are not performing properly. Examples of these uses will be demonstrated as we proceed.

OBJECTIVES

After studying this chapter, you will be able to:
1. Read values of properties from the psychrometric chart.
2. Determine sensible and latent heat changes in air conditioning equipment.
3. Determine mixed air conditions.
4. Determine required supply air conditions.
5. Determine cooling coil performance specifications.
6. Determine reheat requirements.

7.1 PROPERTIES OF AIR

The physical properties of atmospheric air are defined as follows:

Dry Bulb Temperature (DB). The temperature of the air, as sensed by a thermometer. The words *temperature* and *dry bulb temperature* will be used to mean the same thing with regards to air.

Wet Bulb Temperature (WB). The temperature sensed by a thermometer whose bulb is wrapped with a water-soaked wick, in rapidly moving air.

Dew Point Temperature (DP). The temperature at which the water vapor in the air would begin to condense if the air were cooled, at constant pressure.

Humidity Ratio (W). This is also called the moisture content. The weight of water vapor per pound of dry air, in lb/lb dry air, or grains/lb dry air.

Relative Humidity (RH). The ratio of the actual water vapor pressure in the air to the vapor pressure if the air were saturated at that dry bulb temperature. It is expressed in percent.

Specific Volume (v). The volume of air per unit weight of dry air, in ft³/lb dry air.

Specific Enthalpy (h). The heat content of air per unit weight, in BTU/lb dry air.

The dry bulb temperature is the temperature in the conventional meaning that we are all accustomed to using. The meaning and use of *wet bulb temperature* will be explained when the process of evaporative cooling is described.

Note that the *specific* properties, those based on a unit weight, always refer to unit weight of dry air, although the air ordinarily is not dry. This is a convention that is generally used.

The weight unit of *grains* is often used in order to have more convenient numbers:

$$7000 \text{ gr} = 1 \text{ lb}$$

The specific enthalpy of air is the enthalpy of the dry air plus the water vapor, taken above an arbitrary reference temperature at which the enthalpy is given a value of zero. In most data used in air conditioning, the arbitrary zero value point is 0 F for the dry air portion and 32 F for the water vapor portion of the air.

The term *saturated air* is used to describe the condition when air contains the maximum amount of water vapor that it can hold. When the amount of water vapor is less than the maximum possible, the condition is called *unsaturated*. The maximum quantity of water vapor that air can hold depends on the air temperature. This use of the word *saturated* does not have the same meaning as the *saturated state* of a pure substance, explained in Chapter 3.

7.2 DETERMINING AIR PROPERTIES

It is necessary to determine many of the physical properties of atmospheric air for air conditioning work. Some of the relationships between properties will now be explained.

Ideal Gas Laws

Both the dry air and water vapor in atmospheric air behave as ideal gases, and therefore Equation 2.15 applies to each:

$$p_a V = m_a R_a T = 53.3 \, m_a T \qquad (7.1)$$
$$p_w V = m_w R_w T = 85.7 \, m_w T \qquad (7.2)$$

where

p_a = partial pressure of dry air in the mixture, lb/ft²

p_w = partial pressure of water vapor in the mixture, lb/ft²

m_a = weight of dry air, lb

m_w = weight of water vapor, lb

R_a, R_w = gas constants for air and water vapor, ft lb/lb R

V = volume of mixture, ft³

T = temperature of mixture, R

The dry air and water vapor each exert only a part of the total pressure, but each occupies the total volume. A useful principle that applies to the mixture, called *Dalton's Law*, is: *the total pressure equals the sum of the partial pressures:*

$$p = p_a + p_w \qquad (7.3)$$

where

p = total (atmospheric) pressure

p_a = partial pressure of dry air

p_w = partial pressure of water vapor

Humidity Ratio

The ideal gas laws and law of partial pressures can be used to find a relationship for determining the humidity ratio. The definition of humidity ratio expressed as an equation is

$$W = \frac{m_w}{m_a} \qquad (7.4)$$

where

W = humidity ratio, lb water vapor/lb dry air

m_w = weight of water vapor, lb

m_a = weight of dry air, lb

Rearranging the ideal gas law as follows, for both air and water vapor,

$$m_w = \frac{p_w V}{85.7T}$$

$$m_a = \frac{p_a V}{53.3T}$$

Dividing the first equation by the second results in a useful relationship for the humidity ratio:

$$W = \frac{m_w}{m_a} = 0.622\,\frac{p_w}{p_a} \qquad (7.5)$$

In this equation, p_a and p_w are in the same units.

Example 7.1

The partial pressure of the water vapor in the air is 0.20 psia on a day when the barometric (atmospheric) pressure is 14.69 psi. Find the humidity ratio.

Solution

From the law of partial pressures, Equation 7.3,

$$p_a = p - p_w = 14.69 - 0.20 = 14.49 \text{ psia}$$

Using Equation 7.5,

$$W = 0.622\,\frac{p_w}{p_a} = 0.622 \times \frac{0.20}{14.49}$$
$$= 0.0086 \text{ lb w./lb d.a.}$$

or

$$W' = \frac{7000 \text{ gr}}{1 \text{ lb}} \times 0.0086$$
$$= 60.2 \text{ gr w./lb d.a.}$$

Relative Humidity and Dew Point

The relative humidity is defined by the equation

$$\text{RH} = \frac{p_w}{p_{ws}} \times 100 \qquad (7.6)$$

where

RH = relative humidity, %

p_w = partial pressure of water vapor at dry bulb temperature

p_{ws} = saturation pressure of water vapor at dry bulb temperature

The saturation pressure of the water vapor is found from the Steam Tables (Table A.3), at the dry bulb temperature.

The dew point temperature was defined as the temperature at which the water vapor in the air would condense if the air were cooled at constant pressure. Therefore, the water vapor is in a saturated condition at the dew point, and its partial pressure is equal to the saturation pressure at the dew point. Example 7. 2 illustrates this.

Example 7.2

What is the relative humidity and humidity ratio of air at 80 F DB and 50 F DP? The barometric pressure is 14.7 psi.

Solution

Using Table A.3 to find the saturation pressure and partial pressure of the water vapor,

$$\text{at 80 F } p_{ws} = 0.507 \text{ psia}$$
$$\text{at 50 F } p_w = 0.178 \text{ psia}$$

Using Equation 7.6,

$$\text{RH} = \frac{p_w}{p_{ws}} \times 100 = \frac{0.178}{0.507} \times 100 = 35\%$$

Using Equation 7.5, where

$$p_a = p - p_w = 14.7 - 0.178 = 14.52 \text{ psia}$$

$$W = 0.622\,\frac{p_w}{p_a}$$

$$= 0.622 \times \frac{0.178}{14.52}$$

$$= 0.0076 \text{ lb w./lb d.a.}$$

Example 7.3

Find the specific volume (per lb dry air) of the air in Example 7.2.

Solution
The ideal gas law will be used. Rearranging Equation 7.1, remembering that p_a must be in lb/ft^2,

$$V = \frac{53.3 \, m_a T}{p_a}$$

$$= \frac{53.3 \times 1 \times (80 + 460)}{14.52 \times 144}$$

$$= 13.8 \text{ ft}^3/\text{lb d.a.}$$

As the dry air and the water vapor both occupy the same volume, Equation 7.2 could also be used. Note that the weight of water vapor per lb dry air is the humidity ratio, by definition.

$$V = \frac{85.7 \, m_w T}{p_w}$$

$$= \frac{85.7 \times 0.0076 \times (80 + 460)}{0.178 \times 144}$$

$$= 13.9 \text{ ft}^3/\text{lb d.a.}$$

Enthalpy (Heat Content)

The enthalpy of atmospheric air is the sum of the individual enthalpies of the dry air and water vapor. This includes the sensible heat of the dry air and the sensible and latent heat of the water vapor. Using specific heats of air and water vapor of 0.24 and 0.45 BTU/lb-F, and a latent heat value of 1061 BTU/lb for water, the equation for the specific enthalpy of the mixture, per pound of dry air, is

$$h = 0.24t + W(1061 + 0.45t) \tag{7.7}$$

where

h = enthalpy of moist air, BTU/lb d.a.

t = dry bulb temperature of air, F

W = humidity ratio, lb w./lb d.a.

Example 7.4

Find the specific enthalpy of the air in Example 7.2.

Solution
Using Equation 7.7,

$$h = 0.24t + W(1061 + 0.45t)$$

$$= 0.24 \times 80 + 0.0078(1061 + 0.45 \times 80)$$

$$= 27.8 \text{ BTU/lb d.a.}$$

In Example 7.2, the dry bulb temperature (DB) and dew point temperature (DP) were known. The value of p_w was found at the DP temperature, using Table A.3.

Generally, however, it is the wet bulb (WB) temperature and not the DP that is measured. This is partly because the apparatus for measuring the DP is cumbersome, whereas only a thermometer with a wetted wick is needed to measure the WB.

If the WB is known, the value of p_w can be calculated from the following equation, which was developed by Dr. Willis H. Carrier, a pioneer in the field of air conditioning:

$$p_w = p' - \frac{(p - p')(\text{DB} - \text{WB})}{(2830 - 1.43 \times \text{WB})} \tag{7.8}$$

where

p_w = partial pressure of water vapor at dry bulb temperature, psia

p = total air pressure, psia

p' = saturation water vapor pressure at the wet bulb temperature, psia

DB = dry bulb temperature, F

WB = wet bulb temperature, F

The following example illustrates the use of Equation 7.8.

Example 7.5

A measurement of the air DB and WB gives readings of 90 F and 70 F, respectively. The atmospheric pressure is 14.70 psia. What is the RH?

Solution
Equation 7.8 will be used to find p_w, then Equation 7.6 to find the RH.

From Table A.3,

$$\text{at } 70 \text{ F}, p' = 0.363 \text{ psia}$$

$$\text{at } 90 \text{ F}, p_{ws} = 0.698 \text{ psia}$$

From Equation 7.8,

$$p_w = 0.363 - \frac{(14.70 - 0.36)(90 - 70)}{(2830 - 1.43 \times 70)}$$

$$= 0.258 \text{ psi}$$

From Equation 7.6,

$$\text{RH} = p_w/p_{ws} \times 100$$

$$= 0.258/0.698 \times 100$$

$$= 37\%$$

As illustrated by the examples shown in this section, the above equations may be used to determine other unknown properties, after measuring the DB and WB or other conveniently measured properties.

To save repeated calculations, charts have been prepared from the most common values of the properties of air. Their use will be the subject of the next section.

7.3   THE PSYCHROMETRIC CHART

The properties of atmospheric air can be represented in tables or graphical form. The graphical form is called the *psychrometric chart*. It is universally used because it presents a great deal of information very simply and because it is helpful in studying air conditioning processes.

Construction of the Psychrometric Chart

A psychrometric chart is shown in Figure 7.1. There are slight differences in the arrangement of charts furnished by different organizations. This should be studied before using one.

The location of the scales for each of the properties and the constant value lines for those properties are shown in Figure 7.2 on page 168. Each figure is a sketch of the psychrometric charts, but not drawn to the actual scale.

Study these sketches until you are familiar with the scales and the lines of constant values for each property. When reading values or drawing lines, a sharp drafting-type pencil and straight edge must always be used. Values should be read to the best accuracy possible, interpolating between numbered values when necessary.

Example 7.6 _____
Draw a line of 78 F DB on the psychrometric chart.

Solution
The solution is shown in Figure 7.3 on page 169.

Example 7.7 _____
Draw a line of 76.5 F WB on the psychrometric chart.

Solution
The solution is shown in Figure 7.4 on page 169.

Example 7.8 _____
Draw a line of 45% RH on the psychrometric chart.

Solution
The solution is shown in Figure 7.5 on page 169.

Lines of constant enthalpy and constant wet bulb temperature, although actually not exactly parallel, can be considered parallel on the psychrometric chart used here. (See Section 7.10 for a related discussion.) The difference is because the enthalpy values shown on the chart are for saturated air instead of for the actual conditions. The error in using the listed values, however, is less than 2% when calculating enthalpy changes, and on the conservative side. The curved lines on the chart show the corrections to be made to the enthalpy for actual conditions when greater accuracy is necessary.

7.4   LOCATING THE AIR CONDITION ON THE CHART

Any condition of air is represented by a point on the psychrometric chart. The condition can be

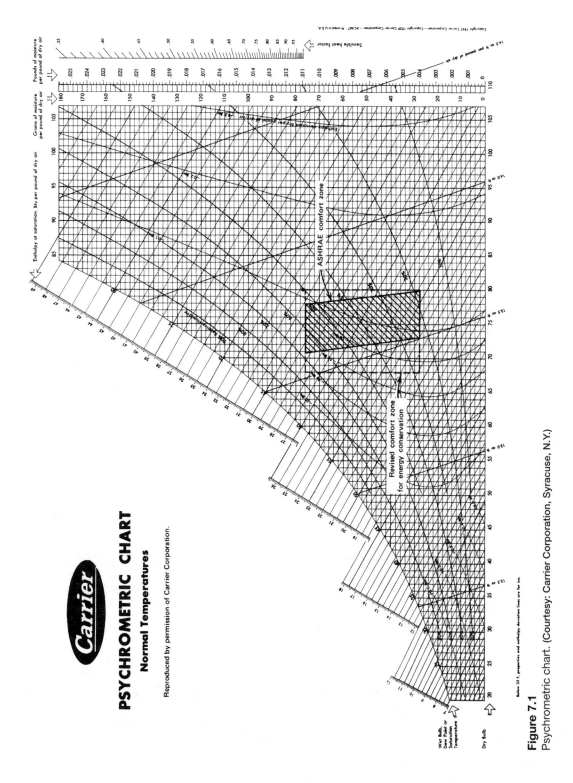

Figure 7.1

Psychrometric chart. (Courtesy: Carrier Corporation, Syracuse, N.Y.)

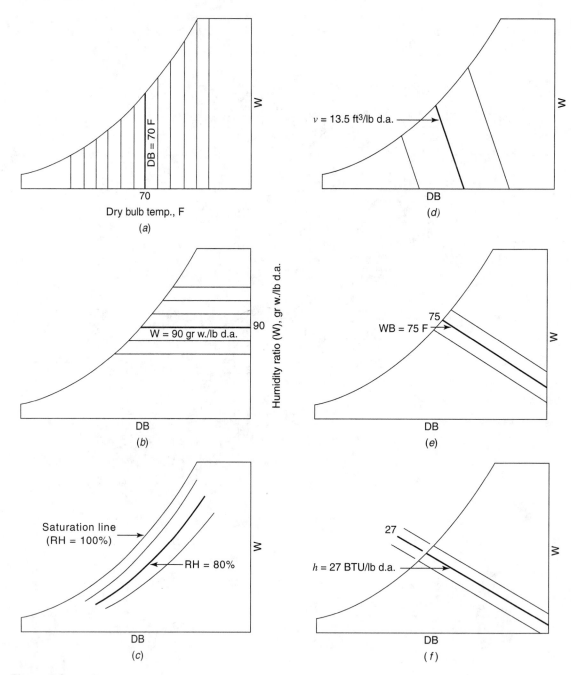

Figure 7.2

Construction of psychrometric chart, showing lines of constant property values. (*a*) Lines of constant dry bulb temperature (DB) on the psychrometric chart. (*b*) Lines of constant humidity ratio (W) on the psychrometric chart. (*c*) Lines of constant relative humidity (RH) on the psychrometric chart. (*d*) Lines of constant specific volume (*v*) on the psychrometric chart. (*e*) Lines of constant wet bulb temperature (WB) on the psychrometric chart. (*f*) Lines of constant enthalpy (*h*) on the psychrometric chart.

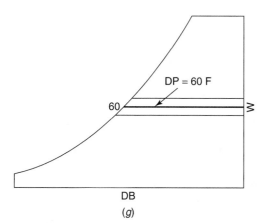

DP = 60 F

60

W

DB

(g)

Figure 7.2
Continued. (*g*) Lines of constant dew point temperature (DP) on the psychrometric chart.

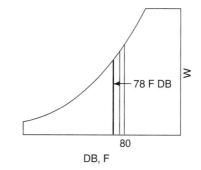

78 F DB

W

80

DB, F

Figure 7.3
Solution to Example 7.6.

located once two independent properties are known. Because each property is represented by a line, *the intersection of the two lines establishes the point representing the condition of the air.* Once the condition is located, any other properties can be read from the chart.

Example 7.9
The weather report reads 90 F DB and 40% RH. What is the WB?

Solution
Using the psychrometric chart, the condition of the air is at the point of intersection of the 90 F DB line and 40% RH line (Figure 7.6 on page 170). Drawing

a line of constant WB from this point, the WB temperature is read as 71.2 F.

Example 7.10
The air leaving a cooling coil is at 60 F DB and 55 F WB. What is its humidity ratio and specific enthalpy?

Solution
On the chart, the condition is found by the point of intersection of the 60 F DB and 55 F WB lines (Figure 7.7 on page 170). From the point, following a line of constant humidity ratio, read $W = 57$ gr w./lb d.a. Following a line of constant enthalpy from the point (parallel to WB lines) read $h = 23.2$ BTU/lb d.a.

Figure 7.4
Solution to Example 7.7.

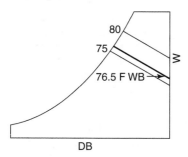

80
75
76.5 F WB
W
DB

Figure 7.5
Solution to Example 7.8.

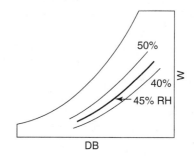

50%
40%
45% RH
W
DB

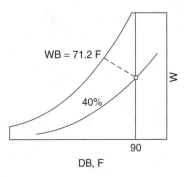

Figure 7.6
Solution to Example 7.9.

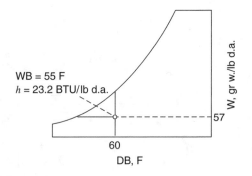

Figure 7.7
Solution to Example 7.10.

Example 7.11 _____

Combustion air enters a furnace at 80 F DB and 23% RH. What is its specific volume?

Solution

The condition is located at the intersection of the 80 F DB and 23% RH lines (Figure 7.8). The specific volume is read as 13.7 ft^3/lb d.a. by interpolation.

_____

The only circumstances under which two properties will not suffice to locate the air condition is when they are not independent properties. When properties are not independent, they are measuring the same thing, even though it may not be apparent by their name or definition. Practically, this means that the property lines are parallel on the psychrometric chart. For example, if we know the DP and humidity ratio of an air sample, we could not establish a point, because this gives only one line. The reader should verify this by studying the chart.

The psychrometric chart of Figure 7.1 shows the properties of air at a pressure of 29.92 in. Hg, the standard atmospheric pressure at sea level. For pressures significantly different, some property readings from the chart will not be correct and it cannot be used. Two solutions are possible—either a chart for the actual pressure can be used, if available, or corrections can be made to the values. These corrections can also be made directly by applying the property equations. Geographical locations at high altitudes (e.g., Denver, Mexico City), which are at lower atmospheric pressures, will require these corrections.

7.5 CONDENSATION ON SURFACES

Who, as a child, does not remember drawing pictures on a fogged windowpane in winter? Moisture on the glass is condensed from the room air when the glass temperature is lower than the room air dew point. Air contacting the glass is cooled below its dew point. From the definition of dew point, the air is saturated with water vapor when cooled to that temperature. When cooled further, it can hold even less water vapor—some is condensed.

Understanding this concept enables us to determine the maximum humidity that can be maintained in a room in winter without condensation occurring on the windows. Condensation should be avoided because the water will stain or damage surfaces. For single-glazed windows, the inside glass surface is only slightly higher than the outside temperature, because the thermal resistance of the glass is low. Example 7.12 illustrates the use of the psychrometric chart in relation to this problem.

Example 7.12 _____

A room with single-glazed windows is at 70 F DB. If the outside temperature is 30 F, what is the maximum RH that should be maintained in the room to avoid condensation on the windows?

Solution

The inside temperature of the glass can be assumed to be at the outside temperature. (The precise

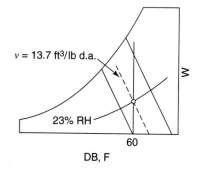

Figure 7.8
Solution to Example 7.11.

temperature can be calculated using the conduction heat transfer equation.)

Room air contacting the glass surface will be cooled to 30 F. Therefore, the dew point of the air must be less than 30 F to avoid condensation. Using the psychrometric chart (Figure 7.9), air with a 70 F DB and 30 F DP has an RH of 23%. This would be the maximum RH that should be maintained. Double glazing will of course increase the inside glass temperature and permissible RH.

The same problem occurs with bare cold water piping running through spaces. Chilled water lines are usually insulated so that the outside surface is well above the air dew point. A vapor barrier covering is necessary, however, to prevent the migration of any water vapor through the insulation to the cold pipe surface, where the water vapor would condense.

Figure 7.9
Solution to Example 7.12.

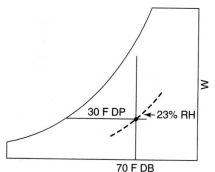

Air Conditioning Processes

7.6 PROCESS LINES ON THE PSYCHROMETRIC CHART

The purpose of air conditioning equipment is to change the condition of the entering air to a new condition. This change is called a *process*. Showing these processes on the psychrometric chart is very helpful in selecting equipment and in analyzing problems.

Processes are shown by drawing a line from the initial air condition to its final condition. The air changes properties along this line. Most processes are represented by straight lines.

Sensible Heat Changes

The *sensible heat change* process is one where heat is added or removed from the air and the DB temperature changes as a result, but there is no change in water vapor content. The direction of the process must therefore be along a line of constant humidity ratio, as shown in Figure 7.10.

Sensible heating (process 1-2) results in an increase in DB and enthalpy.

Sensible cooling process 1-3 (heat removal), results in a decrease in DB and enthalpy.

Figure 7.10
Sensible heating and sensible cooling processes.

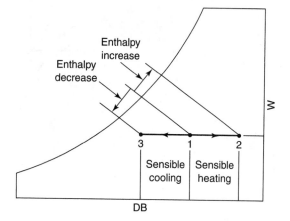

Latent Heat Changes (Humidification and Dehumidification)

The process of adding water vapor to the air is called *humidification,* and removal of water vapor from the air is called *dehumidification,* shown in Figure 7.11.

Process 1-4, humidification, results in an increase in humidity ratio and enthalpy. In humidification, the enthalpy of the air increases due to the enthalpy of the water vapor added. This is why it is called a latent heat change. In dehumidification, process 1-5, removal of water vapor results in a decrease in enthalpy.

These processes—pure humidification and dehumidification without a sensible heat change—do not occur often in practical air conditioning processes. However, the concept is important to understand in analyzing conditions.

Combination Sensible Heat and Latent Heat Change

The following combined sensible and latent processes, shown in Figure 7.12, may occur in air conditioning:

1. Sensible heating and humidification (1-6)
2. Sensible heating and dehumidification (1-7)

3. Sensible cooling and humidification (1-8)
4. Sensible cooling and dehumidification (1-9)

Note that, generally, DB, W, and enthalpy all change. For example, in the cooling and dehumidification process 1-9, both the DB and W are decreased, and the enthalpy decreases due to both sensible and latent heat removal.

It is important to determine the amount of heat and water vapor to be added or removed in the conditioning equipment, and to determine the changes in properties. This can be done by using the sensible and latent heat equations (Chapter 2) with the aid of the psychrometric chart.

7.7 SENSIBLE HEAT CHANGE PROCESS CALCULATIONS (SENSIBLE HEATING AND COOLING)

The sensible heat equation applied to moist air is

$$Q_s = 0.24m_a \times \mathrm{TC} + 0.45m_w \times \mathrm{TC} \qquad (7.9)$$

Figure 7.11
Humidification and dehumidification (latent heat change) processes.

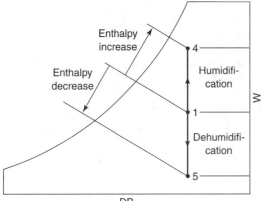

Figure 7.12
Combined sensible and latent heat change processes.

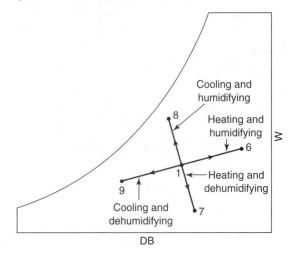

where

Q_s = sensible heat added to or removed from air, BTU/hr

m_a = weight of air, lb/hr

m_w = weight of water vapor, lb/hr

TC = $t_2 - t_1$ = temperature change, F (Specific heats of 0.24 for air and 0.45 for water vapor are used in Equation 7.9.)

The first term in Equation 7.9 expresses the enthalpy change of the dry air and the second term expresses the enthalpy change of the water vapor. For approximate air conditioning calculations, the second term is often small enough so that it can be neglected, and the sensible heat equation is written

$$Q_s = 0.24m_a \times TC \qquad (7.10)$$

Example 7.13 _____

An electric resistance heater is to be installed in a duct to heat 400 lb/hr of air from 60–90 F. What is the required capacity of the heater?

Solution

The electric energy in the resistance heater is converted into the required heat. Using Equation 7.10,

$$Q_s = 0.24m_a \times TC = 0.24 \times 400(90 - 60)$$
$$= 2880 \text{ BTU/hr}$$

Because the capacity of electrical equipment is expressed in kilowatts or watts rather than BTU/hr, units are changed as follows:

$$\text{Capacity} = 2880 \text{ BTU/hr} \times \frac{1 \text{ KW}}{3410 \text{ BTU/hr}}$$
$$= 0.84 \text{ KW}$$

Example 7.14 _____

The air entering the heater in Example 7.13 has an RH of 65%. How much error was there in neglecting the term in the sensible heat equation that included the enthalpy change of the water vapor?

Solution

From the psychrometric chart, the humidity ratio is 0.0072 lb w./lb d.a., and

$$m_w = 0.0072 \text{ lb w./lb d.a.} \times 400 \text{ lb air/hr}$$
$$= 2.9 \text{ lb w./hr}$$

The enthalpy change due to the water vapor is, from Equation 7.9,

$$Q_s = 0.45m_w \times TC = 0.45 \times 2.9 \times 30$$
$$= 39 \text{ BTU/hr}$$

Therefore, the correct amount of heat added is $2880 + 39 = 2919$ BTU/hr. The error from neglecting the enthalpy of the water vapor was about 1% of the total.

_____

A sensible heating or cooling problem can also be solved using the enthalpy values from the psychrometric chart and the Enthalpy Equation 2.13, as seen in Example 7.15.

Example 7.15 _____

Solve Example 7.14 using the psychrometric chart.

Solution

First plot the initial condition, process line, and final condition on the chart. The process and diagrammatic arrangement of the equipment are shown in Figure 7.13 on page 174. The increase in enthalpy of each pound of air is

$$h_2 - h_1 = 29.6 - 22.3 = 7.3 \text{ BTU/lb d.a.}$$

and therefore, using Equation 2.13, the enthalpy increase of the total amount of air is

$$Q_s = m_a(h_2 - h_1) = 400 \text{ lb/hr} \times 7.3 \text{ BTU/lb}$$
$$= 2910 \text{ BTU/hr}$$

which is in close agreement with the previous result. Note how simple and convenient it is to use this method with the aid of the chart.

_____

Always plot the process lines and sketch the equipment arrangement for every job. This will aid in understanding the system performance.

The sensible cooling process problem is handled in the same manner as sensible heating.

The flow rate of air is usually expressed in ft³/min (CFM) rather than lb/hr in air conditioning work, because most instruments read CFM.

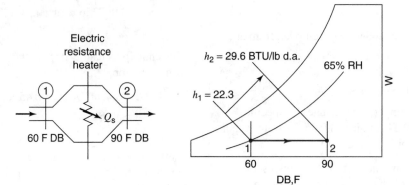

Figure 7.13
Sketch for Example 7.15.

Conversion between these units is therefore often necessary, as illustrated by Example 7.16.

Example 7.16

What is the flow rate of air entering the duct heater in Example 7.15, expressed in CFM?

Solution
From the psychrometric chart, the specific volume of the entering air is 13.25 ft^2/lb d.a. Converting units to CFM,

$$CFM_{in} = 400 \ \frac{lb}{hr} \times \frac{1 \ hr}{60 \ min} \times 13.25 \ \frac{ft^3}{lb}$$

$$= 88.3 \ CFM$$

The specific volume of a gas changes with temperature and pressure, as noted in the Gas Laws in Chapter 2. Therefore, the CFM leaving the duct heater in Example 7.16 will be different from that entering.

Example 7.17

What is the CFM leaving the duct heater in Example 7.16?

Solution
The specific volume leaving, from the chart, is 14.0 ft^3/lb d.a. Therefore,

$$CFM_{out} = 400 \ \frac{lb}{hr} \times \frac{1 \ hr}{60 \ min} \times 14.0 \ \frac{ft^3}{lb}$$

$$= 93.3 \ CFM$$

Of course the same quantity by weight of air is leaving the unit as is entering, but the CFM is greater. This variation in volume can lead to misunderstandings and error, such as selection of the wrong size equipment, unless clearly specified. Problems can be avoided by always indicating the temperature and pressure at which the CFM is specified.

However, when rating the capacity of equipment such as fans, coils, and air handling units, manufacturers do not know the conditions of temperature and pressure that each user will apply. In order to avoid misunderstandings, the CFM of the equipment is often expressed at standard air conditions. *Standard air* is defined as having a specific volume of 13.3 ft^3/lb d.a. (a density of 0.075 lb/ft^3 d.a.). This condition applies at 68 F and 29.92 in. Hg. In many air conditioning applications, the range of temperatures is such that the specific volume is close enough to that of standard air so that no significant error occurs if the specific volume of standard air is used. For heating applications at high temperatures, or where pressures are considerably different from 14.7 psia, as at high altitudes, the actual specific volumes should be used.

The relation between air flow rate expressed in lb/hr and air flow rate expressed as ft^3/min, using standard air conditions, is

$$m_a \ \frac{lb}{hr} = CFM \ \frac{cu \ ft}{min} \times \frac{60 \ min}{1 \ hr} \times \frac{1 \ lb}{13.3 \ ft^3}$$

$$m_a = 4.5 \times CFM \qquad (7.11)$$

Substituting this in the sensible heat equation (7.9) and also assuming a typical average moisture content of air of 0.01 lb w./lb d.a. for air conditioning processes, the result is

$$Q_s = 1.1 \times CFM \times TC$$
$$= 1.1 \times CFM \times (t_2 - t_1) \qquad (7.12)$$

This convenient form of sensible heat equation is commonly used for moist air calculations in air conditioning.

Example 7.18 _____

Determine the capacity of the duct heater in Example 7.16, using Equation 7.12, based on entering air CFM.

Solution

Equation 7.12 is used.

$$Q_s = 1.1 \times 88.3(30) = 2914 \text{ BTU/hr}$$

The result agrees with that found by using the psychrometric chart.

_____

Psychrometrics can be put to good use by the engineer or service technician in troubleshooting, as well as by the system designer. Example 7.19 will illustrate a case.

Example 7.19 _____

A cooling coil with a rated sensible cooling capacity of 50,000 BTU/hr while handling 2000 CFM of air entering at 80 F must be checked to see that it is performing properly. Thermometers at the air entrance and exit of the unit read 80 F and 62 F, and the air flow rate is measured and found to be 2000 CFM. Is the unit performing satisfactorily?

Solution

If the unit is cooling according to its rated capacity, the leaving air temperature will be at least as low as that predicted by the sensible heat equation. Solving for the temperature change in Equation 7.12,

$$TC = \frac{Q_s}{1.1 \times CFM} = \frac{50,000}{1.1 \times 2000} = 23 \text{ F}$$
$$t_2 = TC - t_1 = 80 - 23 = 57 \text{ F}$$

and we see that the unit is not performing as rated. Note: When using equations with temperature changes, the reader should be careful not to make errors resulting from improper use of negative arithmetic signs. In Example 7.19, the temperature leaving, t_2, must be less than t_1, so 23 F is subtracted from 80 F.

_____

7.8 LATENT HEAT CHANGE PROCESS CALCULATIONS (HUMIDIFYING AND DEHUMIDIFYING)

The amount of water vapor added to or removed from air in a humidifying or dehumidifying process is

$$m_w = m_a (W_2 - W_1)$$

where

m_w = water vapor added or removed, lb w./hr

m_a = air flow rate, lb/hr

$W_2 - W_1$ = change in humidity ratio, lb w./lb d.a.

As with the sensible heating process, it is usually acceptable to assume air at standard conditions. If the air flow rate is expressed in CFM, substituting from Equation 7.11 in the above equation gives

$$m_w = 4.5 \times CFM (W_2 - W_1) \qquad (7.13)$$

or, if the humidity ratio is given in gr w./lb d.a., dividing by 7000 gr/lb

$$m_w = \frac{CFM(W_2' - W_1')}{1556} \qquad (7.14)$$

where W' = humidity ratio, gr w./lb d.a.

Example 7.20 _____

A water humidifier in a warm air heating duct handling 3000 CFM increases the moisture content of the air from 30 to 60 gr w./lb d.a. How much water must be supplied?

Solution
The flow diagram is shown in Figure 7.14. Using Equation 7.14,

$$m_w = \frac{\text{CFM}(W_2' - W_1')}{1556} = \frac{3000(60 - 30)}{1556}$$

$$= 58 \text{ lb w./hr}$$

Latent Heat Change

As discussed previously, the evaporation of water requires heat. The latent heat of vaporization of water at typical air conditioning temperatures is approximately 1055 BTU/lb. Using Equation 7.14,

$$Q_l = 1055 \times m_w = 1055 \times \frac{\text{CFM}(W_2' - W_1')}{1556}$$

$$Q_l = 0.68 \times \text{CFM}(W_2' - W_1') \qquad (7.15)$$

where

$$Q_l = \text{latent heat change, BTU/hr}$$

$$W_2' - W_1' = \text{humidity ratio change, gr w./lb d.a.}$$

Example 7.21

How much heat is required in the evaporation process in the humidifier in Example 7.20?

Figure 7.14
Sketch for Example 7.22.

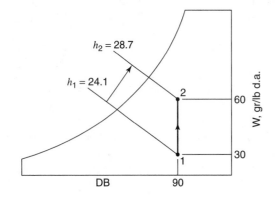

Solution
Using Equation 7.15,

$$Q_l = 0.68 \text{ CFM}(W_2' - W_1') = 0.68 \times 3000(30)$$

$$= 61,200 \text{ BTU/hr}$$

Note: If water were simply evaporated in the air without providing the required heat, the air would cool down. (This process will be described shortly.) Therefore, the heating equipment must provide sufficient heat to prevent cooling of the air.

Another method of humidifying is to generate steam in a separate source and to inject it into the air to be humidified.

Humidification is a desirable process in winter air conditioning. An inspection of the psychrometric chart shows that natural air conditions in the winter have a very low humidity, a condition which is unacceptable for good comfort. A quality environment control system should include winter humidification.

Dehumidification is the reverse of the process described earlier. All of the equations for latent heat change hold true. In this case, heat must be removed from the air to dehumidify it, that is, to condense water vapor from it. Pure dehumidification is not a very commonly used process, and the need arises only in specialized industrial air conditioning. It is usually combined with sensible cooling or heating.

The latent heat change problem can also be solved by using the enthalpy Equation 2.13 and the psychrometric chart, as seen in the following example.

Example 7.22

Find the amount of heat required for the humidifier in Example 7.21, if the air is at 90 F, by using the psychrometric chart.

Solution
From the psychrometric chart (see Figure 7.14),

$$h_2 - h_1 = 28.7 - 24.1 = 4.6 \text{ BTU/lb d.a.}$$

Using Equation 7.11,

$$m_a = 4.5 \times \text{CFM} = 4.5 \times 3000 = 13,500 \text{ lb/hr}$$

The latent heat required is therefore, using Equation 2.13,

$$Q_l = m_a (h_2 - h_1) = 13,500 \qquad (4.6)$$
$$= 62,100 \text{ BTU/hr}$$

The result agrees closely with that found previously in Example 7.21.

7.9 COMBINED SENSIBLE AND LATENT PROCESS CALCULATIONS

In many air conditioning system processes, the air undergoes both sensible and latent heat changes. These changes may take place separately or may occur together. In either case, the procedures for analysis use the sensible and latent heat equations and the psychrometric chart.

The Cooling and Dehumidification Process

Air conditioning for human comfort usually requires a process where both sensible and latent heat are removed from air—that is, the air is cooled and dehumidified.

The sensible heat removed and latent heat removed are found from the Equations 7.12 and 7.15, respectively. The sum, $Q_t = Q_s + Q_l$, is the **total heat** removed for the process.

Example 7.23 _____

An air conditioning unit has a cooling coil that cools and dehumidifies 20,000 CFM of air from 82 F DB and 50% RH to 64 F DB and 61 F WB. Find the sensible, latent, and total capacity of the cooling coil and the amount of moisture condensed.

Solution

The flow diagram is shown in Figure 7.15. The equations developed previously will provide the information. The sensible heat removed (Equation 7.12) is

$$Q_s = 1.1 \times \text{CFM} \times \text{TC}$$
$$= 1.1 \times 20,000 \ (18) = 396,000 \text{ BTU/hr}$$

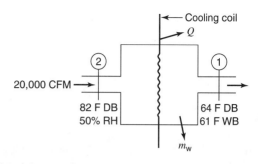

Figure 7.15
Sketch for Example 7.23.

From the psychrometric chart

$$W_2' - W_1' = 82 - 75 = 7.0 \text{ gr w./lb d.a.}$$
$$Q_l = 0.68 \times \text{CFM} \ (W_2' - W_1')$$
$$= 0.68 \times 20,000 \ (7)$$
$$= 95,000 \text{ BTU/hr}$$

The total heat removed is

$$Q_t = Q_s + Q_l = 396,000 + 95,000$$
$$= 491,000 \text{ BTU/hr}$$

or

$$491,000 \text{ BTU/hr} \times \frac{1 \text{ ton}}{12,000 \text{ BTU/hr}} = 41 \text{ tons}$$

The total cooling capacity of the coil required for this job is 491,000 BTU/hr (41 tons) at the conditions specified. The amount of moisture condensed during the process is, using Equation 7.14,

$$m_w = \frac{\text{CFM}(W_2' - W_1')}{1556} = \frac{20,000 \ (7)}{1556}$$
$$= 90 \text{ lb/hr}$$

Provision must be made for draining the water that is continually collecting in the air conditioning unit.

The problem in Example 7.23 can also be solved using Equation 2.13 and the psychrometric chart. Although the sensible heat and latent heat are being removed simultaneously from the air in the conditioner, they can be shown separately on the chart, as seen in the following example.

Example 7.24

Solve Example 7.23 using the psychrometric chart.

Solution

Referring to Figure 7.16, the process line representing the total heat removal is 1-2 (the actual line is slightly curved, as explained later). However, the latent heat removal portion is shown by 1-a and the sensible heat removal is shown by a-2, even though these are not actual process lines.

The flow rate in lb/hr is

$$m_a = 4.5 \times CFM = 4.5 \times 20,000$$
$$= 90,000 \text{ lb/hr}$$

The sensible heat removal is

$$Q_s = m_a(h_a - h_2) = 90,000 \, (31.6 - 27.2)$$
$$= 396,000 \text{ BTU/hr}$$

The latent heat removal is

$$Q_l = m_a(h_1 - h_a) = 90,000 \, (32.7 - 31.6)$$
$$= 99,000 \text{ BTU/hr}$$

Figure 7.16
Sketch for Example 7.24.

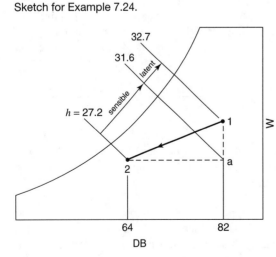

and the total heat removal is

$$Q_t = Q_s + Q_l = 495,000 \text{ BTU/hr}$$

As shown indirectly in Examples 7.22 and 7.24, the enthalpy Equation 2.13 can be expressed in the following form:

$$Q = 4.5 \times CFM \, (h_2 - h_1) \qquad (7.16)$$

where

Q = sensible (Q_s), latent (Q_l), or total (Q_t) heat added or removed, BTU/hr

CFM = volume flow rate of air being processed

$h_2 - h_1$ = sensible, latent, or total enthalpy change, BTU/lb d.a.

Equation 7.16 *can be used for any air conditioning process represented on the psychrometric chart.*

It is advisable to solve air conditioning process problems by both of the methods explained:

1. Using Equations 7.12 and 7.14, the sensible heat and latent heat equations
2. Using Equation 7.16 with the psychrometric chart, applying it to the sensible heat and latent heat parts of the process

When the results are compared, they should substantially agree. If not, an error has been made. (However, agreement does not ensure that there is not a possible error common to both methods.)

The solution of any of the other combined sensible-latent processes is handled in the same manner as the cooling-dehumidification process. Heating and humidification are typical to winter air conditioning systems. The other processes are encountered less often. However, some industrial air conditioning applications may require them.

It should be noted that some combinations of processes may have sensible and latent heat changes opposite in direction. For instance, the heating and dehumidification process has sensible heat added and latent heat removed.

7.10 THE EVAPORATIVE COOLING PROCESS AND THE WET BULB TEMPERATURE

One special cooling and humidification process called *evaporative cooling* requires a more detailed discussion. Referring to Figure 7.17, water is sprayed into the airstream. Some of the water evaporates, increasing the water vapor content of the air. The unevaporated water is recirculated continuously, and no external heat is added to the process.

If the (dry bulb) temperature of the air is measured entering and leaving the conditioning unit, it will be noted that the temperature leaving is lower than that entering, even though no external cooling source is used. This indicates that sensible heat was given up by the air. The important question here is: What caused this?

The evaporation of the water required heat. Because there is no external heat source, unlike the pure humidification process described earlier, this heat must be obtained from the air, lowering its temperature.

The next important fact to note about the evaporative cooling process is that it is a *constant enthalpy* process. This must be so, because there is no heat added to or removed from the air-water vapor mixture. There is simply an exchange of heat. The sensible heat decreases and the latent heat increases by the same amount. A process in which there is no change in total heat content is called an *adiabatic* process. We can now determine the process line on the psychrometric chart for the evaporative cooling process; it is a line of constant enthalpy content, as seen in Figure 7.17.

Referring to the definition of wet bulb temperature—the temperature recorded by a thermometer whose stem is wrapped with a wetted wick, and placed in the airstream—it is seen that evaporative cooling is the process occurring at the thermometer stem. This results in a lower temperature reading of the wet bulb thermometer. The air passing through the wick becomes completely saturated.

Thus we can note that the evaporative cooling process is therefore a constant wet bulb temperature process. If wet bulb thermometers were placed in the airstream entering and leaving the evaporative cooling unit, they would have the same readings.

If the evaporative cooling process can produce air at temperatures low enough for sufficient cooling of spaces (at least as low as 60–65 F DB), it would mean that no refrigeration equipment would be needed, including its operating costs.

However, the evaporative cooling process is practical for air conditioning only in very dry climates. Look at the psychrometric chart at a typical summer outdoor air design condition, in a humid

Figure 7.17
Evaporative cooling process.

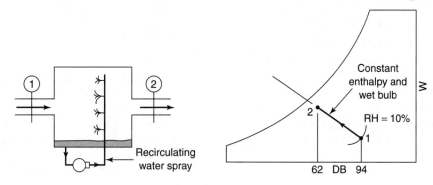

climate, say 90 F DB and 74 F WB. If we follow a constant wet bulb line, evaporative cooling could produce air at only 74 F DB, even with complete saturation. Not only would the DB not be low enough for sufficient cooling, but the high humidity of the supply air would result in extremely uncomfortable humidity conditions.

However, note that if the outdoor air were at 94 F DB and 10% RH (60 F WB), typical of a desert-like climate, as in parts of the southwestern United States, evaporative cooling could produce supply air at about 62 F DB, suitable for air conditioning.

Even in some normally dry climates, there may be some days where the humidity is high enough so that the effective evaporative cooling will not occur. When considering using evaporative cooling type air conditioning units, in these cases the decision must be made as to whether the lack of adequate air conditioning is acceptable for those periods or whether the investment in mechanical refrigeration equipment is wiser.

The evaporative cooling process also occurs in a cooling tower. *Cooling towers* are equipment used to cool water. The water is sprayed into an airstream, and a small portion of the water evaporates. The heat necessary to evaporate the water is taken both from the air and from the water that does not evaporate. The cooled water is then circulated to where it will be used. The use of cooling towers in refrigeration systems will be explained in Chapter 13.

7.11   THE AIR MIXING PROCESS

The air *mixing process* is one where two streams of air are mixed to form a third stream. This process occurs frequently in air conditioning, particularly in mixing outside air with return air from rooms. If the conditions of the two airstreams that are to be mixed are known, the conditions after mixing can be found. Referring to Figure 7.18, the procedures for finding the DB and W will be explained.

According to the Conservation of Energy Principle, the sensible heat content of the air before and after mixing is the same. That is

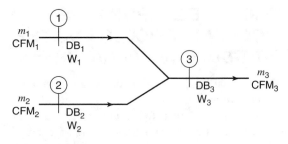

Figure 7.18
Air mixing process.

$$m_3 \times DB_3 = (m_1 \times DB_1) + (m_2 \times DB_2)$$

Solving for DB_3,

$$DB_3 = \frac{(m_1 \times DB_1) + (m_2 \times DB_2)}{m_3} \qquad (7.17)$$

If the specific volumes of the unmixed streams are not widely different, the equation can be written using flow rates in ft^3/min (CFM), rather than lb/hr, without significant loss of accuracy:

$$DB_3 = \frac{(CFM_1 \times DB_1) + (CFM_2 \times DB_2)}{CFM_3} \qquad (7.18)$$

The humidity ratio W_3 of the mixed air is found in a similar manner, applying the principle of conservation of mass—that is, the water vapor content before and after mixing is the same:

$$m_3 \times W_3 = (m_1 \times W_1) + (m_2 \times W_2)$$

Solving for W_3,

$$W_3 = \frac{(m_1 \times W_1) + (m_2 \times W_2)}{m_3} \qquad (7.19)$$

As before, the following approximately correct equation can often be used:

$$W_3 = \frac{(CFM_1 \times W_1) + (CFM_2 \times W_2)}{CFM_3} \qquad (7.20)$$

For determining mixed air conditions, Equations 7.18 and 7.20 are accurate enough if the specific volumes of the unmixed airstreams are within 0.5 ft^3/lb d.a. of each other. This occurs under most outside air (OA) and return air (RA) design conditions. For example, on the psychrometric chart, a RA condition of 78 F DB and 50% shows a specific volume of about 13.8 ft³/lb d.a. An OA condition of 94 F DB and 75 F WB (St. Louis, Missouri) shows a specific volume of about 14.3 ft³/lb d.a. Under these conditions, for most applications, Equations 7.18 and 7.20 would produce satisfactory answers for the mixed air condition.

Example 7.25 _____

Outside air and return air as shown in Figure 7.19 are mixed. Find the mixed air DB and WB.

Solution

The conditions are close enough so that the approximate equations can be used. Using Equation 7.18,

$$DB_3 = \frac{(1000 \times 90) + (2000 \times 75)}{3000} = 80 \text{ F}$$

From the psychrometric chart $W_1' = 89$ and $W_2' = 64$ gr w./lb d.a. Using Equation 7.20,

$$W_3' = \frac{(1000 \times 89) + (2000 \times 64)}{3000}$$

$$= 72 \text{ gr w./lb d.a.}$$

Locating the mixed air condition on the psychrometric chart, at 80 F DB and 72 gr w./lb d.a., the WB = 66 F.

_____

The mixing process can also be solved graphically on the psychrometric chart, from the following two facts:

Figure 7.19
Sketch for Example 7.25.

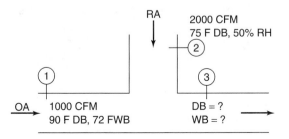

1. The mixed air condition will lie along a straight line connecting the two conditions of the unmixed airstreams.
2. The location of the mixed air condition on this mixing line will be inversely proportional to the quantities of the unmixed airstreams to the total. (This is simply a graphical expression of Equations 7.18 and 7.20).

Example 7.26 _____

Solve Example 7.25 graphically, using the psychrometric chart.

Solution

The mixing process line is drawn between points 1 and 2. The proportion of each airstream to the total is

$$\frac{CFM_1}{CFM_3} = \frac{1000}{3000} = \frac{1}{3}$$

$$\frac{CFM_2}{CFM_3} = \frac{2000}{3000} = \frac{2}{3}$$

Point 3, the mixed air condition, is therefore located one-third of the total distance, starting from point 2.

It is usually convenient to use the DB scale to locate the mixed condition. Any proportional distance on the DB scale is the same as that on the mixing line. The total distance on the DB scale is 90 − 75 = 15 spaces. The one-third distance from point 2 is 75 + ⅓ × 15 = 80 on the DB scale. This distance is projected vertically to the mixing line to locate point 3. The construction is shown in Figure 7.20 on page 182.

Reading from point 3 on the psychrometric chart,

$$DB_3 = 80 \text{ F}, \qquad WB_3 = 66 \text{ F}$$

_____

All of the air processes usually encountered in air conditioning have now been described. Our next task will be to learn how to put this information together in designing an air conditioning system or in analyzing performance problems. This requires determining the conditions of air to be supplied to the rooms. In studying this problem, we

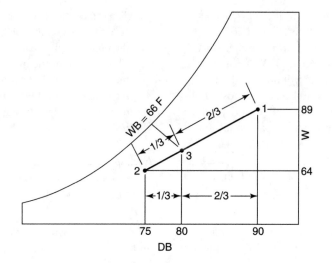

Figure 7.20
Sketch for Example 7.26.

will also learn some new concepts used in psychrometrics—the sensible heat ratio and the coil process line.

Psychrometric Analysis of the Air Conditioning System

In the following discussion, we will use the basic psychrometric processes to analyze a complete air conditioning system, and will also briefly consider some more advanced psychrometric concepts. Some familiarity with types of air handling equipment and systems (Chapter 12) will aid in understanding the following material.

7.12   DETERMINING SUPPLY AIR CONDITIONS

The rooms in a building gain heat in the summer from a number of sources. The procedures for finding these heat gains were discussed in Chapter 6.

The rate at which heat must be extracted from a room to offset these heat gains was given the name *room total cooling load* (RTCL); it is composed of

two parts, the *room sensible cooling load* (RSCL) and *room latent cooling load* (RLCL).

This heat extraction, or cooling effect, is provided by supplying air to the room at a temperature and humidity low enough to absorb the heat gains.

These relationships are shown in Figure 7.21 and are expressed by the sensible and latent heat equations:

$$\text{RSCL} = 1.1 \times \text{CFM}_S \, (t_R - t_S) \tag{7.21}$$
$$\text{RLCL} = 0.68 \times \text{CFM}_S \, (W_R' - W_S') \tag{7.22}$$

where

RSCL = room sensible cooling load, BTU/hr

RLCL = room latent cooling load, BTU/hr

Figure 7.21
Supplying conditioned air to absorb room heat gains.

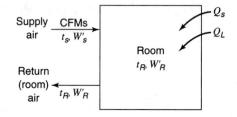

CFM$_S$ = CFM of supply air

t_R, t_S = temperature of room and supply air, F

$W_R' W_S'$ = humidity ratios of room and supply air, gr/lb d.a.

Equations 7.21 and 7.22 are used to find the required conditions of the supply air to offset the sensible and latent loads for each room.

It is usual practice to apply the RSCL equation (7.21) first to determine the supply air CFM$_S$ and t_S, and then apply the RLCL equation (7.22) to determine the supply air humidity ratio W_S'.

In applying Equation 7.21, the RSCL is known from the cooling load calculations (Chapter 6) and t_R and W_R' are selected in the comfort zone (Chapter 1). This still leaves two unknowns, CFM$_S$ and t_S. One of these is chosen according to "good practice" (such as costs and job conditions), and the remaining unknown is then calculated from the equation.

Example 7.27 illustrates the calculation of the supply air conditions.

Example 7.27

The Unisex Hair Salon Shop has a sensible cooling load of 55,000 BTU/hr and latent cooling load of 22,000 BTU/hr. The room conditions are to be maintained at 78 F DB and 50% RH. If 2000 CFM of supply air is furnished, determine the required supply air DB and WB.

Solution

Applying Equation 7.21, solving for the supply air temperature change

$$t_R - t_S = \frac{RSCL}{1.1 \times CFM_S} = \frac{55,000}{1.1 \times 2000} = 25 \text{ F}$$

The supply air temperature is therefore

$$t_R = 78 - 25 = 53 \text{ F}$$

The required humidity ratio of the supply air is then found from Equation 7.22:

$$W_R' - W_S' = \frac{RLCL}{0.68 \times CFM_S}$$

$$= \frac{22,000}{0.68 \times 2000} = 16 \text{ gr w./lb d.a.}$$

From the psychrometric chart $W_R' = 71$ gr w./lb d.a., and therefore

$$W_S' = 71 - 16 = 55 \text{ gr w./lb d.a.}$$

Reading from the chart, WB$_S$ = 52 F.

7.13 SENSIBLE HEAT RATIO

If we recalculate the supply air conditions required in Example 7.27 for other CFM quantities, the conditions found would of course be different. Table 7.1 shows the results for two other assumed values of CFM, as well as the results already found.

TABLE 7.1 SATISFACTORY SUPPLY AIR CONDITIONS FOR EXAMPLE 7.27

Supply Air Condition	CFM	DB, F	W' gr w./lb d.a.
1	2000	53	55
1A	2500	58	60
1B	3200	62.4	62.6

If all three satisfactory supply air conditions are plotted on the psychrometric chart, as shown in Figure 7.22, a surprising fact is noted: All of the points lie on a straight line, and furthermore, this line also passes through the room air condition, *R*.

Figure 7.22
Satisfactory supply air conditions fall along a straight line.

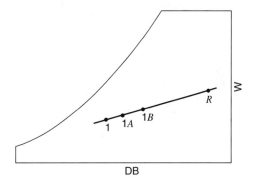

This line has considerable significance. If we were to assume still other air supply rates and then calculate the required supply air conditions, we would find that every one would lie on this same line. This is not a coincidence. Any supply air condition that will satisfactorily remove the proper proportion of room sensible and latent heat gains will be on this line; in addition, any supply air condition that is not on this line will not be satisfactory. Figure 7.23, using the data from Table 7.1, will be used to explain this important fact.

We note from this figure that for air supply at either *A* or *B,* the ratio of sensible to total heat removal, h_s/h_t, is the same for both. It will also be true for any other point on line *RA* (this can be proven by the geometry of similar triangles).

The slope of line *RA,* which is defined as the ratio *Ax/AR,* can also be shown by geometry to be equal to h_s/h_t. To sum up this idea:

$$\frac{Ax}{AR} = \frac{h_s}{h_t} = \frac{RSCL}{RTCL}$$

The ratio RSCL/RTCL is called the *room sensible heat ratio,* RSHR. (It is also called the *room sensible heat factor,* RSHF). That is

$$RSHR = \frac{RSCL}{RTCL} \qquad (7.23)$$

7.14   THE RSHR OR CONDITION LINE

*The **RSHR line** is defined as the line drawn through the room conditions with the room sensible heat ratio slope RSCL/RTCL.*

A scale for sensible heat ratio slopes is shown on most psychrometric charts to make it easier to draw lines with this slope. The following example will illustrate how to plot the RSHR line. Following that, its importance will be explained.

Example 7.28 _____
The Big Boy Hamburger Shop has a sensible cooling load of 45,000 BTU/hr and a latent cooling load of 15,000 BTU/hr. The shop is maintained at 77 F DB and 45% RH. Draw the RSHR line.

Solution
The solution is shown in Figure 7.24. The following steps are carried out:

1. Calculate the RSHR (Equation 7.23):

$$RSHR = \frac{RSCL}{RTCL} = \frac{45,000}{45,000 + 15,000} = 0.75$$

2. On the SHR scale on the psychrometric chart, locate the 0.75 slope. There is also a guide point for the SHR scale, encircled on the chart

Figure 7.23
Sensible and latent heat removal for two different supply air conditions.

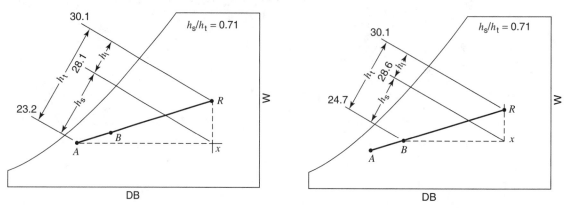

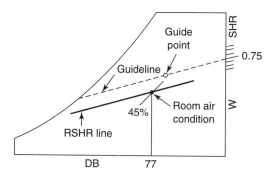

Figure 7.24
Plotting the RSHR line (Example 7.28).

(located at 80 F and 50% RH). Draw a guide line from SHR = 0.75 through the guide point.
3. Draw a line parallel to the guide line through the room condition point. This is the RSHR line, because it has the RSHR slope, and passes through the room condition. (Two drafting triangles will aid in drawing an exact parallel.)

The importance of the RSHR line is that it is *the line on which any satisfactory supply air condition must lie.* The reason for this is that it has the slope representing the correct proportion of sensible and latent heat removal. Therefore, if the supply air condition is on the line, it will remove the correct proportion of the RSHG and RTHG. The RSHR line is the same line that was developed from Table 7.1 by plotting a number of possible supply air conditions. The SHR scale on the chart enables us to plot the RSHR line in a much easier manner than was done there.

In selecting air conditioning equipment, the usual practice is to plot the RSHR line and then choose a supply air condition on this line. This procedure will be discussed later. First, let us look at an example of applying the RSHR line concept in troubleshooting a service problem.

Example 7.29 _____

Mrs. Van Astor, a regular patron of the swank Francais Restaurant, complains on one July day that it feels very "sticky." The manager turns down the

thermostat. Mrs. Van Astor gets so cold she puts on her mink stole. Finally she stalks out. Soon all of the wealthy jet set customers leave. The manager calls I. Fixum, troubleshooter. Fixum looks up the air conditioning system design data, which are

RSCL = 150,000 BTU/hr

RLCL = 53,000 BTU/hr

Room design conditions = 78 F DB, 50% RH

Design supply air = 62 F DB

Solution
The solution is shown graphically in Figure 7.25.

1. The actual supply air conditions are first measured, using instruments. They are 61 F DB and 59 F WB.
2. Using Equation 7.23, the RSHR is calculated.

$$RSHR = \frac{RSCL}{RTCL} = \frac{150,000}{203,000} = 0.74$$

3. The RSHR line is plotted on the psychrometric chart. This is the line with the slope equal to the value of the RSHR (0.74) that passes through the room air point.
4. The actual supply air condition is located on the chart, and it is seen that it does not lie on the RSHR line. Therefore the proper room design conditions will not be maintained.

In the preceding example, turning down the thermostat lowered the supply air dry bulb

Figure 7.25
Sketch for Example 7.29.

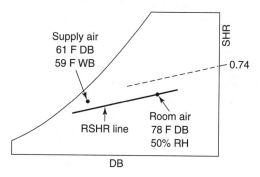

temperature to an approximately satisfactory value (61 F) compared to the design value (62 F). Yet this did not sufficiently reduce the room humidity level.

Fixum knows that the air conditioning unit was not removing enough latent heat (not dehumidifying enough) because the supply air condition is *above* the RSHR line. Turning down the thermostat even further probably would still not solve the problem.

There could be a number of common causes for the uncomfortable conditions existing in the example cited. Perhaps an improper cooling coil was being used. Perhaps the refrigerant temperature was not low enough, or perhaps the amount of outside air used was more than that designed for. These matters should become clearer as we continue our analysis of air conditioning processes with the powerful graphical aid of the psychrometric chart.

7.15   COIL PROCESS LINE

A line can be drawn on the psychrometric chart representing the changes in conditions of the air as it passes over the cooling and dehumidifying coil. This is called the coil process line. This line depends on the coil configuration, air velocity, and refrigerant temperature. It is a curved line, and is difficult to locate. However, it is possible to locate a straight line on the chart that, although it is not the true coil process line, will enable us to select a coil or check the performance of a coil. We will call this the coil process line anyway.

The **coil process line** may then be defined as *the straight line drawn between the air conditions entering and leaving the coil*, as shown in Figure 7.26.

The *capacity* of a coil is defined as the sensible, latent, and total heat that it removes from the air it is conditioning. The required coil capacity, called the *cooling coil load*, can be determined from the coil process line, as illustrated in Example 7.30.

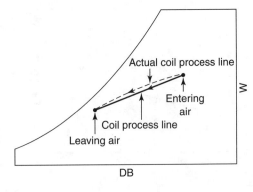

Figure 7.26
Coil process line.

Example 7.30
A cooling coil handles 24,000 CFM of air entering at 86 F DB and 73 F WB. The air leaves the coil at 59 F DB and 56 F WB. Determine the coil capacity.

Solution
The coil process line is drawn on the psychrometric chart (Figure 7.27) from entering condition 1 to leaving condition 2. The sensible, latent, and total heat content change per lb d.a. are as shown. The coil capacity is therefore (Equation 7.16)

$$Q_s = 4.5 \times CFM(h_x - h_2)$$
$$= 4.5 \times 24,000 \, (30.6 - 23.8)$$
$$= 734,000 \text{ BTU/hr}$$
$$Q_l = 4.5 \times CFM(h_1 - h_x)$$
$$= 4.5 \times 24,000 \, (36.8 - 30.6)$$
$$= 670,000 \text{ BTU/hr}$$
$$Q_t = 734,000 + 670,000 = 1,404,000 \text{ BTU/hr}$$
$$= 117 \text{ tons}$$

The total capacity could also have been found directly:

$$Q_t = 4.5 \times CFM(h_1 - h_2)$$
$$= 4.5 \times 24,000 \, (36.8 - 23.8)$$
$$= 1,404,000 \frac{\text{BTU}}{\text{hr}}$$

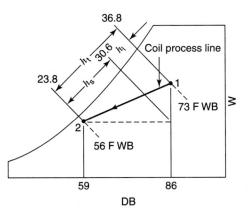

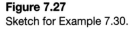

Figure 7.27
Sketch for Example 7.30.

7.16 THE COMPLETE PSYCHROMETRIC ANALYSIS

We are now prepared to determine all of the required supply air conditions and the cooling coil capacity for proper conditioning of the space, based on the following known information (Chapter 6):

1. Room sensible and latent heat gains.
2. Outside and inside design conditions.
3. Ventilation (outside) air requirements.
4. Either CFM or DB temperature of the supply air. One of these is selected and the other is then determined from the sensible heat equation. However, both must be in a range that is considered satisfactory for "good practice."

Supply air temperature values are usually chosen so that the temperature difference between room and supply air is between 15–30 F. Factors such as the type and location of air supply outlets will affect the temperature difference selected (see Chapter 10).

The supply air CFM must neither be too little nor too great, to prevent discomfort from staleness or drafts. Fortunately this is usually not a problem if the supply air temperature and ventilation air quantity are selected within acceptable values.

Example 7.31 will illustrate a complete psychrometric analysis.

Example 7.31
The following design data has been established for the High Life Insurance Company office building:

RSCL = 740,000 BTU/hr, RLCL = 150,000 BTU/hr

Outside design conditions—94 F DB, 75 F WB

Inside design conditions—78 F DB, 50% RH

Outside air required is 6730 CFM

Supply air temperature difference is 20 F

DETERMINE

A. Supply air CFM
B. Supply air conditions
C. Conditions entering cooling coil
D. Cooling coil sensible, latent, and total load

Solution
Each part of the problem will be solved in order. It is advisable to sketch a diagrammatic arrangement of the system and also each process on the psychrometric chart, as shown in Figure 7.28 on page 188. Note that as some outside air (OA) is introduced, the same amount of air leaving the space must be exhausted (EA), and the remaining air is returned (RA) to the air conditioning unit.

A. Using Equation 7.21,

$$CFM_3 = \frac{RSCL}{1.1(t_4 - t_3)} = \frac{740,000}{1.1 \times 20}$$
$$= 33,640 \text{ CFM}$$

B.
$$DB_3 = 78 - 20 = 58 \text{ F}$$

To find the remaining supply air conditions, plot the RSHR line. The slope is

$$RSHR = \frac{Q_s}{Q_t} = \frac{RSCL}{RTCL} = \frac{740,000}{890,000} = 0.83$$

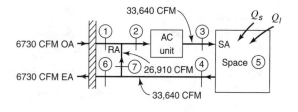

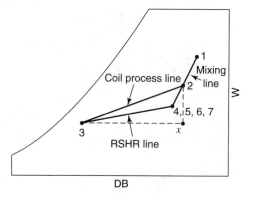

Figure 7.28
Sketch for Example 7.31.

The intersection of the RSHR line and 58 F DB line determines the remaining supply air condition $WB_3 = 56.3$ F.

C. The condition entering the cooling coil will be the mixed air condition of outside air and return air. Using Equation 7.18,

$$DB_2 = \frac{CFM_1 \times DB_1 + CFM_7 \times DB_7}{CFM_2}$$

$$= \frac{6730(94) + 26,910(78)}{33,640} = 81.2 \text{ F}$$

Draw the mixing line 1-7 on the chart. The intersection of this line with the 81.2 F DB line will locate point 2, the condition entering the coil. Read $WB_2 = 67.2$ F.

D. Draw the coil process line 2-3.

Using Equation 7.12, the coil sensible load is

$$Q_s = 1.1 \times CFM_2(DB_2 - DB_3)$$

$$= 1.1 \times 33,640(81.2 - 58)$$

$$= 858,500 \text{ BTU/hr}$$

Using Equation 7.15, the coil latent load is

$$Q_l = 0.68 \times CFM_2 (W_2' - W_3')$$

$$= 0.68 \times 33,640(77.5 - 65.0)$$

$$= 286,000 \text{ BTU/hr}$$

The total cooling coil load (refrigeration load) is

$$Q_t = Q_s + Q_l = 1,145,000 \text{ BTU/hr}$$

$$= 95.4 \text{ tons}$$

E. The results should be checked by using Equation 7.16 and the psychrometric chart

$$Q_s = 4.5 \times CFM(h_x - h_3)$$

$$= 4.5 \times 33,640(29.8 - 24.1)$$

$$= 862,900 \text{ BTU/hr}$$

$$Q_l = 4.5 \times CFM(h_2 - h_x)$$

$$= 4.5 \times 33,640(31.7 - 29.8)$$

$$= 287,600 \text{ BTU/hr}$$

$$Q_t = 4.5 \times CFM(h_2 - h_3)$$

$$= 4.5 \times 33,640(31.7 - 24.1)$$

$$= 1,150,000 \text{ BTU/hr} = 95.4 \text{ tons}$$

which agrees quite well with the results from part D.

It is useful at this time for the students to study Figure 7.28 closely. We have drawn the RSHR line, the mixing line, and the coil process line. Identify these and relate each point to the equipment and duct arrangement.

Note that in the preceding example the coil loads are greater than the room loads. This is because the coil must remove the excess heat from the outside air, as well as remove the room heat gains. The heat removed from the outside air is called the *outside air load*.

Example 7.32 _____

Calculate the outside air load for Example 7.31. Determine the cooling coil load and compare the result with that found previously.

Solution

Using Equation 7.16 and the psychrometric chart to find the total outside air load (we could also use the sensible and latent heat equations),

$$Q_t(OA) = 4.5 \times CFM_{OA}(h_1 - h_5)$$
$$= 4.5 \times 6730(38.6 - 30.1)$$
$$= 257,400\frac{BTU}{hr}$$

The cooling coil load (i.e., the total load on the coil) must be the heat necessary to remove the outside air plus the room loads, or

$$RSCL = 740,000 \text{ BTU/hr}$$
$$RLCL = 150,000$$
$$Q_t(OA) = \underline{257,400}$$
$$\text{Coil load} = 1,147,400 \text{ BTU/hr,}$$

which checks with the previous results.

_____

Note in Example 7.32 the outside air load includes only removing the heat necessary to bring the outside air to room conditions—the excess heat in the outside air.

The psychrometric analysis explained previously provides information on the coil requirements—the coil sensible, latent, and total loads, entering and leaving conditions, and CFM. This data will enable the system designer to select the proper coil from the manufacturer's tables. However, a greater understanding of how a coil performs can be achieved from some further concepts that will now be explained. These ideas are useful in troubleshooting as well as in selecting new equipment.

7.17   THE CONTACT FACTOR AND BYPASS FACTOR

When air passes across the outside surface of a coil, only part of the air actually contacts the surface and is cooled. This would be expected because there is a spacing between tubes.

It can be assumed that only the air that contacts the cooling surface is cooled and dehumidified. The bypass air is untreated—it leaves the coil at the same conditions as it entered.

The *Contact Factor* (CF) is defined as the proportion of air passing through the coil that touches the cooling surface and is thus cooled.

The *Bypass Factor* (BF) is defined as the proportion of air that does not touch the surface, and is therefore not cooled. From this definition, it follows that

$$CF + BF = 1$$

The next section will explain the use of the contact factor and bypass factor.

7.18   THE EFFECTIVE SURFACE TEMPERATURE

The temperature of the outside surface of a cooling coil is not the same at all places along the coil tubing. It will vary due to a number of factors, which need not be discussed here. However, we can think of an average coil surface temperature that will be called the *Effective Surface Temperature* (EST). This can be considered as the temperature to which the air that contacts the surface is cooled. (It is also called the *Apparatus Dew Point.*)

From the definition, it follows that if all the air passing over the coil contacted the surface (CF = 1), the air would leave at a temperature equal to the EST. This air would be saturated when the EST is below the air dew point, because moisture is being removed. Figure 7.29 (page 190) shows this process.

Of course it is not possible for a coil to have a CF = 1, because some air through the unit must bypass the surface. Therefore, the air leaving the coil can never be saturated. The amount of air that bypasses the surface depends on tube size and spacing, air face velocity, and number and arrangement of rows. The CF and BF factors can be measured for a coil at each face velocity. Once this is known, we can predict the performance of the coil, based on the following fact:

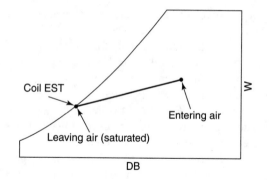

Figure 7.29
Coil process line for a cooling coil with CF = 1.

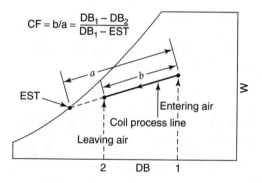

Figure 7.30
Determining CF for a cooling coil.

The CF for a coil is the ratio of the length of the coil process line to the total length of that line extended to the effective surface temperature along the saturation line.

Figure 7.30 illustrates this. The contact factor is CF = b/a.

Note from both Figures 7.29 and 7.30 that the coil effective surface temperature is the intersection of the coil process line with the psychrometric chart saturation line.

The contact factor can also be determined graphically on the psychrometric chart by using the temperatures of the air entering and leaving the cooling coil and the effective surface temperature. As seen in Figure 7.30, the contact factor is equal to the following temperature difference ratio

$$CF = \frac{b}{a} = \frac{DB_1 - DB_2}{DB_1 - EST} \qquad (7.24)$$

where

DB_1 = dry bulb temperature of air entering the cooling coil, F

DB_2 = dry bulb temperature of air leaving the cooling coil, F

EST = effective surface temperature of coil, F

The following example illustrates the use of the contact factor and effective surface temperature concepts.

Example 7.33 _____

Find the required effective surface temperature (EST), contact factor (CF), and bypass factor (BF) for a cooling coil that is to cool air from 85 F DB and 69 F WB to 56 F DB and 54 F WB.

Solution

1. The cooling coil process line is drawn on the psychrometric chart (Figure 7.31) from point 1 (entering air) to point 2 (leaving air).
2. The process line is extended to the saturation line to obtain the effective surface temperature. EST = 50 F.
3. The contact factor is calculated from Equation 7.24.

$$CF = \frac{DB_1 - DB_2}{DB_1 - EST} = \frac{85 - 56}{85 - 50} = 0.83$$

Therefore the bypass factor is

$$BF = 1 - CF = 1 - 0.83 = 0.17$$

Cooling coil selection tables showing CF, BF, and EST values for each coil are used by some manufacturers. After finding the required values of these terms by the procedures just shown, the proper coil can be selected from the tables. Coil selection will be discussed in Chapter 12.

Note from the psychrometric chart that a steep coil process line may not intersect the saturation

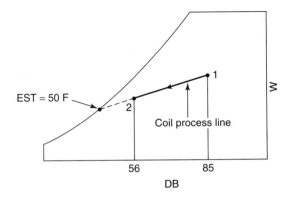

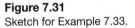

Figure 7.31
Sketch for Example 7.33.

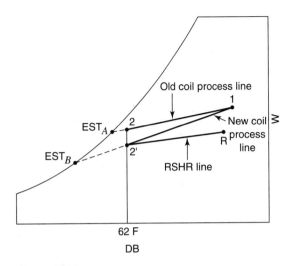

Figure 7.32
Sketch for Example 7.34.

line at all. This means that this required coil process cannot be achieved by any actual coil. We will discuss ways of resolving this problem shortly.

Example 7.34
What might be a solution to the problem that was shown to exist in Example 7.29, where the space humidity was too high?

Solution
Referring to the psychrometric chart in that example, we note that the supply air temperature was not on the RSHR line. The coil entering conditions could be measured and the coil process line drawn. Assume the coil process line 1-2-*A* is as shown in Figure 7.32. This indicates that the EST_A is too high. Lowering the coil refrigerant temperature results in a new coil process line 1-*B* and new EST_B that might result in a satisfactory supply air condition 2', on the RSHR line. (The service engineer would have to check further to see if the refrigeration equipment would allow the lower refrigerant temperature and greater load, and if the coil CF was satisfactory.)

7.19   REHEAT

The term *reheat* refers to the process where, after the warm air is cooled by the cooling coil, it is par-

tially reheated before being supplied to the air-conditioned space. The reheat process may be accomplished with a *reheat coil,* or by using return air, or mixed air. Figure 7.33 on page 192 shows the air handling unit arrangement using a reheat coil.

Sometimes reheat is used because it is difficult to achieve the desired design supply air conditions by a one-step cooling coil process. The most common conditions that may cause this problem are:

1. The room latent cooling load (RLCL) is a high proportion of the room total cooling load (RTCL). Note this results in a steep RSHR line on the psychrometric chart.
2. The air entering the cooling coil is either all outside air, or a large proportion is outside air, and furthermore, this air is at a high humidity level.

Figure 7.33 illustrates the psychrometrics of the situation. The required supply air condition is point 3. The *desired* cooling coil process line is 1-3-*A,* passing through the required supply air condition.

Note that this coil process line does not intersect the saturation line, and that the air leaving the coil is far from being saturated. Commercial cooling

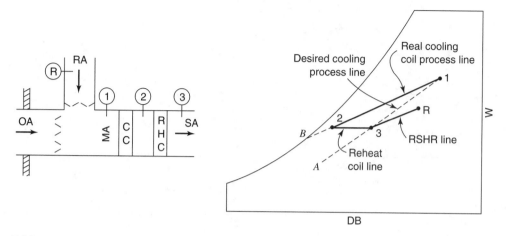

Figure 7.33
Reheat coil used to provide satisfactory supply air condition.

coils will not process air in this manner, because of their heat transfer characteristics. It is approximately safe to say that the air leaving a cooling coil will not have an RH of less than about 85–90% under typical conditions.

It can be further seen from Figure 7.33 that even though the line 1-3-*A* does not have a real effective surface temperature, the closest line we could draw to it that does intersect the saturation line might have a very low EST. This would require increased refrigeration power costs, possible frosting on the air side of the coil, and perhaps freeze-up if a chilled water coil is used.

A solution to the dilemma is to provide an actual cooling coil whose process line is 1-2, followed by a reheat coil whose process line is 2-3. The coil has an $EST_B = 49$ F.

The objections to this solution are the increased capital cost of the reheat coil, and especially the increased energy costs, which are two-fold. First, note that the cooling load is increased (1-2 instead of 1-3), and then there is a heating load 2-3.

Is some cases, a change in the indoor design conditions might avoid the need for reheat, so that a feasible cooling coil can be selected. This should be checked graphically on the psychrometric chart. Of course the conditions must remain in the comfort zone.

Fortunately, applications where reheat may be required at full load design conditions do not occur often. An example is when the latent heat gain is a very high proportion of the total, such as a dance club.

In the next section, we will see that reheat is sometimes used when the air conditioning system is operating at part load.

7.20   PART LOAD OPERATION AND CONTROL

When the cooling load is lower than the design value, the air conditioning equipment must supply less cooling capacity, otherwise the space will be overcooled. This is called part load operation.

The decreased cooling capacity can be achieved by partially reheating the cold air off the coil to the new required supply air temperature. A reheat coil can also be used for this purpose. The psychrometric processes are the same as shown previously in Figure 7.33. Now, however, the excess use of energy is even more objectionable since there are many other less wasteful ways of providing part load capacity.

In smaller commercial equipment, part load capacity is sometimes accomplished by using bypassed return air or mixed air for reheating. In these cases, the space humidity may rise at part loads, because the reheating air is adding humidity. Often the space humidity increase is small, however, so comfortable conditions are maintained.

Part load capacity can also be achieved by reducing the volume flow rate of air to the space, rather than increasing the supply air temperature.

A further discussion of the psychrometrics of the air conditioning processes involved in part load operation is best deferred to Chapter 12, when there is a more in-depth coverage of the equipment involved. We will find then that there are less energy-wasteful methods of operating at part loads than by using reheat.

7.21 FAN HEAT GAINS

The heat gains from the supply and return air fans have not been included in the psychrometric analysis we have explained. If these heat gains are a small proportion of the total, their effect can be neglected. There is no precise rule for determining when they should be considered. The greater the fan pressure, however, the greater the heat gain. Therefore, for small systems with short duct runs the effect can often be neglected. It is best to calculate the gain in each case and then decide if it is significant. Heat gains that raise the air temperature one or more degrees F should usually be included in the analysis. Procedures for calculating this effect have been explained in Chapter 6.

The psychrometric processes with the supply air fan heat gains included are shown in Figure 7.34, for a draw-through type air handling unit (the fan is downstream from the coil). Note that the supply air condition, 1, is at a higher DB than the condition leaving the cooling coil, 7. Therefore, the cooling coil load is greater and the psychrometric analysis should include this.

If the air handling unit has a blow-through fan arrangement, the fan heat gain is imposed on the coil load but does not increase the supply air temperature. If supply or return duct heat gains are significant, these will also affect the process line locations on the psychrometric chart. The system design project in Chapter 17 will provide an opportunity for seeing some of these effects.

Problems

7.1 Using the psychrometric chart on the next page, for conditions (a) to (e), list the properties not shown.

Figure 7.34
Effect of draw-through supply air fan heat gain.

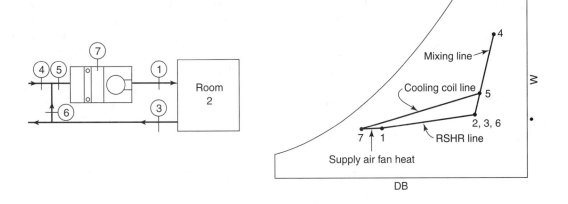

	DB, F	WB, F	DP, F	RH, %	W', gr / lb d.a.	ft³/lb d.a.	BTU/lb d.a.
(a)	80	60					
(b)	75			40			
(c)		65	50				
(d)			50	40			
(e)	70			70			

7.2 Air at 40 F DB and 60% RH is heated by an electric heater to 80 F. Find the DP, WB, and RH of the air leaving the heater. Draw the process line on the psychrometric chart.

7.3 A cold water pipe with a surface temperature of 52 F passes through a room that is at 75 F DB. At what RH in the room will moisture condense on the pipe?

7.4 Air initially at 90 F DB and 70 F WB is cooled and dehumidified to 56 F DB and 54 F WB. Draw the process line on the psychrometric chart and find the sensible, latent, and total heat removed per pound of dry air.

7.5 Using equations, solve Problem 7.4.

7.6 A cooling coil cools 5000 CFM of air from 80 F DB and 70% RH to 58 F DB and 56.5 F WB. Determine the sensible, latent, and total load on the coil, and the GPH of moisture removed.

7.7 An air handling unit mixes 1000 CFM of outside air at 92 F DB and 75 F WB with 4000 CFM of return air at 78 F DB and 45% RH. Determine the mixed air DB, WB, enthalpy, and humidity ratio.

7.8 A space has a RSCL = 83,000 BTU/hr and a RLCL = 31,000 BTU/hr. Determine the RSHR.

7.9 A space to be maintained at 75 F DB and 50% RH has a RSCL = 112,000 BTU/hr and a RLCL = 21,000 BTU/hr. An air supply of 5000 CFM is provided. Determine the supply air DB and WB.

7.10 A room has a RSCL = 20,000 BTU/hr and a RLCL = 9000 BTU/hr. The room design conditions are 77 F DB and 50% RH. Supply air is delivered at 58 F. Determine the required supply air flow rate in CFM. What is the supply air WB?

7.11 Air enters a cooling coil at 80 F DB and 66 F WB and leaves at 60 F DB and 57 F WB. Determine the coil CF, BF, and effective surface temperature.

7.12 Air at 82 F DB and 67 F WB passes through a coil with a CF of 0.91. The effective coil surface temperature is 55 F. Determine the DB and WB of the air leaving the coil.

7.13 An air conditioning unit is supplying 4000 CFM of air at 55 F DB and 53 F WB to a room maintained at 75 F DB and 55% RH. The outside air conditions are 95 F DB and 74 F WB, and 1000 CFM of ventilation air are furnished. Determine the following: RSCL, RLCL, RTCL, outside air load, and required coil CF. Sketch the apparatus arrangement and show conditions at all locations.

7.14 A room has design conditions of 78 F DB and 50% RH and a RSCL = 18,000 BTU/hr and RLCL = 8000 BTU/hr. An air conditioning unit supplies 900 CFM of air to the room at 58 F DB and 56 F WB. Will the unit maintain the room design DB and WB? What are the approximate conditions maintained in the room?

7.15 A space with a RSCL = 172,000 BTU/hr and a RLCL = 88,000 BTU/hr is to be maintained at 76 F DB and 50% RH. Conditioned air is supplied at 56 F DB and 54 F WB. Determine the wasted energy consumed if a reheat coil is used to maintain design conditions.

7.16 The following results have been found from a cooling load calculation for a building in Chicago, Illinois: RSCL = 812,000 BTU/hr., RLCL = 235,0000 BTU/hr, ventilation air = 6000 CFM, supply air = 59 F DB, space design conditions = 77 F DB, 50% RH. Design the air system:

A. Sketch apparatus arrangement

B. Determine supply air CFM and WB

C. Determine mixed air conditions

D. Determine the coil sensible, latent, and total load

E. Determine the outside air sensible, latent, and total load

F. Determine the required coil CF and BF

G. Sketch all psychrometric processes and label all points

7.17 On a day when the barometric pressure is 14.68 psi, the partial pressure of the water vapor in the air is 0.17 psia. Using equations, find the humidity ratio.

7.18 Using equations, find the relative humidity, humidity ratio, and specific volume of air at 70 F DB and 60 F DP, when the barometric pressure is 14.7 psi. Compare the results with those found from the psychrometric chart.

7.19 Using equations, find the specific enthalpy of air in Problem 7.18. Compare the result with that found using the psychrometric chart.

7.20 There is 20,000 CFM at 80 F DB and 60% RH entering an air conditioning unit. Air leaving the unit is at 57 F DB and 90% RH. Determine the

A. Cooling done by the unit in BTU/hr and tons

B. Rate that water is removed from the air in lb/hr

C. Sensible load on the unit in BTU/hr and tons

D. Latent load on the unit in BTU/hr and tons

E. Dew point of the air leaving the air conditioning unit

F. Effective surface temperature (apparatus dew point)

7.21 On a hot September day, a room has a sensible cooling load of 20,300 BTU/hr from occupants, lights, walls, windows, and so on. The latent cooling load for the room is 9000 BTU/hr. The room design conditions are 76 F DB and 50% RH. Supply air will enter the room at 58 F DB.

A. Sketch the equipment and duct arrangement, showing known information.

B. Determine the required air flow rate into the room in CFM.

C. For the above supply air, find the wet bulb temperature, enthalpy, relative humidity, and moisture content in gr/lb and lb/lb.

D. It is known that 260 CFM of outside air is required for ventilation in this room. The outside air is at 94 F DB and 76 F WB. It is mixed with return air from the room before it enters the air conditioning unit. For the mixed air, determine the dry bulb temperature, wet bulb temperature, enthalpy, and moisture content in gr/lb and lb/lb entering the unit.

E. Determine the required size of the refrigeration equipment required to condition this room, in BTU/hr and tons. (Include ventilation cooling load.)

F. Determine the savings in equipment capacity if the outside ventilation air requirement is reduced to 130 CFM. Give the answer in BTU/hr and percent.

7.22 An air-conditioned space has a room sensible cooling load of 200,000 BTU/hr and a room latent cooling load of 50,000 BTU/hr. It is maintained at 76 F DB and 64 F WB. There is 1200 CFM of air vented through cracks and hoods in the space, or through a spill (exhaust) air opening. This means that the outdoor air flow rate is 1200 CFM. The outdoor air, which is at the design conditions of 95 F DB and 76 F WB, is mixed with return air before it enters the air conditioning unit.

A. Sketch the equipment and duct arrangement, showing known information.

B. Calculate the room sensible heat ratio (RSHR).

C. Find the required supply air flow rate in CFM for a supply air temperature of 60 F DB.

D. Determine the cooling load of the outside air in BTU/hr and tons.

E. Calculate the total cooling load in BTU/hr and tons.

F. What is the effective surface temperature (apparatus dew point)?

G. What is the coil CF and BF?

7.23 A refrigeration chiller supplies chilled water to an air conditioning unit. The unit takes in 3000 CFM of outdoor air at 95 F DB and 76 F WB. This outdoor air mixes with 20,000 CFM of return air at 78 F DB and 50% RH. Conditioned air leaves the cooling coil in the air conditioning unit at 52 F DB and 90% RH.

A. What is the load on the chiller due to the coil in the air conditioning unit? Give the answer in BTU/hr and tons.

B. Assume the conditioned air is reheated to 58.5 F DB with electric heaters. What is the operating cost per hour of these heaters if power costs 10 cents per kilowatt hour?

7.24 The supply air for Problem 6.12 is at 58 F DB. Determine

A. The required air flow rate in CFM

B. Supply air WB, RH, enthalpy, and moisture content

C. Mixed air DB, WB, enthalpy, and moisture content

D. Coil sensible, latent, and total load

7.25 The supply air for Problem 6.16 is at 60 F. Determine

A. The required air flow rate

B. Supply air WB, RH, enthalpy, and moisture content

C. Mixed air DB, WB, enthalpy, and moisture content

D. Coil sensible, latent, and total load

CHAPTER 8
Fluid Flow in Piping and Ducts

In planning or servicing an HVAC system, it is often necessary to determine pump and fan pressure requirements and piping or duct pressure losses. These and related problems can be solved by an application of some principles of fluid flow which apply to the flow of water and air in air conditioning systems.

OBJECTIVES

After studying this chapter, you will be able to:

1. Use the continuity equation to find flow rate.
2. Use the energy equation to find pump and fan pressures.
3. Find velocity from total and static pressure.
4. Determine pipe and duct sizes.

8.1 THE CONTINUITY EQUATION

The flow of water through piping and air through ducts in HVAC systems is usually under conditions called *steady flow*. Steady flow means that the flow rate of fluid at any point in a section of pipe or duct is equal to that at any other point in the same pipe or duct, regardless of the pipe or duct's shape or cross section. That is, *the same quantity of fluid is passing through every section at a given moment.*

For example, in Figure 8.1 on page 198, suppose the flow rate of water past section 1 were 10 GPM. If there is steady flow, there must also be 10 GPM flowing past section 2. To see this more clearly, if less than 10 GPM were flowing past section 2, say 4 GPM, ask, what happened to the remaining 6 GPM that left section 1? It cannot disappear or be lost (unless there is a hole in the pipe!). Similarly, there cannot be more flow at section 2 than at section 1 because there was only 10 GPM available initially.

In HVAC systems, the density of the air or water flowing generally does not change significantly. When the density remains constant, the flow is called *incompressible.*

Steady flow is a special case of a general principle called either the *conservation of mass principle* or the *continuity principle.* The continuity principle

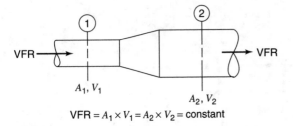

$$VFR = A_1 \times V_1 = A_2 \times V_2 = \text{constant}$$

Figure 8.1
The continuity equation for steady flow of air through a duct or water through a pipe.

can be expressed as an equation, which is called the *continuity equation*. For incompressible steady flow, the continuity equation is

$$VFR = \text{constant} = A_1 \times V_1 = A_2 \times V_2 \qquad (8.1a)$$

where

VFR = volume flow rate of fluid

A_1, A_2 = cross-sectional area of pipe or duct at any points 1 and 2

V_1, V_2 = velocity of fluid at any points 1 and 2

(Figure 8.1 illustrates Equation 8.1a.)

The continuity equation can also be expressed using mass flow rate (MFR) rather than VFR. From Equation 2.1, mass equals density times volume. Then, with constant density (d),

$$\begin{aligned} MFR &= \text{constant} \\ &= d \times VFR = d \times A_1 \times V_1 \qquad (8.1b) \\ &= d \times A_2 \times V_2 \end{aligned}$$

where

MFR = mass flow rate

d = density of fluid

Most flows in HVAC work are incompressible steady flow, so Equations 8.1a and 8.1b can be used. Occasionally unsteady flow exists. (Analysis of this situation is beyond the scope of this book.)

Example 8.1 illustrates uses of the continuity equation.

Example 8.1 _____

A service engineer wishes to check if the proper flow rate is circulating in the chilled water piping on a job. This engineer measures a water velocity of 10 ft/sec. The cross-sectional area of pipe is 2 ft². What is the water flow rate through the pipe in GPM (gal/min)?

Solution
Using Equation 8.1a,

$$VFR = A_1 \times V_1 = 2 \text{ ft}^2 \times 10 \text{ ft/sec}$$
$$= 20 \text{ ft}^3/\text{sec}$$

Converting from ft³/sec to GPM

$$VFR = 20 \frac{\text{ft}^3}{\text{sec}} \times \frac{60 \text{ sec}}{1 \text{ min}} \times \frac{7.48 \text{ gal}}{1 \text{ ft}^3}$$
$$= 8980 \text{ GPM}$$

Area and Velocity Change

The continuity equation may be used to demonstrate how velocity is affected by changes in the pipe or duct size. Since

$$VFR = \text{constant} = A_1 V_1 = A_2 V_2 \qquad (8.1a)$$

Solving for V_1 (or V_2),

$$V_1 = \frac{A_2}{A_1} \times V_2; \qquad V_2 = \frac{A_1}{A_2} \times V_2 \qquad (8.1c)$$

That is, *the velocity changes inversely with the cross-sectional area.* If the pipe or duct size increases, the velocity decreases; if the size decreases, the velocity increases. This is shown in Figure 8.2.

Do not confuse *flow rate* with *velocity*. With steady flow, at any given condition, the flow rate of the fluid (quantity flowing) does not change, regardless of any change in pipe or duct size. The velocity (speed), however, will inevitably change with pipe or duct size.

Examples 8.2 and 8.3 illustrate uses of the continuity equation.

Example 8.2 _____

Air is flowing through a 1 ft × 2 ft (Figure 8.3) duct at a rate of 1200 CFM (ft³/min). The duct

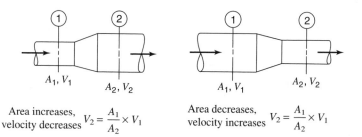

Figure 8.2
Change of velocity with change in cross-sectional area of duct or pipe.

decreases size to 0.5 ft × 1 ft. What is the air velocity in the second section of duct?

Solution
Using Equation 8.1a to find V_1,

$$V_1 = \frac{\text{VFR}}{A_1} = \frac{1200 \text{ ft}^3/\text{min}}{2 \text{ ft}^2} = 600 \text{ ft/min}$$

As the flow rate is constant, $A_1 \times V_1 = A_2 \times V_2$. Solving for V_2,

$$V_2 = \frac{A_1}{A_2} \times V_1 = \frac{2 \text{ ft}^2}{0.5 \text{ ft}^2} \times 600 \text{ ft/min}$$

$$= 2400 \text{ ft/min}$$

Example 8.3 _____
Air is flowing in a duct of 48 in.^{2} cross-sectional area at a velocity of 2400 ft/min. This high velocity results in a disturbing noise. The HVAC contractor wants to reduce the velocity to 1300 ft/min. What size duct should be substituted?

Solution
Using Equation 8.1, solving for A_2,

Figure 8.3
Sketch for Example 8.2.

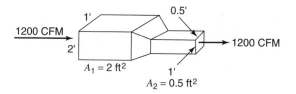

$$A_2 = \frac{V_1}{V_2} \times A_1 = \frac{2400}{1300} \times 48 \text{ in.}^2 = 88 \text{ in.}^2$$

A duct that has this cross-sectional area would be substituted (say 11 in. × 8 in.).

_____

8.2   THE FLOW ENERGY EQUATION

When the energy balance principle (Section 2.11) is applied to flow in a pipe or duct, it may be stated as follows: between any two points 1 and 2 (Figure 8.4 on page 200),

$$E_1 + E_{\text{add}} - E_{\text{lost}} = E_2$$

or

$$E_1 + E_{\text{add}} = E_2 + E_{\text{lost}}$$

where

E_1, E_2 = energy stored in fluid at points 1 and 2

E_{add} = energy added to fluid between points 1 and 2

E_{lost} = energy lost from fluid between points 1 and 2

The energy of the fluid at any point consists of pressure, velocity (kinetic energy), and elevation (potential energy). The energy added may be that of a pump or fan. The energy lost is due to friction. There may be other energy changes (e.g., a temperature change), but they are usually small and may be neglected.

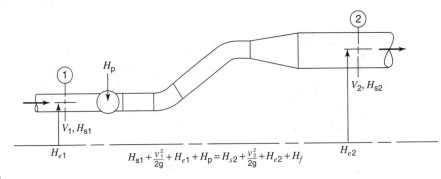

Figure 8.4
The flow energy equation applied to flow in a duct or pipe.

If the energy balance is expressed as an equation using units of head pressure (ft of fluid), as in Figure 8.4, it becomes

$$\underbrace{E_1}_{} + \underbrace{E_{add}}_{} = \underbrace{E_2}_{} + \underbrace{E_{lost}}_{}$$

$$H_{s1} + \frac{V_1^2}{2g} + H_{e1} + H_p = H_{s2} + \frac{V_2^2}{2g} + H_{e2} + H_f$$

$$(8.2a)$$

where

H_s = static pressure of fluid (pressure at rest), ft

V = velocity, ft/sec

g = gravitational constant, 32.2 ft/sec^2

$\dfrac{V^2}{2g}$ = velocity pressure, ft

H_e = elevation, ft

H_p = pressure added by pump or fan, ft

H_f = pressure lost in piping or duct from friction, ft

Equation 8.2a is called the *flow energy equation* or *generalized Bernoulli equation*. It is used often to determine the pressure requirements of pumps and fans and in testing and balancing systems.

Equation 8.2a can be arranged in a useful form by solving for the term H_p and grouping other terms as follows.

$$H_p = (H_{s2} - H_{s1}) + \frac{(V_2^2 - V_1^2)}{2g} + (H_{e2} - H_{e1}) + H_f$$

$$(8.2b)$$

where

H_p = pump or fan pressure, ft

$H_{s2} - H_{s1}$ = change in static pressure, ft

$\dfrac{V_2^2 - V_1^2}{2g}$ = change in velocity pressure, ft

$H_{e2} - H_{e1}$ = change in pressure due to elevation change, ft

H_f = pressure lost in piping or duct from friction, ft

Expressed in the form of Equation 8.2b, the energy equation is used to find the required pump or fan pressure for a system.

Example 8.4
The piping system shown in Figure 8.5 is to deliver water from the basement to the roof storage tank, 180 ft above, in the Abraham Lincoln Apartments. The friction loss in the piping, valves, and fittings is 12 ft w. The water enters the pump at a gage pressure of 10 ft and is delivered at atmospheric pressure (all values are gage pressure). The velocity at the pump suction is 2 ft/sec and at the piping exit is 10 ft/sec. What is the required pump pressure?

Solution
Using Equation 8.2b,

$$H_p = (H_{s2} - H_{s1}) + \frac{(V_2^2 - V_1^2)}{2g}$$

$$+ (H_{e2} - H_{e1}) + H_f$$

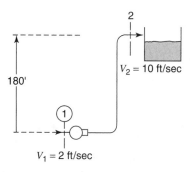

Figure 8.5
Sketch for Example 8.4.

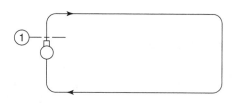

Figure 8.6
Sketch for Example 8.5.

$$(H_{s2} - H_{s1}) = 0 - 10$$
$$= -10 \text{ ft (change in static pressure)}$$
$$\frac{(V_2{}^2 - V_1{}^2)}{2g} = \frac{(10)^2 - (2)^2}{64.4}$$
$$= 1.5 \text{ ft (change in velocity pressure)}$$
$$(H_{e2} - H_{e1}) = 180 \text{ ft (change in elevation)}$$
$$H_f = 12 \text{ ft (friction pressure loss)}$$
$$H_p = -10 + 1.5 + 180 + 12$$
$$= 184 \text{ ft w.g.}$$

The additional pressure required because of the velocity pressure change in Example 8.4 was small. In some cases in piping systems, it is small enough to be neglected.

8.3 PRESSURE LOSS IN CLOSED AND OPEN SYSTEMS

An *open* piping or duct system is one that is open to the atmosphere at some point. Example 8.4 is an open system. Note that any elevation change is included in determining the pump head. A condenser-cooling tower water system is also open.

A *closed* system is one where the water is recirculated continuously and there is no gap or opening in the piping. In a closed system, there is no net change in elevation of the water around the whole circuit, and therefore the change in H_e is 0 in the flow energy equation. A hydronic system is a closed system.

Example 8.5

The pressure loss due to friction in the hydronic system shown in Figure 8.6 is 24 ft w. What is the pump head required?

Solution
This is a closed system. Starting at the pump discharge, point 1, and going around the complete loop back to 1,

$$H_{e1} - H_{e1} = 0 \text{ (no net change in elevation)}$$
$$V_1{}^2 - V_1{}^2 = 0$$
$$H_{s1} - H_{s1} = 0$$

Using Equation 8.2a, and since all the above terms = 0,

$$H_p = H_f = 24 \text{ ft w.}$$

Example 8.5 shows that the *pump head in a closed hydronic system is equal to the pressure loss due to friction around the complete circuit.*

Air duct systems are almost always open systems. When using the energy Equation 8.2a or 8.2b, however, the terms expressing the change in pressure due to elevation change $(H_{e1} - H_{e2})$ is either zero (the duct layout is horizontal) or is usually small enough to be negligible. The velocity change term, however, is sometimes significant, and if so, cannot be neglected.

Air pressure values in ducts are usually measured in inches of water gage (in.w.g.).

Example 8.6

The duct shown in Figure 8.7 has 8000 CFM flowing through it. The friction loss from point 1 to 2 is 0.43 in. w. If the static pressure at 1 is 1.10 in w.g., what is the static pressure at 2?

Solution

Writing Equation 8.2a to solve for H_{s2},

$$H_{s2} = H_{s1} + H_p - H_f + \left(\frac{V_1^2 - V_2^2}{2g}\right) + (H_{e1} - H_{e2})$$

$H_p = 0$ (because there is no fan between 1 and 2)

$H_{e1} - H_{e2} = 0$ (insignificant elevation change)

The static pressure is therefore

$$H_{s2} = H_{s1} + H_f + \left(\frac{V_1^2 - V_2^2}{2g}\right)$$

Finding V_1 and V_2 from Equation 8.1,

$$V_1 = \frac{8000 \text{ ft}^3/\text{min}}{4 \text{ ft}^2}$$

$$= 2000 \text{ ft/min} \times \frac{1 \text{ min}}{60 \text{ sec}}$$

$$= 33.33 \text{ ft/sec}$$

$$V_2 = \frac{8000 \text{ ft}^3/\text{min}}{16 \text{ ft}^2}$$

$$= 500 \text{ ft/min} \times \frac{1 \text{ min}}{60 \text{ sec}}$$

$$= 8.33 \text{ ft/sec}$$

Therefore

$$\frac{V_1^2 - V_2^2}{2g} = \frac{(33.33)^2 - (8.33)^2}{64.4} = 16.2 \text{ ft air}$$

Figure 8.7
Sketch for Example 8.6.

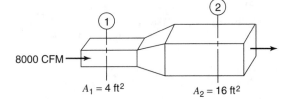

Converting units to in. w.,

$$16.2 \text{ ft air} \times \frac{1 \text{ in. w.}}{69.6 \text{ ft air}} = 0.23 \text{ in. w.}$$

$$H_{s2} = 1.10 + 0 - 0.43 + 0.23 + 0$$

$$= 0.90 \text{ in. w.g.}$$

In Example 8.6, it was found that the pressure decreased from point 1 to 2 because of the pressure loss due to friction, but there was a partial *increase* in pressure because of the velocity *decrease* (0.23 in. w.). This occurrence is of importance in airflow in ducts, as will be explained in the next sections.

8.4   TOTAL, STATIC, AND VELOCITY PRESSURE

The *total pressure* (H_t) of a flowing fluid is defined as

$$H_t = H_s + H_v \qquad (8.3)$$

where

H_t = total pressure

H_s = static pressure

H_v = velocity pressure

The *static pressure* is the pressure the fluid has at rest.

The *velocity pressure* is defined as

$$H_v = \frac{V^2}{2g} \qquad (8.4)$$

Thus the total pressure energy that a fluid has at any point can be considered to consist of two parts, its static pressure energy and its velocity pressure energy.

The velocity pressure concept is useful in measuring velocities and flow rates in piping and ducts. If the velocity pressure can be measured, the velocity can be found by solving Equation 8.4 for V:

$$V = \sqrt{2gH_v} \qquad (8.5)$$

where

V = velocity, ft/sec

g = gravitational constant, ft/sec^2

H_v = velocity head, ft of fluid

Example 8.7

The total pressure and static pressure are measured as 66.5 ft w. and 64.8 ft w., respectively, in the condenser water pipe line from a refrigeration machine. What is the velocity in the line?

Solution

Using Equations 8.3 and 8.5,

$$H_v = H_t - H_s = 66.5 - 64.8 = 1.7 \text{ ft w.}$$

$$V = \sqrt{2gH_v} = \sqrt{2 \times 32.2 \text{ ft/sec}^2 \times 1.7 \text{ ft}}$$

$$= 10.5 \text{ ft/sec}$$

When measuring airflow, using in. w. as the unit of pressure and velocity in ft/min, if the appropriate conversion units are substituted in Equations 8.4 and 8.5, they become

$$H_v = \left(\frac{V}{4000}\right)^2 \tag{8.6}$$

$$V = 4000\sqrt{H_v} \tag{8.7}$$

where

V = air velocity, ft/min

H_v = velocity pressure, in. w.

Many testing and balancing instruments for measuring flow utilize the relationship between total, static, and velocity pressure. Figure 8.8 shows an example. In Figure 8.8(a), a manometer is connected to the duct to read static pressure. In Figure 8.8(b), a manometer reads total pressure because in addition to being exposed to the static pressure, the impact tube at the end of the manometer faces the

oncoming airstream, and therefore receives the velocity pressure energy as well. By connecting the two manometers as shown in Figure 8.8(c), the difference between total and static pressure—the velocity pressure—is read directly.

The *pitot tube* (Figure 8.9 on page 204) is another airflow measuring device that works in the same manner. The probe that is inserted in the duct has two concentric tubes. The opening facing the airstream measures the total pressure and the concentric holes are exposed to static pressure, so the velocity pressure is read directly. A number of readings are usually taken across the duct to get an average velocity.

Example 8.8

A contractor wishes to check the air flow rate in a 28 in. × 16 in. duct. The contractor takes a set of readings with a pitot tube, averaging 0.8 in. w. What is the air flow rate in the duct?

Solution

From Equation 8.7, the air velocity is

$$V = 4000\sqrt{0.8} = 3580 \text{ ft/min}$$

The duct cross-sectional area is

$$A = 28 \text{ in.} \times 16 \text{ in.} \times \frac{1 \text{ ft}^2}{144 \text{ in.}^2} = 3.11 \text{ ft}^2$$

and the flow rate from Equation 8.1 is

$$\text{VFR} = A \times V = 3.11 \text{ ft}^2 \times 3580 \text{ ft/min}$$

$$= 11,100 \text{ CFM}$$

Figure 8.8
Manometer arrangement to read static, total, and velocity pressure.

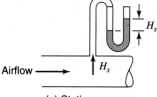

(a) Static pressure

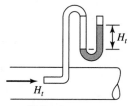

(b) Total pressure

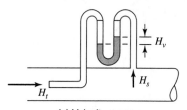

(c) Velocity pressure

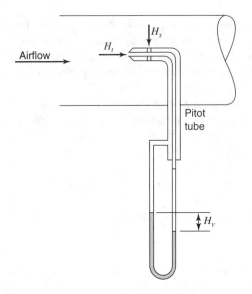

Figure 8.9
Pilot tube used for measuring velocity pressure.

8.5 CONVERSION OF VELOCITY PRESSURE TO STATIC PRESSURE (STATIC REGAIN)

One of the remarkable things that can occur in flow in a duct or pipe is that the static pressure can increase in the direction of flow if the velocity decreases. This is caused by a conversion of velocity energy to static energy, called *static regain*. It is a phenomenon that we have all experienced. If we hold a hand in front of the stream of water from a hose, we feel the pressure that is a result of reducing the velocity energy and converting it to pressure.

Consider the diverging air duct section in Figure 8.10. Using Equation 8.6, the difference in velocity between points 1 and 2 is

$$H_{v1} - H_{v2} = \left(\frac{V_1}{4000}\right)^2 - \left(\frac{V_2}{4000}\right)^2$$

If we now apply the flow energy Equation 8.2a, assuming there is no friction loss H_f and the change in elevation is negligible, then

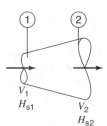

Figure 8.10
Sketch for Equation 8.8.

$$H_{s2} - H_{s1} = H_{v1} - H_{v2} = \left(\frac{V_1}{4000}\right)^2 - \left(\frac{V_2}{4000}\right)^2 \quad (8.8)$$

Equation 8.8 shows that *if the velocity decreases in the direction of flow* (because the pipe or duct size has increased) *then the static pressure increases.* Velocity energy has been converted to pressure energy. This effect is called *static pressure regain.*

Because there is always some friction loss, the actual static pressure regain is never as high as that shown in Equation 8.8. The proportion of static regain that can be recovered, called the *recovery factor R,* depends on the shape of the transition that changes velocity. The actual static pressure regain (SPR) is therefore

$$SPR = R\left(\left[\frac{V_1}{4000}\right]^2 - \left[\frac{V_2}{4000}\right]^2\right) \quad (8.9)$$

Recovery factors of 0.7 to 0.9 can be achieved with reasonably gradual transitions, thereby keeping friction losses low.

Example 8.9 _____
Find the increase in static pressure (regain) from point 1 to 2 in the duct system shown in Figure 8.11, if the recovery factor is 0.7.

Solution
Using Equation 8.9 with $R = 0.7$,

$$H_{s2} - H_{s1} = 0.7\left[\left(\frac{1800}{4000}\right)^2 - \left(\frac{600}{4000}\right)^2\right]$$

$$= 0.13 \text{ in. w.}$$

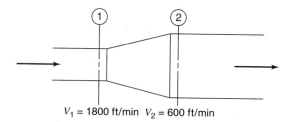

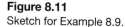

Figure 8.11
Sketch for Example 8.9.

The opposite event to a static pressure regain, a conversion of static pressure to velocity pressure, will occur in a converging transition, resulting in a decrease in static pressure (Figure 8.12). This effect occurs in a nozzle, where the velocity increases.

8.6 PRESSURE LOSS FROM FRICTION IN PIPING AND DUCTS

We have seen from the flow energy equation that one of the effects the pump or fan must overcome is the pressure loss due to friction. Friction is a resistance to flow resulting from fluid viscosity and from the walls of the pipe or duct. In previous examples, we have assumed values of friction pressure loss. Actually we must be able to calculate it.

For the type of flow usually existing in HVAC systems, called *turbulent flow*, the pressure loss or drop due to friction can be found from the following equation (called the Darcy-Weisbach relation):

$$H_f = f \frac{L}{D} \frac{V^2}{2g} \qquad (8.10)$$

where

H_f = pressure loss (drop) from friction in straight pipe or duct

f = a friction factor

L = length of pipe or duct

D = diameter of pipe or duct

V = velocity of fluid

The friction factor f depends on the roughness of the pipe or duct wall. Rougher surfaces will cause increased frictional resistance. This means that by using and maintaining smooth surfaces, friction decreases and less energy is used. The other terms in the equation also indicate useful information. Lower velocities and larger diameters reduce H_f and therefore result in lower energy consumption, although the pipe or duct cost then increases.

Although H_f could be calculated each time from Equation 8.10, charts that are much easier to use and show the same information have been developed for water flow and airflow.

Figure 8.12
Conversion between velocity pressure and static pressure. (*a*) Diverging transition—velocity decreases, static pressure increases. (*b*) Converging transition—velocity increases, static pressure decreases.

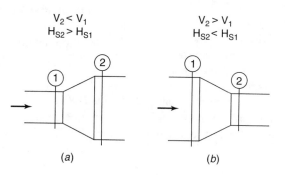

8.7 FRICTION LOSS FROM WATER FLOW IN PIPES

The pressure loss or drop caused by friction with water flow in straight pipe has been put in a convenient chart format for commonly used materials and conditions. Three of such charts are presented in this book.

Figures 8.13 and 8.14 (page 208) are suitable for water at 60 F flowing in steel Schedule 40 pipe. The Schedule number refers to the pipe wall thickness. Schedule 40 pipe is widely used for water under pressure in HVAC installations (see Chapter 9).

Figure 8.13 is suitable when the pipe wall is in a clean condition. This is generally true in a closed hydronic heating and cooling system that is reasonably well maintained.

Figure 8.14 is suitable for open piping systems —that is, systems open to the atmosphere at some point. In such systems, the pipe wall is usually rougher than in closed systems, resulting in a higher friction loss. Figure 8.14 accounts for this. The water piping system between a refrigeration condenser and cooling tower is an example of an open system.

Figure 8.15 on page 209 is suitable for water at 60 F flowing in copper tubing. The Type K, L, and M lines on the chart refer to different tube wall thicknesses. Type K or L copper tubing is widely used for water under pressure in HVAC installations. Figure 8.15 is suitable for both closed and open systems. Copper tube wall will usually not roughen significantly with age in open systems.

For chilled water temperature ranges (40–50 F) and condensing water temperature ranges (80–100 F), Figures 8.13, 8.14, and 8.15 may be used without correction. For hot water systems with temperatures in the vicinity of 200 F, the pressure loss due to friction is about 10% less than shown and should be corrected. This is a result of the change in viscosity and density with temperature.

Pressure drop charts for other piping materials and liquids can be found in appropriate handbooks.

The following example will illustrate use of the friction loss charts.

Example 8.10

What is the pressure loss due to friction and the velocity in 500 ft of 2 in. Schedule 40 steel piping through which 40 GPM of water at 60 F is flowing in a closed system?

Solution

The information can be found from Figure 8.13 (closed systems). The solution is indicated in Figure 8.16 on page 210, at the point of intersection of a 40 GPM flow rate and $D = 2$ in. Note that the chart lists friction loss per 100 ft of pipe, which is then converted to the loss in the actual length of pipe.

At 40 GPM and $D = 2$ in., H_f per 100 ft = 3.2 ft w., therefore

$$H_f = \frac{3.2 \text{ ft w.}}{100 \text{ ft}} \times 500 \text{ ft} = 16.0 \text{ ft w.}$$

The velocity at the intersection point is $V = 3.9$ ft/sec.

Example 8.11

A copper tubing system is to be used to circulate 30 GPM of water at 60 F. The system is to be designed to have a friction pressure drop no greater than 3 ft w. per 100 ft pipe. What is the smallest size tubing that can be used?

Solution

Figure 8.15 will be used. The solution is shown in Figure 8.17 on page 210. The intersection point of 30 GPM and 3 ft w./100 ft pipe lies between a 2 in. and 1½ in. diameter. If a 1½ in. diameter is used, the pressure drop will be greater than 3 ft w./100 ft at 30 GPM, so this is unacceptable. If a 2 in. diameter is used, the pressure drop will be less than the maximum allowed; therefore, this is the correct solution. Note that the actual rather than the allowed pressure drop should be recorded. The solution is

$$D = 2 \text{ in.}, H_f = 2.0 \text{ ft w./100 ft}$$

Example 8.12

A 3 in. steel pipe, Schedule 40, is supposed to circulate 200 GPM in a chilled water system. I. Fixit, a service troubleshooter, is asked to check if the

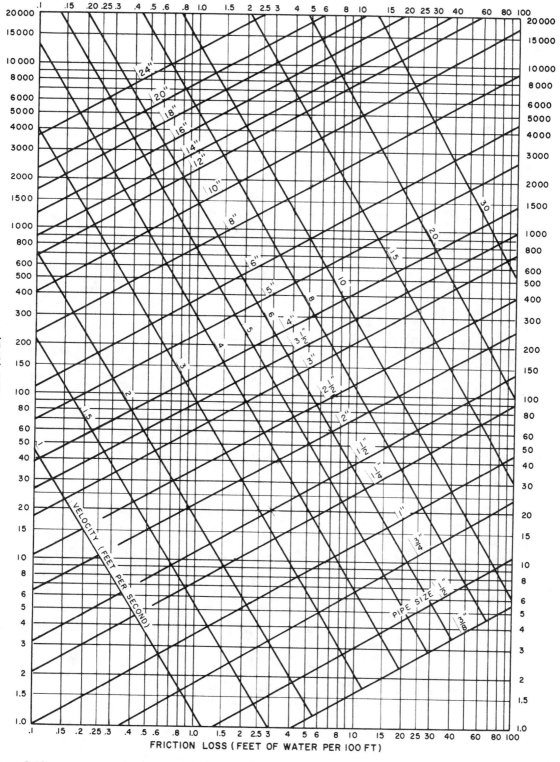

Figure 8.13
Friction loss for water in Schedule 40 steel pipe—closed system. (Courtesy: Carrier Corporation, Syracuse, N.Y.)

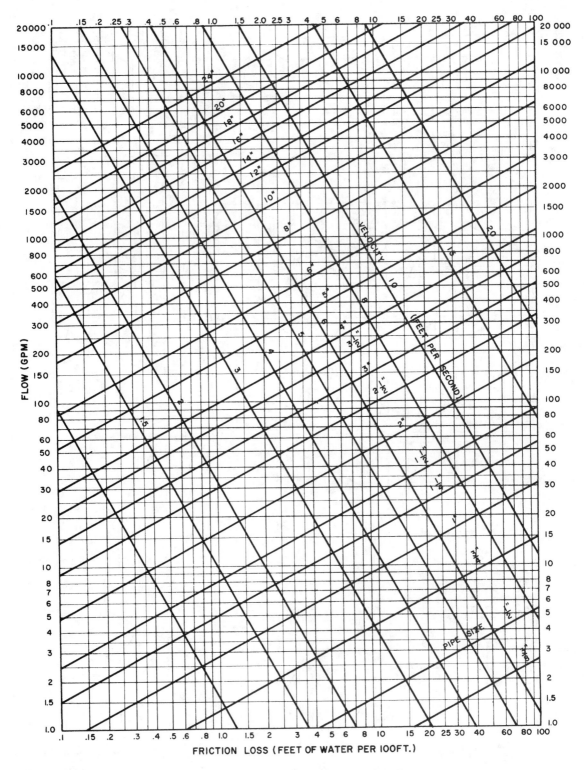

Figure 8.14
Friction loss for water in Schedule 40 steel pipe—open system. (Courtesy: Carrier Corporation, Syracuse, N.Y.)

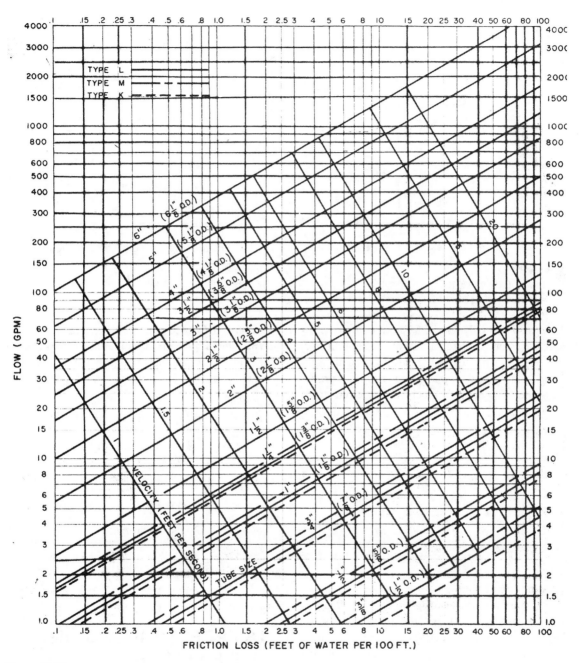

Figure 8.15
Friction loss for water in copper tubing—open or closed system. (Courtesy: Carrier Corporation, Syracuse, N.Y.)

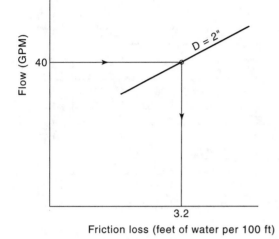

Figure 8.16
Sketch for Example 8.10.

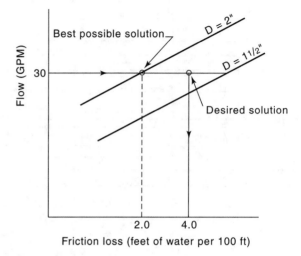

Figure 8.17
Sketch for Example 8.11.

flow rate is actually 200 GPM. Fixit puts two pressure gages on a horizontal run of the straight line, 200 ft apart. The first gage reads 40 ft w., the second, 32 ft w. Is the system delivering the proper flow rate of water? How much is being circulated?

Solution
The actual pressure drop (equal to the friction loss) is

$$H_{s1} - H_{s2} = H_f = 40 - 32$$
$$= 8 \text{ ft w. per 200 ft pipe}$$

or

$$H_f/100 \text{ ft pipe} = \frac{8 \text{ ft w.}}{200 \text{ ft}} \times 100 \text{ ft} = 4 \text{ ft w.}$$

Reading from Figure 8.13, at a friction loss of 4 ft w. in a 3 in. pipe, the actual flow rate is 130 GPM, far less than normal. (Fixit now must look for the cause of the problem. Perhaps a valve is throttled closed too much, or perhaps a pump is not performing properly.)

Note: For hot water systems, the correction of 10% less pressure drop should be made. Otherwise oversized equipment or wasteful energy losses will result.

Example 8.13 _____
What would be the friction pressure drop in 800 ft of 2 in. copper tubing through which 50 GPM of water is flowing in a hydronic heating system?

Solution
From Figure 8.15, the pressure drop for cold water is

$$H_f/100 \text{ ft} = 5 \text{ ft w.}$$

correcting this for hot water

$$H_f/100 \text{ ft} = 0.9 \times 5 = 4.5 \text{ ft w.}$$

For 800 ft, the pressure drop is

$$H_f = \frac{4.5 \text{ ft w.}}{100 \text{ ft}} \times 800 \text{ ft} = 36.0 \text{ ft w.}$$

8.8 PRESSURE LOSS IN PIPE FITTINGS

In addition to the pressure loss in straight pipe, there will be pressure losses from turbulence and change of direction through fittings and valves. These are called dynamic losses.

These pressure losses are shown in Table 8.1. The pressure losses are expressed in this table in a way that is called the *equivalent length*. The listings for a particular fitting of a given size show the equivalent length (E.L.) of straight pipe that would have the same pressure drop as that fitting. After finding the E.L. from Table 8.1, the appropriate friction loss chart is used to find the actual pressure drop through the fitting.

Example 8.14

Find the pressure drop through a 4 in. 90° cast iron (C.I.) standard elbow in a chilled water system though which 300 GPM of water is flowing.

Solution

From Table 8.1, find the equivalent length of the fitting

$$E.L. = 11.0 \text{ ft}$$

Using Figure 8.13, $H_f/100$ ft = 5.2 ft w. The pressure drop through the fitting is

$$H_f = \frac{5.2 \text{ ft w.}}{100 \text{ ft}} \times 11.0 \text{ ft} = 0.6 \text{ ft w.}$$

In addition to the equivalent length method of determining pressure drop through pipe fittings, there is another procedure called the loss coefficient method. A loss coefficient (called C, C_v, or K) for the fitting is determined from an appropriate table listing C-values. The loss coefficient method will not be used for pipe fittings here. It will be used for duct fittings (see Section 8.13).

8.9 PIPING SYSTEM PRESSURE DROP

A common problem is to determine the pressure loss from friction in a closed system, in order to determine the required pump head. The system pressure drop is simply the sum of the losses through each item in one of the paths or circuits from pump discharge to pump suction, including piping,

TABLE 8.1   EQUIVALENT FEET OF PIPE FOR FITTINGS AND VALVES

	Nominal Pipe Size (inches)												
	½	¾	1	1¼	1½	2	2½	3	4	5	6	8	10
45° Elbow	0.8	0.9	1.3	1.7	2.2	2.8	3.3	4.0	5.5	6.6	8.0	11.0	13.2
90° Elbow standard	1.6	2.0	2.6	3.3	4.3	5.5	6.5	8.0	11.0	13.0	16.0	22.0	26.0
90° Elbow long	1.0	1.4	1.7	2.3	2.7	3.5	4.2	5.2	7.0	8.4	10.4	14.0	16.8
Gate valve open	0.7	0.9	1.0	1.5	1.8	2.3	2.8	3.2	4.5	6.0	7.0	9.0	12.0
Globe valve open	17	22	27	36	43	55	67	82	110	134	164	220	268
Angle valve	7	9	12	15	18	24							
Tee—side flow	3	4	5	7	9	12	14	17	22	28	34	44	56
Swing check valve	6	8	10	14	16	20	25	30	40	50	60	80	100
Tee—straight through flow	1.6	2.0	2.6	3.3	4.3	5.5	6.5	8.0	11.0	13.0	16.0	22.0	26.0
Radiator angle valve	3	6	8	10	13								
Diverting tee		20	14	11	12	14	14	14					
Flow check valve		27	42	60	63	83	104	125	126				
Air scoop		2	3	4	5	7	8	13	15				
Boiler (typical)	5	7	9	11	13	17							

fittings, valves, and equipment. Information on pressure drops through equipment is obtained from the manufacturer.

To determine the system pressure loss, the pressure losses *through only one circuit are considered.* This is because the pressure losses are the same through every circuit. This idea is quite similar to that in electric circuits, where the voltage loss through parallel electric circuits is the same. Figure 8.18 illustrates this. The pressure drop from A to D is indicated by the difference in readings on the two pressure gages located at A and D. Therefore, the pressure drop through the longer circuit ABD is the same as that through ACD.

It would seem, from this explanation, that it would not matter which circuit one chooses for actually calculating the system pressure drop. It usually does not work out that way, however. Most piping systems are designed to have equal friction loss per foot of length. It would seem, therefore, using our example above, that in this case the pressure drop ABD would be greater than that through ACD. This certainly is not possible, because each pressure gage has one fixed reading.

What happens in most hydronic systems is that valves are used to "balance the system"; each valve is throttled to a position that results in the correct flow rate in that valve's circuit. By throttling (partially closing) the valve, an additional pressure drop in that circuit is created. In circuit ACD, the valve might be throttled considerably.

Pressure drops in Table 8.1 and other measured results are valid only for fully open valves. Therefore, the pressure loss through a partially open valve in an actual installation cannot be determined.

For this reason, it is customary to select the longest circuit in a system to calculate the pressure drop, and to assume that the valves in this circuit are full-open.

Therefore, to find the total system pressure drop in a multi-circuited system, proceed as follows:

1. Examine the piping layout to determine which of the parallel connected circuits has the longest total equivalent length, bearing in mind the following:
   A. Usually, this is the circuit that has the longest straight pipe length.
   B. Occasionally it is a shorter circuit that has such an exceptionally large number of fittings, valves, and equipment that makes this circuit the one with the longest total equivalent length (TEL).
2. Having decided on the basis of this investigation which circuit has the greatest TEL, *calculate the pressure drop through this circuit and only this circuit, ignoring all others,* using the procedures we have explained.

It will be helpful to draw a sketch of the piping system, labeling each point (intersections, equipment), as well as indicating all flow rates, lengths, and pipe sizes in each section.

Prepare a table listing each section and item in that circuit (and *only* that circuit) chosen for the calculation. List all of the features related to the task in this table. Figure 8.19, Table 8.2, and Example 8.15 illustrate the procedure.

Example 8.15

For the steel piping systems shown in Figure 8.19, determine the required pump head.

Figure 8.18

Pressure drop from points A to D is always the same, regardless of path length.

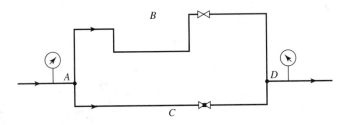

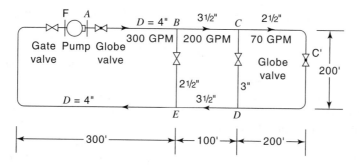

Figure 8.19
Sketch for Example 8.15.

Solution

From the energy equation, the required pump head rise *F–A* is equal to the pressure drop due to friction from *A–F* through the system. But to find this, only the circuit with the greatest pressure drop is chosen. This is *ABCC'DEF*, the longest circuit. Circuits *ABEF* and *ACDF* are ignored. (It is possible in un-

usual cases that one of these two shorter paths may have a greater pressure drop if it has items with great resistance. Such cases must be checked).

Using Figure 8.13 and Table 8.1, the friction pressure loss for each item in circuit *ABCC'DEF* is found and summed up. This information is presented in an organized form in Table 8.2. Review

TABLE 8.2 PIPING PRESSURE DROP CALCULATONS FOR EXAMPLE 8.15

Section	Item	D, in.	GPM	V FPS	E.L., ft	No. of Items	Total Length, ft	Friction Loss H_f ft w. 100 ft	Total ft w.
EFAB	Pipe	4	300	7.8			800	5.2	
EFAB	Gate valve				4.5	1	5		
EFAB	Globe valve				110.5	1	110		
EFAB	90° std ell				11	2	22		
B	Tee				11	1	11		
Subtotal							948	× 5.2/100 =	49.3
BC	Pipe	3½	200	7.0			100	4.8	
C	Tee				9	1	9		
Subtotal							109	× 4.8/100 =	5.2
CD	Pipe	2½	70	4.8			600	3.7	
C'	Globe valve				67	1	67		
CD	90° std ell				6.5	2	13		
D	Tee				6.5	1	7		
Subtotal							687	× 3.7/100 =	25.4
DE		3½	200	7.0			100	4.8	
E	Tee				9	1	9		
Subtotal							109	× 4.8/100 =	5.2
							Pump head = Total H_f =		85.1

each entry carefully, comparing it with the piping diagram and the pressure drop charts, to see if you arrive at the same results.

A quick estimate method sometimes used for determining the system pressure drop is to multiply the straight pipe friction loss (in the longest circuit) by 1.5 to allow for fittings and valves. The author does not recommend this rule of thumb, except for preliminary studies.

8.10 SYSTEM PIPE SIZING

An important task in designing a hydronic piping system is to determine the appropriate size (diameter) of each section of pipe. The most common procedure for doing this is called the *equal friction method*. The steps in this procedure are as follows:

1. Prepare a diagrammatic sketch of the piping system, including each terminal unit.
2. Determine the flow rate (GPM) through each unit, as explained in Chapter 5.
3. Find the flow rate through each section of pipe. In a two-pipe system, the flow progressively decreases in each supply main section downstream from the pump, since some of the water branches off at each unit; in the return main, the flow increases in each section.

   The simplest way to do this is to start from the last terminal unit supplied and progressively add the flow rates to each preceding section of the supply main. The reverse procedure is used for the return main sections.
4. Choose a value of friction loss rate to be used for the system piping, based on all of the following recommendations:
   A. The friction loss rate should be between approximately 1 to 5 ft w./100 ft pipe. Within these limits, values in the higher end are usually used for larger systems, since this reduces pipe sizes, and piping costs are very substantial in large projects. Values

from 1.5–3.5 ft w./100 ft are commonly used in most applications.
   B. The velocity in the largest mains should not exceed 4–6 FPS in small systems, or 8–10 FPS in large systems.

      The velocity in any pipe passing through occupied areas should not exceed 4 FPS, regardless of system size, since excessive noise may result.
   C. The velocity in any pipe section should not be below about 1.5 FPS. At lower velocities, dirt or air may be trapped in the line, blocking water flow. This problem is more common in small branch lines.
5. Based on the friction loss guidelines, select a pipe size for the supply main leaving the pump, based on its flow rate. Often the friction loss rate chosen will result in a selection between two standard sizes. In this case, use one of the two adjacent pipe sizes, usually that which is closet to the originally designed friction loss rate. Judgment is needed here, affected largely by expected system piping costs as well as the guidelines cited.

   Note that a new friction loss rate results from the necessity of selecting a standard pipe size; it may differ slightly from that originally designed. This is usually satisfactory.

   Check the velocity limit guidelines before selecting the desired friction loss rate. If there is a violation, change the friction loss rate used so that the velocity is in conformity with the standards.
6. The value of the new friction loss rate chosen is then used as a desired standard to select the pipe sizes for the rest of the system. This is why the procedure is called the *equal friction method.*

   Continue selecting the size of each section of supply main pipe, based on its flow rate, each with a friction loss rate as close as possible to the desired standard value. These friction loss rates will again not be identical, because the choice of an adjacent standard pipe size is always necessary, as before. Be sure to record the *actual* friction loss rates in each case. Note that not every successive pipe section will change size, until the flow rate change becomes great enough.

7. Branch piping runout sizes to units may also be selected using the equal friction method, but there are potential exceptions:
   A. When using the selection charts, you may find that the minimum velocity requirement determines the pipe size.
   B. The branch pipe size is sometimes chosen to equal the fitting connection size at the terminal unit; this can reduce installation costs.
   C. In two-pipe direct return systems, the branch piping in those circuits with short total lengths is sometimes deliberately undersized to reduce the tendency to excess flow in those circuits. This makes balancing of flow easier.
8. When all the piping in the system has been sized, the circuit with the greatest total length is determined. The pressure drop in this circuit is then calculated, as described previously.

Example 8.16 will illustrate pipe sizing procedures.

Example 8.16

Select the pipe sizes for the chilled water piping system shown in Figure 8.20. Type L copper tubing is used.

Each terminal unit takes 10 GPM. The branches to each terminal unit are a total of 10 ft long. Lengths of mains are shown on the sketch.

Solution

The stepwise procedure explained previously will be used to size the pipe.

1. The piping sketch is first drawn as in Figure 8.20.
2. The flow rate in each section is found by adding the flow rates from each unit, starting with the last as shown.
3. *ABCDEFGHIJ* is clearly the longest circuit.
4. Main *AB* has 40 GPM. Using Figure 8.15, either a 2½ in. or 2 in. size will result in the friction loss rate between 1–5 ft w./100 ft.
5. It is decided that the 2 in. pipe size will be selected to minimize initial costs. The friction loss is 3.3 ft w./100 ft.
6. The tabulation of pipe sizes selected for each remaining section in the longest circuit is shown in Table 8.3 on page 216. Note that the pipe size is decreased gradually as flow rate decreases, to maintain a friction loss reasonably close to the initial friction loss, within limits of available pipe sizes.
7. The piping system shown is a direct return arrangement, which in this case will have some circuits much shorter than others. The water flow will tend to short-circuit through *ABIJ* and other short loops, starving the last units. By increasing the pressure drop in the branches, this unbalance can be improved somewhat. A check should be made that the velocities are not excessive. The pipe size selected for branches is 1 in. rather than 1¼ in.

Figure 8.20
Sketch for Example 8.16.

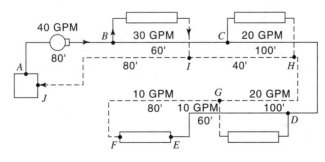

TABLE 8.3   PIPE SIZING PROCEDURES FOR EXAMPLE 8.16

Section	GPM	D, in.	L, ft	Friction, ft w./100 ft	V, FPS
AB	40	2	80	3.3	4.3
BC	30	2	60	2.0	3.2
CD	20	1½	100	3.7	3.7
DE	10	1¼	60	2.5	2.7
FG	10	1¼	80	2.5	2.7
GH	20	1½	100	3.7	3.7
HI	30	2	40	2.0	3.2
IJ	40	2	80	3.3	4.3
BRANCHES					
BI	10	1	20	7.0	4.0
CH	10	1	20	7.0	4.0
DG	10	1	20	7.0	4.0

in order to accomplish this. This is still probably not enough to solve the problem completely, and balancing valves would be required. In reality, a reverse return system would be a better choice of piping arrangements than the own shown.

8. The pressure drop in the longest circuit can now be calculated by the same procedures as used in Example 8.15. This is left as a problem for the student.

8.11   FRICTION LOSS FROM AIRFLOW IN DUCTS

Pressure loss from friction for airflow in straight round ducts is shown in charts in a manner similar to water flow in piping. Figure 8.21 shows this information. This chart is suitable for clean galvanized steel round ducts with about 40 joints per 100 ft, and with air at standard conditions. It can be used for the general range of HVAC temperatures and for altitudes up to 2000 ft.

Example 8.17 _____
A 12 in. diameter round galvanized duct 250 ft long has 100 CFM of air flowing through it. What is the pressure loss due to friction and the velocity in the duct?

Solution
The solution is found from Figure 8.21, as seen in the sketch in Figure 8.22 on page 218:

$$H_f/100 \text{ ft} = 0.20 \text{ in. w.}$$

$$H_f = \frac{0.20 \text{ in. w.}}{100 \text{ ft}} \times 250 \text{ ft} = 0.50 \text{ in. w.}$$

$$V = 1300 \text{ FPM}$$

To find the friction loss in rectangular section ducts, Figure 8.23 on page 218 must first be used. This chart shows *equivalent round duct* sizes.

The equivalent round duct is defined as the round duct that would have the same friction loss as a rectangular duct found in the chart.

Example 8.18 _____
A 30 in. by 19 in. rectangular duct is delivering 7000 CFM of air. What is the friction loss per 100 ft?

Solution
Referring first to Figure 8.23, as seen in the sketch in Figure 8.24 on page 218, the equivalent round diameter to a 30 in. by 19 in. duct is

$$D = 26 \text{ in.}$$

Figure 8.21 can now be used to find the friction loss in the rectangular duct, as shown previously:

$$H_f/100 \text{ ft} = 0.17 \text{ in. w.}$$

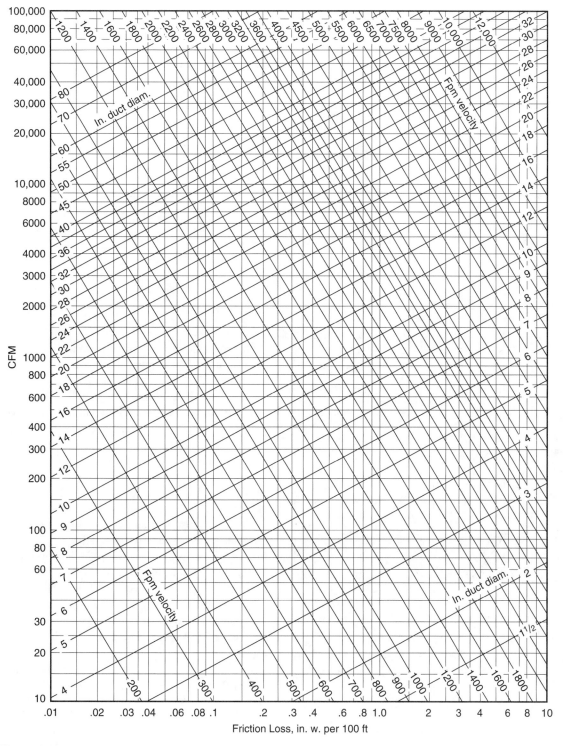

Figure 8.21
Friction loss for airflow in galvanized steel round ducts.

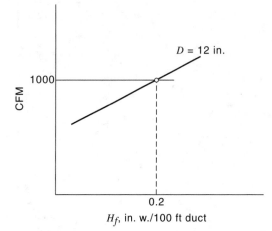

Figure 8.22
Sketch for Example 8.17.

Figure 8.23
Equivalent round duct sizes.

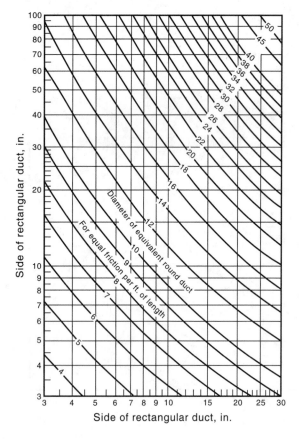

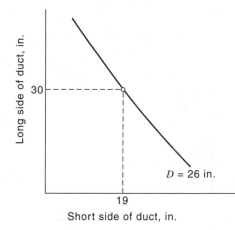

Figure 8.24
Sketch for Example 8.18.

8.12   ASPECT RATIO

At first consideration, it might seem that the equivalent round duct would have the same cross-sectional area as a rectangular duct for the same friction loss. This is not quite true. A rectangular duct with the same friction loss will have a greater area than a round duct. This is because the rectangular shape, with a greater ratio of surface to cross section, causes more friction. This problem becomes worse as the aspect ratio increases. The *aspect ratio* is the ratio of the dimensions of the two adjacent sides of a rectangular duct.

As a general rule, the aspect ratios of rectangular ducts should be as low as possible to keep friction losses reasonably low and thereby avoid excess energy consumption. A high aspect ratio will also mean more sheet metal and therefore a more expensive system. Unfortunately, the height available for horizontal ducts is often limited by the clearance above hung ceilings, resulting in high aspect ratios.

Example 8.19
Ace Sheet Metal, a contractor, wants to install a duct handling 3000 CFM in a hung ceiling that has 12 in. of vertical clear space available for the duct.

The velocity in the duct is not to exceed 1600 FPM to avoid excessive noise. What size duct should Ace install?

Solution

Ace wants to keep the aspect ratio as low as possible, to reduce friction loss and also to save money on sheet metal cost, so they will try to use as much of the 12 in. as possible. Let us say that Ace is going to put 1 in. of insulation on the duct; therefore, the maximum duct depth can be 10 in.

From Figure 8.21, with 3000 CFM at 1600 FPM, a 19 in. round duct is found. From Figure 8.23, for a round duct of 19 in., the equivalent rectangular duct with one side 10 in. is 33 in. by 10 in. This is a reasonably good solution, because the aspect ratio is $33/10 = 3.3$.

The friction loss charts can be used for testing and troubleshooting as well as design and installation, as illustrated in Example 8.20.

Example 8.20 _____

A 20 in. by 11 in. duct is supposed to be handling 3000 CFM. The engineer from Top Testing Co. is assigned to check the performance. The engineer takes pressure readings on manometers 50 ft. apart in the duct and reads 1.75 in. w. and 1.63 in. w. Is the system handling the proper airflow? If not, what is the flow?

Solution

From Figure 8.23, the equivalent round duct diameter to a 20 in. by 11 in. rectangular duct is 16 in. Using Figure 8.21, the friction loss for this duct at 3000 CFM is

$$H_f/100 \text{ ft} = 0.37 \text{ in. w.}$$

and for 50 ft is

$$H_f = \frac{0.37}{100} \times 50 = 0.19 \text{ in. w.}$$

The friction loss is actually

$$H_f = 1.75 - 1.63 = 0.12 \text{ in. w.}$$

and therefore the duct is supplying less than 3000 CFM. The actual conditions are

$$H_f/100 \text{ ft} = \frac{0.12 \text{ in. w.}}{50 \text{ ft}} \times 100 \text{ ft} = 0.24 \text{ in. w.}$$

From Figure 8.21, at this friction loss,

$$\text{Flow rate} = 2400 \text{ CFM}$$

Of course this check is accurate only if the installation is similar to the one on which the friction charts are based, as described previously.

8.13 PRESSURE LOSS IN DUCT FITTINGS

In addition to the pressure loss in straight lengths of duct, there is a pressure loss when the air flows through duct fittings (elbows, tees, transitions). These pressure losses, called dynamic losses, are due to the turbulence and change in direction. They can be expressed in either of two ways. One is the *equivalent length* method, explained in Section 8.8, where it was used for pipe fittings.

Another procedure is called the *loss coefficient* method. With this method, the pressure loss through a duct (or pipe) fitting is expressed as follows:

$$H_f = C \times H_v = C \times \left(\frac{V}{4000}\right)^2 \qquad (8.11)$$

where

H_f = total pressure loss through fitting, in. w.

C = a loss coefficient

H_v = velocity pressure at fitting, in. w.

V = velocity, ft/min

Some values of C for various duct fittings are shown in Tables 8.4–8.8 on pages 220–228.

Example 8.21 _____

A 90° smooth radius elbow without vanes has the dimensions shown in Figure 8.25 on page 223. It has 1500 CFM flowing through it. Find the pressure loss through the fitting.

TABLE 8.4   LOSS COEFFICIENTS, ELBOWS
Use the velocity pressure (H_v) of the upstream section. Fitting loss (H_t) = $C \times H_v$

A. Elbow, Smooth Radius (Die Stamped), Round

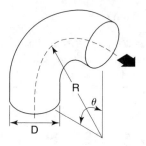

Coefficients for 90° Elbows: (See Note)

R/D	0.5	0.75	1.0	1.5	2.0	2.5
C	0.71	0.33	0.22	0.15	0.13	0.12

Note: For angles other than 90° multiply by the following factors:

θ	0°	20°	30°	45°	60°	75°	90°	110°	130°	150°	180°
K	0	0.31	0.45	0.60	0.78	0.90	1.00	1.13	1.20	1.28	1.40

B. Elbow, Round, 3 to 5 pc — 90°

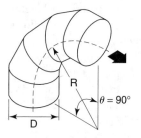

No. of Pieces	Coefficient C				
	R/D				
	0.5	0.75	1.0	1.5	2.0
5	—	0.46	0.33	0.24	0.19
4	—	0.50	0.37	0.27	0.24
3	0.98	0.54	0.42	0.34	0.33

C. Elbow, Round, Mitered

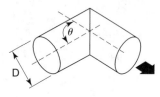

	Coefficient C					
θ	20°	30°	45°	60°	75°	90°
C	0.08	0.16	0.34	0.55	0.81	1.2

D. Elbow, Rectangular, Mitered

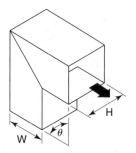

θ	Coefficient C										
	H/W										
	0.25	0.5	0.75	1.0	1.5	2.0	3.0	4.0	5.0	6.0	8.0
20°	0.08	0.08	0.08	0.07	0.07	0.07	0.06	0.06	0.05	0.05	0.05
30°	0.18	0.17	0.17	0.16	0.15	0.15	0.13	0.13	0.12	0.12	0.11
45°	0.38	0.37	0.36	0.34	0.33	0.31	0.28	0.27	0.26	0.25	0.24
60°	0.60	0.59	0.57	0.55	0.52	0.49	0.46	0.43	0.41	0.39	0.38
75°	0.89	0.87	0.84	0.81	0.77	0.73	0.67	0.63	0.61	0.58	0.57
90°	1.3	1.3	1.2	1.2	1.1	1.1	0.98	0.92	0.89	0.85	0.83

TABLE 8.4   (Continued)

E. Elbow, Rectangular, Smooth Radius without Vanes

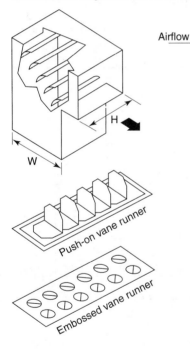

	Coefficients for 90° elbows: (See Note)										
R/W	H/W										
	0.25	0.5	0.75	1.0	1.5	2.0	3.0	4.0	5.0	6.0	8.0
0.5	1.5	1.4	1.3	1.2	1.1	1.0	1.0	1.1	1.1	1.2	1.2
0.75	0.57	0.52	0.48	0.44	0.40	0.39	0.39	0.40	0.42	0.43	0.44
1.0	0.27	0.25	0.23	0.21	0.19	0.18	0.18	0.19	0.20	0.27	0.21
1.5	0.22	0.20	0.19	0.17	0.15	0.14	0.14	0.15	0.16	0.17	0.17
2.0	0.20	0.18	0.16	0.15	0.14	0.13	0.13	0.14	0.14	0.15	0.15

F. Elbow, Rectangular, Mitered with Turning Vanes

SINGLE THICKNESS VANES

*No.	Dimensions, inches			Coeff.
	R	S	L	C
1†	2.0	1.5	0.75	0.12
2	4.5	2.25	0	0.15
3	4.5	3.25	1.60	0.18

*Numbers are for reference only.

†When extension of trailing edge is not provided for this vane, losses are approximately unchanged for single elbows, but increase considerably for elbows in series.

DOUBLE THICKNESS VANES
Coefficient C

*No.	Dimensions, in.		Velocity (V), fpm				Remarks
	R	S	1000	2000	3000	4000	
1	2.0	1.5	0.27	0.22	0.19	0.17	Embossed Vane Runner
2	2.0	1.5	0.33	0.29	0.26	0.23	Push-On Vane Runner
3	2.0	2.13	0.38	0.31	0.27	0.24	Embossed Vane Runner
4	4.5	3.25	0.26	0.21	0.18	0.16	Embossed Vane Runner

*Numbers are for reference only.

TABLE 8.5   LOSS COEFFICIENTS, TRANSITIONS (DIVERGING FLOW)
Use the velocity pressure (H_v) of the upstream section. Fitting loss (H_t) = $C \times H_v$

A. Transition, Round, Conical

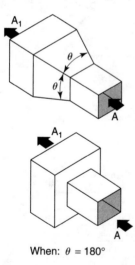

When: $\theta = 180°$

$R_e = 8.56\, DV$

where:

D = Upstream diameter (inches)
V = Upstream velocity (fpm)

R_e	A_1/A	Coefficient C (See Note)							
		\u03b8							
		16°	20°	30°	45°	60°	90°	120°	180°
0.5×10^5	2	0.14	0.19	0.32	0.33	0.33	0.32	0.31	0.30
	4	0.23	0.30	0.46	0.61	0.68	0.64	0.63	0.62
	6	0.27	0.33	0.48	0.66	0.77	0.74	0.73	0.72
	10	0.29	0.38	0.59	0.76	0.80	0.83	0.84	0.83
	≥16	0.31	0.38	0.60	0.84	0.88	0.88	0.88	0.88
2×10^5	2	0.07	0.12	0.23	0.28	0.27	0.27	0.27	0.26
	4	0.15	0.18	0.36	0.55	0.59	0.59	0.58	0.57
	6	0.19	0.28	0.44	0.90	0.70	0.71	0.71	0.69
	10	0.20	0.24	0.43	0.76	0.80	0.81	0.81	0.81
	≥16	0.21	0.28	0.52	0.76	0.87	0.87	0.87	0.87
$\geq 6 \times 10^5$	2	0.05	0.07	0.12	0.27	0.27	0.27	0.27	0.27
	4	0.17	0.24	0.38	0.51	0.56	0.58	0.58	0.57
	6	0.16	0.29	0.46	0.60	0.69	0.71	0.70	0.70
	10	0.21	0.33	0.52	0.60	0.76	0.83	0.84	0.83
	≥16	0.21	0.34	0.56	0.72	0.79	0.85	0.87	0.89

B. Transition, Rectangular, Pyramidal

When: $\theta = 180°$

A_1/A	Coefficient C (See Note 1)							
	\u03b8							
	16°	20°	30°	45°	60°	90°	120°	180°
2	0.18	0.22	0.25	0.29	0.31	0.32	0.33	0.30
4	0.36	0.43	0.50	0.56	0.61	0.63	0.63	0.63
6	0.42	0.47	0.58	0.68	0.72	0.76	0.76	0.75
≥10	0.42	0.49	0.59	0.70	0.80	0.87	0.85	0.86

Note: A = Area (Entering airstream), A_1 = Area (Leaving airstream)

TABLE 8.6   LOSS COEFFICIENTS, TRANSITIONS (CONVERGING FLOW)
Use the velocity pressure (H_v) of the downstream section. Fitting loss (H_t) = $C \times H_v$

A. Contraction, Round and Rectangular, Gradual to Abrupt

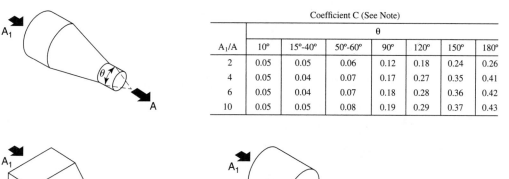

	Coefficient C (See Note)						
				θ			
A_1/A	10°	15°-40°	50°-60°	90°	120°	150°	180°
2	0.05	0.05	0.06	0.12	0.18	0.24	0.26
4	0.05	0.04	0.07	0.17	0.27	0.35	0.41
6	0.05	0.04	0.07	0.18	0.28	0.36	0.42
10	0.05	0.05	0.08	0.19	0.29	0.37	0.43

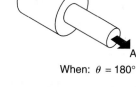

When: $\theta = 180°$

Note: A = Area (Entering airstream), A_1 = Area (Leaving airstream)

Reprinted with permission from the SMACNA "HVAC Systems — Duct Design" manual Second Edition, 1981.

Solution
The loss coefficient is found in Table 8.4e. Referring to Figure 8.25

$$\frac{H}{W} = \frac{12}{8} = 1.5; \quad \frac{R}{W} = \frac{16}{8} = 2.0$$

From Table 8.4e

$$C = 0.14$$

The duct cross-sectional area and velocity are

$$A = 12 \text{ in.} \times 8 \text{ in.} \times \frac{1 \text{ ft}^2}{144 \text{ in.}^2} = 0.667 \text{ ft}^2$$

$$V = 1500 \frac{\text{ft}^3}{\text{min}} \times \frac{1}{0.667 \text{ ft}^2} = 2250 \text{ ft/min}$$

Using Equation 8.11, the pressure loss is

$$H_f = 0.14 \left(\frac{2250}{4000}\right)^2 = 0.04 \text{ in. w.}$$

The pressure loss in transition pieces is calculated in the same manner. With converging transitions, the downstream velocity is used, and with diverging transitions, the upstream velocity is used.

Figure 8.25
Sketch for Example 8.21.

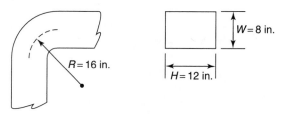

$R = 16 \text{ in.}$

$W = 8 \text{ in.}$

$H = 12 \text{ in.}$

TABLE 8.7   LOSS COEFFICIENTS, CONVERGING JUNCTIONS
Use the velocity pressure (H_v) of the downstream section. Fitting loss (H_t) = $C \times H_v$

A. Converging Tee, Round Branch to Rectangular Main

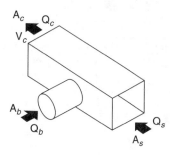

| V_c | Branch, Coefficient C (See Note) | | | | | | | | | |
| | Q_b/Q_c | | | | | | | | | |
	0.1	0.2	0.3	0.4	0.5	0.6	0.7	0.8	0.9	1.0
< 1200 fpm	−.63	−.55	0.13	0.23	0.78	1.30	1.93	3.10	4.88	5.60
> 1200 fpm	−.49	−.21	0.23	0.60	1.27	2.06	2.75	3.70	4.93	5.95

When:

A_b/A_s	A_s/A_c	A_b/A_c
0.5	1.0	0.5

B. Converging Tee, Rectangular Main and Branch

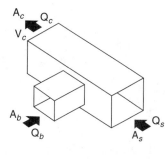

| V_c | Branch, Coefficient C (See Note) | | | | | | | | | |
| | Q_b/Q_c | | | | | | | | | |
	0.1	0.2	0.3	0.4	0.5	0.6	0.7	0.8	0.9	1.0
< 1200 fpm	−.75	−.53	−.03	0.33	1.03	1.10	2.15	2.93	4.18	4.78
> 1200 fpm	−.69	−.21	0.23	0.67	1.17	1.66	2.67	3.36	3.93	5.13

When:

A_b/A_s	A_s/A_c	A_b/A_c
0.5	1.0	0.5

Note: A = Area (sq. in.), Q = Airflow (cfm), V = Velocity (fpm)

C. Converging Tee, 45° Entry Branch to Rectangular Main

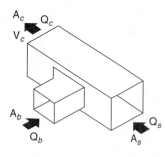

When:

A_b/A_s	A_s/A_c	A_b/A_c
0.5	1.0	0.5

| V_c | Branch, Coefficient C (See Note) | | | | | | | | | |
| | Q_b/Q_c | | | | | | | | | |
	0.1	0.2	0.3	0.4	0.5	0.6	0.7	0.8	0.9	1.0
< 1200 fpm	−.83	−.68	−.30	0.28	0.55	1.03	1.50	1.93	2.50	3.03
> 1200 fpm	−.72	−.52	−.23	0.34	0.76	1.14	1.83	2.01	2.90	3.63

TABLE 8.7   (Continued)

D. Converging Wye, Rectangular

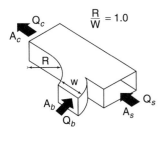

$\frac{R}{W} = 1.0$

		Branch, Coefficient (See Note)								
		Q_b/Q_c								
A_b/A_s	A_b/A_c	0.1	0.2	0.3	0.4	0.5	0.6	0.7	0.8	0.9
0.25	0.25	−.50	0	0.50	1.2	2.2	3.7	5.8	8.4	11
0.33	0.25	−1.2	−.40	0.40	1.6	3.0	4.8	6.8	8.9	11
0.5	0.5	−.50	−.20	0	0.25	0.45	0.70	1.0	1.5	2.0
0.67	0.5	−1.0	−.60	−.20	0.10	0.30	0.60	1.0	1.5	2.0
1.0	0.5	−2.2	−1.5	−.95	−.50	0	0.40	0.80	1.3	1.9
1.0	1.0	−.60	−.30	−.10	−.04	0.13	0.21	0.29	0.36	0.42
1.33	1.0	−1.2	−.80	−.40	−.20	0	0.16	0.24	0.32	0.38
2.0	1.0	−2.1	−1.4	−.90	−.50	−.20	0	0.20	0.25	0.30

		Main, Coefficient C (See Note)								
		Q_b/Q_c								
A_s/A_c	A_b/A_c	0.1	0.2	0.3	0.4	0.5	0.6	0.7	0.8	0.9
0.75	0.25	0.30	0.30	0.20	−.10	−.45	−.92	−1.5	−2.0	−2.6
1.0	0.5	0.17	0.16	0.10	0	−0.08	−.18	−.27	−.37	−.46
0.75	0.5	0.27	0.35	0.32	0.25	0.12	−.03	−.23	−.42	−.58
0.5	0.5	1.2	1.1	0.90	0.65	0.35	0	−.40	−.80	−1.3
1.0	1.0	0.18	0.24	0.27	0.26	0.23	0.18	0.10	0	−.12
0.75	1.0	0.75	0.36	0.38	0.35	0.27	0.18	0.05	−.08	−.22
0.5	1.0	0.80	0.87	0.80	0.68	0.55	0.40	0.25	0.08	−.10

Reprinted with permission from the SMACNA "HVAC Systems — Duct Design" manual Second Edition, 1981.

Example 8.22

The diverging transition piece in Figure 8.26 is handling 12,000 CFM. Find the pressure loss through the fitting.

Solution

From Table 8.5b, with $A_1/A = 2.0$, read $C = 0.25$. Using Equation 8.11,

$$V = 12,000 \ \frac{ft^3}{min} \times \frac{1}{8 \ ft^2} = 1500 \ ft/min$$

$$H_f = 0.25\left(\frac{1500}{4000}\right)^2 = 0.04 \ in. \ w.$$

Figure 8.26
Sketch for Example 8.22.

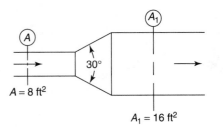

$A = 8 \ ft^2$

$A_1 = 16 \ ft^2$

30°

TABLE 8.8 LOSS COEFFICIENTS, DIVERGING JUNCTIONS
Use the velocity pressure (H_v) of the upstream section. Fitting loss (H_t) = $C \times H_v$

A. Tee, 45° Entry, Rectangular Main and Branch

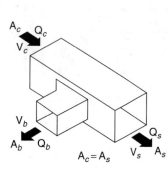

	Branch, Coefficient C (See Note)								
V_b/V_c	Q_b/Q_c								
	0.1	0.2	0.3	0.4	0.5	0.6	0.7	0.8	0.9
0.2	0.91								
0.4	0.81	0.79							
0.6	0.77	0.72	0.70						
0.8	0.78	0.73	0.69	0.66					
1.0	0.78	0.98	0.85	0.79	0.74				
1.2	0.90	1.11	1.16	1.23	1.03	0.86			
1.4	1.19	1.22	1.26	1.29	1.54	1.25	0.92		
1.6	1.35	1.42	1.55	1.59	1.63	1.50	1.31	1.09	
1.8	1.44	1.50	1.75	1.74	1.72	2.24	1.63	1.40	1.17

B. Tee, 45° Entry, Rectangular Main and Branch with Damper

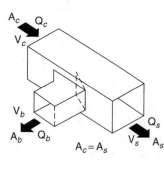

	Branch, Coefficient C (See Note)								
V_b/V_c	Q_b/Q_c								
	0.1	0.2	0.3	0.4	0.5	0.6	0.7	0.8	0.9
0.2	0.61								
0.4	0.46	0.61							
0.6	0.43	0.50	0.54						
0.8	0.39	0.43	0.62	0.53					
1.0	0.34	0.57	0.77	0.73	0.68				
1.2	0.37	0.64	0.85	0.98	1.07	0.83			
1.4	0.57	0.71	1.04	1.16	1.54	1.36	1.18		
1.6	0.89	1.08	1.28	1.30	1.69	2.09	1.81	1.47	
1.8	1.33	1.34	2.04	1.78	1.90	2.40	2.77	2.23	1.92

Note: A = Area (sq. in.), Q = Airflow (cfm), V = Velocity (fpm)

C. Tee, Rectangular Main and Branch

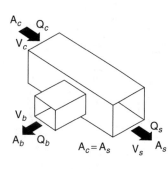

	Branch, Coefficient C (See Note)								
V_b/V_c	Q_b/Q_c								
	0.1	0.2	0.3	0.4	0.5	0.6	0.7	0.8	0.9
0.2	1.03								
0.4	1.04	1.01							
0.6	1.11	1.03	1.05						
0.8	1.16	1.21	1.17	1.12					
1.0	1.38	1.40	1.30	1.36	1.27				
1.2	1.52	1.61	1.68	1.91	1.47	1.66			
1.4	1.79	2.01	1.90	2.31	2.28	2.20	1.95		
1.6	2.07	2.28	2.13	2.71	2.99	2.81	2.09	2.20	
1.8	2.32	2.54	2.64	3.09	3.72	3.48	2.21	2.29	2.57

TABLE 8.8 (Continued)

D. Tee, Rectangular Main and Branch with Damper

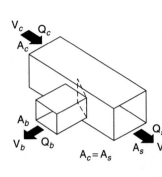

	Branch, Coefficient C (See Note)								
V_b/V_c	Q_b/Q_c								
	0.1	0.2	0.3	0.4	0.5	0.6	0.7	0.8	0.9
0.2	0.58								
0.4	0.67	0.64							
0.6	0.78	0.76	0.75						
0.8	0.88	0.98	0.81	1.01					
1.0	1.12	1.05	1.08	1.18	1.29				
1.2	1.49	1.48	1.40	1.51	1.70	1.91			
1.4	2.10	2.21	2.25	2.29	2.32	2.48	2.53		
1.6	2.72	3.30	2.84	3.09	3.30	3.19	3.29	3.16	
1.8	3.42	4.58	3.65	3.92	4.20	4.15	4.14	4.10	4.05

E. Tee, Rectangular Main and Branch with Extractor

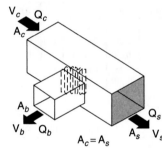

	Branch, Coefficient C (See Note)								
V_b/V_c	Q_b/Q_c								
	0.1	0.2	0.3	0.4	0.5	0.6	0.7	0.8	0.9
0.2	0.60								
0.4	0.62	0.69							
0.6	0.74	0.80	0.82						
0.8	0.99	1.10	0.95	0.90					
1.0	1.48	1.12	1.41	1.24	1.21				
1.2	1.91	1.33	1.43	1.52	1.55	1.64			
1.4	2.47	1.67	1.70	2.04	1.86	1.98	2.47		
1.6	3.17	2.40	2.33	2.53	2.31	2.51	3.13	3.25	
1.8	3.85	3.37	2.89	3.23	3.09	3.03	3.30	3.74	4.11

	Main, Coefficient C (See Note)							
V_b/V_c	0.2	0.4	0.6	0.8	1.0	1.2	1.4	1.6	1.8
C	0.03	0.04	0.07	0.12	0.13	0.14	0.27	0.30	0.25

F. Tee, Rectangular Main to Round Branch

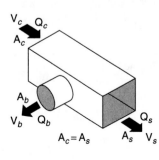

	Branch, Coefficient C (See Note)								
V_b/V_c	Q_b/Q_c								
	0.1	0.2	0.3	0.4	0.5	0.6	0.7	0.8	0.9
0.2	1.00								
0.4	1.01	1.07							
0.6	1.14	1.10	1.08						
0.8	1.18	1.31	1.12	1.13					
1.0	1.30	1.38	1.20	1.23	1.26				
1.2	1.46	1.58	1.45	1.31	1.39	1.48			
1.4	1.70	1.82	1.65	1.51	1.56	1.64	1.71		
1.6	1.93	2.06	2.00	1.85	1.70	1.76	1.80	1.88	
1.8	2.06	2.17	2.20	2.13	2.06	1.98	1.99	2.00	2.07

TABLE 8.8 (Continued)

G. Tee Rectangular Main to Conical Branch (2)

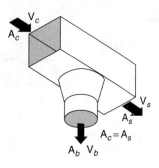

Branch, Coefficient C (See Note)

V_b/V_c	0.40	0.50	0.75	1.0	1.3	1.5
C	0.80	0.83	0.90	1.0	1.1	1.4

H. Wye, Rectangular (15)

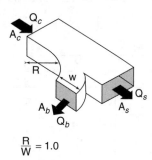

$\dfrac{R}{W} = 1.0$

90° Branch

Branch, Coefficient C (See Note)

A_b/A_s	A_b/A_c	Q_b/Q_c								
		0.1	0.2	0.3	0.4	0.5	0.6	0.7	0.8	0.9
0.25	0.25	0.55	0.50	0.60	0.85	1.2	1.8	3.1	4.4	6.0
0.33	0.25	0.35	0.35	0.50	0.80	1.3	2.0	2.8	3.8	5.0
0.5	0.5	0.62	0.48	0.40	0.40	0.48	0.60	0.78	1.1	1.5
0.67	0.5	0.52	0.40	0.32	0.30	0.34	0.44	0.62	0.92	1.4
1.0	0.5	0.44	0.38	0.38	0.41	0.52	0.68	0.92	1.2	1.6
1.0	1.0	0.67	0.55	0.46	0.37	0.32	0.29	0.29	0.30	0.37
1.33	1.0	0.70	0.60	0.51	0.42	0.34	0.28	0.26	0.26	0.29
2.0	1.0	0.60	0.52	0.43	0.33	0.24	0.17	0.15	0.17	0.21

Main, Coefficient C (See Note)

A_b/A_s	A_b/A_c	Q_b/Q_c								
		0.1	0.2	0.3	0.4	0.5	0.6	0.7	0.8	0.9
0.25	0.25	−.01	−.03	−.01	0.05	0.13	0.21	0.29	0.38	0.46
0.33	0.25	0.08	0	−.02	−.01	0.02	0.08	0.16	0.24	0.34
0.5	0.5	−.03	−.06	−.05	0	0.06	0.12	0.19	0.27	0.35
0.67	0.5	0.04	−.02	−.04	−.03	−.01	0.04	0.12	0.23	0.37
1.0	0.5	0.72	0.48	0.28	0.13	0.05	0.04	0.09	0.18	0.30
1.0	1.0	−.02	−.04	−.04	−.01	0.06	0.13	0.22	0.30	0.38
1.33	1.0	0.10	0	0.01	−.03	−.01	0.03	0.10	0.20	0.30
2.0	1.0	0.62	0.38	0.23	0.13	0.08	0.05	0.06	0.10	0.20

Reprinted with permission from the SMACNA "HVAC System—Duct Design" manual. Second Edition, 1981

In Sections 8.4 and 8.5, we discussed total pressure (H_t), velocity pressure (H_v), and static pressure (H_s). It was noted that for flow through a diverging transition, velocity pressure is converted to static pressure, resulting in an increase in static pressure. If there was no frictional pressure, the decrease in velocity pressure would exactly equal the increase in static pressure. Also, the total pressure would remain the same.

The pressure loss due to friction (H_f), however, causes a decrease in total pressure. The net result of the two effects (conversion of pressure and the frictional pressure loss) on the static pressure is found by algebraically summing them up. The following example illustrates this.

Example 8.23 _____

The total pressure at point 1 for the fitting in Example 8.22 is 2.35 in. w.g. Find the static pressure at point 1 and the total and static pressure at point 2.

Solution

Refer to Figure 8.27, which shows the results of the calculations following, as well as a profile of the pressure changes in the fitting. Using Equation 8.3, the static pressure at point 1 is

$$H_{s1} = H_{t1} - H_{v1} = 2.35 - \left(\frac{1500}{4000}\right)^2 = 2.21 \text{ in. w.g.}$$

Figure 8.27
Sketch for Example 8.23.

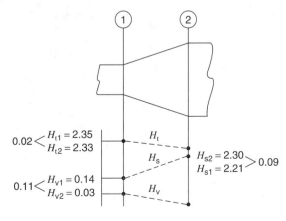

The frictional pressure loss was found in Example 8.22 to be $H_f = 0.03$ in. w. The total pressure at point 2 is therefore

$$H_{t2} = H_{t1} - H_f = 2.35 - 0.03 = 2.32 \text{ in. w.g.}$$

The static pressure at point 2 is

$$H_{s2} = H_{t2} - H_{v2} = 2.32 - \left(\frac{750}{4000}\right)^2$$

$$= 2.28 \text{ in. w.g.}$$

Example 8.23 illustrates a case of static pressure regain (SPR). The actual regain was

$$\text{SPR} = H_{s2} - H_{s1} = 2.28 - 2.21 = 0.07 \text{ in. w.}$$

If there had been no frictional pressure loss, the theoretical SPR would be

$$\text{SPR} = 0.07 + 0.03 = 0.10 \text{ in. w.}$$

The proportion of actual to theoretical SPR is

$$R = \frac{0.07}{0.10} = .70$$

R is the recovery factor, as defined in Section 8.5. To put this another way, 70% of the pressure loss in the fitting is recovered.

The more gradual a transition or change in direction in a fitting, the higher the recovery factor will be; that is, there will be less of a frictional loss. This results in a lower fan power to overcome these losses and a resultant savings in energy. Often however, this more gradual fitting will be more expensive to fabricate.

Where there is a combined transition and branch in a duct system, the pressure loss in the straight main run and in the branch are separate, and the value of each depends on the shape. If it is important to minimize losses, the shape should be as shown in Figure 8.28 on page 230. In this case, the pressure loss through the straight run can usually be neglected, and the branch pressure loss can be calculated as an elbow.

In order to reduce the fabrication cost of fittings, they are often made as shown in Figure 8.29 on page 230. In this case, the pressure loss through the branch may be considerable, particularly at high velocities.

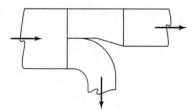

Figure 8.28
Branch with low pressure loss.

A further discussion of recommended duct fitting construction can be found in Chapter 9.

8.14   PRESSURE LOSS AT FAN INLET AND OUTLET

There will also be a pressure loss at the fan inlet and outlet, the value of which depends on the shape of the fan-duct connection. This is called the *system effect*. Some values of the resulting loss coefficient C are shown in Table 8.9. An inspection of the types of connections in Table 8.9 will show the importance of considering the system effect and of installing fans with good connections. A list of system effects can be found in the *Air Moving and Conditioning Association* (AMCA) Manuals.

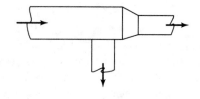

Figure 8.29
Branch with high pressure loss.

Example 8.24

A contractor installs the inlet connection to a fan as shown in Figure 8.30(*a*) instead of as shown in Figure 8.30(*b*). The fan inlet velocity is 2000 ft/min. What is the pressure loss inlet to the fan in each case?

Solution

From Table 8.9, we read the values of $C = 1.2$ and $C = 0.25$ for the poor and good connection. The pressure loss for each, using Equation 8.11, is

$$\text{Poor } H_f = 1.2\left(\frac{2000}{4000}\right)^2 = 0.30 \text{ in. w.}$$

$$\text{Good } H_f = 0.25\left(\frac{2000}{4000}\right)^2 = 0.06 \text{ in. w.}$$

Note the greatly increased pressure loss with the poor connection, resulting in wasted energy.

TABLE 8.9   LOSS COEFFICIENTS (C) FOR STRAIGHT ROUND DUCT FAN INLET CONNECTIONS

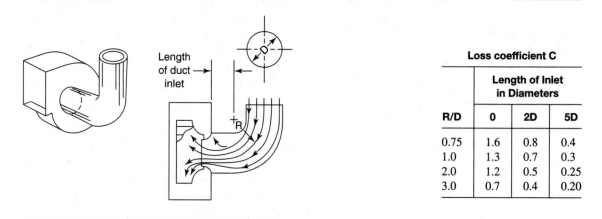

R/D	0	2D	5D
0.75	1.6	0.8	0.4
1.0	1.3	0.7	0.3
2.0	1.2	0.5	0.25
3.0	0.7	0.4	0.20

Loss coefficient C — Length of Inlet in Diameters

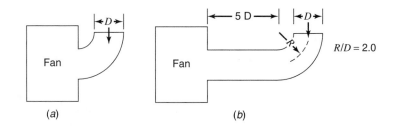

Figure 8.30
Sketch for Example 8.24. (*a*) No straight inlet duct. (*b*) Long straight inlet duct.

8.15 DUCT SYSTEM PRESSURE LOSS

The duct pressure losses must be found in order to determine fan capacity, check equipment performance, and balance air quantities.

The *system total pressure loss* is defined as the *total pressure loss through the duct path that has the largest pressure losses.* This path is often the longest one, but it may be a shorter path that contains an unusual number of fittings or devices with large pressure losses.

It is better to work with total pressure loss rather than static pressure loss when analyzing duct pressure losses. This gives a better understanding of the total pressure available at any point in case problems exist. To find the system total pressure loss, the losses are summed up for each section of straight duct and each fitting in the path chosen.

Pressure losses through any equipment (coils, filters, diffusers) must be included. The manufacturer will furnish this data.

Example 8.25

For the duct system shown in Figure 8.31, determine the system total pressure loss and fan requirements. The fan inlet and outlet connections are not shown, but it has been found that the system effect inlet loss is 0.20 in. w. and the outlet loss is 0.08 in. w. The total pressure required at each air outlet for proper distribution is 0.1 in. w.g.

Solution

From inspection of the duct layout, path *XABCDEF* is the longest. It is also the path with the greatest pressure loss, because none of the shorter paths have unusual pressure losses.

The pressure losses for the straight ducts and fittings are read from the appropriate tables. The results

Figure 8.31
Sketch for Example 8.25.

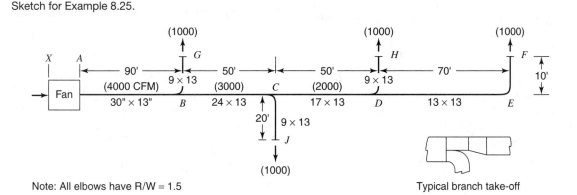

TABLE 8.10 SUMMARY OF RESULTS FOR EXAMPLE 8.25

Section	Item	Flow Rate, CFM	Duct Size, in.	Equivalent Round, in.	V, FPM	Friction Loss/100 ft, in. w.	Length, ft	Loss Coeff, C	Velocity Press., in. w.	Pressure Loss, in. w.
X	Fan inlet	4000								0.20
A	Fan outlet				1477					.08
AB	Duct		30 × 13	21	1477	0.17	90			.15
B	Diverging transitions				1385					0
BC	Duct	3000	24 × 13	19	1385	0.17	50			.09
C	Diverging transitions				1303					0
CD	Duct	2000	17 × 13	16	1303	0.18	50			.09
D	Diverging transitions									0
DEF	Duct	1000	13 × 13	14	852	0.09	80			.07
E	Elbow							.17		.01
F	Outlet									.10

Total system pressure loss = 0.79

are listed in Table 8.10. Each item should be checked by the student. The diverging transitions have a negligible pressure loss due to the gradual transformation and relatively low velocities.

The fan selected for the system would be specified for a total pressure of 0.79 in. w.g.

The method used in Example 8.25 is also valuable in solving balancing problems, because the total pressure can be found anywhere in the duct, as shown in Example 8.26.

Example 8.26
The system in Example 8.25 is installed in Governor Jawbone's offices. There is a complaint about drafts near outlet G. What is the trouble?

Solution
We will find the total pressure at outlet G and check it against the value (0.1 in. w.g.) required at the air outlet. The pressure loss is calculated now through path XABG. The pressure loss XAB has already been found. From Table 8.10 it is

$$XAB\ H_f = 0.20 + 0.08 + 0.15 = 0.43 \text{ in. w.}$$

The pressure loss from B to G is now found. For the transition elbow at B, H/W = 0.7. If R/W = 1.5, C = 0.19

$$\text{Elbow } H_f = 0.19\left(\frac{1700}{4000}\right)^2 = 0.03 \text{ in. w.}$$

For duct BG H_f/100 ft = 0.2 in. w.

$$\text{Duct } H_f = 0.2 \times \left(\frac{10}{100}\right) = 0.02 \text{ in. w.}$$

The total pressure loss in this path is therefore

$$XABG\ H_f = 0.43 + 0.03 + 0.02 = 0.48 \text{ in. w.}$$

The fan total pressure is 0.79 in. w.g. Therefore, the total pressure in the duct at outlet G is

$$H_{tG} = 0.79 - 0.48 = 0.31 \text{ in. w.g.}$$

This pressure is much greater than the pressure required (0.1 in. w.g.) and will result in excess air at uncomfortable velocities being delivered through outlet G.

The solution to the excess pressure at the outlet in Example 8.26 might be handled by partially closing a damper in the branch duct, if one had been installed.

This might create noise problems, however. A better solution would be to design the duct system so that excess pressures are dissipated in duct friction losses. Procedures for doing this will be explained shortly.

8.16   DUCT DESIGN METHODS

In the previous section, we explained how to find the pressure losses in ducts after their sizes were known. In designing a new system, however, the duct sizes must be determined first. Two methods of sizing ducts will be explained here, the *equal friction* method and the *static regain* method.

Equal Friction Method

With this method, *the same value friction loss rate per length of duct is used to size each section of duct in the system.*

The friction loss rate is chosen to result in an economical balance between duct cost and energy cost. A higher friction loss results in smaller ducts but higher fan operating costs.

Duct systems for HVAC installations may be loosely classified into *low velocity* and *high veloc-*ity groups, although these are not strictly separate categories.

Typical ranges of design equal friction loss rates used for low velocity systems are from 0.08 to 0.15 in. w./100 ft of duct. Maximum velocities in the main duct at the fan outlet are limited where noise generation is a problem (see Table 8.11). However, sound attenuation devices and duct sound lining can be used if needed.

High velocity duct systems are designed with initial velocities from about 2500 FPM to as high as about 4000 FPM. The corresponding friction loss rates may be as high as 0.6 in. w./100 ft.

High velocity duct systems are primarily used to reduce overall duct sizes. In many large installations, space limitations (above hung ceilings, in shafts) make it impossible to use the larger ducts resulting from low velocity systems.

The higher pressures result in certain special features of these systems. The ducts and fans must be constructed to withstand the higher pressures. The noise produced at the high velocities requires special sound attenuation.

The following example illustrates duct sizing by the equal friction method.

TABLE 8.11   SUGGESTED VELOCITIES IN LOW VELOCITY AIR CONDITIONING SYSTEMS

Designation	Recommended Velocities, FPM			Maximum Velocities, FPM		
	Residences	Schools, Theaters, Public Buildings	Industrial Buildings	Residences	Schools, Theaters, Public Buildings	Industrial Buildings
Outside air intakes[a]	500	500	500	800	900	1200
Filters[a]	250	300	350	300	350	350
Heating coils[a]	450	500	600	500	600	700
Air washers	500	500	500	500	500	500
Suction connections	700	800	1000	900	1000	1400
Fan outlets	1000–1600	1300–2000	1600–2400	1700	1500–2200	1700–2800
Main ducts	700–900	1000–1300	1200–1800	800–1200	1100–1600	1300–2200
Branch ducts	600	600–900	800–1000	700–1000	800–1300	1000–1800
Branch risers	500	600–700	800	650–800	800–1200	1000–1600

[a] These velocities are for total face area, not the net free area, other velocities are for net free area. Reprinted with permission from the 1967 Systems and Equipment. ASHRAE Handbook & Product Directory.

Example 8.27 _____

Find the size of each duct section for the system shown in Figure 8.32, using the equal friction design method. Use rectangular ducts. The system serves a public building.

Solution

1. Sum up the CFMs backward from the last outlet, to find the CFM in each duct section. The results are shown in Table 8.12.
2. Select a design velocity for the main from the fan, using Table 8.11. A velocity of 1400 ft/min. will be chosen, which should be reasonably quiet for the application.
3. From Figure 8.21, the friction loss rate for the main section *AB* is read as 0.13 in. w./100 ft. The equivalent round duct diameter is read as 20.5 in.
4. The equivalent round duct diameter for each duct section is read from Figure 8.21 at the intersection of the design friction loss rate (0.13 in. w./100 ft) and the CFM for the section.
5. The rectangular duct sizes are read from Figure 8.23. In the actual installation, the duct proportions chosen would depend on space available.
6. The pressure loss in the system can be calculated as shown previously.

Figure 8.32

Sketch for Example 8.27.

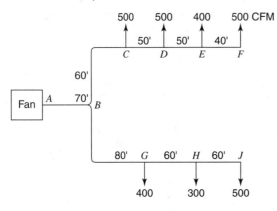

With systems that use package air conditioning units, the available pressure to overcome the friction loss is limited to the external pressure that the fan in the unit can develop.

In this case, the appropriate procedure would be to find the total equivalent length of the system. Dividing the external fan pressure by the length will establish the maximum friction loss that can be used. In reality, this limitation seldom occurs, since package units are mostly used with systems of relatively short duct length.

Return air ducts can be sized by the equal friction method in the same manner as supply air units. Return air ducts are usually in the low velocity category, even if the supply ducts are of the high velocity type.

The equal friction method of designing ducts is quite simple, and is probably the most popular one used. For systems that do not have great distances between the first and the last outlets, it works quite satisfactorily. If there are long distances between the outlets at the beginning and the end of the system, however, those near the fan will be over-pressured. This was demonstrated in Example 8.26. The result may cause difficulties in balancing the flow rates, and possibly excess noise.

If the outlets closest to the fan are on long separate branches, this problem may be overcome by modifying the equal friction design method. The longest run is sized by the design friction loss rate, but some branches are chosen at a higher friction loss rate, thus using up the excess pressure.

To reduce extreme pressure differences throughout the system, the static regain method of duct design may also be used.

Static Regain Method

The *static regain* method of sizing ducts is most often used for high velocity systems with long duct runs, especially in large installations.

With this method, an initial velocity in the main duct leaving the fan is selected, in the range of 2500–4000 FPM.

After the initial velocity is chosen, *the velocities in each successive section of duct in the main run*

TABLE 8.12   SUMMARY OF RESULTS FOR EXAMPLE 8.27

Section	CFM	V, ft/min	Friction Loss, in. w. per 100 ft	Eq. D, in.	Rect. Duct Size, in.
AB	3100	1240	0.13	20.5	24 × 15
BC	1900	1140	0.13	17	20 × 12
CD	1400	1050	0.13	15	16 × 12
DE	900	900	0.13	12.5	16 × 9
EF	500	889	0.13	10	9 × 9
BG	1200	1029	0.13	14	14 × 12
GH	800	914	0.13	12	14 × 9
HJ	500	889	0.13	10	9 × 9

are reduced so that the resulting static pressure gain is enough to overcome the frictional losses in the next duct section.

The result is that the static pressure is the same at each junction in the main run. Because of this, there generally will not be extreme differences in the pressures among the branch outlets, so balancing is simplified.

The following example illustrates how to size ducts by this method.

Example 8.28

Determine the duct sizes for the system shown in Figure 8.33, using the static regain method. Round duct will be used.

Solution

The results of the work are summarized in Table 8.13 on page 236. The steps are as follows:

1. A velocity in the initial section is selected. (This system is a high velocity system, so the noise level will not determine the maximum velocity. Sound attenuating devices must be used.) An initial velocity of 3200 ft/min will be chosen.

2. From Figure 8.21, the duct size and static pressure loss due to friction in section *AB* is determined. The friction loss per 100 ft is 0.56 in. w. and therefore the friction loss in the section is 0.56 × 50/100 = 0.28 in. w.

3. The velocity must be reduced in Section *BC* so that the static pressure gain will be equal to the friction loss in *BC*. There will not be a complete regain, due to dynamic losses in the transition at *B*. We will assume a 75% regain factor for the fittings. A trial-and-error procedure is necessary to balance the regain against the friction loss. Let us try a velocity of 2400 ft/min. in section *BC*. The friction loss is

$$\text{Loss in } BC = \frac{0.32 \text{ in. w.}}{100 \text{ ft}} \times 40 \text{ ft}$$

$$= 0.13 \text{ in. w.}$$

Figure 8.33
Sketch for Example 8.28.

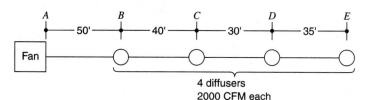

4 diffusers
2000 CFM each

TABLE 8.13   SUMMARY OF RESULTS FOR EXAMPLE 8.28

Section	CFM	V, ft/min	Eq. D, in.	Velocity Pressure, in. w.	Friction Loss, in. w./100 ft	Length, ft	Friction Loss, in. w.	Static Pressure Regain, in. w.
AB	8000	3200	22	0.64	0.56	50	0.28	
B								0.16
BC	6000	2600	21	0.43	0.40	40	0.16	
C								0.09
CD	4000	2200	18	0.30	0.33	30	0.10	
D								0.09
DE	2000	1700	15	0.18	0.26	35	0.09	

The static pressure regain available to overcome this loss, using Equation 8.9, is

$$\text{Regain at } B = 0.75\left[\left(\frac{3200}{4000}\right)^2 - \left(\frac{2400}{4000}\right)^2\right]$$

$$= 0.21 \text{ in. w.}$$

This is too large a regain. Try a velocity of 2600 ft/min.

$$\text{Loss in } BC = \frac{0.40 \text{ in. w.}}{100 \text{ ft}} \times 40 \text{ ft}$$

$$= 0.16 \text{ in. w.}$$

$$\text{Regain at } B = 0.75\left[\left(\frac{3200}{4000}\right)^2 - \left(\frac{2600}{4000}\right)^2\right]$$

$$= 0.16 \text{ in. w.}$$

This trial is satisfactory. The regain at B is precisely enough to overcome the loss in section BC. The duct size of BC is 21 in.

4. Continue the same procedure at transition C. Let us try a velocity of 2200 ft/min. in CD. The results are

$$\text{Loss in } CD = \frac{0.33 \text{ in. w.}}{100 \text{ ft}} \times 30 \text{ ft}$$

$$= 0.10 \text{ in. w.}$$

$$\text{Regain at } C = 0.75\left[\left(\frac{2600}{4000}\right)^2 - \left(\frac{2200}{4000}\right)^2\right]$$

$$= 0.09 \text{ in. w.}$$

This first guess is satisfactory. No further trial is needed. The duct size is 18 in.

5. The trial-and-error process at D results in a duct size of 15 in. for section DE. The reader should check this.

The result of this method is that the static pressure in the duct at outlets B, C, D, and E will be the same. Assuming that these outlets all required the same static pressure for proper air distribution, the static regain procedure provided duct sizes that will reduce air balancing difficulties. On the other hand, if the equal friction method had been used, the static pressure at B would be considerably higher than at E, causing air outlet balancing problems.

One disadvantage of the static regain method of duct design is that it usually results in a system with some of the duct sections larger than those found by the equal friction method. For systems at high velocities, however, this method is recommended.

For return air duct systems, the equal friction duct sizing method is generally used.

Computer software is available for all popular duct sizing methods. These programs can save considerable time, especially if the static regain method is used.

Problems

8.1   The average velocity of air flowing in a 24 × 18 in. duct is 1300 ft/min. What is the volume flow rate of air in the duct in CFM?

8.2   A pipe with a cross-sectional area of 8.4 in.^{2} has 12 GPM of water flowing through it. What is the water velocity in ft/sec?

8.3   A duct is to be installed that will carry 3600 CFM of air. To avoid excess noise the maximum velocity of air allowed in the duct is 1750 ft/min. The depth of the duct is to be 10 in. What is the minimum width of the duct?

8.4   A 2 in. pipe has water flowing through it at 4 FPM. The diameter is increased to 3 in. What is the velocity in the 3 in. section?

8.5   A 42 in. wide by 20 in. deep duct has a flow rate of 18,000 CFM of air. It is desired to reduce the velocity to 1800 ft/min. The depth of the duct is fixed. What should be the width of the new section?

8.6   Cooling water is pumped from a river to a refrigeration machine condenser 80 ft above the pump intake. The condenser requires 920 GPM. The friction loss through all of the piping, valves, fittings, and condenser is 31 ft w. The water leaves the condenser at atmospheric pressure and flows back to the river by gravity. The velocity of the water entering and leaving the system is the same. What is the required pump head?

8.7   For the piping system shown in Figure 8.34, determine the required pump head. The friction loss is 27 ft w.

8.8   For piping system shown in Figure 8.35, determine the pressure drop due to friction between points 1 and 2, if the pressure gages read as shown.

8.9   For the piping system shown in Figure 8.36 on page 238, the friction loss between points 1 and 2 is 18 ft w. The pressure gage at point 1 reads 23 psig. What would be the reading of the pressure gage at point 2?

8.10   A hydronic cooling system has a pressure drop due to friction of 41 ft w. The pump discharge pressure is 83 ft w.g. What would be the reading on a pressure gage at the pump suction, in psig?

8.11   In the duct system shown in Figure 8.37 on page 238, what is the pressure change from point 1 to 2, if the friction loss is 2.1 in. w.g. between the points?

8.12   The average velocity pressure in a 48 by 18 in. duct is 0.5 in. w. Determine the flow rate in CFM.

8.13   The duct transition shown in Figure 8.38 on page 238 has a recovery factor of 0.8. Determine the static pressure regain.

8.14   Find the pressure loss due to friction and the velocity in a 250 ft straight section of 4 in.

Figure 8.34
Sketch to Problem 8.7.

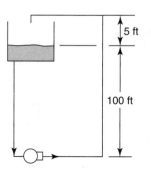

Figure 8.35
Sketch for Problem 8.8.

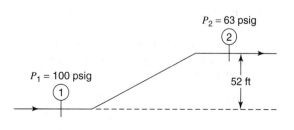

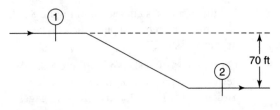

Figure 8.36
Sketch for Problem 8.9.

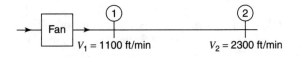

Figure 8.37
Sketch for Problem 8.11.

Schedule 40 steel pipe in a hydronic cooling system through which 200 GPM of water is flowing.

8.15   What would be the pressure loss in the piping in Problem 8.14 if it were in a hydronic heating system?

8.16   Determine the friction loss per 100 ft of pipe for 10 GPM flowing through 1¼ in. Type L copper tubing.

8.17   Water flows through a clean 3 in. Schedule 40 steel pipe at a velocity of 4 ft/sec. What is the flow rate and friction loss per 100 ft?

8.18   A Schedule 40 steel pipe is to be used to deliver 150 GPM to a cooling tower. It shall have a friction pressure loss no greater than 4 ft w. per 100 ft. What is the minimum pipe size that can be used? What is the actual friction loss?

8.19   Determine the pressure drop through a 1½ in. cast iron globe valve through which 40 GPM is flowing.

8.20   Find the pressure drop through the cold water piping system shown in Figure 8.39, composed of Type L copper tubing.

8.21   Find the pressure drop through the condenser-cooling tower water system shown in Figure 8.40, composed of 8 in. Schedule 40 steel pipe, through which 1100 GPM is flowing.

8.22   For the piping arrangement shown in Example 8.16, size the piping by the constant

friction loss method, using a value of about 1.5 ft w. per 100 ft.

8.23   For the piping arrangement shown in Example 8.16, with each unit circulating 6 GPM, size the piping system, using Type L copper tubing, at 4 ft w. per 100 ft.

8.24   Determine the pressure drop in the piping system in Problem 8.23. Assume a typical set of valves and auxiliaries. The boiler has a pressure drop of 3 ft w. The length of each branch is 12 ft.

8.25   Find the equivalent round diameter of a 36 in. by 12 in. rectangular duct.

8.26   A 28 by 14 in. galvanized steel duct has a flow rate of 5000 CFM of air. Find the friction loss per 100 ft and the velocity.

8.27   A straight length of duct 420 ft long has a flow rate of 2000 CFM of air. The friction loss in the duct must be limited to 1.6 in. w. What is the smallest round duct size that could be used?

Figure 8.38
Sketch for Problem 8.13.

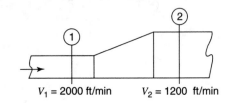

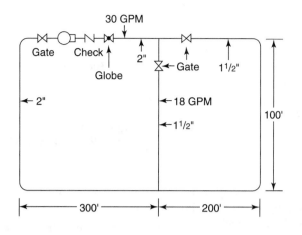

Figure 8.39
Sketch for Problem 8.20.

8.28 Find the loss coefficient of a 30 in. by 12 in. 90° duct elbow with a mean radius of 12 in., without vanes. If the elbow is carrying 5000 CFM, what is the pressure drop?

8.29 Find the static pressure change from A to B in the duct shown in Figure 8.41, assuming a recovery factor of 0.75.

8.30 A 24 by 12 in. duct has a flow rate of 4000 CFM. The static pressure is 3 in. w.g. at a certain location. What is the total pressure at this location?

8.31 A 24 in. by 14 in. galvanized duct is supposed to be delivering 5000 CFM. Test readings on static pressure manometers 80 ft apart in a straight section of the duct read 2.10 in. w.g. and 1.90 in. w.g. Determine if the duct is supplying the proper air quantity.

8.32 Find the total pressure loss in the duct system shown in Figure 8.42 (page 240) from the fan outlet at A to F and also to G. The pressure loss through each air outlet is 0.20 in. w.g.

Figure 8.40
Sketch for Problem 8.21.

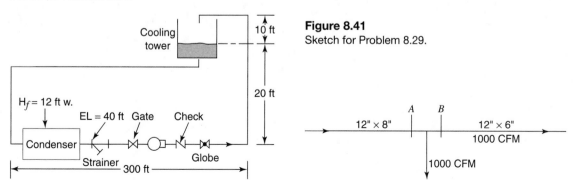

Figure 8.41
Sketch for Problem 8.29.

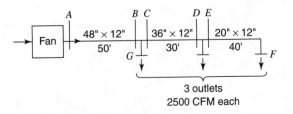

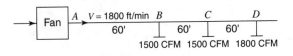

Figure 8.42
Sketch for Problem 8.32.

Figure 8.43
Sketch for Problem 8.33.

8.33 Size the rectangular ducts in the system shown in Figure 8.43 by both the equal friction method and by the static region method. All ducts shall be 10 in. deep.

8.34 Find the pressure drop in the duct system in Example 8.27. Outlets have a pressure loss of 0.10 in. w. Assume reasonable shapes of transitions and elbows.

8.35 Size the ducts in Example 8.27 by the static regain method, starting with the same initial velocity. Compare the duct sizes and pressure drop found by the two methods.

8.36 Determine the pipe sizes and system pressure drop for the house in Problems 3.20 and 5.14.

8.37 Determine the pipe sizes and system pressure drop for the building in Problems 3.21 and 5.15.

CHAPTER 9
Piping, Valves, Ducts, and Insulation

The selection of the correct types of piping, valves, ducts, and insulation for a particular HVAC system is an important task. In addition, the proper method of installation should be understood, considering such problems as expansion, hanging, supporting, anchoring, and vibrations.

OBJECTIVES

After studying this chapter, you will be able to:

1. Specify the appropriate materials, fittings, and values for an HVAC piping installation.
2. Determine the need for pipe expansion and vibration treatment.
3. Use correct practices for pipe installations.
4. Specify the appropriate materials for HVAC duct installations.
5. Use correct practices for duct installations.

9.1 PIPING MATERIALS AND SPECIFICATIONS

Piping is made of many possible materials, and the proper choice depends on the *service* for which the piping is intended. The service includes:

1. Properties of the fluid being carried
2. Temperature
3. Pressure
4. Exposure to oxidation or corrosion

In addition, cost and availability also affect choice of materials. Finally, codes and regulations usually limit the choice of materials for a given use. It is the responsibility of the HVAC specialist to be aware of the codes that apply to each installation before selecting materials. The piping material most commonly used in hydronic systems is either low carbon ("black") steel pipe or copper tube. For severe problems of oxidation or corrosion, other materials may be necessary.

Physical specifications of steel pipe and copper tubing are standardized by the *American Society of Testing Materials* (ASTM). The term *black steel pipe* is used in the trade, but this actually refers to

ASTM A-120 or ASTM A-53 low carbon steel. The engineer should always specify the pipe intended by the ASTM number. For the interested reader, the chemical composition of those materials can be found in the ASTM publications.

Some specifications for black steel pipe are shown in Table 9.1. The wall thickness is referred to by a *Schedule number,* such as 20, 30, 40, or 80. These numbers supersede a former description called standard, extra strong, and double extra strong.

The choice of the correct Schedule number of piping depends on the pressure and temperature service. The allowable pressures can be calculated from formulas established in the *American Standard Code for Pressure Piping.* In hydronic systems at pressures commonly encountered, Schedule 40 pipe is usually specified, except in very large diameters, where Schedule 30 or 20 is sometimes used.

The allowable pressure is only part of the story, however. The engineer should recognize that corrosion and erosion may reduce the pipe wall thickness over a period of years. Therefore, selecting piping with a substantial wall thickness may mean a longer life system.

The wall thickness of copper tubing is specified by a lettering system, Type K, L, M, and DWV. Type K has the thickest wall, and is used with high pressures and refrigerants. Type L has an intermediate thickness wall. It is usually adequate for hydronic system piping. Type M is used for low pressure plumbing. The outside diameter (OD) is the same for any size for all three types; the inside diameter (ID) changes. The pressure drop will therefore be greatest for Type K. Figure 8.15 confirms this. The specifications for Type L tubing are shown in Table 9.2. Type DWV is used for drainage.

Hard temper as opposed to *soft temper* tubing has greater rigidity and will not sag as much as soft tubing when hung horizontally.

The decision to choose between steel piping or copper tubing for an installation is based primarily on cost. Copper is more expensive than steel, but in smaller sizes the labor cost for copper is often less.

TABLE 9.1   SPECIFICATIONS OF STEEL PIPE

Pipe Size (inches)	Schedule	Outside Diameter (inches)	Inside Diameter (inches)	Wall Thickness (inches)	Weight, lbs/ft	Volume, gal/ft
½	40	0.840	0.622	0.109	0.850	0.0158
¾	40	1.050	0.824	0.113	1.130	0.0276
1	40	1.315	1.049	0.133	1.678	0.0449
1¼	40	1.660	1.380	0.140	2.272	0.0774
1½	40	1.900	1.610	0.145	2.717	0.106
2	40	2.375	2.067	0.154	3.652	0.174
2½	40	2.875	2.469	0.203	5.790	0.248
3	40	3.500	3.068	0.216	7.570	0.383
3½	40	4.000	3.548	0.226	9.110	0.513
4	40	4.500	4.026	0.237	10.790	0.660
5	40	5.563	5.047	0.258	14.620	1.039
6	40	6.625	6.065	0.280	18.970	1.501
8	40	8.625	7.981	0.322	28.550	2.597
10	40	10.750	10.020	0.365	40.480	4.098
12	30	12.750	12.090	0.330	43.800	5.974
14	30	14.000	13.250	0.375	54.600	7.168
16	30	16.000	15.250	0.375	62.400	9.506

TABLE 9.2 SPECIFICATIONS OF COPPER TUBING (TYPE L)

Nominal Size (inches)	Outside Diameter (inches)	Inside Diameter (inches)	Wall Thickness	Weight, lbs/ft	Volume, gal/ft
⅜	½	0.430	0.035	0.198	0.00753
½	⅝	0.545	0.040	0.285	0.0121
⅝	¾	0.660	0.042	0.362	0.0181
¾	⅞	0.785	0.045	0.455	0.0250
1	1⅛	1.025	0.050	0.655	0.0442
1¼	1⅜	1.265	0.055	0.884	0.0655
1½	1⅝	1.505	0.060	1.140	0.0925
2	2⅛	1.985	0.070	1.750	0.1610
2½	2⅝	2.465	0.080	2.480	0.2470
3	3⅛	2.945	0.090	3.330	0.3540
3½	3⅝	3.425	0.100	4.290	0.4780
4	4⅛	3.905	0.110	5.380	0.6230
5	5⅛	4.875	0.125	7.610	0.9710
6	6⅛	5.845	0.140	10.200	1.3900
8	8⅛	7.725	0.200	19.300	2.4300
10	10⅛	9.625	0.250	30.100	3.7900
12	12⅛	11.565	0.280	40.400	5.4500

It is common to see larger installations in steel and smaller ones in copper.

Copper tubing has two advantages that should be noted. First, the frictional resistance is less than for steel, resulting in the possibility of smaller pumps and less power consumption. Second, it is not subject to oxidation and scaling to the same extent as steel. On the other hand, steel is a stronger material and therefore does not damage as easily. Sometimes the larger piping is made of steel and smaller branches to units are copper. When this is done, a plastic bushing should be used to separate the copper and steel, because otherwise corrosion may occur at the joint due to electrolytic action.

In open piping systems, such as piping to a cooling tower, oxidation may occur if black steel is used. *Galvanized* steel pipe is sometimes used in these applications. This is black steel pipe that has a coating of a tin alloy which resists oxidation. In very severe corrosion applications, galvanized piping is not adequate, and much more costly *wrought iron* or *cast iron* pipe is used.

Note that pipe and tubing diameters are specified by a *nominal* size, which does not always correspond exactly to either the inside diameter or outside diameter.

The specification tables contain much useful information, as shown in Example 9.1.

Example 9.1
A 60 ft long, 5 in. Schedule 40 chilled water steel pipe is to be hung horizontally from a floor slab above. The structural engineer asks the HVAC contractor to determine how much extra weight the floor will have to carry.

Solution
The weight includes the pipe and the water it carries. Using Table 9.1,

$$\text{Pipe weight} = 14.6 \text{ lb/ft}$$
$$\text{Water weight} = 1.04 \text{ gal/ft} \times 8.3 \text{ lb/gal}$$
$$= 8.6 \text{ lb/ft}$$
$$\text{Total weight} = (14.6 + 8.6) \text{ lb/ft} \times 60 \text{ ft}$$
$$= 1392 \text{ lb}$$

9.2 FITTINGS AND JOINING METHODS FOR STEEL PIPE

In hydronic systems, joining of steel pipe is usually done with either *screwed, welded,* or *flanged* fittings. Specifications for fittings are established by the *American National Standards Institute (ANSI)* for both steel pipe and copper tube.

Screwed fittings for steel piping are generally made of cast or malleable iron. For typical hydronic systems, fittings with a 125 psi pressure rating are usually adequate. If in doubt, the system pressure should be checked. Typical pipe fittings are shown in Figure 9.1.

Elbows (ells), used for changing direction, are available in 30°, 45°, and 90° turns. *Long radius* ells have a more gradual turn than *standard* ells, and thus have a lower pressure drop. Sometimes, however, tight spaces require standard ells. *Tees* are used for branching, and *couplings* are used to join straight lengths of threaded pipe. When joining to equipment, however, *unions* should be used so that the connection may be disconnected for service. With welded fittings, mating welding flanges on

Figure 9.1
Steel pipe fittings. (*a*) A 90° elbow, threaded. (*b*) Tee, threaded. (*c*) Coupling, threaded. (*d*) Bushing, threaded. (*e*) Nipple. (*f*) Union. (*g*) A 90° elbow, flanged. (*h*) Tee, welding. (*i*) A 90° elbow, welding. **(Courtesy: Grinnell Corporation, Providence, R.I.)**

(*a*)

(*b*)

(*c*)

(*d*)

(*e*)

(*f*)

(*g*)

(*h*)

(*i*)

the pipe and equipment serve the same purpose as unions. *Bushings* are used when connecting from a pipe of one size to a piece of equipment that has a different size opening.

Welding is a process where the two metal ends to be joined are melted and then fused together with a metal welding rod that also liquifies and fuses. The heat may come from either a gas torch or an electric arc. Welding makes a very strong joint. Welding fittings similar to screwed fittings are available. Straight lengths of pipe may be butt welded directly together without couplings.

In hydronic installations, screwed steel pipe joints are commonly used up to about 2–3 in., and welded joints are used for larger sizes. This generally results in the lowest cost of labor plus materials in most U.S. locations. Furthermore, it is more difficult to make pressure-tight threaded joints in very large sizes.

9.3 FITTINGS AND JOINING METHODS FOR COPPER TUBING

Copper tubing joints in hydronic systems are made either by soldering (also called *sweating*) or by *flaring*. Typical solder fittings are shown in Figure 9.2.

Soldering is a process where a metal alloy called solder is melted (between 400 and 1000 F) and when it solidifies, it forms a pressure-tight joint between the two parts to be joined. Solder fittings are made to slip over the tubing with enough clearance for the solder to flow in the annular space between the fitting and the tube. Surfaces must be clean of all oxidation. A chemical coating called *flux* is then used to prevent further oxidation. When the joints must withstand high temperatures and pressures or severe vibrations, a soldering process called *brazing* is used. Basically it is no different from lower temperature soldering, except that a different soldering material is used which melts at a higher temperature (above 1000 F) and makes a stronger joint.

Flared joints are made by flaring out the end of the copper tubing and using a flare fitting union

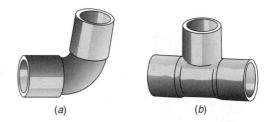

Figure 9.2
Copper tube solder fittings. (*a*) A 90° copper elbow. (*b*) Copper tee.

that will make a pressure-tight seal when tightened against the flare (Figure 9.3 on page 246). Flared fittings are expensive but are removable, and therefore should be used when access to equipment is required for service or maintenance.

Strainers are used to remove solid particles from the circulating system. The water passes through a perforated plate or wire mesh in which the particles are trapped. The strainer is cleaned at regular intervals. In small pipe sizes, the Y-type is generally used, and in large sizes, a basket type is used (Figure 9.4 on page 246). Strainers are usually installed at the suction side of pumps and before large automatic control valves.

9.4 VALVES

There are many types and uses of valves. We will discuss mainly *general service* valves, types that are used widely in piping systems. Automatic control valves will be discussed in Chapter 14. Valves for controlling flow may be grouped into three classes according to their function.

Stopping Flow
Valves in this group are used only to shut off flow. This procedure is useful in isolating equipment for service or in isolating sections of a system so that it may be serviced, yet allowing the rest of the system to operate. *Gate* valves (Figure 9.5 on page 247) are used for this purpose. Note that a gate valve has a straight through flow passage, resulting in a low pressure loss.

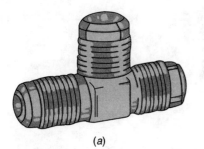

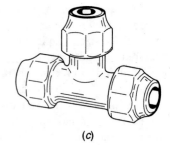

(a)   (b)   (c)

Figure 9.3
Copper tube flaring fittings. (*a*) Tee. (*b*) Nut. (*c*) Tee with nuts assembled.

Regulating Flow Rate

Valves in this group are used to adjust flow rate manually. This is desirable in setting proper flow rates through equipment and different circuits in a system. *Globe* valves (Figure 9.6), *angle* valves, *plug* valves, *needle* valves, *ball* valves, and *butterfly* valves can be used to regulate flow.

Gate valves should not be used to regulate flow, and they must be closed or left completely open. Their internal construction is not suitable for throttling flow. If partially closed, *wiredrawing* (erosion of the valve seat) may occur. However, most flow regulation valves can be used to stop flow. This should only be done in an emergency, as the system must then be balanced again.

Limiting Flow Direction

Valves that allow flow in only one direction are called *check* valves. In water circulating systems, reverse flow could occur when the system is not operating, particularly if there is a static head of water. Reverse flow may damage equipment or empty out a line or equipment unintentionally. Figure 9.7 (page 248) shows some types of check valves. The *swing check* can be installed only in horizontal lines. A *vertical lift check* or *spring-loaded check* can be used in vertical lines. Check valves are usually installed at a pump and other critical points in a system.

9.5   PRESSURE REGULATING AND RELIEF VALVES

Where water pressure may exceed safe limits for equipment, a *pressure regulating valve* (PRV) is used. This valve limits the discharge pressure to a preset value. These valves are often used in the make-up water supply line to a system where the make-up is from a city water supply at high pressures.

Figure 9.4
Y-type strainer. (Courtesy: Grinnell Corporation, Providence, R.I.)

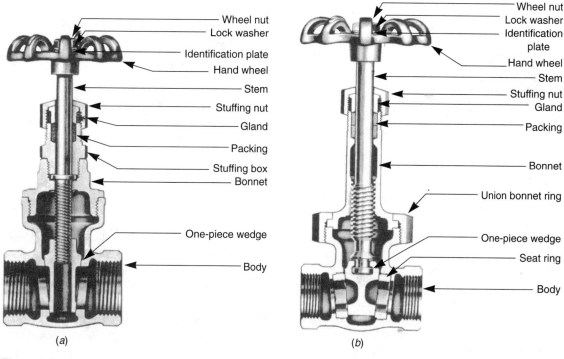

Figure 9.5
Gate valves. (*a*) Screwed bonnet, nonrising stem. (*b*) Union bonnet, rising stem.

Figure 9.6
Globe valve.

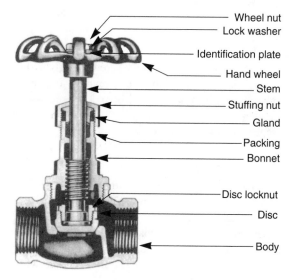

Pressure relief valves are not the same as a PRV. A relief valve opens when the valve inlet pressure exceeds a preset value. Relief valves are used as safety devices to relieve excess pressure in boilers and other equipment (Chapter 4).

9.6 VALVE CONSTRUCTION

The construction of gate and globe valves will be discussed in more detail here. Refer to Figures 9.5 and 9.6 to see what each part looks like and how it is assembled. Knowledge of valve construction will enable you to select the correct valve for each application and to understand how to service it.

Disc or Wedge

The part that closes off flow is called the disc or wedge. A *solid* wedge is simplest and is often used in

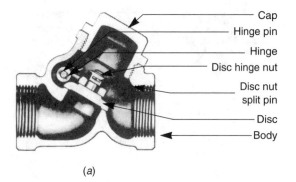

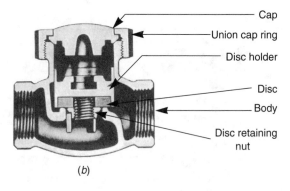

Figure 9.7
Check valves. (*a*) Swing check. (*b*) Horizontal lift check.

gate valves. A two-piece *split* wedge is less subject to sticking and is often used at high temperatures and pressures. Globe valve discs made in bevel or plug shapes are best for throttling service. Flat *composition discs* are not recommended for close throttling, but they have the advantage of being replaceable without removing the valve. Bevel and plug discs can usually be reground in place when they wear.

Stem

The stem lifts and lowers the disc or wedge. The valve may be constructed with a *rising* or *nonrising* stem. In the rising stem type (Figure 9.6), the disc and stem rise together. The length of exposed stem therefore provides a quick visual indication of whether the valve is open or closed. This is advantageous to the operating engineer, particularly if the service is one where an incorrect position of the valve might result in immediate harm to some process or even danger to people. In the nonrising stem type, the disc travels on the stem, and the stem stays in one position. There is therefore no visual indication of whether the valve is open or closed. This arrangement takes up less space, however, because the valve stem does not travel.

The valve may be constructed so that the stem has an inside screw or *outside screw and yoke* (OS & Y). In the inside screw arrangement, the stem threads are inside the valve body. In the OS & Y type (Figure 9.8), the threads are outside the valve body and are held by a yoke. With the inside arrangement, the threads are exposed to the system fluid. The OS & Y arrangement might be used when corrosive fluids or extreme temperatures and pressures exist to prevent damage to the threads.

Packing Nut and Stuffing

The valve must have a means of sealing around the stem to prevent the fluid from leaking out under high pressure. This is accomplished with the *stuffing* or *packing*, usually made of a soft material impregnated with graphite. The packing is held in place and compressed against the stem by tightening the packing nut.

Bonnet

The bonnet connects the nut to the body of the valve. *Screwed bonnets*, which screw directly to the valve body (Figure 9.6), are common in small valves. A screwed *union* type bonnet is used when frequent disassembly is expected. Bolted bonnets are used on larger valves.

Valve Materials

For hydronic service, either *all bronze* valves or *iron body* with bronze parts are generally used. For many applications, valves with a 125 psi pressure rating are suitable, but this should be checked before selection. At extremely high pressures, steel valves may be required. Valves are available with screwed, flanged, or welded ends.

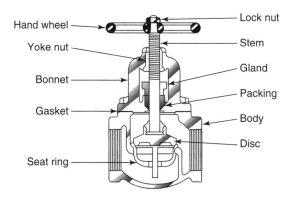

Figure 9.8
OS & Y globe valve.

9.7 VALVE SELECTION

Although the types of valves were described previously, specific recommendations on choosing the proper valve for an application will be discussed here.

Gate valves should be used only for stopping flow, not for throttling. Use them for isolating equipment and sections of a system.

Regulating or throttling flow can be accomplished with globe, angle, plug, needle, or butterfly valves. Use these valves in any section, branch, or unit where flow needs to be manually adjusted or balanced. For large diameters, butterfly valves have become popular in hydronic systems because they cost considerably less than globe valves. However, they should not be used where extremely tight shut-off or very close regulation is required.

9.8 PIPE EXPANSION AND ANCHORING

Most materials (unless constrained) will change their length when their temperature is changed. Pipe lines in hydronic heating systems will therefore tend to expand from their initial lengths when brought up to system temperature. This fact must be accounted for in the piping design and installation. Failure to do so may result in broken piping and damaged equipment.

The engineer provides for *expansion* where desirable, and for proper *anchoring* of piping where expansion is undesirable. Each system layout is unique and must be analyzed to determine the correct solution. Some guidelines will be suggested here. If pipe expansion is completely prevented, considerable forces and stresses may result in the piping, which could rupture it. Therefore, it is usual to provide for some expansion. This may be done by the following methods:

1. Using expansion *loops* or *offsets* (Figure 9.9) allows the pipe to bend at the loop or offset, thus accommodating the expansion. Of course, the bending itself results in stresses on the pipe, so the size of the loop or offset must be adequate, and depends on the length, size, material, and temperature change. Often a run of piping will have enough natural offsets to accommodate expansion, particularly in small installations. Cast iron fittings should not be used in expansion loops, because cast iron is brittle and may crack.

Figure 9.9
Expansion loop and offset.

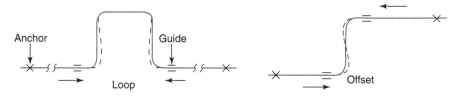

2. Expansion joints, which are manufactured items, may be used. There is a *slip* type where the pipe slides inside the joint, and there is a *bellows* type where the joint is a movable bellows. Expansion joints are subject to wear and leakage and must be periodically inspected and maintained. Therefore, they should not be installed in inaccessible locations. Access doors must be provided if they are located in shafts or otherwise closed in.

The effect that expansion of a long run has on branch piping must be provided for. Branch connections must be provided with sufficient flexibility so as not to break. This is done by offsets—changing directions at a branch connection. Movement at the offset then prevents a break. This type of offset is called a *swing joint* (Figure 9.10).

The provisions for expansion described above will result in reduced forces at points where the piping is anchored, but does not exclude the need to provide proper anchoring methods. Equipment should not be used as anchoring points. Do not allow rigid connections to equipment where expansion occurs; use offsets or flexible connections. Anchoring connections must be made so that any force is transmitted to a part of the building structure adequate to take the force. It is sometimes necessary to solve this problem with the aid of the structural engineer. Location of anchoring points can be determined only by studying the particular installation. It may be best to anchor at both ends, or only in the middle, or at a number of points, depending on the length of the run, where expansion loops are provided, and where equipment connections are located.

Pipe *supports* are necessary to carry the weight of the piping and water. Vertical piping may be supported at the bottom, or at one or more points along the height. The supports also may or may not serve as anchors. Horizontal piping is supported by *hangers*. Supports must be provided at frequent enough intervals to not only carry the weight, but to prevent sag. The hanger usually consists of a *rod* and a *cradle*.

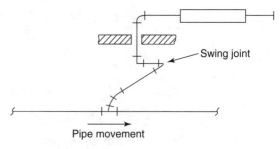

Figure 9.10
Swing joint.

There are various types of cradles, as shown in Figure 9.11. Where considerable movement occurs, *roll-type* hangers should be used. If hung from the concrete floor slab above, hanger inserts are installed before the concrete is poured. This requires careful planning and coordination between the HVAC and structural engineers. If extra hangers are needed after the construction is completed, inserts can be driven into the concrete slab with a gun-like tool.

9.9   VIBRATION

Consideration must be given to possible vibrations occurring in the piping system. Pumps and compressors usually are the source of vibrations. Reciprocating machinery generally creates more vibrations than rotating machinery. Vibrations may be transmitted to the building structure or to piping. Both of these problems must be examined.

In some cases, the intensity of vibration produced by machinery may not be great enough to result in significant transmission to the piping or structure, and no further consideration is necessary. When vibration transmitted to the structure requires treatment, it may be reduced by use of heavy concrete foundations and by suitable machinery locations. There are cases, however, where prevention of any transmission of vibration to the building structure is critical. An example is where machinery is located on a lightweight penthouse floor above office spaces. In this case, the equipment is mounted on

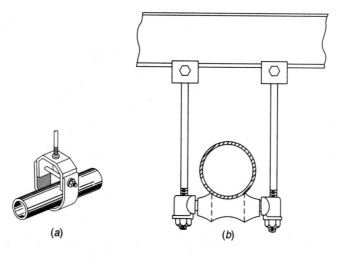

Figure 9.11
Pipe supports. (*a*) Clevis support and hanger. (*b*) Roller support and hanger.

vibration isolators. Isolating supports may be made of cork, rubber, or steel springs (Figure 9.12).

One or more of the following procedures may be used when vibrations are transmitted to the piping:

1. Use of isolation hangers. Using an isolation material (cork, felt) between the cradle and pipe may sometimes be adequate. For a more serious problem, spring hangers can be used.

2. Use of flexible pipe connections to the troublesome machinery. However, flexible connections tend to become inflexible at high pressures.

Difficult vibration problems may require the aid of a specialist in these fields, and the HVAC engineer should not hesitate to call on such help when necessary.

Figure 9.12
Vibration isolation mountings. (*a*) Rubber pad. (*b*) Spring. **(Courtesy: Vibration Mountings and Controls, Inc.)**

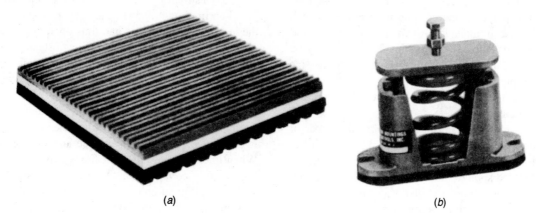

(*a*)   (*b*)

9.10 PIPE INSULATION

Thermal insulation should be used on all cold or hot hydronic system piping. On both hot and chilled water systems, thermal insulation serves two purposes:

1. To reduce energy waste and possible increased size of heating or refrigerating equipment.
2. To reduce incorrect distribution of heat. Uninsulated piping may result in the water being at an unsatisfactory temperature when it reaches the conditioned spaces.

In chilled water systems, it is also necessary to prevent condensation of moisture from the air on the outside of the cold piping, which could both damage the insulation and drip onto surrounding surfaces. The prevention of condensation is achieved by covering the insulation with a material that serves as a *vapor barrier*. It is impervious to the flow of the water vapor in the air.

There are a great many materials from which pipe insulation may be made. A good insulation should have the following characteristics:

1. Low thermal conductivity
2. Noncombustible
3. Not subject to deterioration or rot
4. Adequate strength

Pipe insulation may be made from natural materials such as wool, felt, rock, or glass fibers; cork; and rubber. In recent years, synthetic materials such as polyurethane have been developed, which have an extremely low thermal conductivity and other excellent properties. Pipe insulation may be furnished in blanket form or premolded to the size of the pipe to be covered. The latter is preferable because it requires less labor and will have a neater appearance.

Vapor barriers are made from treated paper or aluminum foil. Usually the manufacturer furnishes it already wrapped on the insulation (Figure 9.13).

Molded sponge rubber insulation is very popular on small diameter chilled water lines. The rubber serves as both a thermal insulation and vapor

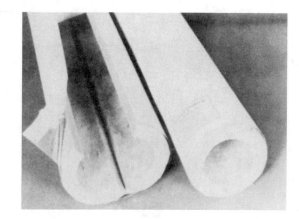

Figure 9.13
Pipe insulation with vapor barrier covering. (Courtesy: Owens Corning Fiberglas Corporation)

barrier. It is very easy to cut and install, thus resulting in low labor costs.

When exposed, the surface of insulation is often provided with a canvas cover. The surface is then *sized*, that is, painted with a material which makes the surface smooth and stiff.

Fittings and valves are also insulated. This is usually done by trowelling on an insulating cement mixture. Premolded, shaped insulation is also available for some typical fittings. The installer must take special care not to cover operating parts of valves or removable flanges when applying the insulation.

The question of what thickness of insulation to use is an important one. The greater the thickness, the less the energy losses and thus operating costs are reduced. However, the insulation costs increase with thickness. Therefore, the correct insulation thickness is generally that which provides the minimum owning and operating cost.

Exposed piping and equipment, whether insulated or not, is usually painted when the installation is completed, both for appearance and protection. When lines carrying different fluids exist, the piping is often *color-coded*, that is, each system is painted a different color. In any case, lines should be stenciled with their proper names and direction of flow at reasonable intervals, to make operation and maintenance easier.

Brass name tags should always be furnished and attached to valves. These are usually numbered and a key list is made identifying them by cross-referencing. This list should be mounted in a highly visible place, such as the operating engineer's office.

9.11 THE PIPING INSTALLATION

Some good general practices for installing the piping system will be listed here.

1. Piping should generally be parallel to building walls.
2. Direction changes should be minimized to reduce the number of fittings.
3. The installation should provide simple access to and maintenance of equipment. For example, do not run piping in front of a control panel.
4. Piping should avoid penetration of beams or other structural members. Where this is unavoidable, the structural engineer must be consulted.
5. The piping must not interfere with installations of other trades. This must be checked with the plans of ducts, lighting, and so on.
6. The piping location must not affect the building function. An obvious example is running piping across a door opening.
7. Install horizontal piping with a slight pitch and take all branch connections from the top so that any entrapped air will flow to high points.
8. Provide air vent devices at all high points (see Chapter 11).
9. Provide a short pipe connection and gate valve at all low points in order to drain the system.

This list does not include special features peculiar to each project, nor safety and code requirements.

9.12 DUCT CONSTRUCTION

The most commonly used material for general HVAC ducts is galvanized steel sheet metal. In re-

cent years, molded glass fiber ducts have also come into use. When the air being carried is corrosive, more corrosion-resistant materials are used, such as stainless steel, copper, or aluminum. Exhausts from kitchens and chemical laboratories are examples where special materials would be required.

SMACNA (Sheet Metal and Air Conditioning Contractors National Association) has established standards for construction of ductwork. These standards specify the sheet metal thickness (gage), methods of bracing and reinforcing the duct to prevent collapse or sagging, and methods of joining sections. The standards depend on the air pressure in the duct. High pressure systems require stronger construction. The details of recommended duct construction can be found in *SMACNA* publications. Glass fiber ducts are recommended only for low pressure systems.

Rectangular-shaped sheet metal duct is most commonly used in lower pressure HVAC applications (up to 3 in. w.g. static pressure). Machine-made round sheet metal duct is popular in high velocity, high pressure systems, although heavy gage rectangular duct is also used. Round flexible duct is often used to make final connections to air diffusers, because this permits the contractor to make small adjustments in the location of the diffuser.

Rectangular duct is usually made to order for each job. Round duct is fabricated by machinery in standard diameters. Both for this reason and because round duct is lighter in gage, it is less expensive than rectangular duct for high pressure systems.

Rectangular duct fittings are very expensive because of the labor cost involved, and should be as simple as possible, unless minimum pressure loss is important. Figure 9.14 (page 254) shows a transition and branch fitting for minimum pressure loss. Figure 9.15 (page 254) shows a simpler lower cost fitting, but it will have a greater branch pressure loss.

When changing duct shapes, the transition should preferably have a slope of 7:1, and should not be less than 4:1, to minimize pressure loss (Figure 9.16 on page 254).

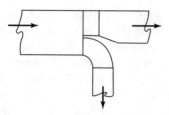

Figure 9.14
Branch connection with low pressure loss.

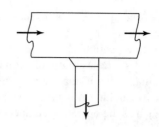

Figure 9.15
Branch connection with high pressure loss.

When changing direction, round elbows with a wide sweep radius should be used to keep pressure loss low. If short radius or square elbows are necessary to save space, turning vanes should be installed in the elbow (Figure 9.17).

Duct joints should be made as tight as possible to reduce air leakage, using a sealant if necessary. It is not unusual to find installations losing 10% or more of the design airflow due to poor installation.

Small horizontal ductwork is supported by sheet metal straps. Heavier ducts require angle iron support cradles suspended from rods (Figure 9.18).

The ductwork standards described here apply to commercial applications. Ducts for residential use are simpler in construction and will not be described here. For further information, see the *ASHRAE Systems Volume*. Duct connections to fans and air distribution devices will be discussed in Chapter 10.

9.13   DUCT INSULATION

Ducts carrying hot or cold air are covered with thermal insulation to reduce heat loss. In addition, the insulation is covered with a vapor barrier to prevent condensation of water on cold ducts. Glass fiber or similar material with a high thermal resistance is used for insulation. The vapor barrier is usually aluminum foil. Insulation comes in either *rigid board* or *blanket* form (Figure 9.19). The rigid board costs considerably more, and is used only when the duct is exposed and appearance is important or abuse is likely.

Ducts are frequently lined internally with acoustical insulation to absorb sound. In this case, the acoustical lining often also serves as thermal insulation.

Figure 9.16
Recommended slope for duct transition.

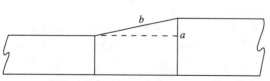

Slope *b:a* of 7:1 preferred
Slope *b:a* of 4:1 minimum recommended

Figure 9.17
Square elbow with turning vane.

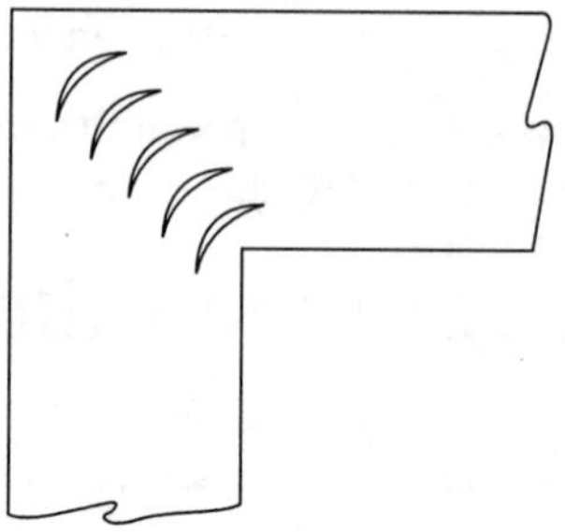

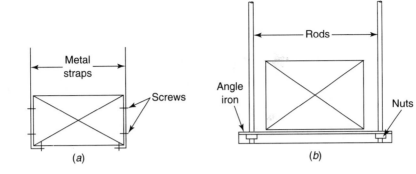

Figure 9.18
Duct hangers. (*a*) Strap hanger. (*b*) Trapeze hanger.

Review Questions

1. Describe the systems for specifying wall thicknesses of both steel pipe and copper tubing.

2. List the pipe fittings described in this chapter and explain their uses.

3. Describe the joining methods used for copper tubing and steel pipe.

4. List the major types of valves and their uses.

5. List the types of materials used for HVAC ducts and their applications.

6. Discuss the features and uses of rectangular and round ducts.

7. List the good practices for duct design and installation.

8. What are the two forms of duct insulation? What is a vapor barrier?

9. List recommended good practices for pipe installation.

Figure 9.19
Duct insulation. (*a*) Blanket. (*b*) Rigid board. (**Courtesy: Owens Corning Fiberglas Corporation**)

(*a*)

(*b*)

CHAPTER 10
Fans and Air Distribution Devices

Fans are necessary to distribute air through equipment and through ductwork to spaces that are to be air conditioned. In the first part of this chapter, we will study types of fans and their performance, selection, application, and construction, and some installation and energy conservation recommendations. After that, we will discuss air distribution devices and their selection, and sound control in air distribution systems.

OBJECTIVES

After studying this chapter, you will be able to:

1. Distinguish the types of fans and their characteristics.
2. Select a fan.
3. Use the fan laws to determine the effect of changed conditions.
4. Distinguish the types of air distribution devices and their applications.
5. Select an air distribution device.
6. Analyze the sound conditions in an air distribution system.

10.1 FAN TYPES

Fans may be classified into two main types, *centrifugal* fans and *axial flow* fans, which differ in the direction of airflow through the fan. In a centrifugal fan, air is pulled along the fan shaft and the blown radially away from the shaft. The air is usually collected by a scroll casing and concentrated in one direction (Figure 10.1). In an axial flow fan, air is pulled along the fan shaft and then blown along in the same direction (Figure 10.2).

Centrifugal Fans

Centrifugal fans may be subclassified into *forward curved, radial, backward curved,* and *backward inclined* types, which differ in the shape of their impeller blades (Figure 10.3). In addition, backward curved blades with a double-thickness blade are called *airfoil* blades.

Axial Fans

Axial fans may be subclassified into *propeller, tubeaxial,* and *vaneaxial* types. The propeller fan

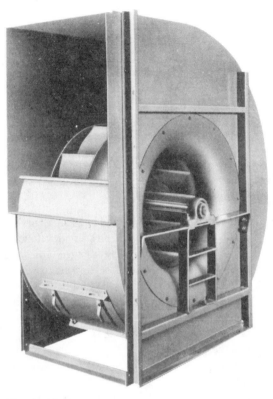

Figure 10.1
Centrifugal fan, airfoil blade type. (Courtesy: Buffalo Forge Company.)

Figure 10.2
Vaneaxial fan. (Courtesy: Buffalo Forge Company.)

plied to the air. This is accomplished by the rotating fan impeller, which exerts a force on the air, resulting in both flow of the air and an increase in its pressure.

The volume flow rate of air delivered and the pressure created by the fan are called *performance characteristics*. Other performance characteristics of importance are *efficiency* and *brake horsepower* (BHP).

shown in Figure 10.4(*a*) on page 258 is like a window fan, consisting of a propeller-type wheel mounted on a ring or plate. The tubeaxial fan shown in Figure 10.4(*b*) has a vaned wheel mounted in a cylinder. The vaneaxial fan in Figures 10.4(*c*) and 10.2 is similar to the tubeaxial type, except that it also has guide vanes behind the fan blades which improve the direction of airflow through the fan.

10.2   FAN PERFORMANCE CHARACTERISTICS

In the general discussion on fluid flow (Chapter 8), we noted that there is a resistance, caused by friction, to the flow of air through ducts. To overcome this resistance, energy in the form of pressure must be sup-

Figure 10.3
Types of centrifugal fan impeller blades.

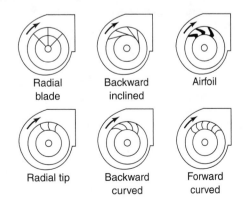

Radial blade   Backward inclined   Airfoil

Radial tip   Backward curved   Forward curved

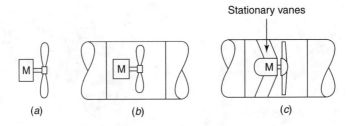

Figure 10.4
Types of axial flow fans. (*a*) Propeller. (*b*) Tubeaxial. (*c*) Vaneaxial.

Knowledge of the fan performance is useful for correct fan selection and proper operating and troubleshooting procedures.

The following symbols and definitions will be used in discussing fan performance.

CFM = volume flow rate, ft³/min

H_s = static pressure, inches of water gage (in. w.g.)

H_v = velocity pressure, in. w.g.

H_t = total pressure, in. w.g.

BHP = brake horsepower input

N = speed, revolutions per min (RPM)

d = air density, lb/ft³

ME = mechanical efficiency

= air horsepower output/BHP input

Fan performance is best understood when presented in the form of curves. Figures 10.5 and 10.6 are typical performance curves for forward and backward curved bladed centrifugal fans. Some important features seen are:

1. For both forward curved and backward curved blade centrifugal fans, the pressure developed has a slight peak in the middle range of flow, then the pressure drops off as flow increases.
2. The BHP required for the forward curved blade fan increases sharply with flow, but with the backward curved blade fan, the BHP increases only gradually, peaks at a maximum, and then falls off.
3. Efficiency is highest in the middle ranges of flow.
4. A higher maximum efficiency can often be achieved with a backward curved blade fan.

10.3 FAN SELECTION

The choice of the best type of fan to be used for a given application depends on the fan performance

Figure 10.5
Typical performance characteristics of a forward curved blade centrifugal fan.

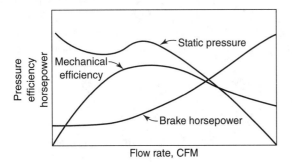

Figure 10.6
Typical performance characteristics of a backward curved blade centrifugal fan.

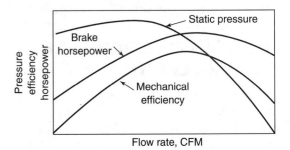

characteristics and other features that will be discussed.

Propeller fans cannot create a high pressure, and are thus used where there is little or no ductwork. They are low in cost, and typical applications are as wall- or window-installed exhaust fans.

Centrifugal fans are the most commonly used type in ducted air conditioning systems. Forward curved blade centrifugal fans are usually lower in initial cost than backward blade types for the same performance. The operating cost will often be higher, however, due to lower efficiency. The rising BHP characteristic curve could result in overloading the motor if operated at a condition beyond the selected CFM. These fans are often used in packaged air conditioning units because of low cost.

Backward (curved or inclined) blade centrifugal fans are generally more expensive than forward curved types, but usually have lower operating costs due to high efficiency. The limiting horsepower characteristic reduces the possibility of overloading the motor or electrical distribution system if the fan is delivering more air than it was designed for. Airfoil bladed fans have the highest efficiency of any type.

Tubeaxial and vaneaxial fans can be used in ducted systems. The air distribution from tubeaxial fans is uneven, thus making them undesirable for air conditioning systems. Vaneaxial fans are suitable for ducted air conditioning systems. They usually produce a higher noise level than centrifugal fans and therefore may require greater sound reduction treatment. Their compact physical construction is useful when space is limited.

10.4 FAN RATINGS

After the best type of fan is selected for an application, the next task is to determine the proper size to be used.

Manufacturer's fan ratings are presented as either performance curves (Figure 10.7) or tables (Table 10.1 on pages 260–262) for each fan size. Curves and tables each have their good and bad features.

Performance curves enable the engineer to visualize changes in static pressure, BHP, and efficiency easily. Note that each fan curve represents the performance at a specific fan speed and air density. Fans are usually rated with air at *standard conditions:* a density of 0.075 lb/ft³ at 70 F and 29.92 in. Hg. Performance curves at different air conditions may be available from the manufacturer, but if not, they may be predicted from the *fan laws* to be described in a later section.

Tables list fan performance at different speeds, and therefore replace a large number of curves. For this reason, tables are used more often than curves for fan selection. However, the operating condition of maximum efficiency is not apparent when using tables. Some manufacturers resolve this by noting the point of maximum efficiency on their tables (usually in boldface).

Operating near maximum efficiency generally results in the lowest noise output by a fan.

To select a fan, the duct system static pressure resistance (duct H_s) is first calculated, using the procedures explained in Chapter 8. Manufacturer's data are then used to select a fan that will produce the required CFM against the system static pressure resistance.

In effect, the fan must develop a static pressure (fan H_s) and CFM equal to the system requirements. (We will discuss the fan–system interaction in more detail shortly, but for now we will focus on the fan selection.)

Figure 10.7
Performance curves of a 33 in. diameter backward inclined blade centrifugal fan at 1440 RPM.

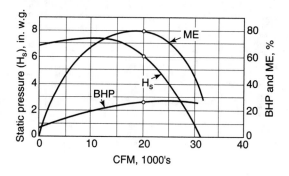

TABLE 10.1   PERFORMANCE CHARACTERISTICS OF TYPICAL AIRFOIL BLADE CENTRIFUGAL FANS (27 in. WHEEL DIAMETER)

CFM	OV	1/4" SP		3/8" SP		1/2" SP		5/8" SP		3/4" SP		1" SP		1 1/4" SP		1 1/2" SP		1 3/4" SP		2" SP	
		RPM	BHP	RPM	BHP	RPM	BHP	RPM	BHP	RPM	BHP	RPM	BHP	RPM	BHP	RPM	BHP	RPM	BHP	RPM	BHP
2085	500	325	.10	376	.15																
2502	600	351	.13	395	.18	438	.24	481	.30												
2919	700	382	.16	421	.22	458	.28	495	.35	532	.42										
3336	800	414	.19	450	.26	484	.33	516	.40	548	.47	613	.63								
3753	900	447	.23	481	.31	513	.38	542	.46	572	.54	629	.70	686	.89	744	1.09				
4170	1000	482	.28	514	.36	543	.45	571	.53	598	.62	650	.79	702	.98	753	1.18	805	1.40		
4587	1100	518	.34	547	.42	575	.52	602	.61	627	.70	676	.89	723	1.08	770	1.29	816	1.51	863	1.74
5004	1200	555	.40	582	.50	609	.59	634	.69	658	.79	703	1.00	748	1.20	791	1.42	834	1.64	876	1.87
5421	1300	592	.47	618	.57	642	.68	667	.79	689	.89	733	1.11	775	1.33	815	1.56	855	1.79	895	2.03
5838	1400	629	.56	654	.66	677	.78	700	.89	722	1.00	765	1.23	804	1.47	842	1.71	880	1.96	917	2.21
6255	1500	668	.65	691	.76	714	.88	734	1.00	755	1.12	796	1.37	834	1.62	870	1.88	906	2.13	941	2.40
6672	1600	707	.76	728	.87	749	1.00	770	1.13	789	1.25	828	1.51	866	1.78	900	2.05	934	2.32	967	2.60
7089	1700	745	.87	765	1.00	786	1.13	806	1.26	824	1.40	861	1.67	897	1.95	932	2.24	964	2.53	996	2.81
7506	1800	785	.99	804	1.14	823	1.27	842	1.41	860	1.55	895	1.84	930	2.13	963	2.43	995	2.73	1025	3.05
7923	1900	823	1.13	842	1.29	860	1.43	878	1.57	896	1.72	930	2.02	963	2.34	995	2.64	1027	2.96	1056	3.28
8340	2000	863	1.28	881	1.45	898	1.61	915	1.75	933	1.91	965	2.23	997	2.55	1028	2.87	1058	3.19	1088	3.53
8757	2100	903	1.45	920	1.62	936	1.80	952	1.94	969	2.11	1000	2.44	1031	2.77	1061	3.12	1090	3.45	1119	3.79
9174	2200	942	1.65	960	1.81	975	2.00	990	2.16	1006	2.31	1036	2.66	1066	3.01	1095	3.37	1123	3.72	1151	4.08
9591	2300	982	1.86	998	2.01	1014	2.21	1028	2.40	1043	2.55	1073	2.91	1102	3.28	1130	3.63	1157	4.02	1184	4.38
10008	2400	1023	2.09	1037	2.23	1053	2.43	1068	2.64	1080	2.81	1110	3.17	1137	3.54	1165	3.93	1191	4.32	1217	4.71
10425	2500	1062	2.33	1077	2.47	1092	2.68	1106	2.89	1119	3.09	1147	3.45	1173	3.84	1200	4.25	1226	4.63	1251	5.05
10842	2600	1102	2.59	1117	2.74	1131	2.94	1145	3.16	1158	3.38	1184	3.74	1210	4.16	1235	4.56	1261	4.98	1285	5.40
11259	2700	1142	2.85	1156	3.04	1170	3.22	1184	3.45	1196	3.68	1222	4.07	1247	4.49	1271	4.90	1296	5.35	1320	5.76
11676	2800	1183	3.13	1196	3.36	1209	3.51	1223	3.76	1236	4.00	1259	4.40	1284	4.83	1308	5.28	1331	5.72	1355	6.16
12093	2900	1223	3.44	1237	3.69	1249	3.84	1262	4.09	1274	4.33	1297	4.75	1321	5.20	1344	5.67	1367	6.11	1390	6.59
12510	3000	1264	3.78	1277	4.04	1289	4.21	1301	4.43	1313	4.69	1335	5.14	1359	5.60	1381	6.07	1403	6.54	1425	7.01
12927	3100	1304	4.14	1316	4.42	1328	4.60	1340	4.79	1353	5.07	1373	5.55	1397	6.02	1419	6.49	1440	6.99	1461	7.46
13344	3200	1345	4.52	1356	4.81	1368	5.02	1379	5.19	1392	5.47	1412	5.98	1434	6.44	1456	6.95	1477	7.45	1498	7.95
13761	3300	1385	4.92	1396	5.21	1408	5.46	1419	5.63	1430	5.88	1450	6.45	1472	6.92	1494	7.43	1514	7.94	1535	8.47
14178	3400	1426	5.35	1437	5.62	1449	5.92	1459	6.10	1469	6.32	1490	6.91	1510	7.41	1531	7.94	1552	8.45	1571	8.99
14595	3500	1466	5.80	1477	6.07	1489	6.40	1499	6.61	1509	6.79	1529	7.40	1549	7.94	1569	8.44	1590	8.99	1608	9.54

CFM	OV	1/4" SP		3/8" SP		1/2" SP		5/8" SP		3/4" SP		1" SP		1 1/4" SP		1 1/2" SP		1 3/4" SP		2" SP	
		RPM	BHP	RPM	BHP	RPM	BHP	RPM	BHP	RPM	BHP	RPM	BHP	RPM	BHP	RPM	BHP	RPM	BHP	RPM	BHP
2575	500	292	.13	339	.19																
3090	600	316	.16	356	.22	394	.29	433	.37												
3605	700	344	.19	379	.27	412	.34	445	.43	478	.51										
4120	800	373	.24	405	.32	435	.41	465	.49	494	.58	551	.78								
4635	900	403	.29	433	.38	461	.47	488	.57	514	.67	566	.87	617	1.09	669	1.35				
5150	1000	434	.35	463	.45	489	.55	514	.66	538	.76	585	.98	631	1.20	678	1.45	725	1.73		
5665	1100	466	.42	492	.52	518	.64	542	.75	565	.86	608	1.10	651	1.34	693	1.59	735	1.86	777	2.15
6180	1200	499	.49	524	.61	548	.73	570	.85	592	.97	633	1.23	673	1.49	712	1.75	750	2.02	789	2.31
6695	1300	533	.58	556	.71	578	.84	600	.97	620	1.10	660	1.37	697	1.64	734	1.92	770	2.21	805	2.51
7210	1400	566	.69	589	.82	609	.96	630	1.09	650	1.24	688	1.52	723	1.82	758	2.11	792	2.42	825	2.72
7725	1500	601	.81	622	.94	642	1.09	661	1.23	680	1.39	716	1.69	751	2.00	783	2.32	815	2.63	847	2.96
8240	1600	636	.93	655	1.08	674	1.23	693	1.39	710	1.54	745	1.87	779	2.20	810	2.53	841	2.87	871	3.21
8755	1700	671	1.07	689	1.24	707	1.39	725	1.56	742	1.73	775	2.07	807	2.40	839	2.76	868	3.12	896	3.47
9270	1800	707	1.23	723	1.41	741	1.56	758	1.74	774	1.92	806	2.28	837	2.63	867	3.00	895	3.37	923	3.76
9785	1900	741	1.39	758	1.59	774	1.76	791	1.94	807	2.13	837	2.50	867	2.88	896	3.26	924	3.45	951	4.05
10300	2000	777	1.58	793	1.79	808	1.99	824	2.16	839	2.36	869	2.75	897	3.15	925	3.54	952	3.94	979	4.36
10815	2100	812	1.79	828	2.00	843	2.22	857	2.40	872	2.60	900	3.01	928	3.42	955	3.85	981	4.26	1007	4.69
11330	2200	848	2.03	864	2.24	878	2.47	891	2.67	906	2.86	933	3.29	960	3.72	986	4.16	1011	4.60	1036	5.04
11845	2300	884	2.30	899	2.48	913	2.73	926	2.96	939	3.15	966	3.59	992	4.05	1017	4.49	1041	4.96	1065	5.41
12360	2400	921	2.58	934	2.75	948	3.01	961	3.26	973	3.48	999	3.92	1023	4.38	1049	4.85	1072	5.33	1095	5.82
12875	2500	956	2.88	969	3.05	983	3.31	996	3.57	1007	3.82	1032	4.26	1056	4.74	1080	5.24	1103	5.72	1126	6.24
13390	2600	992	3.20	1005	3.38	1018	3.64	1030	3.91	1043	4.18	1066	4.62	1089	5.14	1112	5.63	1135	6.15	1157	6.66
13905	2700	1028	3.53	1041	3.75	1053	3.98	1066	4.26	1077	4.54	1100	5.02	1122	5.54	1144	6.06	1167	6.61	1188	7.12
14420	2800	1065	3.87	1077	4.15	1088	4.34	1101	4.65	1112	4.94	1134	5.44	1156	5.97	1177	6.52	1198	7.06	1220	7.61
14935	2900	1101	4.25	1113	4.56	1124	4.75	1136	5.05	1147	5.35	1168	5.87	1190	6.43	1210	7.00	1231	7.55	1251	8.14
15450	3000	1138	4.67	1149	5.00	1160	5.20	1171	5.47	1182	5.79	1202	6.35	1224	6.92	1243	7.50	1263	8.08	1283	8.66
15965	3100	1174	5.12	1185	5.46	1196	5.69	1206	5.92	1218	6.26	1236	6.86	1257	7.44	1277	8.02	1297	8.64	1315	9.2
16480	3200	1211	5.58	1221	5.95	1232	6.20	1242	6.41	1253	6.75	1271	7.39	1291	7.96	1311	8.58	1330	9.21	1348	9.83
16995	3300	1247	6.08	1257	6.44	1268	6.74	1277	6.95	1288	7.26	1306	7.96	1325	8.55	1345	9.18	1363	9.80	1382	10.46
17510	3400	1284	6.60	1294	6.95	1304	7.31	1313	7.54	1323	7.80	1341	8.53	1360	9.16	1379	9.81	1397	10.43	1415	11.11
18025	3500	1320	7.16	1330	7.50	1340	7.91	1349	8.16	1358	8.39	1377	9.14	1394	9.81	1413	10.42	1431	11.11	1448	11.78

TABLE 10.1 (Continued) (33 in. WHEEL DIAMETER)

CFM	OV	¼" SP		⅜" SP		½" SP		⅝" SP		¾" SP		1" SP		1¼" SP		1½" SP		1¾" SP		2" SP	
		RPM	BHP	RPM	BHP	RPM	BHP	RPM	BHP	RPM	BHP	RPM	BHP	RPM	BHP	RPM	BHP	RPM	BHP	RPM	BHP
3130	500	**261**	**.15**	302	.22	....	....	....	....	....	....	....	....	....	....	....	....	....	....	....	....
3756	600	282	.19	**317**	**.26**	351	.35	385	.44	....	....	....	....	....	....	....	....	....	....	....	....
4382	700	306	.23	337	.32	**367**	**.41**	397	**.50**	426	.61	....	....	....	....	....	....	....	....	....	....
5008	800	332	.28	360	.38	388	.48	414	.58	**440**	**.69**	491	.93	....	....	....	....	....	....	....	....
5634	900	359	.35	386	.45	410	.56	435	.68	458	.79	**504**	**1.03**	550	1.30	595	1.60	....	....	....	....
6260	1000	388	.42	412	.54	436	.65	457	.77	479	.90	521	1.16	**563**	**1.43**	604	1.73	645	2.06	....	....
6886	1100	416	.50	440	.63	461	.76	482	.89	502	1.02	541	1.30	580	1.59	**617**	**1.89**	655	2.21	692	2.57
7512	1200	446	.59	468	.74	488	.87	508	1.02	527	1.16	564	1.46	599	1.76	634	2.08	**669**	**2.40**	703	2.75
8138	1300	476	.70	496	.86	516	1.01	535	1.16	553	1.31	587	1.62	621	1.95	653	2.28	686	2.63	**718**	**2.98**
8764	1400	506	.82	526	.99	544	1.15	562	1.31	579	1.48	612	1.81	644	2.15	675	2.51	705	2.87	735	3.23
9390	1500	536	.95	555	1.13	573	1.32	590	1.49	607	1.66	638	2.01	668	2.37	698	2.74	726	3.13	754	3.51
10016	1600	567	1.10	585	1.30	602	1.49	618	1.68	634	1.86	665	2.24	693	2.61	721	2.99	749	3.40	776	3.81
10642	1700	598	1.27	615	1.47	631	1.68	647	1.89	663	2.08	692	2.48	720	2.87	746	3.28	772	3.68	798	4.12
11268	1800	630	1.45	645	1.67	661	1.89	676	2.11	691	2.32	719	2.73	746	3.16	772	3.57	797	4.00	822	4.44
11894	1900	661	1.66	676	1.88	691	2.11	706	2.34	720	2.58	747	3.01	773	3.46	799	3.90	822	4.35	846	4.80
12520	2000	693	1.90	707	2.12	721	2.36	735	2.61	749	2.84	776	3.33	801	3.76	825	4.25	849	4.71	871	5.19
13146	2100	724	2.13	738	2.38	752	2.63	766	2.88	779	3.13	804	3.65	829	4.12	852	4.60	875	5.11	897	5.59
13772	2200	756	2.39	770	2.64	783	2.91	796	3.17	808	3.46	833	3.99	857	4.50	880	4.98	902	5.52	924	6.04
14398	2300	787	2.69	801	2.95	814	3.23	826	3.51	838	3.79	862	4.34	886	4.90	908	5.41	929	5.93	950	6.50
15024	2400	819	3.02	833	3.30	845	3.58	857	3.85	868	4.13	892	4.73	914	5.32	936	5.87	957	6.40	977	6.97
15650	2500	851	3.36	864	3.66	876	3.92	888	4.21	899	4.53	922	5.15	943	5.74	965	6.35	985	6.90	1005	7.45
16276	2600	884	3.73	896	4.01	908	4.29	919	4.64	930	4.94	951	5.59	973	6.20	993	6.85	1013	7.44	1033	8.01
16902	2700	917	4.13	928	4.40	939	4.74	950	5.07	960	5.36	980	6.01	1002	6.69	1022	7.35	1042	8.00	1061	8.60
17528	2800	950	4.57	959	4.85	971	5.22	981	5.50	992	5.84	1012	6.51	1032	7.22	1051	7.88	1071	8.58	1089	9.22
18154	2900	981	5.03	991	5.34	1002	5.69	1013	5.97	1023	6.37	1044	7.04	1061	7.77	1081	8.45	1099	9.16	1118	9.86
18780	3000	1013	5.51	1023	5.86	1034	6.16	1044	6.51	1054	6.89	1074	7.58	1090	8.30	1111	9.05	1129	9.77	1146	10.52
19406	3100	1046	6.02	1055	6.37	1066	6.68	1076	7.10	1085	7.40	1104	8.14	1122	8.89	1140	9.72	1158	10.43	1175	11.19
20032	3200	1078	6.55	1088	6.95	1097	7.26	1107	7.70	1117	7.98	1136	8.78	1154	9.55	1169	10.36	1188	11.11	1204	11.89
20658	3300	1111	7.13	1122	7.54	1129	7.89	1139	8.28	1148	8.65	1169	9.50	1184	10.23	1198	11.00	1217	11.86	1234	12.63
21284	3400	1144	7.74	1154	8.19	1161	8.57	1171	8.89	1180	9.37	1201	10.24	1214	10.90	1230	11.72	1247	12.64	1263	13.40
21910	3500	1177	8.42	1186	8.87	1193	9.27	1203	9.57	1211	10.08	1231	10.97	1245	11.63	1262	12.51	1275	13.37	1293	14.23

CFM	OV	¼" SP		⅜" SP		½" SP		⅝" SP		¾" SP		1" SP		1¼" SP		1½" SP		1¾" SP		2" SP	
		RPM	BHP	RPM	BHP	RPM	BHP	RPM	BHP	RPM	BHP	RPM	BHP	RPM	BHP	RPM	BHP	RPM	BHP	RPM	BHP
3830	500	**235**	**.18**	273	.27	....	....	....	....	....	....	....	....	....	....	....	....	....	....	....	....
4596	600	255	.23	**287**	**.32**	317	.43	348	.54	....	....	....	....	....	....	....	....	....	....	....	....
5362	700	277	.28	305	.39	**332**	**.50**	359	**.62**	385	.75	....	....	....	....	....	....	....	....	....	....
6128	800	300	.35	326	.46	350	.59	374	.71	**398**	**.84**	444	1.13	....	....	....	....	....	....	....	....
6894	900	325	.42	349	.55	371	.68	393	.83	414	.96	**456**	**1.26**	491	1.59	538	1.96	....	....	....	....
7660	1000	350	.51	373	.65	394	.80	414	.94	433	1.10	471	1.42	**509**	**1.75**	546	2.12	583	2.52	....	....
8426	1100	377	.61	397	.77	417	.96	436	1.09	454	1.25	489	1.59	524	1.94	**558**	**2.31**	592	2.71	625	3.14
9192	1200	403	.73	423	.90	442	1.07	460	1.24	477	1.42	510	1.78	541	2.16	573	2.54	**605**	**2.94**	636	3.37
9958	1300	430	.86	449	1.05	467	1.23	484	1.42	500	1.60	531	1.98	562	2.39	590	2.79	620	3.21	**649**	**3.64**
10724	1400	457	1.00	475	1.21	492	1.41	508	1.60	524	1.81	553	2.21	582	2.63	610	3.07	637	3.51	665	3.96
11490	1500	485	1.17	502	1.39	518	1.61	534	1.82	548	2.03	577	2.46	604	2.90	631	3.36	657	3.82	682	4.30
12256	1600	513	1.34	529	1.59	544	1.82	559	2.05	574	2.28	601	2.74	627	3.19	652	3.66	677	4.16	701	4.66
13022	1700	541	1.56	556	1.80	571	2.05	585	2.31	599	2.55	625	3.03	651	3.51	674	4.01	698	4.51	722	5.04
13788	1800	569	1.77	584	2.04	598	2.31	612	2.58	625	2.84	651	3.33	675	3.87	698	4.37	720	4.90	743	5.43
14554	1900	598	2.03	611	2.30	625	2.59	638	2.87	651	3.16	676	3.69	699	4.23	722	4.78	743	5.32	765	5.87
15320	2000	626	2.32	640	2.59	652	2.88	665	3.19	678	3.48	701	4.07	724	4.61	746	5.21	767	5.76	787	6.34
16086	2100	655	2.61	668	2.92	680	3.22	692	3.53	704	3.83	727	4.47	749	5.04	771	5.63	791	6.26	811	6.84
16852	2200	684	2.92	696	3.23	708	3.56	719	3.88	731	4.23	753	4.88	775	5.51	796	6.10	815	6.76	835	7.3
17618	2300	712	3.29	725	3.61	736	3.95	747	4.29	758	4.64	780	5.31	801	6.00	821	6.62	840	7.25	859	7.96
18384	2400	741	3.69	753	4.04	764	4.38	775	4.71	785	5.06	807	5.78	827	6.51	846	7.18	865	7.83	884	8.53
19150	2500	770	4.12	782	4.47	792	4.80	803	5.15	813	5.55	834	6.30	853	7.03	872	7.77	890	8.44	909	9.11
19916	2600	799	4.57	810	4.91	821	5.26	831	5.67	841	6.05	860	6.84	880	7.59	898	8.38	916	9.10	934	9.80
20682	2700	830	5.06	839	5.39	849	5.73	859	6.21	868	6.56	886	7.36	906	8.18	924	8.99	942	9.79	960	10.52
21448	2800	859	5.59	867	5.94	878	6.38	887	6.73	897	7.15	915	7.96	933	8.84	951	9.64	968	10.50	985	11.28
22214	2900	887	6.16	896	6.53	906	6.96	916	7.30	925	7.79	944	8.62	959	9.51	977	10.34	994	11.21	1011	12.0
22980	3000	916	6.75	925	7.17	935	7.54	944	7.97	953	8.43	971	9.28	986	10.15	1004	11.07	1020	11.96	1037	12.88
23746	3100	946	7.37	954	7.80	964	8.17	973	8.69	981	9.06	998	9.96	1014	10.89	1031	11.89	1047	12.76	1063	13.70
24512	3200	975	8.02	984	8.50	992	8.89	1001	9.42	1010	9.77	1027	10.74	1043	11.69	1057	12.68	1074	13.60	1089	14.55
25278	3300	1005	8.73	1014	9.23	1021	9.66	1030	10.14	1038	10.59	1057	11.63	1071	12.52	1084	13.46	1101	14.51	1116	15.46
26044	3400	1034	9.48	1043	10.02	1050	10.49	1059	10.88	1067	11.46	1086	12.53	1098	13.35	1112	14.35	1127	15.47	1142	16.40
26810	3500	1064	10.30	1072	10.86	1078	11.34	1087	11.72	1095	12.34	1114	13.43	1126	14.24	1141	15.32	1153	16.36	1169	17.42

TABLE 10.1   (Continued) (40 1/4 in. WHEEL DIAMETER)

CFM	OV	1/4" SP		3/8" SP		1/2" SP		5/8" SP		3/4" SP		1" SP		1 1/4" SP		1 1/2" SP		1 3/4" SP		2" SP	
		RPM	BHP	RPM	BHP	RPM	BHP	RPM	BHP	RPM	BHP	RPM	BHP	RPM	BHP	RPM	BHP	RPM	BHP	RPM	BHP
4655	500	**213**	**.22**	**248**	**.33**	....	....	....	....	....	....	....	....	....	....	....	....	....	....	....	....
5586	600	230	.28	259	.39	**288**	**.52**	318	.66	....	....	....	....	....	....	....	....	....	....	....	....
6517	700	248	.34	276	.47	301	.61	**325**	**.75**	350	.91	....	....	....	....	....	....	....	....	....	....
7448	800	268	.42	293	.56	317	.72	339	.87	**360**	**1.03**	404	1.38	....	....	....	....	....	....	....	....
8379	900	289	.50	313	.67	335	.84	355	1.01	375	1.17	**413**	**1.54**	452	1.94	492	2.37	....	....	....	....
9310	1000	312	.61	333	.78	354	.97	373	1.16	392	1.35	427	1.72	**461**	**2.13**	496	2.58	532	3.05	....	....
10241	1100	335	.73	354	.92	373	1.12	392	1.32	409	1.53	443	1.94	475	2.36	**505**	**2.81**	537	3.30	570	3.81
11172	1200	358	.87	376	1.07	394	1.28	412	1.50	428	1.72	460	2.18	490	2.62	519	3.09	**547**	**3.58**	**577**	**4.11**
12103	1300	382	1.03	399	1.25	416	1.47	432	1.70	448	1.94	479	2.43	507	2.92	535	3.40	561	3.91	587	4.43
13034	1400	406	1.20	422	1.45	438	1.68	453	1.92	468	2.17	497	2.70	525	3.22	552	3.75	577	4.26	602	4.81
13965	1500	431	1.40	446	1.67	460	1.91	475	2.17	489	2.43	517	2.99	544	3.54	569	4.10	594	4.67	617	5.22
14896	1600	455	1.63	470	1.91	484	2.18	497	2.44	511	2.72	537	3.29	563	3.89	588	4.48	611	5.08	634	5.69
15827	1700	480	1.87	494	2.17	508	2.47	520	2.74	533	3.04	558	3.62	583	4.26	606	4.89	630	5.52	652	6.16
16758	1800	504	2.13	519	2.46	532	2.78	544	3.08	555	3.37	580	3.99	603	4.63	626	5.32	648	5.99	670	6.65
17689	1900	529	2.43	543	2.78	555	3.11	568	3.44	579	3.75	602	4.40	624	5.05	646	5.76	667	6.48	688	7.18
18620	2000	554	2.76	567	3.13	580	3.47	591	3.83	602	4.17	624	4.83	645	5.52	666	6.22	687	6.99	707	7.74
19551	2100	579	3.13	592	3.51	604	3.87	615	4.24	626	4.61	647	5.28	667	6.02	687	6.74	707	7.51	727	8.31
20482	2200	605	3.52	616	3.90	629	4.31	639	4.68	650	5.08	670	5.81	689	6.56	709	7.30	728	8.07	747	8.90
21413	2300	630	3.95	641	4.34	653	4.77	664	5.16	674	5.57	693	6.36	712	7.09	731	7.91	749	8.69	767	9.52
22344	2400	656	4.43	666	4.81	677	5.27	688	5.69	698	6.09	717	6.94	735	7.71	753	8.56	771	9.37	788	10.19
23275	2500	682	4.94	691	5.34	702	5.78	713	6.25	722	6.66	741	7.57	758	8.40	776	9.19	793	10.09	810	10.92
24206	2600	707	5.50	716	5.91	727	6.34	737	6.85	747	7.29	765	8.21	782	9.09	798	9.91	815	10.84	832	11.72
25137	2700	733	6.09	742	6.50	751	6.95	761	7.47	771	7.95	789	8.88	806	9.83	821	10.72	838	11.58	854	12.56
26068	2800	758	6.70	768	7.13	776	7.60	786	8.11	796	8.65	813	9.61	830	10.62	845	11.54	860	12.41	876	13.41
26999	2900	784	7.35	793	7.83	802	8.31	811	8.80	820	9.38	838	10.37	854	11.42	869	12.39	883	13.35	899	14.27
27930	3000	809	8.06	819	8.58	827	9.07	836	9.56	845	10.15	861	11.19	878	12.25	893	13.32	907	14.31	921	15.23
28861	3100	835	8.85	845	9.39	852	9.86	861	10.36	869	10.92	886	12.08	902	13.14	917	14.27	931	15.28	944	16.30
29792	3200	861	9.69	870	10.23	878	10.68	886	11.23	894	11.77	911	12.99	926	14.09	941	15.24	955	16.34	968	17.40
30723	3300	887	10.59	896	11.12	903	11.59	911	12.16	919	12.69	935	13.94	950	15.07	965	16.24	979	17.44	992	18.51
31654	3400	913	11.54	921	12.04	929	12.57	936	13.13	944	13.66	960	14.93	974	16.14	989	17.31	1003	18.56	1016	19.70
32585	3500	939	12.51	947	12.98	955	13.59	962	14.12	969	14.70	985	15.96	999	17.26	1013	18.46	1027	19.71	1040	20.96

The fan may also be selected on the basis of total pressure rather than static pressure. Either basis is satisfactory for low velocity systems. For high velocity systems, it is sometimes more accurate to use total pressure (see Chapter 8). In our examples, static pressure will be used.

The system static pressure resistance is often called the external static pressure, and is frequently abbreviated SP or ESP in manufacturers' literature (as in Table 10.1).

Examples 10.1 and 10.2 illustrate the use of fan manufacturers' curves and tables.

Example 10.1

What static pressure (H_s) will the fan whose performance curves are shown in Figure 10.7 develop at a delivery of 20,000 CFM? What will be the brake horsepower (BHP) and mechanical efficiency (ME) at this condition?

Solution

Using Figure 10.7, at the intersection of 20,000 CFM and the H_s curve,

$$H_s = 6 \text{ in. w.g.}$$

AT this CFM, we also note

$$BHP = 27 \text{ HP}$$
$$ME = 80\%$$

This example does not indicate if there are better choices of fans to deliver 20,000 CFM at 6 in. w.g. static pressure. Perhaps a more efficient choice exists, or if the emphasis is on initial cost, a smaller fan might be found. Other fan performance curves could be studied to determine these possibilities. However, it is simpler to use fan tables for selection, as Example 10.2 illustrates.

Example 10.2

Select an airfoil blade centrifugal fan to supply 8400 CFM at 1½ in. w.g. static pressure.

Solution

The selection will be made from Table 10.1. Assume that energy conservation is important. The following possible selections are noted from the data in the table. (Interpolation between listed values is carried out where necessary.)

Wheel Size, in.	CFM @1½ in.	BHP
27	8400	3.0
30	8400	2.6
33	8400	2.4
36½	8400	2.3
40¼	8400	2.4

The best selection is probably a 33 in. fan. The saving on energy use is negligible with the 36½ in. fan. The 30 in. fan uses 10% more energy, but if initial cost were the most important consideration, it might be selected. It will be noisier, however, because it is less efficient.

10.5 SYSTEM CHARACTERISTICS

In a manner similar to considering fan performance characteristics of CFM versus pressure developed, we can examine the duct *system characteristics* of CFM versus pressure loss (H_f). The pressure loss due to frictional resistance in a given duct system varies as the CFM changes, as follows:

$$H_{f2} = H_{f1}\left(\frac{CFM_2}{CFM_1}\right)^2 \tag{10.1}$$

Equation 10.1 can be used to find the changed pressure loss in a duct system for a changed CFM flow, if the pressure loss is known at some other flow rate.

Example 10.3

The ductwork in a certain ventilating system has a pressure loss of 2 in. w.g. with 5000 CFM of air flowing. What would be the pressure loss if the airflow were 7000 CFM?

Solution
Using Equation 10.1,

$$H_{f2} = 2\left(\frac{7000}{5000}\right)^2 = 3.9 \text{ in. w.g.}$$

By plotting a few of such H_f versus CFM points, a system characteristic curve can be determined. Note that the pressure loss rises sharply with CFM for any duct system, as shown in Example 10.4.

Example 10.4

Plot the system CFM versus H_f curve for the duct system of Example 10.3.

Solution
Using Equation 10.1 to plot a few points, the results are shown in Table 10.2 and plotted in Figure 10.8.

TABLE 10.2 RESULTS FOR EXAMPLE 10.4

CFM	0	2500	4000	5000	6000	7000
H_f, in. w.g.	0	0.5	1.3	2.0	2.9	3.9

Figure 10.8
Sketch for Example 10.4 (system characteristic curve).

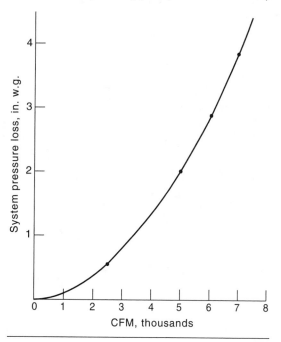

10.6 FAN–SYSTEM INTERACTION

By plotting both the fan and system characteristic pressure versus flow curves together, we can find the condition of operation of the fan and system (Figure 10.9). Because the fan can only perform at conditions on the fan curve, and the system can only perform at conditions on the system curve, the following important principle is always true: *The point of intersection of the system and fan curves is the operating condition of the system.*

Example 10.5 _____

For the fan whose performances characteristics are shown in Figure 10.10, what will be the operating conditions when used with the duct system whose characteristic curve is also shown in Figure 10.10?

Solution

The intersection point of the fan and system pressure characteristic curves is the operating condition, 25,000 CFM and a static pressure of 5.3 in. w.g. The fan BHP = 35, ME = 60% at this condition.

_____

Examining the fan and system curves is not only useful for selecting the operating condition, but

Figure 10.9

Fan and system curves plotted together—intersection is point of operation.

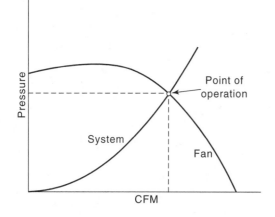

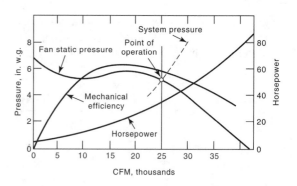

Figure 10.10

Fan and system curves for Example 10.5.

aids in analyzing changed conditions and in finding causes of operating difficulties. A common occurrence in air conditioning systems is that the actual system resistance for a design CFM is different from that calculated by the designer. Some reasons this may happen are:

1. An error in calculating pressure loss.
2. Allowance of an extra resistance as a "safety factor" by the designer.
3. The contractor installs the ductwork in a manner different from that planned.
4. Filters may have a greater than expected resistance due to excess dirt.
5. An occupant may readjust damper positions.

The result of this type of condition is that the duct system has a different characteristic than planned. An examination of the fan and system curves will aid in analyzing these situations.

Example 10.6 _____

For the fan whose performance is shown in Figure 10.11, the system required CFM is 5000 and calculated pressure loss is 4.4 in. w.g. The design operating condition is therefore point 1. The system design performance curve *A,* calculated from Equation 10.1, is also shown.

The system designer has allowed an extra 1.0 in. w.g. as a "safety factor" in calculating the pressure loss. Assuming that there is no real extra pressure loss, what will the actual operating conditions be?

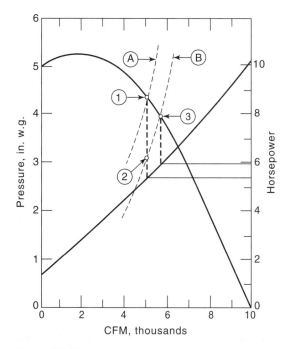

Figure 10.11
Sketch for Example 10.6.

Solution
The actual system pressure loss is 4.4 − 1.0 = 3.4 in. w.g. at 5000 CFM (point 2). The real system pressure curve can be plotted (curve *B*). The actual operating condition is the intersection of the fan and real system pressure curves (point 3).

At this point, the flow rate is 5500 CFM. The system will be delivering too much air to the spaces. Also note that the BHP = 6.0 instead of 5.5, as would have occurred with the design condition. This is a considerable waste of energy and money. Of course, the actual flow rate could be throttled by using dampers. Another solution might be to change the fan speed.

10.7   SYSTEM EFFECT

The fan that is chosen for a given application must develop, at the design CFM, a pressure at least equal to the duct system pressure loss. The manu-facturer's fan rating curves or tables are used to find a suitable fan.

However, the fan ratings are based on testing the fans in a manner prescribed by the *Air Movement and Control Association (AMCA)*. Because every fan installation is unique, each arrangement of inlet and outlet connections to the fan will probably differ from that of the laboratory arrangement in which the fan was tested. The result will be that there are additional pressure losses at the fan inlet and fan outlet, which must be added to the system pressure loss before selecting the required fan pressure.

The precise loss at fan inlet and outlet depends on the shape, size, and direction of the connections to the fan. These are called *system effects,* which we have shown how to determine in Chapter 8. In any case, connections should be made to minimize losses, as described later.

For fans that are part of a packaged unit, the manufacturer usually has allowed for the system effect pressure loss for the unit.

10.8   SELECTION OF OPTIMUM FAN CONDITIONS

Often a number of fans of different sizes, or operating at different speeds, will satisfy the pressure and CFM requirements; therefore the next step in selecting fans is to decide what further criteria should be used in selecting the "best" choice. Some of these factors will now be examined.

1. Fans should be chosen for close to maximum efficiency, particularly with high energy costs. This will fall in the middle ranges of the pressure–CFM curve. Avoid the temptation of selecting a fan far out on the CFM curve, near maximum delivery. This temptation is great, because it may mean a smaller fan and therefore lower initial cost, but the operating cost will be high.

2. Fans should not be selected to the left of the peak pressure on the fan curve (Figure 10.12 on page 266). At these conditions, the system operation may be unstable—there may be pressure fluctuations and excess noise generated.

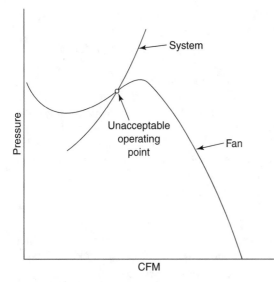

Figure 10.12
Unstable operating condition.

3. When using forward curved blade centrifugal fans, check the system to see if it might operate at significantly greater than design CFM. If so, the motor horsepower required will increase, and a larger motor may be necessary.

4. Allow for system effect according to the duct inlet and outlet connections as explained previously.

5. Fans may have pressure curves of varying steepness (Figure 10.13). If it is expected that there will be considerable changes in system resistance, but constant CFM is required, a fan with a steep curve is desirable. For variable air volume (VAV) systems (Chapter 12), where the CFM varies considerably, a flat curve type is desirable.

10.9 FAN LAWS

There are a number of relationships between fan performance characteristics for a given fan operating at changed conditions, or for different size fans of similar construction; these are called *fan laws*. These relationships are useful for predicting performance if conditions are changed. We will present some of these relationships and their possible uses:

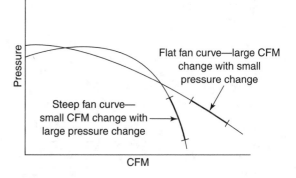

Figure 10.13
Steep and flat fan pressure curves.

$$CFM_2 = CFM_1 \times \frac{N_2}{N_1} \qquad (10.2)$$

$$H_{t2} = H_{t1}\left(\frac{N_2}{N_1}\right)^2 \text{ and } H_{s2} = H_{s1}\left(\frac{N_2}{N_1}\right)^2 \qquad (10.3)$$

$$BHP_2 = BHP_1\left(\frac{N_2}{N_1}\right)^3 \qquad (10.4)$$

$$H_{t2} = H_{t1} \times \frac{d_1}{d_2} \text{ and } H_{s2} = H_{s1} \times \frac{d_1}{d_2} \qquad (10.5)$$

(The symbols above are defined in Section 10.2.)

Example 10.7

A ventilating fan is delivering 8000 CFM while running at a speed of 900 RPM and requiring 6.5 BHP. The operating engineer wants to increase the air supply to 9000 CFM. At what speed should the fan be operated? What must be checked first before making such a change?

Solution
Using Equation 10.2 to find the new speed,

$$N_2 = \frac{CFM_2}{CFM_1} \times N_1$$

$$= \left(\frac{9000}{8000}\right) \times 900 = 1010 \text{ RPM}$$

Equation 10.4 shows that the horsepower will increase, so this must be checked:

$$BHP_2 = \left(\frac{1010}{900}\right)^3 \times 6.5 = 9.2 \text{ HP}$$

This is a considerable increase in power. If the fan had a 7.5 HP motor, which would be likely because it originally required only 6.5 HP, the motor would now be overloaded. The motor would have to be changed, as might the wiring, and the fan capability itself should be checked with the manufacturer.

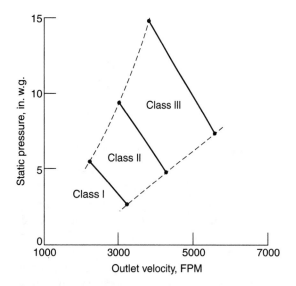

Figure 10.14
Fan construction classes for allowable pressures, backward inclined type. (**Courtesy: Air Movement & Control Association, Inc. [AMCA].**)

10.10 CONSTRUCTION AND ARRANGEMENT

The *AMCA* has established standards of centrifugal fan construction and arrangement that are generally followed in the United States.

Pressure Rating
In order to construct fans of sufficient strength to withstand air pressures to which they will be subjected, and yet not to overdesign the fan, resulting in unnecessary cost, centrifugal fans are classified into groups of different allowable maximum pressures (Figure 10.14). The engineer should select and specify fans in the pressure class required for the job. A different pressure classification has been established for fans mounted in cabinets (see *AMCA* standards).

Inlet
Single width single inlet (SWSI) fans have the air inlet on one side; double width double inlet (DWDI) fans have air inlets on both sides (Figure 10.15). DWDI fans would thus be suitably installed in a plenum-type cabinet.

Arrangement
Centrifugal fans are available in nine different arrangements of bearings and air inlets. Arrange-

ments 1, 2, and 3 (Figure 10.16 on page 268) are often used in HVAC applications because of lower cost and convenience, particularly Arrangement 3.

Rotation, Discharge, and Motor Position
Direction of rotation is described when viewed from the opposite side of an inlet, and is referred to

Figure 10.15
Single width single inlet and double width double inlet fans.

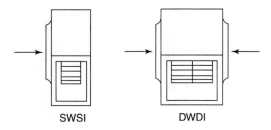

SWSI   DWDI

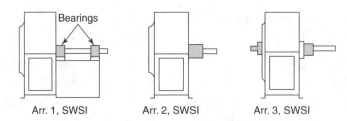

Figure 10.16
Examples of bearing arrangements commonly used in HVAC system fans.

as clockwise (CW) or counterclockwise (CCW). Several discharge arrangements are available, as shown in Figure 10.17. The standard motor positions available are shown in Figure 10.18.

Centrifugal fans are usually belt driven. Different size pulleys make it possible to change speeds. Adjustable pulleys are available so that a limited speed adjustment can be made on the job.

10.11   INSTALLATION

The above specifications of fan construction, rotation, discharge, and motor position must be decided upon when planning the HVAC system installation. In addition, there are other general installation procedures that should be followed:

1. Inlet and discharge connections to the fan should be made to create airflow with minimum pressure loss and equal velocity across the duct section. Some good and poor examples are shown in Figure 10.19.
2. Connection between the inlet and discharge duct and fan should be made with canvas, to reduce vibration transmission. Fans should be mounted or hung on vibration isolators.
Spring and rubber type isolators are available. The manufacturer should be consulted about correct choice of isolator.

Figure 10.17
Examples of discharge arrangements. (*Note:* Rotation direction determined from drive side of fan.)

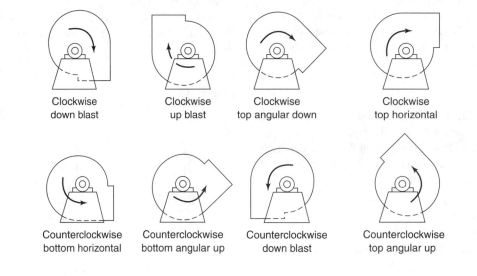

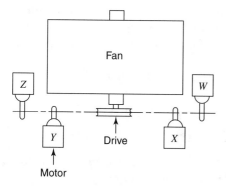

Figure 10.18
Motor positions. (*Note:* Specify motor position by letter, from drive side.)

3. Access openings should be provided if needed, for inspection and service of bearings.
4. A belt guard with a mesh front should be provided so that the belts can be seen without removing the guard.
5. Adequate clearance for inspection and removal should be provided on all sides of the fan. Allow space for the motor installation.

10.12 ENERGY CONSERVATION

1. Airfoil blade centrifugal fans have the highest efficiency and therefore use the least power.
2. Do not allow extra pressure loss as a "safety factor" in the duct system.
3. Select fans in the mid-range of total flow, where efficiency is highest.
4. If volume control of the fan is to be used, inlet guide vane dampers (Figure 10.20 on page 270) are preferable to outlet dampers; less power will be used when the volume flow rate is reduced.
5. Reducing fan speed to reduce flow is the most efficient method for reduced power consumption. However, multispeed fan drives are expensive.
6. Inlet and discharge connections should be arranged to provide minimum pressure loss.

Air Distribution Devices

The conditioned air that is being supplied to each room must be distributed throughout the space in a

Figure 10.19
Examples of good and poor inlet and discharge connections.

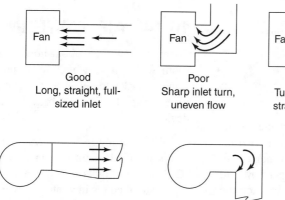

Good	Poor	Better
Long, straight, full-sized inlet	Sharp inlet turn, uneven flow	Turning vanes straighten flow

Good	Poor	Better
Long straight discharge, 15° maximum spread	Sharp discharge turn, uneven flow	Turning vanes straighten flow

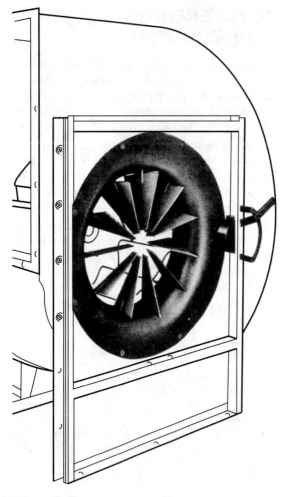

Figure 10.20
Inlet guide vanes for volume control. (Courtesy: Buffalo
Forge Company.)

certain manner; if it is not, uncomfortable conditions will result. This is an aspect of the environmental control system that is often neglected, because it seems simple. Air at the proper flow rate and conditions may be supplied, and yet often the occupants are quite uncomfortable. This is because the air is not distributed properly in the room. We

will investigate some of the principles of air distribution, and then will look at some of the devices (terminal units) that are used to provide proper air distribution. We will consider the types available, their features, and selection.

10.13   ROOM AIR DISTRIBUTION

Good room air distribution requires the following characteristics for comfort:

1. Temperatures throughout the *occupied zone* of the room within ± 2 F (1 C) of the design temperature. Temperature fluctuations greater than this will usually result in discomfort. The occupied zone of most spaces is considered to be from the floor to an elevation of 6 ft. Above this height, greater temperature fluctuations are permissible.
2. Air velocities throughout the occupied zone (called *residual velocities*) between 25–35 FPM for applications where people are seated. Higher velocities (drafts) cause discomfort. Lower velocities also result in discomfort, usually a feeling of stuffiness or staleness. In applications where people are moving around and occupancy is for a short period, as in department stores, higher air velocities are acceptable (50–70 FPM).

10.14   AIR PATTERNS

There are few facts about how an air supply to a room will behave which are important to understand in order to select and locate air supply devices properly and to balance and adjust the devices.

1. When air lower in temperature than room air is supplied (as in summer), it will drop.
2. When air higher in temperature than room air is supplied (as in winter), it will rise.
3. When air is supplied parallel to and near a ceiling, it will tend to "hug" the ceiling for

some distance. This is called the *ceiling* or *surface effect.*

4. The supply air to the room (called the *primary air*) when distributed from an air supply device, will *induce* room air (called *secondary air*) into the airstream, thus rapidly mixing the supply air and the room air.

There are certain other terms used in studying air distribution that need to be defined. Figure 10.21 illustrates these terms. The *throw* from a supply air device is the distance that the supply air travels before reaching a relatively low velocity, called the *terminal velocity.* Terminal velocities of 75–200 FPM are recommended, resulting in residual velocities of 20–70 FPM. The *drop* is the vertical distance the (cold) supply air drops by the end of its throw. The *temperature differential* is the temperature difference between the supply air and the room air. The *spread* is the horizontal divergence of the airstream.

10.15 LOCATION

The location of air distribution devices in the room is an important consideration in achieving good air distribution.

1. *High wall* (Figure 10.22 on page 272). This is a good location for cooling, because the cold air will drop naturally and adequate air circulation throughout the occupied zone will occur. It is not a good location for heating, because the warm air will rise, leaving a stagnant zone in the occupied area. Separate heating (under the window) should be used in this case.

2. *Ceiling* (Figure 10.23 on page 272). This is an excellent location for cooling, because the cold air will drop naturally. It is not a very good location for heating, because the warm air will rise, unless forced down at a high velocity.

3. *Low wall.* This is a good location for heating, because the warm air will rise naturally, but is not desirable for cooling, because the cold air will tend to remain near the floor.

4. *Floor or sill* (Figure 10.24 on page 272). This is an excellent location for heating if located under windows, because it counteracts the cold air downdraft that would otherwise result near the glass. It can also be used for cooling if an adequate outlet velocity is achieved, forcing the cold air to rise and circulate.

Beams and ceiling mounted lighting fixtures create a problem for ceiling or high wall air outlets. The primary air hugs the ceiling due to the ceiling effect, and then bounces off the obstruction, sending a cold draft down to the occupied zone (Figure 10.25 on page 273). In this case, a ceiling outlet should be mounted below the obstruction. For a high wall outlet, the air should be directed to clear the obstruction.

Figure 10.21
Description of terms used in air distribution.

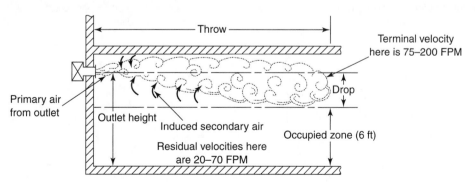

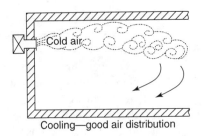

Figure 10.22
High wall outlet location.

For very high ceilings, it is usually better to install ceiling and high wall outlets below or at the level of lighting. In this way, cooling of unused space near the ceiling is reduced, saving energy.

When locating floor or sill outlets, care must be taken not to let drapes or furniture block the airflow.

10.16   TYPES OF AIR SUPPLY DEVICES

There are four types of air supply devices used for creating proper air distribution in the conditioned space:

1. *Grilles and registers*
2. *Ceiling diffusers*
3. *Slot diffusers*
4. *Plenum ceilings*

Grilles and Registers

These devices consist of a frame and parallel bars, which may be either fixed or adjustable. The bars serve to deflect the supply air in the direction the bars are set, and if the bars are adjustable, to adjust the throw and spread of air. Grilles with two sets of bars at right angles to each other are available and are called *double deflection* grilles (Figure 10.26). They enable control of the air distribution in both directions, if needed. Grilles with volume control dampers mounted behind the grille are called *registers*.

Ceiling Diffusers

These devices usually consist of a series of separate concentric rings or louvers with a collar or neck to connect to the duct (Figure 10.27). They may be round, square, or rectangular in shape. In addition to those that distribute air equally in all directions,

Figure 10.23
Ceiling outlet location for cooling provides good distribution.

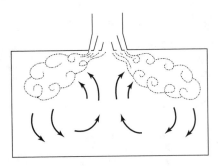

Figure 10.24
Floor or sill location under window for heating provides good air distribution.

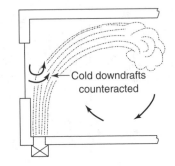

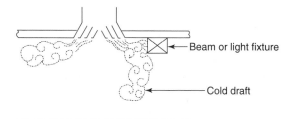

Figure 10.25
Effect of obstruction at ceiling.

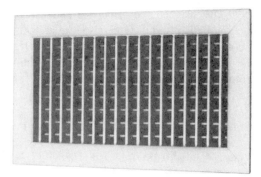

Figure 10.26
Supply register, double deflection type. (Courtesy: Tuttle & Bailey®. Division of Interpace Corporation.)

Figure 10.27
Ceiling diffusers. (Courtesy: Tuttle & Bailey®. Division of Interpace Corporation.)

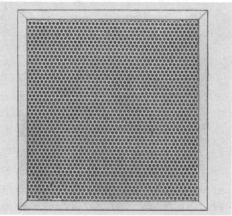

they can be designed to distribute air in any desired direction. Ceiling diffusers are also available in the form of perforated panels. This type is sometimes used because it blends architecturally with the appearance of a suspended panel ceiling.

Slot Diffusers

This is a long strip-shaped outlet with one or more narrow openings, depending on the number of bars or vanes (Figure 10.28). It is also called a *linear* or *strip* diffuser.

A combination fluorescent lighting fixture and slot diffuser is also available. (The slot openings are at the long edges of the fixture.) The fluorescent tubes are therefore cooled somewhat, resulting in more light output per unit of power input, providing a significant energy savings. Another version also has the return air openings in the fixture, as well as the supply diffuser.

Plenum Ceilings

Suspended (hung) ceilings are available with slots or perforations throughout most or all of the ceiling, which can serve as supply air outlets. The space above the ceiling is used as a large plenum through which the supply air is delivered. In this way, air can be distributed evenly throughout the whole space to be conditioned. The design and balancing of plenum ceiling systems is a specialized procedure and will not be discussed further here. Manufacturers of these ceilings will aid the interested designer or contractor.

10.17 APPLICATIONS

Grilles

When used for cooling, a high sidewall location is one of the preferred locations. The air can be directed slightly arched upward; it then will follow the ceiling due to the ceiling effect, mixing well with induced secondary air. In this way, the mixed air temperature will not be unacceptably lower than room temperature before it drops to the occupied zone. When used for warm air heating, the high sidewall outlet may result in stratification of the warm air. However, a careful selection of outlet throw, vertical spread, and return air locations could make the installation satisfactory in mild winter climates. In any case, caution is urged.

Grilles and registers can also be used at ceilings with results similar to high sidewall locations. They are not installed in ceilings as often, however, because their appearance is not considered aesthetically pleasing in a ceiling. Adjustable deflection vanes are used to set proper air direction.

For warm air heating, a perimeter location under windows discharging vertically upward from the floor is ideal in cold climates. This not only provides good mixing of primary and secondary air, but blankets the glass with warm air, offsetting cold downdrafts. This location is popular in residential installations. It usually results in low installation costs in this application because the ductwork in the basement below is relatively simple. When used for cooling, the discharge air velocity must be adequate to overcome the gravity effect of the denser air.

Ceiling Diffusers

These are usually located at the ceiling. The air is discharged horizontally when used for cooling, and

Figure 10.28
Slot diffuser. (Courtesy: Tuttle & Bailey®. Division of Interpace Corporation.)

follows the ceiling for some distance due to the ceiling effect. They are also often installed in the bottom of horizontal ductwork below the ceiling when a suspended ceiling is not used.

Round and square diffusers that have equal openings all around are used to cover a square floor area. A part of the outlet can be blanked off with a piece of sheet metal to get directional air patterns. However, diffusers are available with 1-, 2-, or 3-way blow to cover rectangular-shaped room areas (Figure 10.29).

Some types of ceiling diffusers can be used for heating by adjusting the air pattern to discharge vertically downward at a high velocity. This is more common in industrial applications.

Slot Diffusers

These are available in arrangements enabling them to be used either at ceilings or sidewalls. In addition, they are popular in perimeter applications discharging vertically up from the floor under sills. This popularity is due to the use of low sills and long expanses of glass in many modern buildings.

10.18 SELECTION

The air outlets chosen for a project depend on the following:

Figure 10.29
Use of 1-, 2-, or 3-way blow diffusers.

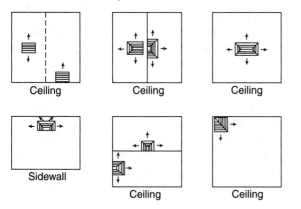

1. *Architectural requirements.* The architect usually wishes the space to have a certain appearance, which may guide the type used and their location.
2. *Structural requirements.* The building structure behind walls, ceilings, and floors may restrict the location of ductwork and thus the air outlets. The structural engineer must be consulted on this.
3. *Temperature differentials.* For large cooling temperature differentials (TD) between supply and room air, the danger of unacceptable air temperatures entering the occupied zone increases. Ceiling diffusers generally have a high induction ratio of room air, therefore lessening this problem when large temperature differential is needed. Some types can be used up to 35 F TD. Grilles are generally limited to 25 F TD.
4. *Location.* When distributing cold air, supply outlets may be located at the ceiling or high on sidewalls. If there are exposed beams, they may deflect the air down to the occupied zone. In this case, the diffuser may have to be located below the beam line. An alternate location for cold air supply is a perimeter location discharging vertically upward from below the windowsill. In this case, the return air location becomes important—it should be located in the interior of the room, preferably at a low elevation, to avoid short circuiting. Warm air supply outlets are preferably located at the perimeter discharging vertically upward.
5. *Quantity.* Often more than one supply outlet is located in a room. This choice depends on a number of factors—air quantity, cost, and architectural requirements. Generally, in a large space, better air distribution is achieved by using a number of diffusers. However, this will increase the cost of the installation. The architect will also set certain requirements, and will want the ceiling to have a certain appearance in regard to location of diffusers. (This is called the *reflected ceiling plan.*) If there is a likelihood of rearranging partitions

in the future, diffuser location and quantity may be chosen to allow changes without having to move ducts and diffusers.

6. *Size.* When the above decisions regarding selection have been made, the proper sizes of outlets can be chosen. This is done with the aid of manufacturers' rating tables. The variable performance characteristics of ceiling diffusers that are of major importance are CFM, throw, mounting height, and sound level. The CFM is the quantity previously determined as required to condition the space. The throw of radius of diffusion is the horizontal distance that the diffuser projects the air. The maximum throw allowable is the distance to a wall or to the edge of the zone of the next diffuser. The minimum throw for adequate circulation is recommended by the manufacturer (usually ¾ of maximum). A selection should never be made with a throw greater than the maximum, for this will result in drafts bouncing off walls.

Most manufacturers give ratings acceptable for a mounting height between 8 and 10 ft, and show corrections for other heights. The sound level produced by a diffuser depends on the air velocity. Therefore, for a given required CFM, a smaller diffuser will mean a higher sound level, but a lower cost. The engineer must balance these needs according to the applications. The diffuser manufacturer usually lists sound ratings of the diffusers by NC (*noise criteria*) levels. This is a weighted perceived sound level. Recommended NC levels are shown in Table 10.3. Ratings for one type of round ceiling diffuser are shown in Table 10.4 on page 278.

Example 10.8

Select a single round ceiling diffuser for Betina's Boutique (Figure 10.30). The RSHG is 18,000 BTU/hr. The supply air temperature differential is 25 F.

Solution
The required CFM must first be determined. Using the sensible heat equation (3.7),

$$CFM = \frac{18,000}{1.1 \times 25} = 650 \text{ CFM}$$

The maximum radius of diffusion permitted is 10 ft (from the center of the room to the wall). The NC level suggested for a small store is from 40 to 50 (Table 10.3). From Table 10.4, a Size 10 diffuser has the following listed rating:

650 CFM, 7 – 16 ft radius of diffusion,

NC-39 sound level

This is a satisfactory selection.

Note that the pressure requirements are also given. This information is needed when balancing the airflow.

Figure 10.30
Sketch for Example 10.8.

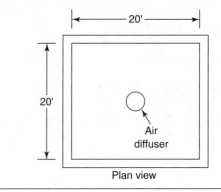

Plan view

Example 10.9

Determine suitable diffuser locations and throws for a room with a 60 ft by 30 ft floor plan if two diffusers are to be used.

Solution
The diffusers will have equal throws in all directions. Figure 10.31 on page 279 shows the floor plan divided into squares. The diffusers will be located in the center of each square. The maximum throw therefore is 15 ft. Acceptable throws are from 12–15 ft when the diffuser size is selected.

TABLE 10.3 RANGES OF INDOOR DESIGN GOALS FOR AIR CONDITIONING SYSTEM SOUND CONTROL

Type of Area	Range of A-Sound Levels, Decibels	Range of NC Criteria Curves	Type of Area	Range of A-Sound Levels, Decibels	Range of NC Criteria Curves
RESIDENCES			**CHURCHES AND SCHOOLS** (Con't)		
Private homes (rural and suburban)	25–35	20–30	Laboratories	40–50	35–45
Private homes (urban)	30–40	25–35	Recreation halls	40–55	35–50
Apartment houses, 2- and 3-family units	35–45	30–40	Corridors and halls	40–55	35–50
			Kitchens	45–55	40–50
HOTELS					
Individual rooms or suites	35–45	30–40	**PUBLIC BUILDINGS**		
Ballrooms, banquet rooms	35–45	30–40	Public libraries, museums, courtrooms	35–45	30–40
Halls and corridors, lobbies	40–50	35–45	Post offices, general banking areas, lobbies	40–50	35–45
Garages	45–55	40–50	Washrooms and toilets	45–55	40–50
Kitchens and laundries	45–55	40–50			
			RESTAURANTS, CAFETERIAS, LOUNGES		
HOSPITALS AND CLINICS			Restaurants	40–50	35–45
Private rooms	30–40	25–35	Cocktail lounges	40–55	35–50
Operating rooms, wards	35–45	30–40	Night clubs	40–50	35–45
Laboratories, halls and corridors			Cafeterias	45–55	40–50
Lobbies and waiting rooms	40–50	35–45			
Washrooms and toilets	45–55	40–50	**STORES, RETAIL**		
			Clothing stores		
OFFICES			Department stores (upper floors)	40–50	35–45
Boardroom	25–35	20–30	Department stores (main floor)		
Conference rooms	30–40	25–35	Small retail stores	45–55	40–50
Executive office	35–45	30–40	Supermarkets	45–55	40–50
Supervisor office, reception room	35–50	30–45			
General open offices, drafting rooms	40–50	35–45	**SPORTS ACTIVITIES, INDOOR**		
Halls and corridors	40–55	35–50	Coliseums	35–45	30–40
Tabulation and computation	45–65	40–60	Bowling alleys, gymnasiums	40–50	35–45
			Swimming pools	45–60	40–55
AUDITORIUMS AND MUSIC HALLS					
Concert and opera halls			**TRANSPORTATION (RAIL, BUS, PLANE)**		
Studios for sound reproduction	20–30	15–25	Ticket sales offices	35–45	30–40
Legitimate theaters, multipurpose halls	30–35	25–30	Lounges and waiting rooms	40–55	35–50
Movie theaters, TV audience studios					
Semi-outdoor amphitheaters	35–45	30–35	**EQUIPMENT ROOMS**		
Lecture halls, planetarium			8 hr/day exposure	<90	
Lobbies	40–50	35–45	3 hr/day exposure	<97	
			(or per OSHA requirement)		
CHURCHES AND SCHOOLS					
Santuaries	25–35	20–30			
Libraries	35–45	30–40			
Schools and classrooms	35–45	30–40			

Note: These are for unoccupied spaces with all systems operating.

Reprinted with permission from the *1976 ASHRAE Handbook & Product Directory.*

TABLE 10.4 PERFORMANCE DATA FOR TYPICAL ROUND CEILING DIFFUSERS

Size			400	500	NC 600	20 700	800	30 900	1000	40 1200	1400	1600
	Neck Velocity, fpm		400	500	600	700	800	900	1000	1200	1400	1600
	Vel. Press. in W.G.		.010	.016	.023	.031	.040	.051	.063	.090	.122	.160
	Total Press	Horizontal	.021	.034	.048	.065	.084	.107	.132	.189	.256	.346
		Vertical	.027	.044	.063	.085	.109	.139	.172	.246	.333	.437
6"	**Flow Rate, cfm** **Radius of Diff., ft.** NC		80 1-2-3 –	100 2-3-4 –	120 2-3-5 15	140 2-4-6 20	160 3-4-7 24	180 3-5-7 27	200 4-5-8 31	235 4-6-10 36	275 5-7-11 41	315 6-8-11 45
8"	**Flow Rate, cfm** **Radius of Diff., ft.** NC		140 2-3-4 –	175 2-3-5 –	210 3-4-7 16	245 3-5-8 21	280 4-5-9 26	315 4-6-10 29	350 5-7-11 32	420 5-8-13 38	490 6-9-15 43	560 7-11-17 47
10"	**Flow Rate, cfm** **Radius of Diff., ft.** NC		220 2-3-5 –	270 3-4-7 –	330 3-5-8 17	380 4-6-9 22	435 4-7-11 27	490 5-8-12 30	545 6-8-14 33	655 7-10-16 39	765 8-12-19 44	870 9-13-22 48
12"	**Flow Rate, cfm** **Radius of Diff., ft.** NC		315 3-4-7 –	390 3-5-8 –	470 4-6-10 18	550 5-7-11 23	630 5-8-13 27	705 6-9-15 31	785 7-10-16 34	940 8-12-19 40	1100 9-14-23 45	1255 11-16-26 50
14"	**Flow Rate, cfm** **Radius of Diff., ft.** NC		425 3-5-8 –	530 4-6-9 13	635 5-7-11 19	745 5-8-13 24	850 6-9-15 28	955 7-11-17 32	1060 8-12-19 35	1270 9-14-22 41	1490 11-16-26 46	1695 13-19-30 60
16"	**Flow Rate, cfm** **Radius of Diff., ft.** NC		560 4-5-9 –	700 5-7-11 14	840 5-8-13 19	980 6-9-15 25	1120 7-11-17 29	1260 8-12-20 32	1400 9-14-22 36	1680 11-16-26 41	1960 13-19-30 46	2240 14-22-35 51
18"	**Flow Rate, cfm** **Radius of Diff., ft.** NC		710 4-6-10 –	885 5-8-12 15	1060 6-9-15 20	1240 7-11-17 26	1420 8-12-20 30	1590 9-14-22 33	1770 10-15-24 36	2120 12-18-29 42	2480 14-21-34 47	2830 16-24-39 52
20"	**Flow Rate, cfm** **Radius of Diff., ft.** NC		875 4-7-11 –	1100 6-9-14 15	1310 7-10-16 21	1530 8-12-19 26	1750 9-14-22 30	1970 10-15-24 34	2190 11-17-27 37	2610 13-19-32 43	3060 16-24-38 48	3500 18-27-43 52
24"	**Flow Rate, cfm** **Radius of Diff., ft.** NC		1260 5-8-13 –	1570 7-10-16 16	1880 8-12-19 22	2200 9-14-23 27	2510 11-16-26 31	2820 12-18-29 35	3140 14-20-32 38	3770 16-24-39 44	4400 19-28-45 49	5020 22-32-52 53
30"	**Flow Rate, cfm** **Radius of Diff., ft.** NC		1960 7-10-16 –	2450 8-13-20 17	2940 10-15-24 23	3430 12-18-28 27	3920 13-20-32 32	4410 15-23-36 36	4900 17-25-41 39	5880 20-30-49 45	6860 24-35-57 50	7840 27-40-65 54
36" **	**Flow Rate, cfm** **Radius of Diff., ft.** NC		2820 8-12-20 –	3520 10-15-24 18	4230 12-18-29 24	4930 14-21-34 28	5630 16-24-39 33	6340 18-27-44 37	7040 20-30-49 40	8450 24-36-58 46	9850 28-42-68 51	11,260 32-48-78 55

Note: 1. Minimum radii of diffusion are to a terminal velocity of 150 fpm, middle to 100 fpm, and maximum° to 50 fpm. If diffuser is mounted on exposed duct, multiply radii of diffusion shown by 0.70. 2. The NC values are based on a room absorption of 18 dB, Re 10^{-13} watts or 8dB, Re 10^{-12} watts.
Values shown are for a horizontal pattern, add 1 dB for a vertical pattern.
Reprinted with permission from Environmental Elements Corporation, Baltimore, Maryland.

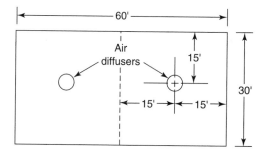

Figure 10.31
Sketch for Example 10.9.

Of course the space could also have been divided into eight areas, each 15 ft square, and eight diffusers could be used. This would greatly increase the cost.

Selection procedures for supply grilles and slot diffusers are similar. The interested student can refer to manufacturers' catalogs.

10.19 ACCESSORIES AND DUCT CONNECTIONS

There are a number of accessories that are used with air supply devices to control or improve air distribution.

Equalizing Grids

When an air outlet is connected to a duct as shown in Figure 10.32(*a*), the air may flow unevenly from the outlet, resulting in poor room air distribution. An equalizing grid installed in the duct collar, as shown in Figure 10.32(*b*), can be used to equalize the airflow pattern to the outlet.

Splitter Dampers

These are sometimes used to direct air into the outlet and to control volume, but they can cause both uneven flow and additional noise (Figure 10.33 on page 280).

Control Damper

These are used to adjust the volume rates of flow to the desired quantity (Figure 10.34 on page 280). Opposed blade dampers are preferable to those that rotate in the same direction, because they will not result in uneven flow. The dampers can usually be adjusted from the face of the outlet with a special key.

Anti-Smudge Rings

A strip of dirt on the ceiling surrounding a ceiling diffuser is a common sight. A ring that surrounds the diffuser is available which will reduce this problem.

Figure 10.32
Use of equalizing grid in duct collar. (*a*) Poor air distribution in duct collar. (*b*) Equalizing grid evens flow.

(*a*)   (*b*)

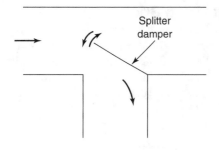

Figure 10.33
Splitter damper to control flow-poor air distribution.

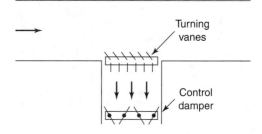

Figure 10.34
Control damper.

Cushioned Head

For a diffuser that is at the end of a duct, it is advisable to extend the duct beyond the outlet neck (about one-half the take-off diameter, with a 6-in. minimum); this results in more even air distribution to the outlet.

10.20   RETURN AIR DEVICES

All of the devices used for air supply are suitable for return air. Grilles are the most commonly used, however, because of their lower cost. The location of return air devices is not as critical as supply devices because the air will not short-circuit in most cases. It is nevertheless a wise rule to locate return air inlets far from outlets, and, if this is not possible, to check the possibility of short-circuiting.

When air is being returned to corridors or adjacent spaces, transfer grilles or louvers may be located in partitions or doors, or the door may be undercut.

The selection of return air devices is usually quite simple. Recommended face inlet velocities that generally provide acceptable noise levels are shown in Table 10.5. The face velocity refers to the velocity calculated by using the overall dimensions of the grille. The actual velocity is higher, because of the grille bars.

Example 10.10

A return air grille is to be located high on a sidewall in a room, exhausting 1500 CFM. What should be the grille face area?

Solution

Referring to Table 10.5, a velocity of 800 FPM is acceptable. The required face area is therefore

$$\text{Area} = \frac{1500 \text{ ft}^3/\text{min}}{800 \text{ ft/min}} = 1.9 \text{ ft}^2 \text{ (use 2 ft}^2\text{)}$$

10.21   SOUND

Air conditioning systems generate sound, which may be objectionable in some cases. It is therefore the responsibility of the designer and contractor to provide adequate sound control when necessary. We will examine only a small part of this complex subject, but will present enough information so that the student will have an understanding of the problems that occur and how they may be resolved.

The magnitude of sound is measured in a unit called the *decibel* (dB). *Sound power* is the sound level generated by a noise source. We are interested

TABLE 10.5   RECOMMENDED RETURN AIR INLET FACE VELOCITIES

Location	Velocity, FPM
Above occupied zone	800
Within occupied zone, not near seats	600–800
Within occupied zone, near seats	400–600
Door or wall louvers	200–300
Door undercuts	200–300

in the sound transmitted and received, however, particularly by humans. This is called *sound pressure.* For our purposes, we do not need to be overly concerned with distinctions between these terms.

In addition to *magnitude,* sound also has *frequency.* Most sound generated has a range of frequencies. The audible range is from about 20–20,000 Hz (cycles per second). We are particularly interested in frequency because the human ear has less sensitivity to lower frequencies (low pitch) than to higher pitch sounds. That is, a higher pitch sound of the same dB level as a lower one seems louder to the individual.

In studying an actual sound problem, the sound levels at each frequency sometimes need to be analyzed. However, an average of the levels at each frequency is often sufficient to work with. A *weighted average* is used to account for the change in sensitivity of the ear to different frequencies.

The weighted average that corresponds well to human response to sound is called the *A-scale network* (dB-A). Sound level measuring meters are available that read dB-A levels. This provides a simple means of measuring effective surrounding sound levels. Table 10.3 lists recommended dB-A levels as well as NC-levels. Both are used in setting standards.

Example 10.11 _____

What are the average recommended dB-A and NC sound levels for a hotel room?

Solution

From Table 10.3, the recommended range of dB-A is 35–45 and of NC is 30–40. The average dB-A is 40 and NC is 35. This would be a suitable sound level in a hotel room.

When two sources produce sound, the combined level is found from Table 10.6.

Example 10.12 _____

The sound power level in a duct approaching an air diffuser is 52 dB. The sound power level of the diffuser is 49 dB. What is the sound power exiting into the room?

TABLE 10.6 EFFECT OF COMBINING TWO SOUND LEVELS

Difference Between Levels, dB	0–1	2–4	5–9	10 or more
dB addition to highest level to obtain combined level	3	2	1	0

Solution
Using Table 10.6,

$$\text{Difference} = 52 - 49 = 3 \text{ dB}$$

$$\text{dB to be added to higher level} = 2 \text{ dB}$$

$$\text{Combined sound power level} = 52 + 2 = 54 \text{ dB}$$

10.22 SOUND CONTROL

The main sources of sound generation in an air conditioning system are the fan and the noise generated by air in the ductwork. Often the resultant sound levels in the rooms are satisfactory and no special treatment is necessary. In any case, the system design and installation should be carried out to minimize sound problems. Some general recommendations are:

1. Select fans near their most efficient operating point. Some of the wasted energy is otherwise converted into noise.
2. Isolate fans from their supports by using vibration isolators, and from the ductwork by using flexible connections.
3. Make duct connection transitions as gradual as possible.
4. Use duct velocities recommended for quietness.
5. Avoid abrupt changes in direction in ducts. Use wide radius elbows or turning vanes.
6. Avoid obstructions in the ductwork. Install dampers only when required.
7. Balance the system so that throttling of dampers is minimized.

8. Select air outlets at sound levels as recommended by the manufacturer.

In many applications, special sound treatment must be carried out, particularly in high velocity systems where considerable noise is generated. In a thorough sound analysis, a series of calculations are made. This involves first determining the sound level generated at each frequency and then the amount of sound attenuation (reduction) required to meet the sound level required in the room.

Ductwork, branches, and elbows provide some natural sound attenuation, and the attenuation varies with the sound frequency. Tables listing this information can be found in the *ASHRAE Systems Volume.* Tables that list average attenuation for all frequencies are also available (Table 10.7). They are not precise, but are suitable for noncritical applications.

TABLE 10.7   NATURAL DUCT ATTENUATION

Ducts		Radius Elbow	
Size, in.	dB	Size, in.	dB
(small) 6 × 6	0.10	10 × 10	1
(med.) 24 × 24	0.05	20 × 20	2
(large) 72 × 72	0.01	over 20	3
Branches			

Ratio of branch to main						
CFM, %		5	10	20	30	50
dB attenuation		13	10	7	5	3

Example 10.13 _____
Determine the natural attenuation in the duct system shown in Figure 10.35.

Solution
From Table 10.7, the attenuation is

$$\text{Duct } 0.05 \text{ dB/ft} \times 500 \text{ ft} = 2.5 \text{ dB}$$

$$\text{Elbow} = 3$$

$$\text{Natural attenuation} = \overline{5.5 \text{ dB}}$$

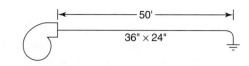

Figure 10.35
Sketch for Example 10.13.

Rooms also have sound-absorbing characteristics, which depend on the size and sound-absorbing qualities of the surfaces and furnishings. Values range from 0 up to 20 or 25 dB. Tables for determining this effect are also available.

The following procedure can be used for predicting sound levels and required sound treatment:

1. Determine the sound level generated by the fan. This information is usually available from the manufacturer.
2. Determine the amount of natural sound attenuation from ducts and fittings, using tables.
3. Subtract item 2 (natural sound attenuation) from item 1 (sound generated). This will be the sound level in the ducts to the air outlet.
4. Determine the sound level of the air outlet from the manufacturer.
5. Combine the sound power level in the duct (item 3) to that of the outlet (Table 10.6). This will be the sound generated at the outlet exit to the room.
6. Determine the room effect from tables. This is the sound absorption from room materials.
7. Subtract item 6 from item 5. This will be the room sound level if no special sound treatment is used.
8. Determine the recommended dB and NC level for the type of room, using tables.
9. If item 7 is less than item 8 (the desired room sound level), no treatment is required. If item 7 is greater, then the difference between them is the amount of additional sound attenuation that must be added.

As an exercise in understanding the procedure, we will assume some figures in the following example, which would be taken from the tables recommended.

Example 10.14

Find the additional average sound attenuation required (if any) for an air conditioning system for a private office, with the following data:

1.	Fan sound power level	= 72 dB
2.	Attenuation in ducts	= 17 dB
3	Sound power to diffuser (item 1 less 2)	= 55 dB
4.	Diffuser sound power level	= 51 dB
5.	Sound power level from diffuser, using Table 10.6, is 55 + 2	= 57 dB
6.	Room effect attenuation	= 10 dB
7.	Room sound level (item 5 less 6)	= 47 dB
8.	Recommended room sound level (executive office)	= 40 dB
9.	Required additional sound attenuation	= 7 dB

Solution

For more accurate results, the above analysis would be carried out at each sound frequency in order to determine how much additional sound attenuation is needed.

There are a few methods for achieving sound attenuation. It can be accomplished by lining ducts internally with a sound-absorbing material. It is also very effective to internally line the air handling unit casing. Manufactured sound traps can also be used. These devices have special internal configurations of sound-absorbing materials and perforated plates. They are quite effective and are used frequently in high velocity systems. Often a combination of these sound attenuation methods is used on a system.

Vibrations from fans, pumps, and compressors can also transmit sound if not isolated. This subject is discussed in Chapter 9.

Review Questions

1. List the types of centrifugal fans and sketch their blade positions.

2. What is the difference between a vaneaxial and tubeaxial fan?

3. Sketch the three performance curve shapes for a backward and forward curved blade centrifugal fan.

4. How is the point of operation of a duct system determined?

5. What criteria should be used in selecting a fan?

6. Describe the effect of changing fan speed on the CFM, BHP, and total pressure.

7. What are the main features of centrifugal fan construction?

8. What good practices should be followed when installing fans?

9. What energy conservation practices should be considered with fans?

10. Describe what is meant by the terms *throw, drop, spread, terminal velocity, residual velocity, ceiling effect,* and *primary* and *secondary air.*

11. List the considerations in choosing the location of air supply devices.

12. List the types of air supply devices and their applications.

13. Describe the use of equalizing grids, splitter dampers, control dampers, anti-smudge rings, and cushion heads.

14. List the recommended procedures to minimize sound generated by a duct system.

Problems

10.1 Find the static pressure developed, BHP, and ME of the fan whose performance curves are shown in Figure 10.7, at a flow of 25,000 CFM.

10.2 A 30 in. centrifugal fan of the type shown in Table 10.1 is specified to deliver 10,000 CFM at 1 in. SP (static pressure). At what speed should the TAB technician set it? What would be the expected motor BHP?

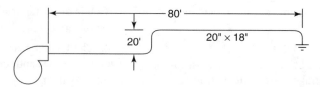

Figure 10.36
Sketch for Problem 10.8.

10.3   A fan is delivering 8100 CFM at 1½ in. SP while running at a speed of 650 RPM and using 2.3 BHP. The fan speed is changed to 700 RPM. Find the new CFM, SP, and BHP.

10.4   Select a ceiling diffuser to deliver 2000 CFM in a 30 ft by 30 ft classroom.

10.5   A 100 ft by 50 ft pharmacy requires 8000 CFM of air. Locate and select three ceiling diffusers.

10.6   A louver is to be installed in a door to return 400 CFM of air to a corridor. What is the recommended louver area?

10.7   What are the recommended NC-level and dB-A level for a classroom?

10.8   The sound lower level of the fan shown in Figure 10.36 is 44 dB, and 40 dB for the diffuser. Determine the duct attenuation and sound level exiting from the diffuser.

10.9   If the room attenuation in Problem 10.8 is 8 dB, what additional attenuation would be required for a conference room?

10.10 Select air diffusers for the warehouse shown in Figure 6.12.

   A. Use 1.1 CFM/ft^2. First select two, then four diffusers.

   B. Use the results of your solution to Problem 7.24. Select two, then four diffusers.

10.11 Select air diffusers for the office building shown in Figure 6.13.

   A. Use 1.5 CFM/ft^2.

   B. Use the results of your solution to Problem 7.25.

CHAPTER 11
Centrifugal Pumps, Expansion Tanks, and Venting

A pump is a device that circulates liquids through piping systems. The centrifugal pump is the type most widely used in circulating water in HVAC systems. In this chapter, we will discuss the principles of operation, selection, construction, in-stallation, and maintenance of centrifugal pumps. The subject of controlling and venting air from the circulating water system will also be discussed, including use of the expansion tank, because this subject is closely related to how the pump is used.

OBJECTIVES

After studying this chapter, you will be able to:

1. Identify the basic parts and construction of a centrifugal pump.
2. Use pump characteristic curves for rating and selecting a pump.
3. Use pump similarity laws to find the effect of changing speed.
4. Determine how to locate and size expansion tanks.

11.1 TYPES OF PUMPS

A pump provides the pressure necessary to overcome the resistance to flow of a liquid in a piping system. Pumps can be classified into two groups according to the way they develop this pressure—either by positive displacement or centrifugal force. In the first group are included reciprocating, gear, vane, screw, and rotary pumps. They are used only in specialized cases in HVAC work and will not be discussed further here. The centrifugal pump is generally used in both hydronic and cooling tower water systems. It is very reliable, rugged, and efficient.

11.2 PRINCIPLES OF OPERATION

The centrifugal pump increases the pressure of the water by first increasing its velocity, and then converting that velocity energy to pressure energy. Figure 11.1 shows the operating elements of a centrifugal pump.

The *impeller* is the part that transmits energy to the water. Flowing from the pump suction line, water enters the opening in the center of the impeller called the *eye*. The impeller rotates, driven by a motor or other prime mover. The water is forced in a centrifugal direction (radially outward) by the motion of the impeller vanes. The velocity of the water is increased considerably by this action. The pump casing contains and guides the water toward the discharge opening.

The action of the impeller has increased the velocity of the water, but not its pressure. The velocity energy is converted into pressure energy by decreasing its velocity. This is accomplished by increasing the flow area in the *volute* and *diffuser*

section of the pump casing. Refer to Chapter 8 where this principle is explained.

11.3   PUMP CHARACTERISTICS

The items of major importance in the performance of a pump are the pressure (*head*) it will develop, the *flow rate* it will deliver, the *horsepower* required to drive the pump, and its *efficiency*. These are called the *pump characteristics*. The characteristics are usually presented in the form of curves or tables; these are used to select the correct pump for an application. The general shape of the curves is similar for all centrifugal pumps. Analyzing these curves is often quite useful in troubleshooting operating problems. Three curves are usually presented:

Figure 11.1
Operating elements of a centrifugal pump.

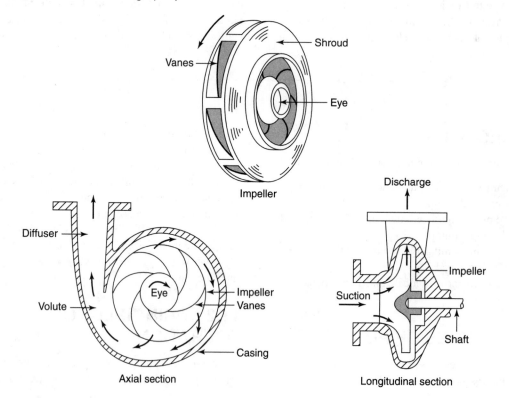

1. Flow rate versus head
2. Flow rate versus brake horsepower
3. Flow rate versus efficiency

The curves for a typical centrifugal pump are shown in Figure 11.2.

Notice that the flow–head curve indicates that a centrifugal pump develops less head at greater flow. The condition of no flow is called *shut-off*, at which the head is at or close to a maximum.

The power required to drive the pump is called the *brake horsepower* (BHP). The flow rate–BHP curve indicates that the BHP increases with flow rate for a centrifugal pump.

The BHP is the power input to a pump. The power output is the power transmitted to the water, given by the following equation:

$$\text{WHP} = \frac{\text{GPM} \times H \times s.g.}{3960} \qquad (11.1)$$

where

WHP = water horsepower (output), HP

GPM = flow rate, GPM

  H = total pump head, ft of liquid

 s.g. = specific gravity of liquid,

  (*s.g.* = 1 for water)

The power input to a pump is always greater than the power output because of friction and other unavoidable losses. The efficiency (E) of a pump is defined as

$$E = \frac{\text{power output}}{\text{power input}} \times 100 = \frac{\text{WHP}}{\text{BHP}} \times 100 \quad (11.2)$$

Figure 11.2 also shows the flow rate efficiency curve for the same centrifugal pump. Note that at shut-off efficiency is zero because there is no flow, then it rises to a maximum, and then decreases again at the pump's maximum flow rate.

Example 11.1 _____
A chilled water pump for the air conditioning system in the Five Aces Casino is delivering 200 GPM at a total head of 36 ft of water. The manufacturer lists the pump efficiency as 60% at this condition. What is the minimum size motor that should be used to drive the pump?

Solution
We must find the required power input (BHP). Using Equation 11.1 to find the power output,

$$\text{WHP} = \frac{200 \times 36}{3960} \times 1 = 1.82 \text{ HP}$$

Using Equation 11.2,

$$\text{BHP} = \frac{\text{WHP}}{E} \times 100 = \frac{1.82}{60} \times 100 = 3.0 \text{ HP}$$

This would be the minimum power needed for a motor. In reality, a larger nominal size motor might be used to prevent possible overloading of the motor, especially for reasons that will be discussed shortly.

The pump performance characteristics, also called ratings, are shown for a particular pump in Figure 11.3 on page 288. They are determined by the manufacturer by testing the pump. The performance depends on the speed at which the pump is operated. The ratings of the pump in Figure 11.3

Figure 11.2
Typical performance characteristic curves for a centrifugal pump.

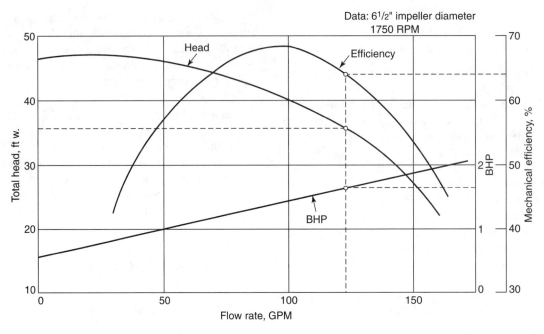

Figure 11.3
Performance curves for a 6½ in. pump at 1750 RPM.

are for a speed of 1750 RPM. (Its performance would be different at other speeds.) This speed and 3500 RPM are the most commonly used in the United States because they are the natural speeds resulting from direct connection to a 60 Hz motor. (In countries that use 50 Hz current, 1450 RPM and 2900 RPM would be the usual speeds.)

The performance of a given pump is found from its curves, as Example 11.2 illustrates.

Example 11.2 _____
If the pump whose ratings are shown in Figure 11.3, operating at 1750 RPM, is delivering 120 GPM, find the head it is developing, the BHP it uses, and its efficiency.

Solution
Using Figure 11.3, at 120 GPM, proceed vertically up to the intersection with the head, BHP, and efficiency curves. Reading horizontally across,

$H = 35.5$ ft w.   BHP = 1.6 HP $E = 64\%$

To conserve data space, a manufacturer may show the performance curves for a number of different size pumps together. Figure 11.4 is a set of flow–head curves for a number of small pumps. BHP and efficiency are not indicated. Each pump is furnished with a motor large enough to handle the maximum BHP.

Another form of presenting pump curves is shown in Figure 11.5 on page 290. In this case, the flow–head curves are shown for a few pumps with impeller sizes ranging from 5–7 in. in diameter, all using the same casing. Instead of BHP and efficiency curves, lines of constant BHP and constant efficiency are shown.

Example 11.3 _____
A 6 in. pump of the type shown in Figure 11.5 is operating at 1750 RPM. Suction and discharge gages at the pump read 30 psig and 45 psig, respectively. How much water is the pump circulating, what BHP is it using, and what is its efficiency?

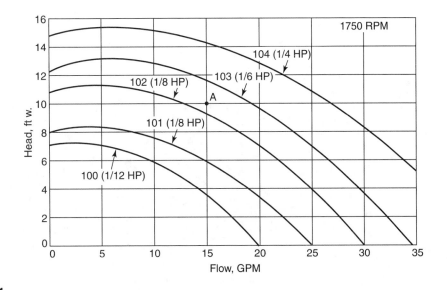

Figure 11.4
Performance curves for a group of small in-line pumps.

Solution

The pump head is the difference between the suction and discharge pressures. Converting this to ft of water

$$H = (45 - 30) \text{ psig} \times \frac{2.3 \text{ ft water}}{1 \text{ psi}} = 34.5 \text{ ft w.g.}$$

Using the flow–head curve for the 6 in. pump from Figure 11.5, at a pump head of 34.5 ft of water, the flow rate is 78 GPM. At this flow, we read the BHP and efficiency, interpolating if necessary, as

$$BHP = 1.0 \text{ HP} \qquad E = 67\%$$

11.4 PUMP SELECTION

In the previous section, we learned how to determine the performance of a pump from its curves. This is useful for the operator or TAB engineer in testing an existing pump. Another situation is the need to select a pump for a system. The pump must have a capacity equal to the system flow rate and a head equal to the system pressure loss. These two system characteristics are the primary ones in selecting a pump.

Example 11.4 _____

For the group of pumps shown in Figure 11.4, which one would be the best choice, based on size, for a system with a required flow rate of 15 GPM and 10 ft water pressure loss?

Solution

Locating the system point *A* in the figure, at the required flow rate we see that the smallest pump that will provide the required head is a size 103, and the head it develops is 11.5 ft water, which is more than enough. The size 102 pump does not develop adequate head at the required flow rate, and the larger size 104 develops much more than required. That pump might be used by throttling flow rate, but it would be unnecessarily expensive and would use more energy.

In Example 11.4, there was only one suitable pump for the application. Usually, however, there are a number of factors that should be considered in selecting the most appropriate pump.

1. A pump that is operating near the point of maximum efficiency should be selected. This generally falls in the mid-range of pump flow capacity.

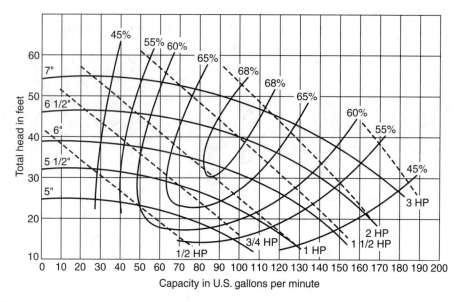

Figure 11.5
Performance curves for a group of pumps with impeller sizes ranging from 5–7 in. at 1750 RPM. (Courtesy: ITT Fluid Handling Division.)

2. For hydronic systems, it is preferable to select a pump operating at 1750 RPM rather than 3500 RPM. At 3500 RPM, a smaller pump can be used, but the higher speed results in higher noise levels that may be disturbing in occupied areas.

3. It is not advisable to select a pump operating near its maximum capacity, even though a smaller pump results from this choice. If the system flow rate actually required is greater than designed for, the pump will not have the extra needed capacity. Select a pump in the vicinity of 50–75% of maximum flow.

4. The steepness of the flow–head curves varies among centrifugal pumps, depending on their design. Figure 11.6 shows examples of a *flat* head curve and *steep* head curve pump. It is recommended that pumps with flat head characteristic curves be used for hydronic systems. If there is a large change in flow rate, there will be a corresponding small change in pump head. This makes balancing and controlling flow rates easier. A steep head curve pump

might be used in a system where the system pressure resistance is expected to gradually increase with time, yet where it is desired to maintain reasonably constant flow rate. A cooling tower circuit might be an example, where the pipe will roughen with age, increasing frictional resistance, and therefore also increasing the required pump head.

Figure 11.6
Flat versus steep pump head characteristics.

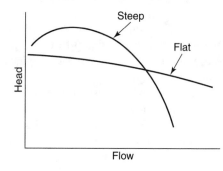

11.5 SYSTEM CHARACTERISTICS

The pressure loss in a piping system changes with the flow rate through the system. The pressure loss–flow rate relationship is called the *system characteristic*, which can be determined from the following equation:

$$\frac{H_{f2}}{H_{f1}} = \left(\frac{GPM_2}{GPM_1}\right)^2 \qquad (11.3)$$

where

H_{f2}, H_{f1} = pressure loss in piping due to friction at conditions 2 and 1.

GPM_2, GPM_1 = flow rates at conditions 2 and 1.

If the pressure loss is calculated at one flow rate, it can be found at any other flow rate, using Equation 11.3.

Example 11.5

A piping system has a pressure loss due to friction of 30 ft water when the flow rate is 60 GPM. If the flow rate were 80 GPM, what would be the pressure loss due to friction?

Solution
Using the system characteristic Equation 11.3,

$$H_{f2} = H_{f1}\left(\frac{GPM_2}{GPM_1}\right)^2 = 30\left(\frac{80}{60}\right)^2 = 53 \text{ ft w.}$$

A system characteristic curve can be plotted for any piping system by calculating the pressure loss at a few different conditions (Figure 11.7). Note that the system frictional resistance rises very sharply with increased flow rate.

This characteristic curve includes frictional pressure loss only, not any static head, if there is any. Therefore, it applies to a closed circuit only (see Chapter 8). If the circuit is open, then to find the total system resistance, the static head would be added to account for the net height the water is lifted.

Some manufacturers offer computer software to the system designer for pump selection. Although

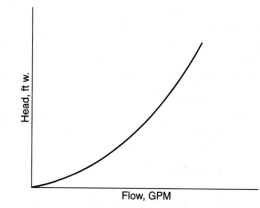

Figure 11.7
Typical system characteristic curve.

this may be a significant convenience, one must be careful not to become locked into using only that manufacturer's product.

11.6 SYSTEM CHARACTERISTICS AND PUMP CHARACTERISTICS

The system and pump characteristic curves can both be plotted together (Figure 11.8). This is very useful in analyzing operating problems.

Figure 11.8
System and pump head characteristic curves indicating point of operation.

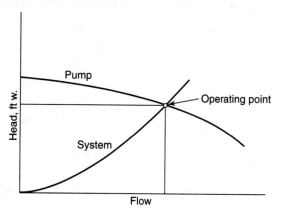

The head developed by the pump must be exactly equal to the system pressure loss. The only point where this is true is where the system and pump head curves intersect. Therefore, the following important statement always holds true:

The point of intersection of the system characteristic and pump characteristic flow–head curves is always the actual operating condition for the system–pump combination.

It is not usually necessary to plot both curves to make a pump selection, but many types of problems encountered in balancing and operating systems can be understood by studying the curves together. For instance, consider the situation where the actual system pressure loss is less than the design pressure loss (Figure 11.9).

The system design pressure loss and flow rate is given by point 1. A pump is selected to develop this head. Point 1 therefore represents the expected point of operation. Curve *A*, the expected system characteristic curve, could also be plotted. Suppose, however, that the actual system pressure loss at the design GPM was only that indicated by point 2. This might occur because the system designer allowed for a "safety factor" in calculating the piping friction loss. Point 2 of course cannot be the system operating point because it is not a point of intersection with the pump curve. Then what is the

operating condition? To find it, we plot a new system characteristic curve, the real curve *B*, through point 2, using Equation 11.3. The real operating condition must be where this curve intersects the pump curve, point 3. But notice what has happened. The pump is actually delivering more flow than is desired. This may overcool or overheat the building. Furthermore, we know that the real operating point is farther out on the pump curve than expected, and the pump will thus use more power than expected. This is a waste of energy, and if the motor has not been oversized, we may have a burned out motor. Instead of the safety factor, the condition may be less safe!

The problem could be resolved after installation by adding resistance in the circuit, say by throttling a balancing valve. This would bring us back to point 1. However, if the excess and unnecessary pressure loss had not been allowed for originally, a smaller pump might have been chosen (at point 2), using less power.

Even though it is proper to select a pump with head close to the actual system pressure loss, it is often advisable to select a motor for nonoverloading conditions, with a capacity greater than the BHP at maximum flow. The extra motor cost is a nominal part of the total cost, and there are many variables that might cause operation at higher than design flow.

Figure 11.9

Illustration of excess power use and incorrect operation condition by use of "safety factor."

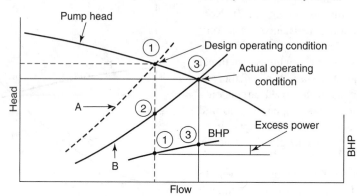

11.7 PUMP SIMILARITY LAWS

There are a number of relations concerning flow rate, speed, power, and head for any given centrifugal pump that are sometimes useful to the HVAC engineer. A few of these are:

$$\frac{\text{GPM}_2}{\text{GPM}_1} = \frac{N_2}{N_1} \tag{11.4}$$

$$\frac{H_2}{H_1} = \left(\frac{N_2}{N_1}\right)^2 \tag{11.5}$$

$$\frac{\text{BHP}_2}{\text{BHP}_1} = \left(\frac{N_2}{N_1}\right)^3 \tag{11.6}$$

where

H = pump head, ft w.

N = pump speed, RPM

BHP = brake horsepower

1, 2 = any two operating conditions

Example 11.6

The Pumpernickel Pump Co. ships a centrifugal pump to Argentina, where 50 Hz electric current is used. The pump is rated at 380 GPM and 40 ft head, using 6 BHP, at a speed of 1750 RPM. What will be the pump's rating in Argentina?

Solution
The pump will operate at 1450 RPM on the 50 Hz current. Using the similarity laws,

$$\text{GPM}_2 = 380 \text{ GPM} \times \frac{1450}{1750} = 315 \text{ GPM}$$

$$H_2 = 40 \text{ ft} \times \left(\frac{1450}{1750}\right)^2 = 27.5 \text{ ft w.}$$

$$\text{BHP}_2 = 6 \text{ BHP} \times \left(\frac{1450}{1750}\right)^3 = 3.4 \text{ HP}$$

There are also pump similarity laws for determining the effect of a change in impeller diameter. It is advisable, however, to consult the manufacturer if a change in impellers is being considered.

Note that the centrifugal pump similarity laws are identical to those for a centrifugal fan (see Chapter 10).

11.8 PUMP CONSTRUCTION

Centrifugal pumps are available in varied arrangements and features of construction, each having different applications.

The *in-line pump* (Figure 11.10) is used for small, low head applications. The pump and motor are mounted integrally, and the pump suction and discharge connections are in a straight line (in-line). Because of this arrangement and the relatively light weight, the pump can be supported directly by the piping and is inexpensive and simple to install. In-line pumps, sometimes called booster pumps or *circulators*, are popular for small hydronic heating systems.

The *close-coupled* pump (Figure 11.11 on page 294) has the impeller mounted on and supported by the motor shaft. The motor has a mounting flange

Figure 11.10
In-line type pump. (Courtesy: ITT Fluid Handling Division.)

Figure 11.11
Close-coupled pump. (Courtesy: ITT Fluid Handling Division.)

Figure 11.12
Pump and motor connected by flexible coupling and mounted on common baseplate. (Courtesy: ITT Fluid Handling Division.)

for supporting the motor–pump combination from a suitable base. The pump has an end suction connection. The close coupled pump is relatively inexpensive and is available from small to medium capacities and heads.

In addition to the in-line and close-coupled pump and motor combinations, pumps are also furnished as separate items. The pump and motor shafts are connected by a flexible coupling. In the medium size range, the pump, coupling, and motor may be preassembled by the manufacturer on a common baseplate (Figure 11.12) for convenience of installation.

With larger pumps the contractor mounts the pump and motor and connects them together through the coupling.

The flexible coupling aids in alignment of the two shafts and helps to reduce vibrations.

Centrifugal pumps can have either *single-* or *double-suction* construction, in which water enters either through one or both sides of the pump. Larger pumps are constructed with double-suction inlets.

The pump casing can be one cast piece or can be split—manufactured in two halves that are bolted together. The split casing makes repairs more convenient. The pump can be opened on the job for ac-

cess to bearings or other internal parts. Both *horizontal split case* and *vertical split case* construction are used. The horizontal split case is used on very large pumps so that the very heavy upper part of the casing can be lifted vertically by a mechanical hoist.

If the impeller has walls (called *shrouds*) on both sides, it is called a *closed* impeller; if it has shrouds on one side, it is called *semi-open*; if it has no shrouds, it is called an *open* impeller. Open-type impellers are not generally used in HVAC applications because their purpose is to permit handling of liquids containing solids, such as sewage.

The *bronze-fitted* pump is generally the combination of materials used for hydronic systems. The casing is cast iron and the impeller is bronze or brass.

Sleeve- and *ball*-type bearings are both used. Sleeve bearings are lubricated with oil. One arrangement has a reservoir of oil and an oil ring that flings the oil around as it rotates. Another arrangement uses cotton waste packing that is impregnated with oil. Large pumps may have an oil pump for forcing the oil to the bearings.

Ball bearings are lubricated with grease. Some motor ball bearings are sealed and cannot be lubricated in the field (Figure 11.13).

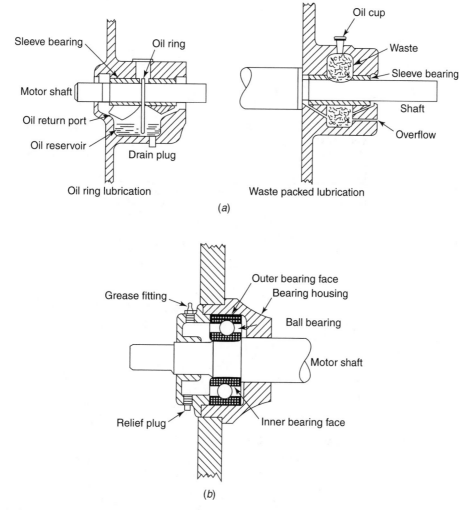

Figure 11.13
Sleeve and ball bearings. (*a*) Sleeve bearings. (*b*) Ball bearings.
(Courtesy: ITT Fluid Handling Division.)

Ball bearings are generally used on smaller pumps. Larger pumps may have either sleeve or ball bearings. Sleeve bearings are quieter and are therefore recommended for hydronic service with larger pumps.

Seals are required to prevent leakage of water under pressure. Either *packing* or *mechanical* seals are used (Figure 11.14 on page 296). Packed seals use a soft material that presses against the shaft, which has a tightness that is adjustable. The packing will wear and must be inspected and replaced at intervals. A small leakage is expected and normal, and a drip pan and drain lines should be provided to handle this. Mechanical seals have two hard, very smooth mating surfaces, one that is stationary and one that rotates. Properly applied, they will prevent any significant leakage. They cannot be used where there are any solid particles in the system, because

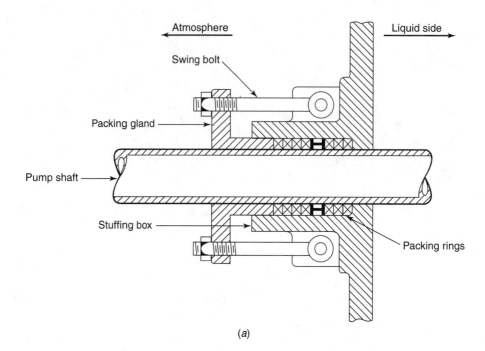

(a)

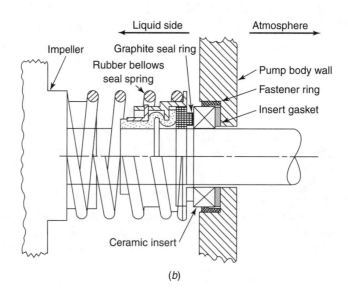

(b)

Figure 11.14
(a) Packed seal. (b) Mechanical seal. (**Courtesy: ITT Fluid Handling Division.**)

the surfaces will become scored and the seal will be lost. They are very popular in hydronic systems because no maintenance is required and they can last many years.

11.9 NET POSITIVE SUCTION HEAD

Under certain conditions in circulating water systems, a phenomenon called *cavitation* may occur in the pump suction, causing operating problems and possible damage to the pump. It results when the water pressure at the pump suction is too low. If this happens, the vapor pressure of the water in the pump may fall below its corresponding saturation temperature (Chapter 2) and the water will flash into steam. The steam bubbles formed may collapse in the pump, momentarily leaving pockets or cavities. The water will rush into these cavities at great force, causing erratic operation, noise, and possible damage to the pump.

To avoid this, a minimum pressure must be maintained at the pump suction, called *net positive suction head* (NPSH). The *required* NPSH for a pump can be obtained from the manufacturer. The *available* NPSH is calculated from examining the suction system arrangement of pressure loss, lift, and temperature. The available NPSH must be greater than the pump requires. If it is too low, the piping arrangement must be changed.

The possibility of cavitation is usually of concern in an open system where there is suction lift to the pump and where the temperature is high. It can be a problem in cooling tower systems if the pump is elevated to a location requiring a high suction lift. If this occurs, the pump or tower may have to be relocated. The problem is not usually encountered in hydronic systems because the static head in a closed system acts on the suction. The placement of the compression tank is an important factor relating to this, however, and will be explained later.

Boiler condensate return systems are also subject to cavitation if they are not designed and installed in accordance with the NPSH requirements.

The available NPSH can be determined from the following equation:

$$H_n = H_a \pm H_z - H_f - H_v \qquad (11.7)$$

where

H_n = available NPSH

H_a = absolute pressure at surface of liquid where pump takes suction (atmospheric pressure, if open)

H_z = elevation of the liquid suction above (+) or below (−) the centerline of the impeller

H_f = friction and velocity head loss in the suction piping

H_v = absolute vapor pressure of water corresponding to the temperature

The units in Equation 11.7 are in feet of water.

Example 11.7
A pump takes water at 180 F from an open tank that is 8 ft below the pump centerline. Friction and velocity head loss in the piping is 2.5 ft w. Atmospheric pressure is 14.7 psi. Determine the available NPSH.

Solution
The vapor pressure of water at 180 F is 7.5 psia (Table A.3). Changing all units to ft w.,

$$H_v = 7.5 \text{ psi} \times \frac{2.31 \text{ ft w.}}{1 \text{ psi}} = 17.3 \text{ ft w.}$$

$$H_a = 14.7 \times 2.31 = 34 \text{ ft w.}$$

Substituting in Equation 11.7,

$$H_n = H_a \pm H_z - H_f - H_v$$
$$= 34 - 8 - 2.5 - 17.3 = 6.2 \text{ ft w.}$$

The pump used must have a required NPSH less than 6.2 ft w.

11.10 THE EXPANSION TANK

Water expands when its temperature increases, unless it is restricted. In a hydronic system, an allowance must be made for this. If the piping system is completely filled and there is no space for the water to expand, the piping or equipment might break. An *open expansion* tank can be provided at the highest point in the system to solve this problem. Figure 11.15(*a*) shows that as the water temperature increases, the total volume of water in the system increases, the effect being a rise in the water level in the tank. Because the tank is open to the atmosphere, however, the system has some of the defects of open hydronic systems. Particularly undesirable is the continual exposure to air and its possible corrosive effects. A much better solution is to use a closed expansion tank containing a gas (air or nitrogen). When the water expands, it partially fills the tank, compressing the gas. For this reason, the closed expansion tank such as the one shown in Figure 11.15(*b*) is usually called a *compression tank*.

The compression tank serves an additional purpose beyond that of providing for the water expansion—it aids in controlling system pressure. For these reasons, compression tanks have largely replaced open expansion tanks in hydronic systems.

11.11   SYSTEM PRESSURE CONTROL

The pressure in a hydronic system must be controlled within certain maximum and minimum limits. This is a subject that is often not understood correctly, leading to operating difficulties and possible equipment damage.

The maximum allowable pressures are usually based on the permissible equipment pressures. In a low temperature hydronic heating system, for example, the boiler relief valve is often set at 30 psig, which would therefore be the maximum allowable pressure at the boiler.

The minimum pressure requirement is based on two factors:

1. The pressure at any location must not be lower than the saturation pressure of the water. If this happens, the water will boil and the steam created will cause operating problems. As mentioned previously, this may occur particularly at the pump suction.

Figure 11.15
Expansion- and compression-type tanks.

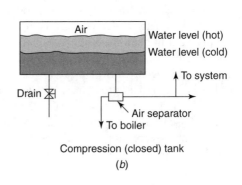

Open expansion tank
(*a*)

Compression (closed) tank
(*b*)

2. The pressure at any location should not be lower than atmospheric pressure. If this happens, air may enter the system.

Control of maximum and minimum pressures to ensure that none of these problems occur is achieved by proper sizing and location of the compression tank and by correctly pressurizing the system when filling. To know how to accomplish this, we must understand how the compression tank functions.

The compression tank acts similarly to a spring or an air cushion. The water in the tank will be at the same pressure as the gas in the tank. The value of this pressure will depend on how much the gas in the tank is compressed. Once the system is filled with water and the water is heated to operating temperature, the total volume of water in the system remains constant. The volume of gas in the tank therefore also remains constant, and its pressure does not change. This holds true regardless of where the tank is located and whether or not the pump is operating. The following statement summarizes this fact:

The point at which the compression tank is connected to the system is the point of no pressure change.

By assuming two different tank locations and utilizing the above principle, we can see what effect the tank's location has on controlling system pressure. Consider first the compression tank located at the discharge side of the pump for the system shown in Figure 11.16. All of the piping is on the same level. Assume that the pressure throughout the system initially is 10 psig (25 psia) without the pump running as seen in Figure 11.16(*a*). Figure 11.16(*b*) shows what happens to the pressures when the pump runs. Assume the pump has a head of 20 psi. The pressure at the tank must be the point of no pressure change, so when the pump runs, the pressure at this point is still 25 psia. The pressure at the pump suction must therefore be 20 psi less than this value, or 5 psia, because the pump adds 20 psi. But 5 psia is −10 psig, which is far below atmospheric pressure. Air would undoubtedly leak into the system at the pump suction. Cavitation in the pump might also occur in a heating system, because the boiling point of water at 5 psia is only 160 F.

Let us see what happens if the tank is located at the suction side of the pump. In Figure 11.17(*a*) on page 300, the initial pressure is at 25 psia throughout the system, as before. When the pump runs, the pressure at the tank location remains at 25 psia. As the pump adds 20 psi, the pressure at the pump discharge must be 45 psia (30 psig), as seen in Figure 11.17(*b*). The pressure throughout the system is well above atmospheric.

This example shows that *the compression tank should be connected to the system at the pump suction*, not at the pump discharge.

If the pump head is low and there is a static head of water above the pump and tank elevation, the pressure at the pump suction might not fall below atmospheric even if the tank is located at the pump discharge. However, this arrangement is usually still not advisable, except for small residential systems.

Figure 11.16
Effect of compression tank location at pump discharge. (*a*) Tank at pump discharge, pump not operating. (*b*) Tank at pump discharge, pump operating.

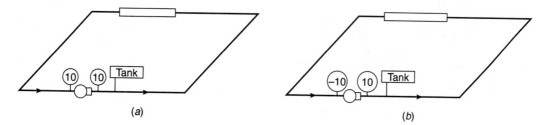

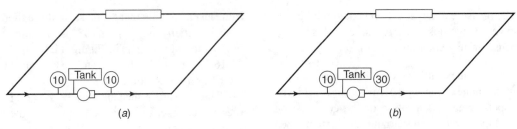

Figure 11.17
Effect of compression tank location at pump suction. (a) Tank at pump suction, pump not operating. (b) Tank at pump suction, pump operating.

To keep the pressure exerted on the boiler as low as possible, the pump should be located to discharge away from the boiler, so that the boiler is not subject to the pump discharge pressure. On a high-rise building, the static head at the boiler, if it is in the basement, might be above the maximum pressure. In this case, the boiler could be located in a penthouse, or a steam boiler and hot water heat exchanger might be used (Chapter 5).

(The recommended arrangement of pump, boiler, and compression tank is shown in Figure 11.18. Accessories are not shown.)

11.12   COMPRESSION TANK SIZE

The size of the compression tank for a system must be adequate to receive the increased volume of water from expansion and also to keep the pressures within minimum and maximum limits. The size depends on the following sources of pressure:

1. *Static pressure.* This is the pressure due to the height of water above any point. Usually the critical point is the boiler, which is often at the bottom of the system.
2. *Initial fill pressure.* If the system were initially filled without pressure, the pressure at the highest point in the system would be atmospheric. In order to provide a safety margin to prevent the pressure from going below atmospheric and thus leaking air in, the contractor should fill the system under pressure. A pres-

sure of 4–5 psig at the top of the system is adequate for hydronic systems.
3. *Pressure–temperature increase.* After the system is filled with cold water and pressurized, when the temperature is raised in a hydronic heating system, the pressure will increase further due to expansion of the water compressing air in the tank.
4. *Pump pressure.* When the pump is operated, the pressures change in the system by the value of the pump head. As explained earlier, this depends on where the compression tank is located. If the tank is connected at the pump

Figure 11.18
Sketch for Example 11.8.

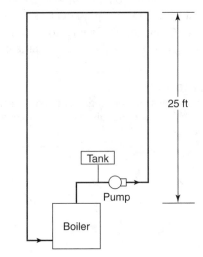

suction the head is added at every point. If the tank is connected at the pump discharge, the pump head is subtracted from the pressure at every point.

These factors have been combined in the following formula developed by the ASME for determining tank size:

$$V_t = \frac{(0.00041t - 0.0466)V_s}{H_a/H_t - H_a/H_0} \qquad (11.8)$$

where

V_t = required volume of compression tank, gallons

V_s = volume of system, gallons

t = design average water temperature, F

H_a = atmospheric pressure, ft water absolute

H_t = minimum pressure at tank, equal to the fill pressure plus static pressure at tank, ft water absolute

H_0 = maximum pressure at tank, ft water absolute

The term in the parentheses in the equation represents the expansion of water. The system volume is determined from the pipe sizes and from equipment volumes, information that can be obtained from manufacturers. In using the formula, pressure loss from friction is usually neglected. Example 11.8 will illustrate how the compression tank is sized.

Example 11.8

Determine the required size of the closed expansion tank for the hydronic heating system shown in Figure 11.18. The system volume is 600 gal. The high point is 25 ft above the boiler. The pump head is 20 ft of water. The pressure relief valve on the boiler is set at 30 psig. The design average water temperature is 200 F. Fill pressure is 5 psig. The tank is located at the boiler elevation.

Solution

Equation 11.8 will be used. From the previous information, the terms in the equation are

t = 200 F

V_s = 600 gallons

H_a = 14.7 psi × 2.3 ft w./1 psi = 34 ft w.

H_t = 5 × 2.3 + 25 = 38 ft w.g. + 34 = 72 ft w. absolute

H_0 = 30 psig = 45 psia × 2.3

     = 104 ft w. absolute

Substituting in the equation

$$V_t = \frac{[0.00041(200) - 0.0466]600}{34/72 - 34/104} = 146 \text{ gal}$$

A compression-type expansion tank of approximately 150 gal capacity would be used on this system, if connected at the suction side of the pump. If the tank is connected to the discharge side of the pump, the minimum pressure at the tank must be increased by the amount of pump head. That is,

$$H_t = 72 + 20 = 92 \text{ ft w. absolute}$$

The required tank size in this case becomes

$$V_t = \frac{21.2}{34/92 - 34/104} = 493 \text{ gal}$$

Note that the tank size must be increased greatly because of its location.

For residential and other small hydronic heating systems, it is usually not necessary to calculate the size of the compression tank. Tables are available from manufacturers that list the appropriate tank size according to the building heating load.

For small systems, a convenient flexible diaphragm-type compression tank is often used (Figure 11.19 on page 302). The flexing of the diaphragm allows for the expansion and contraction of the system water.

11.13   AIR CONTROL AND VENTING

When the system is initially filled, air unavoidably enters the system. Dissolved air is in the fill water and the compressed air is in the tank. Eventually, further air will enter even a carefully designed and

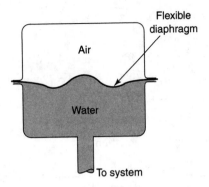

Figure 11.19
Diaphragm-type compression tank.

Figure 11.20
Automatic air venting valve for terminal units.
(Courtesy: Taco, Inc.)

installed system, through makeup water and when the system is opened for maintenance and repair. Control of air in the system is necessary for two reasons: The presence of air will block the flow of water, and the air and water together may promote corrosion.

When the water in the system is first heated to operating temperature, dissolved air is released from the water as air bubbles. Much of this air will find its way to the compression tank. Sometimes an *air separator* device (Figure 11.15) is located at the tank connection to divert air to the tank. However, some air probably will not be collected in the tank, but will find its way to other high points in the piping system. This air must be *vented* from the system or it will block flow through terminal units. *Air vents*, which are small valves, must be provided at all high points in the system. Vent valves may be manual or automatic (Figure 11.20).

Ideally, a system should have only one high point, but often there are many rises and drops in piping, each resulting in a high point. It is helpful to pitch horizontal piping slightly up toward high points when installing it. It is also advisable to install a vent at each terminal unit. After the system is filled and put into operation, the contractor opens each vent and bleeds air until none are left. Automatic air vents are convenient, but if they stick in an open position, they could cause considerable water damage to a building.

An eccentric type reducing fitting (Chapter 9) is recommended when changing pipe size, which often occurs at the pump connections. This prevents the trapping of air at the top of the fitting.

11.14 ENERGY CONSERVATION

1. Select pumps in their range of greatest efficiency, which is usually in the range of 50–70% of their maximum capacity.
2. Do not allow an extra pressure loss in the piping as a "safety factor."
3. Provide for venting air from the system in design, installation, and maintenance. Air will block flow and prevent proper operation, therefore it will indirectly reduce efficiency.

Review Questions

1. List the basic parts of a centrifugal pump and their functions.
2. On one sketch, show the typical centrifugal pump characteristics curves.
3. What factors should be considered in selecting a pump for a hydronic system?

4. What is the relationship between the pump and system characteristics?

5. Explain the following terms: *in-line*, *close-coupled*, *single-*, or *double-suction*, and *closed* or *open impeller*.

6. Name the two types of pump bearings for hydronic systems and their features.

7. List the methods of bearing lubrication and the types of bearings to which they apply.

8. List the two types of pump seals and their features.

9. What is the importance of NPSH?

10. What are the functions of a compression tank?

11. What considerations are important for minimum and maximum pressure control?

12. Where should air vents be installed?

13. How should the piping be installed to improve air venting?

Problems

11.1 A pump for a hydronic solar heating system is circulating 60 GPM of an antifreeze solution that has a specific gravity of 1.14, against a total head of 23 ft w. The pump efficiency is 55%. What is the minimum size motor required to drive the pump?

11.2 The pump whose characteristic curves are shown in Figure 11.3 is circulating 100 GPM. What is the head, BHP, and efficiency?

11.3 Pump 6½ (Figure 11.5) is operating with a total head of 38 ft w. Find the GPM it is circulating, the power it is using, and its efficiency.

11.4 A hydronic system requires 20 GPM of water. The system pressure loss is 6 ft w. at this flow rate. Select a pump from Figure 11.4 for this application.

11.5 The pressure loss due to friction in a hydronic system is 10 ft w. when the flow rate is 15 GPM. Pump 103 (Figure 11.4) is used. It is decided to increase the flow rate to 18 GPM. Will pump 103 be satisfactory?

11.6 For a system requiring 110 GPM, the engineer calculates the pressure loss at 37 ft w., selecting the pump in Figure 11.3. The actual pressure drop is only 25 ft w. Determine the actual operating conditions of GPM and BHP if no changes are made in the system.

11.7 For Problem 11.6 how could the proper flow rate be achieved? Determine the BHP and head at this condition.

11.8 A centrifugal pump has a rating of 120 GPM and 36 ft w. total head at a speed of 1150 RPM. It requires 2 BHP at these conditions. If its speed is increased to 1450 RPM, what will be its expected rating and required BHP?

11.9 A hydronic heating system has a design average water temperature of 220 F. The high point is 20 ft above the boiler. The pump head is 30 ft w. The system volume is 290 gal. The pressure relief valve on the boiler is set at 30 psig. The system is filled under a pressure of 4 psig. Find the required compression tank volume if it is located at the (a) pump suction, (b) pump discharge, and (c) high point of the system.

CHAPTER 12
Air Conditioning Systems and Equipment

There are a large number of variations in the types of air conditioning systems and the ways they can be used to control the environment in a building. In every application, the planner must consider the features of each type of system and decide which is the best choice. Load changes, zoning requirements, space available, and costs are some of the variables that determine which type of system is to be used.

OBJECTIVES

After studying this chapter, you will be able to:

1. Identify the components of single zone central system air conditioning equipment and their functions.
2. Identify the types of zoned air conditioning systems and their features.
3. Describe the features of all-water and air-water systems.
4. Describe the features of the different types of unitary equipment.
5. Select a cooling coil.
6. Identify the types and performance characteristics of air cleaners.
7. Describe the causes of and solutions to poor indoor air quality.

12.1 SYSTEM CLASSIFICATIONS

Air conditioning systems can be classified in a number of ways, such as

A. The cooling/heating fluid that is used.

There are three possible groups in regards to the fluid used:

1. **All-air** systems. These systems use only air for cooling or heating.
2. **All-water (hydronic)** systems. These systems use only water for cooling or heating.
3. **Air-water** combination systems. These systems use both water and air for cooling and heating.

B. **Unitary** or **central** systems

A *unitary system* uses packaged equipment. That is, most, if not all, of the system components (fans, coils, refrigeration equipment) are furnished as an assembled package from the manufacturer.

A *central* or *built-up* system is one where the components are furnished separately and installed and assembled by the contractor.

C. **Single zone** or **multiple zone** systems

A *single zone* system can satisfactorily air condition only one zone in a building. A *multiple zone* system can satisfactorily air condition a number of different zones.

12.2 ZONES AND SYSTEMS

The amount of heating or cooling that the air conditioning equipment delivers to a space must always match the space load or requirements. This load is continually changing, because of the variations in outside air temperature, solar radiation and internal loads.

The room thermostat controls the air conditioning equipment so that it responds properly to the changing load. For example, if the equipment is cooling too much, the thermostat either stops the unit or reduces its cooling output, thus maintaining the desired room temperature.

This presents a serious problem if there are other rooms or spaces that do not have the same load change behavior as the room where the thermostat is located. For instance, if the thermostat is in room A, where the cooling load decreases, the cooling output of the unit is reduced and the room temperature remains approximately constant. But suppose room B still needs full cooling. Now it will get insufficient cooling and the room temperature will rise, creating discomfort.

This situation occurs in any building where the load changes behave differently among rooms. Consider rooms on different exposures. The solar heat gain may increase in rooms on one side and decrease in those on another side. Internal loads also are frequently not uniform in their changes.

Lights may be switched off in one space and not another. People change locations.

When these situations exist, a single zone air conditioning system is unsatisfactory. It will satisfy only one or a group of zones whose heat gains vary in unison.

One solution to this problem is to use a separate air conditioning unit for each differently behaving zone. Often however, an air conditioning system of the multiple zone type is used. The various ways that this can be accomplished will be part of our discussion.

*An **air conditioning zone** is a room or group of rooms in which comfortable conditions can be maintained by a single controlling device.*

12.3 SINGLE ZONE SYSTEM

A *single zone* air conditioning system has one thermostat automatically controlling one heating or cooling unit to maintain the proper temperature in a single room or a group of rooms constituting a zone. A window air conditioner is an example of a single zone air conditioning unit. Our focus in this section, however, will be on a central type, all-air single zone system, as shown in Figure 12.1 on page 306.

The central unit, called an air handling unit (AHU), cools or heats air that is then distributed to one or a group of rooms that constitute a single zone.

The equipment shown in Figure 12.1 provides a complete year-round air conditioning system to control both temperature and humidity. Not all of the components are used in all circumstances.

The *supply air fan* is necessary to distribute air through the unit, ductwork, and air distribution devices to the rooms.

The *cooling coil* cools and dehumidifies the air in summer. It receives chilled water or refrigerant from a remote refrigeration unit.

The *reheat coil* partially reheats the cooled air when the room heat gain is less than maximum, thus providing humidity control in summer. If no reheat coil is used, temperature, but not humidity can be controlled in summer (see Chapter 7). This

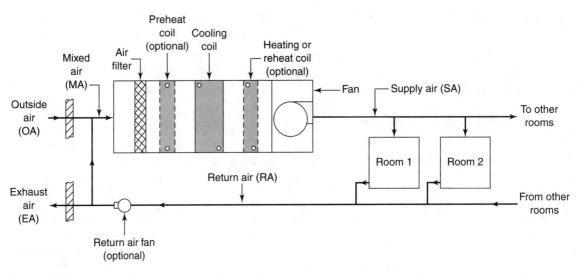

Figure 12.1
Arrangement of single zone central system air conditioning equipment.

coil may alternately be a full capacity heating coil, capable of handling the winter heating needs.

The ductwork is generally arranged so that the system takes in some outside ventilation air (OA), the rest being return air (RA) recirculated from the rooms. The equivalent amount of outside air must then be exhausted (EA) from the building. Provisions are often made in the arrangement of dampers (Figure 12.2) so that 100% outside air can be drawn in and exhausted. This would be

Figure 12.2
Arrangement of ducts and dampers to vary proportion of outside and return duct air.

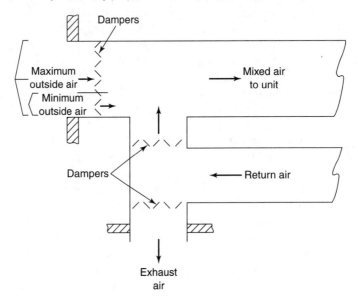

done in intermediate season cool weather to obtain cooling if needed without operating refrigeration machinery. Some systems use 100% outside air and no return air at all times, even though this increases the refrigeration load considerably. Examples might be operating rooms or laboratories where contaminated exhaust air often cannot be recirculated.

The *return air fan* takes the air from the rooms and distributes it through return air ducts back to the air conditioning unit or to the outdoors. In small systems with little or no return air ducts, the return air fan is not required, because the supply fan can be used to draw in the return air.

The *preheat coil* is required in cold climates (below freezing) to temper air so that chilled water cooling coils cannot freeze up. It is optional in milder climates and when DX (dry expansion) cooling coils are used. The preheat coil may be located either in the outside air or the mixed airstream.

When the system is used for winter heating, both the preheat and reheat coils can be utilized.

The *filters* are required to clean the air.

Bypassing air around the cooling coil (Figure 12.3) provides another method of controlling humidity (see Chapter 7), but does not give as good a humidity control in the space as with a reheat coil.

A room thermostat will control the cooling coil capacity to maintain the desired room temperature. If control of room humidity is required, a room humidistat is used. The interaction of these controls is explained in Chapter 14, Automatic Controls.

To achieve satisfactory temperature and humidity control in different zones, individual single zone units can be used for each zone. This may unacceptably increase costs and maintenance. However, there are a number of schemes that require only one air handling unit to serve a number of zones.

Four basic types of multiple zone all-air units and systems are available:

1. *Reheat system*
2. *Multizone system*
3. *Dual duct system*
4. *Variable air volume (VAV) system*

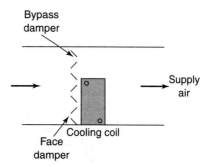

Figure 12.3
Arrangement of face and bypass dampers to provide reheat for humidity control.

The reheat, multizone, and dual duct systems are all constant air volume (CAV) type systems. That is, the air quantity delivered to the rooms does not vary. The variable air volume (VAV) varies the quantity of air delivered to the rooms.

Each of these types of systems will now be explained.

12.4 REHEAT SYSTEM

In the *reheat system*, separate single ducts from the air handling unit are distributed to each zone or room that is to be controlled separately (Figure 12.4 on page 308). A reheat coil is used in each of these ducts. In this way, separate control of both temperature and humidity can be achieved in each zone. The basic air handling unit is the same as with a single zone system, except perhaps the main reheat coil can be eliminated.

The reheat system provides good control of each zone. As seen in Figure 12.4, however, it is very wasteful of energy because the air must always be completely cooled to C and then often reheated (to S1 as shown for zone 1)—a double waste of energy.

Room thermostats located in each zone control their respective reheat coils to maintain the space set point temperature.

The use of the basic zone reheat system as described is often restricted by local energy codes,

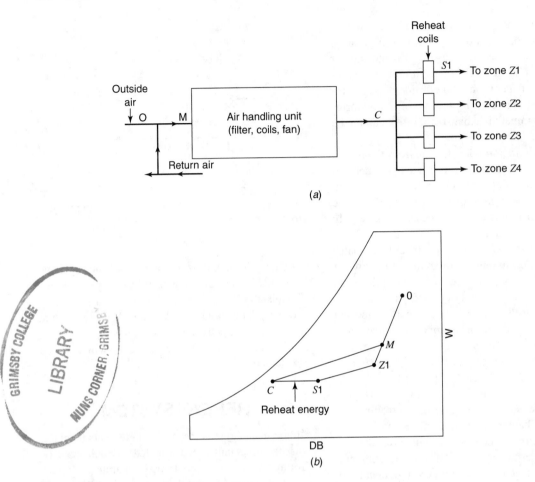

Figure 12.4
Reheat system with individual reheat coils. (a) Equipment arrangement. (b) Psychrometric processes for reheat system.

except for special applications, because of its inherently inefficient use of energy.

12.5 MULTIZONE SYSTEM

The *multizone system* uses an air handling unit that has a heating coil (*hot deck*) and cooling coil (*cold deck*) in parallel (Figure 12.5). Zone dampers are provided in the unit across the hot and cold deck at the outlet of the unit.

Separate ducts are run from each set of dampers to each zone (Figure 12.6). Cold and hot air are

mixed in varying proportions by the dampers according to zone requirements.

The psychometric processes for the multizone system are the same as that for the dual duct system, described in the next section.

The multizone system can provide good zone temperature control, but because mixed air is bypassed around the dehumidifying coil, humidity control may not be satisfactory in applications with high proportions of outside air. Because of the limit on the size of units available, each air handling unit is limited to about 12–14 zones. It is a relatively

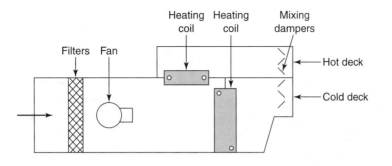

Figure 12.5
Multizone unit.

inexpensive system for small- and medium-size applications where a few separate zones are desired and humidity conditions are not critical.

The energy use features of the multizone system are similar to those of the dual duct system, to be discussed in the next section.

12.6   DUAL DUCT SYSTEM

In the *dual duct system* arrangement, separate hot and cold main ducts are run from heating and cooling coils in the air handling unit (Figure 12.7 on page 310).

Mixing boxes (Figure 12.8 on page 311) are provided in each zone, tapping air from the hot and cold ducts. Dampers in the mixing box respond to a room thermostat to mix the proper proportion of hot and cold air delivered to the zone.

The psychometric processes for summer cooling zone control area shown in Figure 12.7. Mixed air at M is heated to H from the fan heat. Chilled air leaves the cooling coil at C. This air mixes with air from the hot duct to produce supply air at S. RSHR line RS is an average condition for all zones, not an actual room line. Line $Z1$-$S1$ is an actual room line for a zone $Z1$, with a less-than-peak sensible cooling load and high latent load. Warm air and cold air are supplied in the correct proportion from the zone mixing box to provide zone supply air $S1$. Note that the room humidity is higher than the average. In most applications, the humidity increase is not great enough to be uncomfortable. $Z2$ is an example of a room condition with a higher sensible and lower latent load.

As the outside air temperature falls, it may be necessary for the reheat coil to operate to maintain an adequate hot duct temperature, so that humidity

Figure 12.6
Duct arrangement for multizone system.

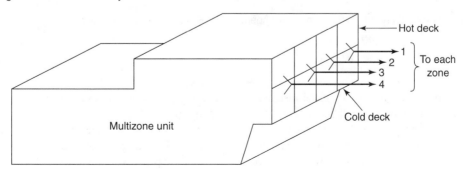

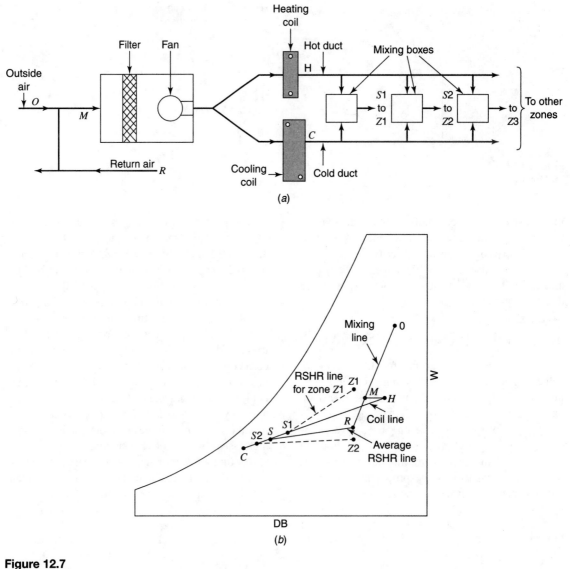

Figure 12.7
Dual duct system arrangement. (*a*) Equipment arrangement. (*b*) Psychrometric processes.

does not rise too high. This is one of the instances where extra energy may be used. In any case, the hot duct temperature control should be set at the minimum required to provide comfort. Many installations have been designed or operated with hot duct temperatures that result in considerable excess energy use.

In order to simplify the explanation, duct and return air fan heat gains are not shown in the psychrometric processes that have been described.

Figure 12.8
Mixing box for dual duct system. (Courtesy: Environmental Elements Corporation, Baltimore, Maryland.)

Dual duct systems are usually designed as high-velocity air systems in order to reduce duct sizes. The mixing boxes therefore have a sound attenuating section built into them. The air downstream from the mixing box is run at conventional low velocities.

The availability of cold and warm air at all times in any proportion gives the dual duct system great flexibility in handling many zones with widely varying loads. The installed cost of the dual duct system is usually quite high, and fan horsepower requirements are high because large volumes of air area moved at high pressure.

Both the dual duct and multizone systems are inherently energy wasteful, since, during part load cooling for a zone, overcooled air is reheated by mixing warm air with it, a double waste of energy. As with reheat systems, the use of constant volume multizone and dual duct systems is restricted. In some situations, they are prohibited in new installations.

Where these systems are allowed, they must have controls that reset the cold deck (duct) temperature at the highest value needed for cooling at all times. Similarly, there must be controls that reset the hot deck (duct) temperature at the lowest value needed for heating at all times. This minimizes the excess energy expended from the reheating or recooling.

12.7 VARIABLE AIR VOLUME (VAV) SYSTEM

The types of air conditioning systems we have already discussed are all constant air volume (CAV) systems. The air quantity delivered from the air handling unit to each zone remains constant. The temperature of this air supply is changed to maintain the appropriate room temperature.

The variable air volume or *VAV system* varies the air quantity rather than temperature to each zone to maintain the appropriate room temperature.

The basic VAV system arrangement is shown in Figure 12.9(*a*) on page 312. A single main duct is run from the air handling unit. Branch ducts are run from this main through *VAV boxes* to each zone. The VAV box has an adjustable damper or valve so that the air quantity delivered to the space can be varied. Room thermostats located in each zone control the dampers in their respective zone VAV boxes to maintain the desired room set point temperature.

The psychometric processes for a VAV system are shown in Figure 12.9(*b*) for summer cooling. The average room conditions are point *R*. Zone Z1 is shown at part load, when its sensible load has decreased, but its latent load has not. Its RSHR line is therefore steeper, as shown. To maintain the design room DB temperature, the air flow rate to zone Z1 is throttled, and the room DB in zone Z1 is the same as *R*, as desired. Notice, however, that the humidity in zone Z1 is higher (point Z1) than desired.

However, for applications that do not have high latent loads, this increase in room humidity conditions at part load is not enough to cause discomfort. Examples of installations where there may be a problem are conference rooms and auditoriums. In such cases, a solution is to limit throttling in the VAV box, and then apply reheat for further cooling load reduction.

There are other potential problems that may occur with VAV systems. Since the total supply air quantity is reduced at low loads, the quantity of outside air would also be reduced. This might be partially offset by providing some means of increasing the proportion of outside air, but even if

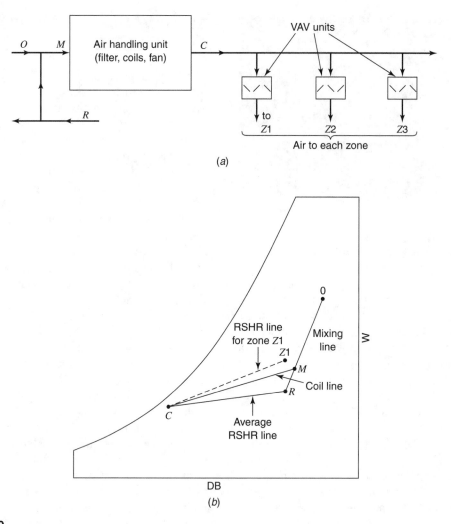

Figure 12.9
Variable air volume (VAV) system arrangement. (*a*) Equipment arrangement. (*b*) Psychrometric processes.

such an arrangement were practical, there often is still some limit of minimum air flow rate, below which there would be inadequate outside air.

One solution to this problem is to use a reheat VAV box, which has a built-in reheat coil. A control limits the minimum air quantity. If the cooling load continues to decrease, the reheat coil is activated. This type of VAV box also can be used to handle the problem with high latent loads, described previously.

Another potential problem at low loads and resulting low air flow rates is poor air distribution in the air conditioning spaces. Air supply diffusers are generally selected to give good coverage at maximum design air quantity. If the air flow rate decreased too much, the air circulation in the room will not be satisfactory, and uncomfortable conditions will result.

There are a few ways this problem may be solved. One is to use the reheat VAV box. When air quantity is reduced to the minimum for good air distribution, the reheat coil takes over.

Another possible solution is to use *variable diffusers*. These diffusers have a variable sized opening. As the air flow rate decreases, the opening narrows, resulting in better air distribution.

A further solution is to use *fan powered* VAV boxes. This type of VAV box has a small fan. In addition to the supply air quantity, this fan draws in and recirculates some room air, thus maintaining a high total air flow rate through the diffuser.

In spite of these potential problems and their special solutions, which is a feature of VAV systems, they are still very popular. This is because of the significant energy savings as compared to the other (CAV) multiple zone central systems. This energy-saving characteristic is partially that cited previously—it does not mix hot and cold air and does not reheat (except as noted).

There is also another significant energy saving. Whenever there is a part load, the air supply quantity is reduced, and there is a saving of fan power. Since a typical air conditioning system operates at part load up to 95% of the time, this saving is considerable.

12.8 ALL-WATER SYSTEMS

The basic concept of all-water systems, that is, hydronic systems, was introduced in Chapter 5. However, further material will be discussed here. Hydronic systems distribute hot or chilled water from the central plant to each space. No air is distributed from the central plant. Hydronic terminal units such as fan-coil units heat or cool the room air. Ventilation air can be brought through the outside wall and the terminal unit.

All-water systems for commercial use can be considerably less expensive and take up much less space than all-air systems (this is not necessarily true for residential use). Water has a much higher specific heat and density than air. This means that considerably less volume of water needs to be circulated for the same amount of heat transfer. The result is that the cross-sectional area of piping is

much smaller than the ductwork would be for the same job.

A hydronic cooling system is therefore useful when space is extremely limited, particularly space in shafts and ceilings. An important example is installation of air conditioning systems in existing large buildings that were not originally designed to include air conditioning.

The lack of need for ductwork and central air handling equipment, and the saving on using much valuable building space, all result in the fact that hydronic systems are often less expensive initially than all-air systems for large jobs, particularly in high-rise buildings. On the other hand, all-water systems have certain disadvantages. The multiplicity of fan-coil units means a great deal of maintenance work and costs. Control of ventilation air quantities is not precise with the small fans in the units. Control of humidity is limited. All-water systems are popular as low cost central systems in multiroom high-rise applications.

12.9 AIR-WATER SYSTEMS

Combination air-water systems distribute both chilled and/or hot water and conditioned air from a central system to the individual rooms. Terminal units in each room cool or heat the room.

Air-water systems utilize the best features of all-air systems and all-water systems. Most of the energy is carried in the water. Often the air quantities distributed are only enough for ventilation. Therefore, the total shaft and ceiling space required is small. In addition, the air is usually carried at high velocities.

One type of air-water system uses *fan-coil units* as the room terminal units. Chilled or hot water is distributed to them from the central plant. Ventilation air is distributed separately from an air handling unit to each room.

Another type of air-water system uses room terminal units called *induction units*; these were described in Chapter 5. It receives chilled or hot water from the central plant, as well as the ventilation air from a central air handling unit. The central air delivered to each unit is called *primary air*. As it flows through the unit at high velocity, it induces room air (*secondary air*) through the unit and across the water coil. Therefore, no fans or motors are required in this type of unit, reducing maintenance greatly.

The induction unit air-water system is very popular in high-rise office buildings and similar applications. Its initial costs are relatively high.

The primary air quantity in the induction system may be only about 25% or less than the total of the air volume rate of a conventional all-air system. Because of this, it is often not adequate for outside air cooling in mild or even cold weather. At these times, chilled water must be supplied to the room unit coils. This is particularly true on southern exposures. There are buildings with air-water induction systems requiring refrigeration at outdoor temperatures as low as 30 F. In some cases, this energy inefficient situation may be improved by utilizing another source of chilled water, such as an outside air heat exchanger (see Chapter 15).

Any of the two-, three-, and four-pipe hydronic system arrangements described in Chapter 5 can be applied to air-water systems. The factors in choosing a particular arrangement are discussed in that chapter.

12.10   UNITARY VERSUS CENTRAL SYSTEMS

As stated previously, air conditioning systems can also be classified into either *unitary* or *central* systems. This classification is not according to how the system functions, but how the equipment is arranged. A unitary system is one where the refrigeration and air conditioning components are factory selected and assembled in a package. This includes refrigeration equipment, fan, coils, filters, dampers, and controls. A central or remote system is one where the components are all separate. Each is selected by the designer and installed and connected by the contractor. Unitary equipment is usually located in or close to the space to be conditioned. Central equipment is usually remote from

the space, and each of the components may or may not be remote from each other, depending on the desirability.

Unitary or central systems can both in theory be all-air, all-water, or air-water systems, but practically, unitary systems are generally all-air systems, and limited largely to the more simple types such as single zone units with or without reheat or multizone units. This is because they are factory assembled on a volume basis. Unitary systems and equipment can be divided into the following groups:

1. *Room units*
2. *Unitary conditioners*
3. *Rooftop units*

These names are not standardized in the industry. For example, *unitary* conditioners are also called *self-contained* units or *packaged* units.

12.11   ROOM UNITS

Room units (Figure 12.10) are available in two types; *window* units and *through-the-wall* units. The window unit fits in the sash opening of an existing window, resting on the sill. The through-the-wall unit fits in an outside wall opening, usually under the windowsill.

Compressor, evaporator cooling coil, condenser, filter, motors, fan and controls are assembled in the unit casing. Dampers can be adjusted so that only room air is used, or so that some outside ventilation air can be brought through the conditioner. Room units are available up to about 3 tons of refrigeration capacity. Their advantages are low cost and simplicity of installation and operation. Window units are particularly applicable to existing buildings. Through-the-wall units are often used in new apartment houses where low cost is primary. In existing buildings, of course, electrical services may have to be increased to take the added electrical load.

Room units have no flexibility in handling high latent heat gains or changed sensible heat ratios, and therefore do not give good humidity

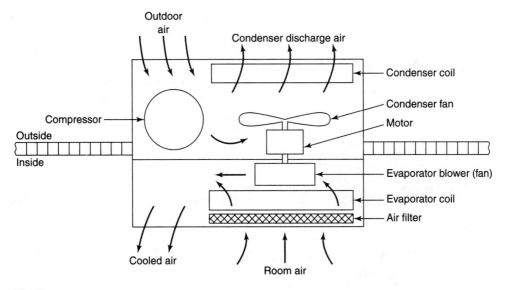

Figure 12.10
Room air conditioner equipment arrangement.

control. Sound levels are higher than with remote equipment. Air cleaning quality is minimal because the filters remove only large particles, in order that the resistance to airflow be low. When used in multiroom buildings, maintenance of the large number of units can be very burdensome and expensive. These units are inherently energy wasteful in multiple use, because they cannot modulate capacity.

12.12 UNITARY AIR CONDITIONERS

This type of unit is designed to be installed in or near the conditioned space (Figure 12.11 on page 316). The components are contained in the unit. Heating components are sometimes included.

Unitary conditioners are available in vertical or horizontal arrangements, according to the space available for the equipment. Although they often discharge air directly into the space, a limited amount of ductwork can be connected if air distrib-

ution with outlets is desired. These units are popular in small commercial applications.

Units are available that have all components packaged except the condenser. This is a popular arrangement in private residential and small commercial applications. In one arrangement, the condenser is located outdoors and compressor, coil, and fan package in an attic or basement.

Another common arrangement is called a *split system*. The condenser and compressor are in one package, located outdoors, and the fan and cooling coil are another package located indoors.

The split system arrangement has distinct advantages in many applications. The compressor-condenser package is located outdoors where its noise is less objectionable and it is more accessible for maintenance. Furthermore, there is no problem of finding a suitable and adequate space in the building.

Unitary conditioners have the same advantages and disadvantages as room units. In larger units, multiple compressors are used. Units are available in sizes up to about 50 tons.

Figure 12.11
Unitary air conditioner.

12.13   ROOFTOP UNITS

This type of unitary equipment (Figure 12.12) is designed to be located outdoors, and is generally installed on roofs. Usually, all of the refrigeration, cooling, and air handling equipment comes in sections that are assembled together, although the compressor and condenser may be remote. Heating equipment may also be incorporated in the unit.

Rooftop units may be used with ductwork and air outlets. They must have weatherproofing features not required with equipment located indoors. All electrical parts must be moistureproof, and the casing and any other exposed parts must be corrosion protected.

The advantages of rooftop units are that they do not use valuable building space and they are relatively low in cost. Units are available with multizone arrangement, thereby offering zone controls, but humidity control is limited. Rooftop systems are extremely popular in low cost, one-story building applications, such as supermarkets and suburban commercial buildings.

12.14   AIR HANDLING UNITS

The central system air handling unit (AHU) consists of the cooling/heating coils, fan, filters, air mixing section, and dampers, all in a casing (Figures 12.1, 12.5, and 12.13). There are basically two arrangements: single zone units and multizone units. The distinction between these has been discussed in Sections 12.3 and 12.5. In small- and medium-capacity systems, air handling units are factory made in sections—fan section, coil sections, mixing box, filter section—in numerous sizes. Those parts that are required are selected by the user. For large systems, separate coils, filters, and fans are selected by the engineer and casings are fabricated by the contractor to suit the equipment.

Casings are usually made of galvanized sheet metal. The casing should be insulated to prevent energy losses. When cooling and dehumidifying, drain pans must be included under the coil to collect condensed moisture, and a piping drain connection must be provided, which is run to a waste drain. The pipe should have a deep seal trap so that a water seal always exists (Figure 12.13).

The dehumidification effect of the cooling coil frequently results in water collecting on the coil. The water may then be carried as droplets into the moving airstream. To prevent this water from circulating into the air conditioning ductwork, *eliminators* are provided downstream from the coil. This consists of vertical Z-shaped baffles that trap the droplets, which then fall into the condensate pan.

Access doors should be provided to permit maintenance. They should be located on both sides of coils and filters. In large equipment, lights should be provided inside each section.

Figure 12.12
Rooftop unit. (Courtesy: McQuay Group, McQuay-Perfex, Inc.)

When the fan is located downstream of the cooling coil, the unit is called a *draw-through* type. When the fan is upstream of the coils, it is called a *blow-through* type. Draw-through is preferable because the air will flow more uniformly across the coil section when drawn through by the fan. Multizone units are blow-through types. To aid in distributing the air more evenly across the heating and cooling coil in blow-through units, a perforated plate is sometimes located between the fan and coils.

12.15   COOLING AND HEATING COILS

Cooling coils may use either chilled water or evaporating refrigerant. The latter are called *dry expansion* (DX) coils.

Figure 12.13
Maintenance accessories in large air handling unit.

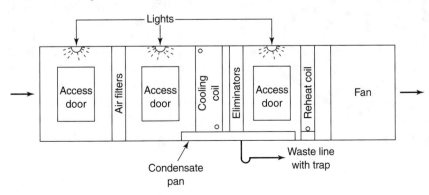

Cooling coils are usually made of copper tubing with aluminum fins, but copper fins are sometimes used. The coils are arranged in a serpentine shape, in a number of rows, depending on the need (Figure 12.14).

The fins increase the effective surface area of tubing, thus increasing the heat transfer for a given length of tube. The coil may be constructed either with tubes in series or in parallel to reduce water pressure drop.

When cooling coils have a number of rows, they are usually connected so that the flow of water and air are opposite to each other, called *counterflow* (Figure 12.15). In this way, the coldest water is cooling the coldest air and fewer rows may be needed to bring the air to a chosen temperature than if parallel flow were used, and the chilled water temperature can be higher.

The water inlet connection should be made at the bottom of the coil and the outlet at top, so that any entrapped air is carried through more easily. In addition, an air vent should be located at the outlet on top.

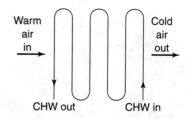

Figure 12.15
Counterflow arrangement of air and water flow for cooling coil.

12.16   COIL SELECTION

Coil selections are made from manufacturers' tables or charts based on the required performance. The performance of a cooling coil depends on the following factors:

1. The amount of sensible and latent heat that must be transferred from the air.
2. Conditions of air entering and leaving, DB and WB.
3. Coil construction—number and size of fins size and spacing of tubing, number of rows.
4. Water (or refrigerant) velocity.
5. Air face velocity. The face velocity is the air flow rate in CFM divided by the projected (face) area of the coil.

Water velocities from 1–8 FPS are used. High water velocity increases heat transfer but also results in high pressure drop, and therefore requires a larger pump and increased energy consumption. Velocities in the midrange of about 3–4 FPS are recommended.

High air velocities also result in better heat transfer and also more CFM handled. However, if the coil is dehumidifying, the condensed water will be carried off the coil into the airstream above 500–550 FPM face velocity, and *eliminator baffles* must be used to catch the water droplets.

The form in which manufacturers present their coil rating data varies greatly one from another. Using these ratings does not give much insight into how a coil performs. For this reason, we will not

Figure 12.14
Cooling coil (chilled water type).

present any rating data. However, the procedures described in Chapter 7 give all the basic data necessary to select a coil. Indeed, that procedure has the advantage of being suitable for any manufacturer's coils, as seen in Example 12.1.

Table 12.1 lists typical contact factors (CF) for finned cooling coils. With this type of table, the number of rows of coil needed for given entering and leaving air conditions can be directly determined.

Example 12.1

A cooling coil has 3200 CFM of air flowing across it at a face velocity of 400 FPM. Air enters the coil at 85 F DB and 69 F WB and leaves at 56 F DB and 54 F WB. Determine the required number of rows and face area of an 8 fin/in. coil.

Solution

The required CF is 0.83, as worked out in Example 7.32. From Table 12.1, for that type of coil, a four-row coil will do the job. The face area needed is

$$3200 \text{ CFM}/400 \text{ FPM} = 8 \text{ ft}^2$$

TABLE 12.1 TYPICAL CONTACT FACTORS FOR HELICAL FINNED COOLING COILS

| | Face Velocity, FPM | | | | | |
| | 8 fins/in. | | | 14 fins/in. | | |
No. of Rows	400	500	600	400	500	600
2	0.60	0.58	0.57	0.73	0.69	0.65
3	0.75	0.73	0.71	0.86	0.82	0.80
4	0.84	0.82	0.81	0.93	0.90	0.88
6	0.94	0.93	0.92	0.98	0.97	0.96
8	0.98	0.97	0.96			

12.17 AIR CLEANING DEVICES (FILTERS)

Air conditioning systems that circulate air generally have provisions for removing some of the objectionable air contaminants. Most systems have devices that remove particles commonly called

dust or dirt, which result largely from industrial pollution. Occasionally gases that have objectionable odors are also removed from the air.

The need for proper air cleaning is often treated casually when designing and operating an air conditioning system. The incorrect type of filter may be chosen, or the filters may not be maintained properly. This is a serious neglect, because we are dealing with a question of air pollution and human health. Proper air cleaning is necessary for the following reasons:

1. *Protection of human health and comfort.* Dust particles are related to serious respiratory ailments (emphysema and asthma).
2. *Maintaining cleanliness of room surfaces and furnishings.*
3. *Protection of equipment.* Some equipment will not operate properly or will wear out faster without adequate clean air. Some manufacturing processes are particularly sensitive.
4. *Protection of the air conditioning machinery.* For example, lint collecting on coils will increase their resistance to heat transfer.

12.18 METHODS OF DUST REMOVAL

Air cleaners can remove dust in three major ways:

1. *Impingement.* The dust particles in the airstream strike the filter media and are therefore stopped.
2. *Straining.* The dust particles are larger than the space between adjacent fibers and therefore do not continue with the airstream.
3. *Electrostatic precipitation.* The dust particles are given an electric charge. The filter media is given the opposite charge, and therefore the particles are attracted to the media.

A filter may remove particles by one or more of the above methods. This will be discussed when specific types of filters are described. Figure 12.16 on page 320 shows each of the methods.

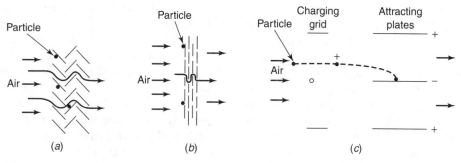

Figure 12.16
Methods of removing particles from air. (*a*) Impingement. (*b*) Straining.
(*c*) Electrostatic precipitation.

12.19   METHODS OF TESTING FILTERS

Understanding how air filter performance is evaluated is important because only in this way can a proper filter be selected. Only in recent years have standard test methods developed. Without standard procedures, filters cannot be compared with each other. The problem is complicated because filter performance depends on the concentration and sizes of dust particles in the air. This varies greatly from one location to another and at different times.

The following tests are generally accepted and recommended in the industry:

1. *Weight*. The weight of dust captured by the air filter is measured. A standard dust of fixed concentration and particle sizes is used. This test is useful in comparing ability to remove larger particles. It does not indicate ability to remove small particles, because the small particles comprise such a small proportion of the total weight of atmospheric dust.

2. *Dust spot discoloration*. In this test, air is first passed through the air cleaning device and then a white filter paper. The degree to which the filter paper is discolored is an indicator of the amount of smaller dust particles not removed by the air cleaner. This test is important because these particles cause soiling of room surfaces.

3. *DOP penetration*. This test is used to measure the ability of air cleaners to remove extremely small particles. A cloud of particles of a substance called DOP is chemically generated. The size of these particles is 0.3 microns in diameter (one micron is about 1/25,000th of one inch). A cloud of DOP particles in an airstream is passed through the air cleaner. The concentration of particles not removed is measured downstream of the cleaner by using a light-scattering technique. In this way, the filter's effectiveness in removing very small particles is tested. As an example, bacteria range from about 0.3–30 microns in diameter, and cigarette smoke particles from 0.01–1 micron. The DOP test is used only on air cleaners that are designed to have a high efficiency in removing very small particles.

4. *Dust holding capacity*. The above three tests all measure efficiency of an air cleaner in removing particles. What they do not measure is how much the filter air resistance will increase with dust accumulation. A filter that will hold a considerable amount of dust before resistance increases considerably is preferable to one that has a lesser capacity before buildup up to a given resistance. The dust holding capacity test compares weight of dust collected with increase in air resistance through the filter.

12.20 TYPES OF AIR CLEANERS

Air cleaners can be classified in a number of ways.

Type of Media

The *viscous impingement* air filter has a media of coarse fibers that are coated with a viscous adhesive. Glass fibers and metal screens are two commonly used media materials. Air velocities range from 300–600 FPM. The pressure drop when clean is low, around 0.1 in. w.g.; the filter should be serviced when the resistance reaches 0.5 in. w.g. This type of filter will remove larger dust particles satisfactorily but not small particles. It is low in cost (Figure 12.17).

The *dry-type* air filter uses uncoated fiber mats. Glass fibers and paper are two commonly used materials. The media can be constructed of either coarse fibers loosely packed or fine fibers densely packed. By varying density, dry-type air filters are available that have good efficiency only on larger particles, as with the viscous impingement type, or are also available with medium or high efficiency for removing very small particles.

The HEPA filter (High Efficiency Particulate Air) is a very high efficiency dry-type filter for removing extremely small particles (Figure 12.18). For example, it is the only type of filter that will effectively remove viruses as small as 0.05 micron (1/500,000th of an inch!). Air face velocities through HEPA filters are very low, about 50 FPM, and resistance rises to about 2.0 in. w.g. before servicing. They are quite expensive.

The media in air filters can be arranged in the form of random fiber mats, screens, or corrugated sinuous strips.

Permanent or Disposable

Air filters may be designed so that they are discarded (*disposable* type) when filled with dust, or are cleaned and reused. *Permanent* types have

Figure 12.17
Viscous impingement disposable filter. (Courtesy: American Air Filter Co., Louisville, Kentucky.)

Figure 12.18
High efficiency dry-type (HEPA) filter. (Courtesy: American Air Filter Co., Louisville, Kentucky.)

metal media that will withstand repeated washings, but they cost more than disposable types.

Stationary or Renewable

Stationary air filters are manufactured in rectangular panels that are placed alongside each other and stacked, according to the size needed. The panels are removed and either replaced or cleaned when dirty. *Renewable-type* air filters consist of a roll mounted on a spool that moves across the airstream (Figure 12.19). The media is wound on a take-up spool, driven by a motor. The movement of the media is often controlled by a pressure switch which senses the pressure drop across the media. When the resistance increases to a set value because of the dirt collected, the motor moves the curtain, exposing clean media. Renewable air filters are considerably more expensive than the sta-

Figure 12.19
Automatic renewable filter. (Courtesy: American Air Filter Co., Louisville, Kentucky.)

tionary types, but maintenance costs are greatly decreased. Either fibrous materials or metal screens are used as media.

Electronic Air Cleaners

In this type, there is no fibrous media to entrap dust (Figure 12.20). Dust particles are given a high voltage charge by an electric grid. A series of parallel plates are given the opposite electric charge. As the dust-laden airstream passes between the plates, the particles are attracted to the plates. The plates may be coated with a viscous material to hold the dust. After an interval of time the air cleaner must be removed from service in order to clean the plates and remove the dirt. Electronic air cleaners are expensive, but are very efficient for removing both large and very small particles.

12.21   SELECTION OF AIR CLEANERS

The selection of the proper air cleaner depends on the degree of contamination of the air to be cleaned and the cleanliness requirements.

For applications that require only minimum cleanliness and low cost, inexpensive viscous impingement type disposable air filters would be used. A private residence or apartment house might be an example. For applications that require a greater degree of cleanliness, and where contamination is greater, perhaps intermediate efficiency dry-type filters would be used. Another choice might be electronic air cleaners, particularly where smoking is heavy. Often electronic cleaners are used in conjunction with a prefilter, a coarse viscous impingement cleanable filter that removes the large particles first, so that they do not cause quick build-up of dirt on the electronic air cleaner. This arrangement is popular in large commercial buildings.

Where removal of extremely small particles is critical, such as viruses, bacteria, or radioactive particles, HEPA filters are used. These are also usually backed up with a coarse prefilter to remove large particles.

Figure 12.20
Electronic air cleaner. (Courtesy: American Air Filter Co., Louisville, Kentucky.)

For removing gases with objectionable odors from the air, *activated carbon* (charcoal) filters are used. The carbon absorbs the gas molecules. These filters are sometimes used in restaurants to remove odorous gases resulting from cooking.

12.22   INDOOR AIR QUALITY

It has become evident that the poor quality of air inside some buildings is contributing to health problems. Since many people spend up to 90% of their time indoors, this subject is of major concern. The HVAC system is connected with Indoor Air Quality (IAQ), in sometimes contributing to the problem and as a part of the potential solution.

Air contaminants from sources inside buildings are the main cause of poor IAQ, but outdoor air pollutants that enter a building can also contribute to the problem.

The emphasis in this discussion will be on IAQ problems in the commercial working environment, rather than private residences, although many of the problems and solutions are similar.

Health Effects

Short-term effects from indoor air pollutants may include eye, nose, and throat irritation, nausea, irritability, headaches, and fatigue. Symptoms of diseases such as asthma may be increased.

Humidifier fever is a respiratory illness caused by exposure to microorganisms found in humidifiers and air conditioners.

Hypersensitivity pneumonitis is a respiratory illness caused by the inhalation of organic dusts.

Long-term effects that may show up after a period of years are respiratory diseases, heart disease, and cancer.

The term *sick building syndrome* (SBS) refers to a set of symptoms that may affect occupants only during the time they are in the building and cannot be traced to a specific pollutant.

Indoor Pollutants

Volatile Organic Compounds (VOC) These are organic substances emitted as gases from building materials, plywood, particle board, adhesives, carpets, cleaning materials, paints, air fresheners, copying machines, and pesticides. *Formaldehyde* is the best known and most common of VOC pollutants.

Biological Contaminants These include bacteria, viruses, molds, mites, pollens, and fungi.

Environmental Tobacco Smoke (ETS) Also called "passive smoking," this is the mixture of substances emitted from burning tobacco, breathed in by occupants.

Radon This is a radioactive gas emitted by soil. It may enter a building through underground walls or floors, sumps, and drains. It is more commonly a problem in private residences than in commercial buildings.

Asbestos This is a mineral substance used in a fibrous form for insulation, tiles, and acoustic material and fireproofing in buildings.

Glass fibers Materials made of glass fibers are used as thermal and sound insulation in HVAC systems. These fibers sometimes peel off and are carried into the occupied spaces. In addition, the fiber lining itself can serve as a breeding place for molds and fungi.

Carbon dioxide (CO)$_2$ A natural constituent of atmospheric air, this gas is not toxic but is sometimes used as a measure of adequate ventilation. Outdoor concentrations of CO_2 are about 300 ppm (parts per million). In indoor spaces that are not well ventilated and that are densely occupied, the CO_2 concentration will increase considerably. The ASHRAE Standard recommends a threshold level of 1000 ppm above which the CO_2 level indicates possibly poor indoor air quality. There is a difference of opinion on this question however—further information and research is needed.

Solutions

There are three general approaches to improving air quality in buildings.

1. *Source control.* This involves avoidance of the use of pollutant source materials or chemicals. If they are already in place, removal or containment may be done.
2. *Ventilation.* In efforts to conserve energy use, air infiltration has been decreased by reducing or sealing crack openings in both existing buildings and in the design of new ones. This coupled with using minimum outside ventilation rates in HVAC systems has amplified the effect of indoor air pollutants, since they are kept in the building longer, and in greater concentrations.

   The concentration of indoor air pollutants can be decreased by supplying a substantial amount of outside ventilation air from the HVAC systems. Table 6.17 lists ventilation requirements typical of present state codes.
3. *Cleaning.* The level of indoor air contaminants can be reduced by both air cleaning and good housekeeping.

Air filters in the HVAC system should be of the proper type and efficiency to reduce the indoor pollutant level as needed. It may be found that more efficient filters than used previously are required; in some cases HEPA filters, electronic cleaners, and activated carbon filters may be desirable. If the outside air is sufficiently contaminated, OA filters may be necessary.

Regular and good housekeeping maintenance is an important part of insuring a satisfactory indoor air quality. Among the items to be considered are

A. Cleaning and replacement of air filters. This is often an area of serious neglect.
B. Elimination or reduction of moist areas. For instance, care should be taken that drain pans in HVAC equipment drain freely.
C. Vacuum cleaning of areas, where dirt may accumulate. An example is the ductwork system.
D. Appropriate application of biocidal cleaners to areas where biological growths are expected, such as cooling towers, humidifiers, coils, and drain pans. A serious illness called Legionnaire's disease sometimes has its origins in building HVAC systems, such as cooling towers. Care must be taken that the cleaning agents themselves are not pollutants that may enter the occupied spaces.

For good indoor air quality, space temperatures and humidity should also be within the range recommended in Section 1.6. The relative humidity (RH) should be maintained below 60% to discourage growth of molds and fungi.

12.23 ENERGY REQUIREMENTS OF DIFFERENT TYPES OF AIR CONDITIONING SYSTEMS

A comparative energy use analysis of some of the major types of air conditioning systems will be made in this section. This will illustrate the opportunities for energy conservation by the proper choice of a system for a given application.

A comparison will be made of energy requirements for constant volume reheat, dual duct, and variable air volume systems. A further comparison will be made with an air-water system such as the induction or fan-coil type. A typical office building will be specified. Some simplifications will be made to avoid unnecessary details that would detract from following the analysis. Of course, in an actual energy study, every factor would be included.

The following are the design specifications:

Office building of 200,000 ft² area.

N, E, S, W zones are 25,000 ft² each. Interior zone is 100,000 ft².

Outside condition is 97 F DB, 74 F WB. Inside 78 F DB, 45% RH.

Ventilation air is 45,000 CFM.

Lights and power are 12 BTU ft².

Occupants 3300.

Building peak load is in July at 4 PM.

Air off cooling coils is at 54 F DB.

Supply fan temperature rise is 4 F. Fan static pressure is 6 in. w.g. Duct and return fan heat gains are neglected; RSHR is 0.88.

1. *Constant volume reheat system.* The air handling unit psychrometric processes are shown in Figure 12.4. Supply air temperature is 58 F if the air off cooling coil is 54 F DB. The air supply rate must satisfy the sum of each zone peak.

$$CFM = \frac{5,510,000}{1.1 \times (78 - 58)} = 250,000$$

The refrigeration capacity of the reheat system must satisfy the sum of the zone peaks:

Sum of zone sensible peaks, BTU/hr	= 5,510,000
People latent	= 660,000
Outside air 45,000 × 45(377 − 29.0)	= 1,760,000
Fan heat 250,000 × 1.1 × 4	= 1,100,000
Refrigeration load, BTU/hr	= 9,030,000
	= 752 tons

The reheat system must furnish heat from the heating coil for all zones except those at peak loads. The heating required at design conditions is

$$5,510,000 - 4,690,000 = 870,000 \text{ BTU/hr}$$

2. *Dual duct system.* The air handling unit psychrometric processes are shown in Figure 12.7. The total air supply rate must satisfy the sum of each zone peak load.

PEAK SENSIBLE HEAT GAINS FOR EACH ZONE, BTU/hr.

	N	E	S	W	I
Solar + trans.	150,000	800,000	450,000	1,000,000	
Lights	300,000	300,000	300,000	300,000	1,200,000
People	90,000	90,000	90,000	90,000	350,000
Totals	540,000	1,190,000	840,000	1,390,000	1,550,000
	Sum = 5,510,000 BTU/hr				

BUILDING PEAK SENSIBLE HEAT GAINS

	N	E	S	W	I
Solar + trans.	150,000	180,000	200,000	1,000,000	
Lights	300,000	300,000	300,000	300,000	1,200,000
People	90,000	90,000	90,000	90,000	350,000
Totals	540,000	570,000	590,000	1,390,000	1,550,000
	Sum = 4,640,000 BTU/hr				

Furthermore, additional air is required because of leakage through closed dampers in the zone mixing box. The equation for finding total CFM is

$$CFM_t = \frac{\text{sum of zone peaks}}{1.1(t_r - t_c)}$$

$$\times \frac{1}{1 - x - x\left(\dfrac{t_w - t_r}{t_r - t_c}\right)}$$

where

CFM_t = total air supply rate, hot and cold ducts

t_r = average room temperature

t_c = cold air supply temperature at mixing dampers

t_w = warm air supply temperature at mixing dampers

x = fraction of air leakage through closed mixing damper

Using a leakage rate of 5%, and noting that t_w = 85.3 F DB (Figure 12.7),

$$CFM_t = \frac{\frac{5,510,000}{1.1(78 - 54)}}{}$$

$$\times \frac{1}{1 - 0.05 - 0.05\left(\dfrac{85.3 - 78}{78 - 54}\right)}$$

$$= 223,000$$

Although the dual duct system must supply an air rate to satisfy each zone peak, the refrigeration capacity must only satisfy the building peak:

Building sensible peak	= 4,640,000 BTU/hr
People latent	= 660,000
Outside air	= 1,760,000
Fan heat 223,000 × 1.1 × 4	= 980,000
Refrigeration load	= 8,040,000 BTU/hr
	= 670 tons

3. *Variable air volume system.* The air handling

unit psychrometric processes are shown in Figure 12.9. The maximum air supply rate must satisfy only the building peaks:

$$CFM = \frac{4,640,000}{1.1(78 - 58)} = 211,000$$

The refrigeration capacity must satisfy only the building peak, 670 tons. (We will neglect the slight difference in fan heat.)

The difference in the energy requirements of the three systems at full load is summarized as follows:

	Reheat	Dual duct	VAV
Tons of refrigeration	752	670	670
Refrigeration KW	677	603	603
CFM	250,000	223,000	211,000
Fan KW	200	178	169
Heating BTU/hr	870,000		
Total KW	877	781	772
Extra cost $/hr	16.23	0.72	0

In the above estimates, figures of 0.9 KW/ton, $0.08/KWH, $9.00/$10^6$ BTU, and 0.8 KW/1000 CFM at 6 in. w.g. were used.

Note the huge extra expense and energy waste from using the reheat system. If we were to assume this energy difference for a full load equivalent of 1200 hours a year, the yearly extra cost would be

Reheat	$19,480
Dual duct	860

This is still not the actual situation, however. The energy consumption differences are even much greater at part load, both because the VAV system throttles airflow with load reduction and the reheat system must add even more heat. We will make a rough analysis of this situation. Assume a mild day with no solar, outside air, or transmission loads.

1. *Reheat system.* The cooling coil still cools all the air to set temperature, 54 F. The outside air

load is the only one not required. The refrigeration load is therefore

Full load for reheat system	9,030,000 BTU/hr
Outside air design load	− 1,760,000
Reheat system part load	7,270,000 BTU/hr
	= 606 tons

The actual building part load is

Peak design load	8,040,000 BTU/hr
Outside air design load	− 1,760,000
Solar + trans. design load	− 1,530,000
Building part load	4,750,000 BTU/hr
	= 396 tons

The reheat system must provide external heat source for all of the difference between the building peak and sum of zone peaks at part load:

$$7,270,000 - 4,750,000 = 2,520,000 \text{ BTU/hr}$$

2. *Dual duct system.* The refrigeration capacity required is the building peak part load (396 tons) since the cold air is throttled as needed in each zone.
3. *VAV system.* The refrigeration capacity required is 396 tons, because the air supply rate is throttled as required. The total air supply rate is

$$\text{CFM} = \frac{\text{room sensible heat}}{1.1 \times 20} = \frac{3,110,000}{1.1 \times 20}$$

$$= 141,000$$

The differences in energy requirements at the part load condition are summarized as follows:

	Reheat	Dual duct	VAV
Tons of refrigeration	606	396	396
Refrigeration KW	545	356	356
CFM	250,000	223,000	141,000
Fan KW	200	178	113
Total KW	745	534	469
Heating BTU/hr	2,270,000		
Extra cost $/hr	42.50	5.20	0

Assuming an operating period of 2000 hours a season with this part load condition as an average, the yearly extra energy costs would be

Reheat	$85,000
Dual duct	10,400

The part load condition selected does not necessarily represent the average of an actual installation, of course. To carry out an accurate yearly energy use, hourly weather and equipment performance data are needed. However, the general conclusions from our analysis hold true.

We have not attempted to show variations in energy use due to types of controls selected. Other factors such as diversity or operation of reheat coils in the dual duct and VAV systems have not been included.

If the systems analyzed were used for winter heating, further sharp differences in energy consumption would appear for similar reasons. Many buildings have a separate radiation heating system for perimeter zones, however, which reduces the penalty of excessive use of air heating. This is particularly true during unoccupied hours, where otherwise fans would have to be operated continually.

We have not made a comparative analysis of the energy use of an air-water system. At full load this system has about the same energy requirements as the VAV system, because the room units handle only the load for their zones. There may be a small advantage in auxiliary energy use for the air-water system because only the primary (ventilation) air is moved. The additional pump energy is usually less than that of the greater air quantity. At lower outdoor temperatures, however, the situation may change significantly. All-air systems of course have as much air for cooling as needed in cold weather. For the air-water system, however, the primary air quantity is not adequate to cool zones with large heat gains, and chilled water must be used.

Consider the south side of the building on a sunny November day with an outdoor temperature of 40 F. With fan heat, the primary air supply temperature is 44 F. The peak refrigeration load is

Solar gain	= 1,200,000 BTU/hr
People sensible	= 90,000
Lighting	= 300,000
Primary air $1.1 \times 5000 \times 30$	= −165,000
Transmission losses	= −255,000
Refrigeration load	= 1,170,000 BTU/hr
	= 98 tons

This is a heavy penalty to pay for using this type of system without heat recovery, because there will be many hours in the heating season when refrigeration is needed. Chilled water and condenser water pumps will also have to be operated.

12.24 ENERGY CONSERVATION

1. Systems that mix hot and cold air (dual duct or multizone) or water (three pipe) may result in energy waste, although they can be designed to minimize the loss.
2. When using air-water systems, care should be taken that they are not producing opposite effects and therefore wasting energy. For example, in an induction system, the primary air may be warm while at the same time chilled water is being distributed to the induction units. This should be avoided where possible by proper design and operation.
3. Reheating is unavoidably wasteful and should be avoided except for special applications, unless the reheating would come from otherwise wasted energy (see Chapter 15).
4. Systems should be designed and operated to use all outside air for cooling when it is adequate (see Chapter 15).
5. Replace or clean filters on a regular schedule to limit pressure losses to those recommended, thereby avoiding excessive fan power.
6. Clean coils regularly, thereby maintaining maximum heat transfer.

Review Questions

1. Sketch and label all elements of a single zone air handling unit and a typical duct arrangement.
2. Sketch and label all elements of a reheat system arrangement and air handling unit.
3. Sketch and label all elements of a multizone system arrangement and air handling unit.
4. Sketch and label all elements of a dual duct system arrangement and air handling unit.
5. Sketch and label all elements of a VAV duct system arrangement and air handling unit.
6. Prepare a list of advantages and disadvantages of reheat, multizone, dual duct, VAV, and unitary systems.
7. Explain the following terms:
   A. *Split system*
   B. *Draw-through unit*
   C. *Blow-through unit*
8. List four purposes of air cleaning devices.
9. Explain the following terms related to air cleaners:
   A. *Impingement*
   B. *Straining*
   C. *Electrostatic precipitation*
10. List and explain the four methods of testing and rating air cleaners.
11. What is a HEPA air filter? What are its applications?
12. What type of air cleaner is used to remove undesired gases?

Problems

12.1 Select a 14 fin/in. cooling coil to cool 12,000 CFM of air from 82 F DB and 70 F

WB to 55 F DB and 54 F WB. The coil face velocity is 600 FPM.

12.2   Select an 8 fin/in. coil for the same requirements as described in Problem 12.1.

12.3   A four-row, 8 fin/in. cooling coil is handling air at a face velocity of 600 FPM. Air enters at 87 F DB and 72 F WB, and leaves at 59 F DB. What is the leaving air WB?

12.4   If the air velocity for the coil in Problem 12.3 is reduced to 400 FPM, assuming the same leaving DB, what is the leaving air WB?

CHAPTER 13
Refrigeration Systems and Equipment

An environmental control system that includes cooling and dehumidification will require a means of removing heat from the conditioned spaces. Because heat flows only from a higher to a lower temperature, a fluid with a temperature lower than the room design temperature must be made available, to which the excess room heat can be transferred. Refrigeration produces this low temperature fluid.

Occasionally a natural low temperature fluid is available. The ancient Roman rulers had slaves transport snow from the high mountains to cool their food and beverages. Cold well water has often been used in modern air conditioning systems. Many communities now restrict the use of well water for air conditioning, however, because of the depleted supply. Furthermore, well water temperatures are often 50–60 F, which is too high to accomplish adequate dehumidification. Another natural heat sink that is used occasionally for cooling water is atmospheric air. In climates where the humidity is extremely low, evaporative cooling of air may reduce both the water and air temperature low enough so that either can be used for cooling (Chapter 7). Well water or evaporative cooling should be considered for refrigeration when available.

Usually there is no natural heat sink at a temperature lower than the desired space temperature when cooling is required. In this case, refrigeration systems that require machinery are used to provide a cold fluid for cooling or dehumidification.

Vapor compression and absorption refrigeration systems are both used widely for producing refrigeration required for air conditioning. In this chapter, we will explain how each system functions, the types of equipment used, and some equipment selection procedures. We will not discuss the calculations related to the thermodynamic cycle or to compressor performance. Our purpose here will be to relate the refrigeration equipment to the complete air conditioning system. However, a further understanding of refrigeration theory is necessary for the well-trained air conditioning practitioner. Textbooks on refrigeration should be consulted, such as the author's text *Refrigeration Principles and Systems: An Energy Approach.*

Other refrigeration methods, such as the air cycle, thermoelectric cooling, and steam jet refrigeration are not widely used in commercial air conditioning and will not be discussed.

OBJECTIVES

After studying this chapter, you will be able to:

1. Describe and sketch the vapor compression refrigeration system.
2. Identify the types of compressors, condensers, evaporators, and flow control devices.
3. Select packaged refrigeration equipment.
4. Describe and sketch the absorption refrigeration system.
5. Describe and sketch the heat pump system.
6. Describe environmental effects of refrigerants.

Vapor Compression Refrigeration System

13.1 PRINCIPLES

A schematic flow diagram showing the basic components of the *vapor compression refrigeration* system is shown in Figure 13.1. To aid in understanding it, some typical temperatures for air conditioning applications are indicated. Refrigerant fluid circulates through the piping and equipment in the direction shown. There are four processes (changes in the condition of the fluid) that occur as it flows through the system:

PROCESS 1-2. *At point (1), the refrigerant is in the liquid state at a relatively high pressure and high temperature. It flows to (2) through a restriction, called the* flow control device *or* expansion device. *The refrigerant loses pressure going through the restriction. The pressure at (2) is so low that a small portion of the refrigerant* flashes *(vaporizes) into a gas. But in order to vaporize, it must gain heat (which it takes from that portion of the refrigerant that did not vaporize), thus cooling the mixture and resulting in a low temperature at (2).*

PROCESS 2-3. *The refrigerant flows through a heat exchanger called the* evaporator. *This heat exchanger has two circuits. The refrigerant circulates in one, and in the other, the fluid to be cooled (usually air or water) flows. The fluid to be cooled is at a slightly higher temperature than the refrigerant, therefore heat is transferred from it to the*

Figure 13.1
The vapor compression refrigeration system.

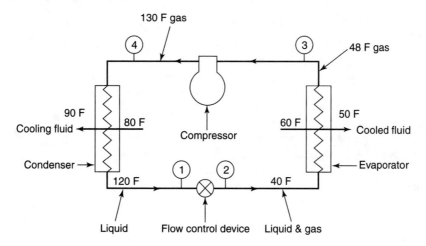

refrigerant, producing the cooling effect desired. The refrigerant boils because of the heat it receives in the evaporator. By the time it leaves the evaporator (4), it is completely vaporized.

PROCESS 3-4. *Leaving the evaporator, the refrigerant is a gas at a low temperature and low pressure. In order to be able to use it again to achieve the refrigerating effect continuously, it must be brought back to the conditions at (1)—a liquid at a high pressure. The first step in this process is to increase the pressure of the refrigerant gas by using a* compressor. *Compressing the gas also results in increasing its temperature.*

PROCESS 4-1. *The refrigerant leaves the compressor as a gas at high temperature and pressure. In order to change it to a liquid, heat must be removed from it. This is accomplished in a heat exchanger called the* condenser. *The refrigerant flows through one circuit in the condenser In the other circuit, a cooling fluid flows (air or water) at a temperature lower than the refrigerant. Heat therefore transfers from the refrigerant to the cooling fluid, and as a result, the refrigerant condenses to a liquid (1).*

The refrigerant has returned to its initial state and is now ready to repeat the cycle. Of course the processes are actually continuous as the refrigerant circulates through the system.

13.2   EQUIPMENT

As noted in the explanation of how the vapor compression refrigeration system functions, the major equipment components are the *compressor, evaporator, condenser,* and *flow control device,* some types of which will now be described. The complete refrigeration plant has many other additional components (e.g., valves, controls, piping) that will not be discussed in detail here.

13.3   EVAPORATORS

These may be classified into two types for air conditioning service—*dry expansion* (DX) evaporators or *flooded* evaporators. In the dry expansion type, refrigerant flows through tubing, and there is no liquid storage of refrigerant in the evaporator. In the flooded type of evaporator, a liquid pool of refrigerant is maintained.

Dry expansion (DX) evaporators exist in two types—DX *cooling coils* or DX *chillers.* Cooling coils are used for cooling air and chillers for cooling water or other liquids.

When cooling air, dry expansion (DX) cooling coils are used. The tubing is arranged in a serpentine coil form, and is finned to produce more heat transfer from a given length. The air flows across the coils. Cooling coils are discussed in more detail in Chapter 12.

Evaporators for cooling water or other liquids are called *chillers.* In the *shell and tube* type, a bundle of straight tubes is enclosed in a cylindrical shell. The chiller may be either the *flooded* type, with water circulating through the tubes and refrigerant through the shell (Figure 13.2), or *dry expansion,* with the reveres arrangement (Figure 13.3). The shell can be made in one piece or can be constructed with bolted removable ends, called *heads.* In the latter case, mechanical cleaning and replacement of individual tubes is possible. This construction is more expensive, however. Flooded chillers are generally used on the larger systems.

Figure 13.2
Flooded chiller. (Reprinted with permission from the *1979 Equipment ASHRAE Handbook & Products Directory.*)

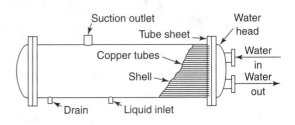

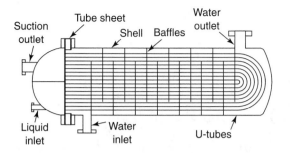

Figure 13.3
Dry expansion chiller. (Reprinted with permission from the *1979 Equipment ASHRAE Handbook & Products Directory*.)

13.4 TYPES OF COMPRESSORS

Positive displacement compressors function by reducing the volume of gas in the confined space, thereby raising its pressure. *Reciprocating, rotary,* and *screw* compressors are positive displacement types. *Centrifugal* compressors function by increasing the kinetic energy (velocity) of the gas, which is then converted to an increased pressure by reducing the velocity.

13.5 RECIPROCATING COMPRESSOR

This is the most widely used type, available in sizes from fractional horsepower and tonnage up to a few hundred tons. Construction is similar to the reciprocating engine of a vehicle, with *pistons, cylinders, valves, connecting rods,* and *crankshaft* (Figure 13.4). The suction and discharge valves are usually a thin *plate* or *reed* that will open and close easily and quickly.

Open compressors have an exposed shaft to which the electric motor or other driver is attached externally. *Hermetic* compressors are manufactured with both compressor and motor sealed in a

Figure 13.4
Reciprocating compressor construction. (Courtesy: Dunham-Bush, Inc.)

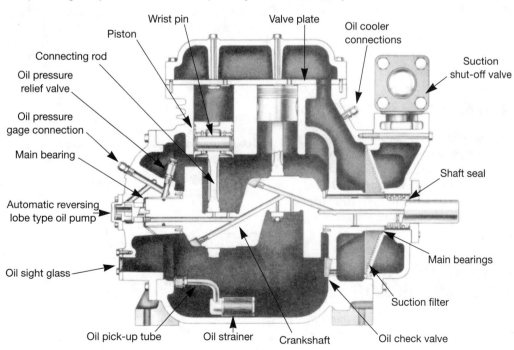

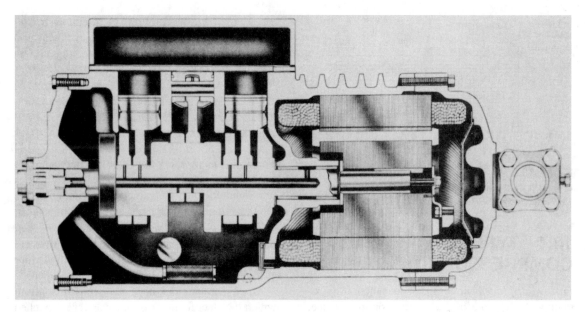

Figure 13.5
Cutaway view of hermetic reciprocating compressor. **(Courtesy: Dunham-Bush, Inc.)**

casing (Figure 13.5). In this way, there is no possibility of refrigerant loss from leaking around the shaft. The motor is cooled by refrigerant in a hermetic compressor. Most modern open compressors use *mechanical seals,* rather than *packing seals,* to reduce refrigerant leakage. These seals are similar to those used in pumps, as discussed in Chapter 11.

13.6   ROTARY COMPRESSOR

This type has a *rotor* eccentric to the casing; as the rotor turns it reduces the gas volume and increases its pressure (Figure 13.6). Advantages of this compressor are that it has few parts, is of simple construction, and can be relatively quiet and vibration-free. Small rotary compressors are often used in household refrigerators and window air conditioners.

13.7   SCREW (HELICAL ROTARY) COMPRESSOR

Two meshing helical shaped *screws* rotate and compress the gas as the volume between the screws de-

creases toward the discharge end. This type of compressor has become popular in recent years due to its reliability, efficiency, and cost. It is generally used in the larger size ranges of positive displacement compressors, in capacities up to about 1000 tons. A screw compressor is shown in Figure 13.7.

Figure 13.6
Sectional view of rotary compressor. **(Reprinted with permission from the** *1979 Equipment ASHRAE Handbook & Products Directory.***)**

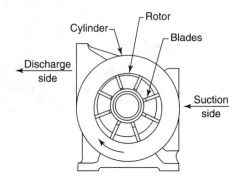

Figure 13.7
Cutaway view of screw compressor. (Courtesy: Dunham-Bush, Inc.)

13.8 SCROLL COMPRESSOR

This type of compressor has two spiral-shaped scrolls, one set inside the other. (These are each shaped somewhat like a pinwheel toy or a spinning spiral firework.)

One scroll rotates and the other is stationary. The refrigerant suction gas is drawn in from the perimeter. The volume decreases as the gas moves to the center, increasing its pressure, and the gas is then discharged.

The scroll compressor has a number of beneficial features. It has few moving parts. It has no suction or discharge valves. Its motion is rotary, reducing vibration. It has a high efficiency and low noise level. It is available as a hermetic compressor, in small and medium sizes.

13.9 CENTRIFUGAL COMPRESSOR

This type of compressor has vaned *impellers* rotating inside a casing, similar to a centrifugal pump. The impellers increase the velocity of the gas, which is then converted into a pressure increase by decreasing the velocity. The nature of the centrifugal compressor makes it suitable for very large capacities, up to 10,000 tons. The impellers can be rotated at speeds up to 20,000 RPM, enabling the compressor to handle large quantities of refrigerant. Hermetic centrifugal compressors as well as open compressors are available. Figure 13.8 shows a complete *hermetic centrifugal refrigeration water chiller*, with compressor, condenser, and evaporator.

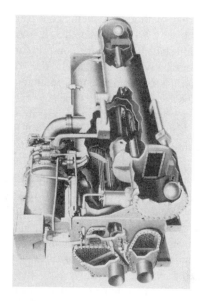

Figure 13.8
Hermetic centrifugal refrigeration water chiller. (Courtesy: Machinery & Systems Division, Carrier Corp., Syracuse, N.Y.)

13.10 CAPACITY CONTROL OF COMPRESSORS

The capacity of a compressor must be regulated to meet the load demand. Control is usually from a signal received from a thermostat or pressurestat (see Chapter 14). In a small reciprocating compressor, capacity is often changed simply by starting and stopping the compressor. In larger multicylinder compressors, a number of steps of capacity can be achieved. In one method, the refrigerant gas is *bypassed* around the compressor when less capacity is called for. This method requires a relatively high power input for low capacity. A more efficient load reduction method is accomplished by holding the suction valve open when a reduction in capacity is called for. The cylinder is then simply idling, and a significant reduction in power input results. Mechanical devices called *unloaders,* automatically controlled from a load signal, are used to open the suction valves.

A reliable method of reducing centrifugal compressor capacity is to use *inlet guide vanes*. This is a set of adjustable vanes in the compressor suction that are gradually closed to reduce the volume of refrigerant gas compressed, thus reducing the capacity. The use of inlet guide vanes lessens a problem of centrifugal compressor operation called *surging*. If the gas flow rate is reduced by throttling with a butterfly-type discharge damper, a point will be reached where instability occurs in which the gas is constantly surging back and forth through the compressor. This is a very serious event that could damage the machine. Inlet guide vanes avoid this by curving the flow direction of the gas in an efficient manner that permits capacity reduction down to about 15% of full load without surging.

For centrifugal compressors that are driven by variable speed prime movers, speed reduction is a convenient method of capacity reduction. Both inlet guide vane and speed control are relatively efficient methods of capacity control, the power input decreasing considerably with capacity. Below about 50% capacity, however, the efficiency falls off rapidly. This is one reason why it is desirable to use multiple centrifugal machines in an installation, if practical.

13.11   PRIME MOVERS

Compressors can be driven by electric motors, reciprocating engines, or by steam or gas turbines. Electric motors are most commonly used because of the convenience and simplicity. However, on very large installations, particularly with centrifugal compressors, steam or gas turbines are often used. The high rotating speed of the turbine often matches that of the compressor, whereas expensive speed-increasing gears may be needed when motors are used. The relative energy costs of electricity, steam, or gas often determine which prime mover will be used. In the Middle East, natural gas from the well (which might otherwise be wasted) is often used in gas turbines that drive large centrifugal machines.

13.12   CONDENSERS

The condenser rejects from the system the energy gained in the evaporator and the compressor. Atmospheric air or water are the two most convenient heat sinks to which the heat can be rejected.

In the *air-cooled condenser* (Figure 13.9), the refrigerant circulates through a coil and air flows across the outside of the tubing. The air motion may be caused by natural convection effects when the air is heated, or the condenser can include a fan to increase the air flow rate, resulting in greater capacity. Air-cooled condensers are normally installed outdoors. They are available in sizes up to about 50 tons.

Water-cooled condensers are usually of shell and tube construction, similar to shell and tube evaporators. Water from lakes, rivers, or wells is sometimes used when available. Usually, however, natural sources of water are not sufficient, and the water must be recirculated through a *cooling tower* to recool it.

Evaporative condensers (Figure 13.10) reject heat to the atmosphere as do air cooled condensers, but by spraying water on the coils some heat is transferred to the water as well as the air, increasing the capacity of the condenser. A pump, piping, spray nozzles, and collection sump are required for the water circulating system. Fans are used to force

Figure 13.9
Air-cooled condenser. (**Courtesy:** Dunham-Bush, Inc.)

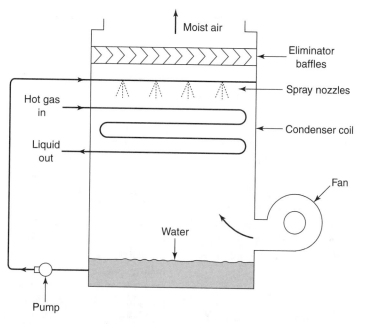

Figure 13.10
Evaporative condenser.

the air through the unit. Evaporative condensers can be installed indoors as well as outdoors by using ductwork to discharge the exhaust air outside.

The capacity of condensers must be controlled to maintain the condensing pressure within certain limits. Higher condensing pressures result in more power use, and extremely high pressures can damage the equipment. On the other hand, if the pressure is too low, the flow control device will not operate satisfactorily. An automatic valve regulating water flow rate is a convenient way of controlling capacity of water-cooled condensers. Air-cooled condensers are often controlled by reducing airflow across the coils, through use of dampers or cycling the fan. The control is usually in response to a change in condensing pressure.

Proper water treatment is important for maintaining the capacity of water-cooled condensers. Manufacturers rate water-cooled condenser and chiller capacity on the basis of a water *fouling factor*—a number that represents the thermal resis-

tance of the water film on the tubes. A value of 0.0005 is considered clean water, and ratings are often based on this value. The water treatment should prevent formation of scale that will increase the thermal resistance, resulting in a decrease in refrigeration capacity and an increase in energy required.

13.13 FLOW CONTROL DEVICES

The restricting device that causes the pressure drop of the refrigerant also regulates the refrigerant flow according to the load. Some of the devices available are the *capillary tube, thermostatic expansion valve,* and the *low side float valve.* The first two are used with dry expansion evaporators; the low side float valve is used in flooded chiller evaporators.

The *capillary tube* is a very small diameter tube of considerable length, which thus causes the required pressure drop. It is used often in small units (e.g., domestic refrigerators and window air conditioners) because of its low cost and simplicity.

The *thermostatic expansion valve* (TEV), shown in Figure 13.11, is widely used in dry expansion systems. The small opening between the valve seat and disc results in the required pressure drop. It also does an excellent job of regulating flow according to the need. The operation of a TEV is shown in Figure 13.12. A bulb filled with a fluid is strapped to the suction line and thus senses the suction gas temperature. This bulb is connected to the valve by a tube in a manner so that the pressure of the fluid in the bulb tends to open the valve more, against a closing spring pressure. If the load in the system increases, the refrigerant in the evaporator picks up more heat and the suction gas temperature rises.

Figure 13.11
Cutaway view of thermostatic expansion valve, internally equalized type. (Courtesy: Sporlan Valve Co.)

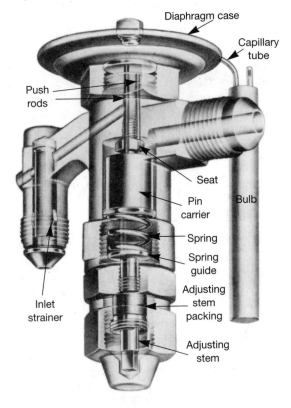

The pressure of the fluid in the bulb increases as its temperature rises, and it opens the valve more. This increases the refrigerant flow needed to handle the increased load. The reverse of all these events occurs when the refrigeration load decreases.

It is important that the refrigerant vapor leaving the evaporator be a few degrees above the saturation temperature (called *superheat*) to ensure that no liquid enters the compressor, which might result in its damage. This is achieved by adjusting the spring pressure to a value that prevents the bulb pressure from opening the valve more unless the gas leaving the evaporator is superheated.

The *internally equalized* TEV has a port connecting the underside of the diaphragm chamber to the valve outlet (Figures 13.11 and 13.12). This neutralizes the effect of any change in evaporator pressure on the balance between spring and bulb pressure. If there is a larger pressure drop in the evaporator, however, this would result in a reduction in superheat. This problem is solved by using an *externally equalized* valve which has a connection to the evaporator outlet rather than the inlet.

A *low side float valve,* as shown in Figure 13.8, is a flow control device that is used with flooded chillers. If too much liquid refrigerant accumulates because flow is not adequate, the float rises and a connecting linkage opens the valve, allowing more flow.

13.14 SAFETY CONTROLS

All refrigeration systems include a number of safety control devises to protect the equipment. The devices required for each system must be determined in each case according to the need. A brief listing of some of the available safety control devices follows.

A **high pressure cut-out** stops the compressor when the refrigerant discharge pressure exceeds a safe limit.

A **low pressure cut-out** stops the compressor when the refrigerant suction pressure is below a safe limit. Usually this is intended as a temperature safety device. The pressure setting on the

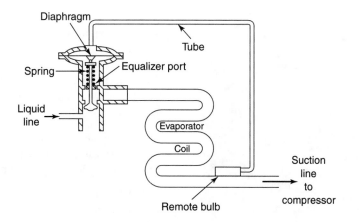

Figure 13.12
Operation of internally equalized thermostatic expansion valve.

device corresponds to a temperature at which water freeze-up might occur.

- A **low temperature cut-out** senses refrigerant temperature on the low side directly and serves to protect against freeze-up.
- A **low oil pressure cut-out** stops the compressor when lubricating oil pressure is inadequate.
- A **flow switch** will stop the compressor when chilled water (or condenser water) flow is inadequate.

When the compressor stops in response to a thermostat, refrigerant may continue to flow to the evaporator due to a vapor pressure difference between the condenser and evaporator. It is not desirable to have the evaporator filled with liquid refrigerant during shutdown because this increases the likelihood of liquid entering the compressor. It also increases the amount of refrigerant absorbed in the crank case oil, thus reducing the lubricating effectiveness of the oil. The problem is solved by using *pump down control.* Instead of having a thermostat control the compressor operation, it controls a *solenoid valve* in the liquid line. This cuts off flow to the compressor. The compressor continues to operate for a time, pumping out the refrigerant from the evaporator. The compressor stops when its low pressure cut-out setting is reached.

13.15 PACKAGED REFRIGERATION EQUIPMENT

Compressors, condensers, evaporators, and accessories are each available separately from manufacturers for selection, purchase, and installation. However, these components may also be available already assembled (packaged) in the factory. There are a number of advantages of using packaged equipment. The components are selected and matched in capacity by the manufacturer, so that they will perform properly together. Installation costs are reduced, as each component does not have to be installed and aligned separately. Controls and interconnecting piping are factory installed, further reducing field costs. The assembled equipment is usually factory tested, reducing the likelihood of operating problems that would have to be corrected on the job. Packaged equipment is available in various combinations, some of which are mentioned below.

Condensing Units

The package of compressor and condenser with interconnecting piping and controls is called a *condensing unit* (Figure 13.13 on page 340). Both water- and air-cooled condensing units are available. Air-cooled units are installed outdoors. When

Figure 13.13
Water-cooled condensing unit. (Courtesy: Dunham-Bush, Inc.)

the air-cooled condensing unit is located outdoors and the air handling unit and evaporator coil are located indoors, the air conditioning system is called a *split system.* This arrangement is popular for residential air conditioning systems.

Compressor-Chiller Unit

This unit consists of compressor, water chiller, interconnecting piping, and controls. It is often used with a remote air-cooled condenser.

Packaged Chiller

This unit, shown in Figure 13.14, contains the complete refrigeration package: compressor, condenser, water chiller, piping and controls, ready to operate when put in place and when external connections are made.

13.16   SELECTION

Refrigeration equipment is selected from manufacturers' ratings after performance requirements are determined. Although the compressor, condenser, and evaporator can be chosen separately, one of the packaged combinations is often used, so we will limit our explanation to these. In any case, selection of individual components is a similar process.

Air-Cooled Condensing Unit

The following data are needed for selection of an air-cooled condensing unit:

1. Refrigeration capacity required (load)
2. Condenser ambient temperature
3. Saturated suction gas temperature

The load is a result of the cooling load calculations. The condenser ambient temperature is usually the outdoor design temperature in summer. Sometimes the condenser is located where the ambient temperature may be even greater than design temperature, and this should be considered. The compressor saturated suction temperature will be equal to the evaporator temperature minus an allowance. This allowance accounts for the pressure drop in the suction line, expressed as an "equivalent temperature drop," usually 2 F. Table 13.1 on pages 342–343 is an example of air-cooled condensing unit ratings.

Example 13.1 _____

The refrigeration load for the air conditioning system of a branch of the Big Bank in Tulsa, Oklahoma, is 12 tons. The system uses refrigerant R-22, evaporating at 42 F. Select an air-cooled condensing unit.

Figure 13.14
Packaged chiller. (Courtesy: Dunham-Bush, Inc.)

Solution

A unit will be selected from Table 13.1. Required capacity = 12 tons. Ambient temperature = 98 F (Table A.6). Allow a friction loss in the suction line equivalent to 2 F. Therefore, the saturated suction temperature = 42 − 2 = 40 F.

From Table 13.1, the unit selected is a Model RCU-0155SS air-cooled condensing unit. Capacity = 12.4 tons at saturated suction temperature = 40 F, ambient temperature = 100 F. Power input = 14.6 KW. (The capacity at 98 F will be slightly higher than at 100 F ambient, as noted from the table.)

Packaged Water Chiller

The following data are needed for selection of a packaged water chiller:

1. Refrigeration load
2. Condenser water temperature leaving unit
3. Condenser water temperature rise
4. Chilled water temperature leaving unit
5. Chilled water temperature drop
6. Fouling factor

The load is determined from cooling load calculations. Condenser water leaving temperature is usually about 5–15 F above ambient wet bulb temperature if a cooling tower is used. The water temperature rise is usually selected between 8–15 F. The leaving chilled water temperature will depend on the cooling coil selection for the air handling equipment. Chilled water temperature ranges of 8–12 F are common.

No exact figures on temperature changes are given because many choices are possible. The designer must frequently try different combinations of values, selecting the equipment each time, to find which will result in the best choice. Computer programs for equipment selection provided by the manufacturer are very useful for this task. Table 13.2 on pages 344–345 is an example of packaged water chiller ratings.

Example 13.2

A package water chiller is required for the air conditioning system of the Royal Arms Apartments.

The load is 27 tons. Chilled water is cooled from 55 F to 45 F. Condenser water enters at 85 F and leaves at 95 F. The condenser and chilled water fouling factors are 0.0005. Select a suitable unit.

Solution

A unit will be selected from Table 13.2. The fouling factor is a number that describes the cleanliness of the water. The size of the condenser required will depend on this. Table 13.2 is based on a water fouling factor of 0.0005, as noted, so no correction for this will be necessary. If the fouling factor is different, tables from the manufacturer show corrections to the selection. The required conditions are:

Capacity = 27 tons

Leaving chilled water temperature = 45 F

Chilled water temperature drop = 10 F

Entering condenser water temperature = 85 F

Condenser water temperature rise = 10 F

The unit chosen for these requirements is a Model PCW 030T water chiller, which has a capacity of 28.1 tons at required conditions. Power input is 25.1 KW.

Note that Table 13.2 indicates both a reduction in refrigeration capacity and an increase in power required, if the fouling factor of the condenser increases to 0.001. There is a loss of 2% in refrigeration and an increase of 3% in power, resulting in a net increase of 5% in energy use for a given capacity. The fouling factor number reflects the effect of dirt on the heat transfer surface. This points up the importance of maintaining a clean condenser to conserve energy.

Additional information about the equipment is usually required, such as dimensions and weight, and water pressure drop through the chiller and condenser. This can be found in the manufacturer's catalog.

13.17 ENERGY EFFICIENCY

When selecting refrigeration equipment, it will often be found that more than one unit will have

TABLE 13.1   AIR-COOLED CONDENSING UNIT RATINGS

Capacity Data* (60 Hz.)**
Condensing Units-R22

Model	Suction Temp °F	Ambient Temperature °F												EER @ ARI Base Rating Cond.
		90°F		95°F		100°F		105°F		110°F		115°F		
		Tons	K.W.	Tons	K.W.	Tons	K.W.	Tons	K.W.	Tons	K.W.	Tons	K.W.	
	30	5.8	7.4	5.6	7.5	5.3	7.6	5.1	7.7	4.7	7.8	4.6	7.9	
RCU-008S	35	6.6	7.8	6.5	8.0	6.1	8.1	5.7	8.2	5.5	8.3	5.3	8.5	11.2
	40	7.4	8.2	7.0	8.4	6.8	8.6	6.6	8.7	6.3	8.9	5.9	9.2	
	45	**8.0**	**8.6**	7.9	8.9	7.6	9.1	7.4	9.3	7.0	9.6	6.7	9.8	
	30	6.4	7.0	6.2	7.1	5.8	7.2	5.6	7.4	5.5	7.6	5.2	7.7	
RCU-008SS†	35	7.0	7.4	6.8	7.6	6.6	7.8	6.4	7.9	6.2	8.0	5.8	8.2	12.4
	40	7.7	7.7	7.5	7.9	7.0	8.1	6.9	8.3	6.7	8.5	6.5	8.7	
	45	**8.4**	**8.1**	8.0	8.4	7.8	8.6	7.6	8.8	7.4	8.9	7.0	9.1	
	30	7.7	8.4	7.5	8.6	7.3	8.7	6.9	8.9	6.7	9.1	6.4	9.3	
RCU-010SS†	35	8.5	8.9	8.3	9.1	8.0	9.2	7.8	9.4	7.5	9.6	7.3	9.8	12.2
	40	9.2	9.3	9.0	9.5	8.8	9.8	8.6	10.0	8.3	10.2	8.0	10.4	
	45	**10.0**	**9.8**	9.8	10.1	9.6	10.3	9.4	10.5	9.0	10.8	8.8	11.0	
	30	8.5	11.5	8.2	11.7	7.9	11.9	7.5	12.0	7.2	12.1	7.0	12.2	
RCU-010T	35	9.4	12.4	9.1	12.6	8.7	12.8	8.4	13.0	8.0	13.1	7.8	13.2	9.7
	40	10.3	13.2	10.0	13.4	9.6	13.6	9.3	13.8	8.9	13.9	8.6	14.1	
	45	**11.3**	**14.0**	10.9	14.2	10.5	14.6	10.2	14.7	9.9	14.8	9.5	15.1	

the capacity needed. In this case, it is useful to know which will give the "best" performance. The most desirable choice is the unit that would produce the most refrigeration with the lowest power input. This can be measured by a performance factor called the *Coefficient of Performance* (COP). The COP is defined as:

$$COP = \frac{\text{refrigeration capacity}}{\text{equivalent power input to compressor}}$$

The higher the COP of a refrigeration unit, the less power is required for a given refrigeration requirement. The COP is thus a useful figure in comparing equipment to minimize energy consumption.

TABLE 13.1 (Continued)

Capacity Data* (60 Hz.)**
Condensing Units-R22

Model	Suction Temp °F	Ambient Temperature °F												EER @ ARI Base Rating Cond.
		90°F		95°F		100°F		105°F		110°F		115°F		
		Tons	K.W.	Tons	K.W.	Tons	K.W.	Tons	K.W.	Tons	K.W.	Tons	K.W.	
RCU-015SS†	30	10.9	12.6	10.6	12.8	10.2	13.0	9.7	13.4	9.5	13.6	9.4	13.8	
	35	12.1	13.2	11.8	13.7	11.4	13.9	11.1	14.2	10.8	14.6	10.5	14.8	11.8
	40	13.3	14.1	12.8	14.2	12.4	14.6	12.0	14.9	11.9	15.4	11.4	15.6	
	45	**14.5**	**14.8**	14.5	15.2	13.8	15.4	13.3	16.0	12.9	16.2	12.5	16.4	
RCU-015T	30	11.2	14.8	10.8	15.1	10.2	15.4	9.7	15.5	9.2	15.7	8.8	15.9	
	35	12.7	15.7	12.1	16.1	11.6	16.4	11.1	16.7	10.7	17.0	10.1	17.2	10.3
	40	14.1	16.8	13.5	17.2	13.0	17.7	12.4	17.9	11.9	18.2	11.3	18.6	
	45	**15.5**	**18.0**	14.9	18.4	14.3	18.8	13.8	19.2	13.3	19.6	12.8	19.9	
RCU-020T	30	14.5	16.8	14.1	17.1	13.5	17.5	13.0	17.9	12.4	18.3	11.9	18.7	
	35	16.1	17.8	15.6	18.1	15.0	18.5	14.5	19.0	14.0	19.2	13.4	19.6	11.6
	40	17.4	18.5	16.9	19.0	16.3	19.5	15.8	20.0	15.4	20.3	14.6	20.4	
	45	**19.0**	**19.7**	18.5	21.0	17.8	20.6	17.3	21.0	16.8	21.4	16.4	21.9	
RCU-020SS†	30	15.7	19.1	15.2	19.5	14.9	19.9	14.3	20.2	14.0	20.5	13.4	21.0	
	35	17.4	20.0	16.8	20.5	16.4	20.9	15.8	21.4	15.5	21.9	14.9	22.4	11.1
	40	19.0	21.3	18.5	21.7	17.9	22.4	17.4	22.8	17.1	23.2	16.3	23.7	
	45	**20.7**	**22.4**	20.1	22.8	19.6	23.5	19.0	24.1	18.5	24.5	17.8	25.0	

Notes: ARI Base rating conditions 90° ambient, 45° suction temperature. These are indicated by boldface type.
† All models with the suffix 'SS' denotes single D/B-Metric accessible Hermetic compressors.
* For capacity ratings at 85°F ambient temperature, multiply the ratings of 90°F ambient by 1.03 × Tons and .97 × KW.
** For 50 hertz capacity ratings, derate above table by .85 multiplier.
(Courtesy of Dunham-Bush, Inc.)

This will be discussed in more detail in Chapter 15, together with another efficiency measure called the *Energy Efficiency Ratio* (EER).

Example 13.3 _____

Determine the coefficient of performance for the package chiller of Example 13.2.

TABLE 13.2   PACKAGED WATER CHILLER RATINGS

| | Lvg. Chilled Water Temp. °F | Condenser Entering Water Temp. °F | | | | | | | | | | | | | | |
| | | 75° | | | 80° | | | 85° | | | 90° | | | 95° | | |
Model		Cap. Tons	KW	Cond. GPM	Cap. Tons	KW	Cond. GPM	Cap. Tons	KW	Cond. GPM	Cap. Tons	KW	Cond. GPM	Cap. Tons	KW	Cond. GPM
	42	10.3	8.2	30.0	10.1	8.5	29.7	9.9	8.8	30.3	9.7	9.2	29.6	9.5	9.6	28.9
	44	10.7	8.3	30.9	10.5	8.6	30.7	**10.3**	**9.0**	**30.5**	10.0	9.3	29.4	9.9	9.7	29.4
PCW010T	45	10.9	8.3	31.8	10.7	8.7	31.3	10.5	9.0	31.0	10.2	9.3	30.0	10.1	9.7	29.4
	46	11.1	8.9	32.0	10.9	8.7	31.8	10.7	9.1	31.5	10.4	9.4	30.8	10.2	9.8	29.7
	48	11.5	8.5	33.1	11.3	8.8	32.8	11.1	9.3	32.6	10.8	9.6	31.6	10.6	9.9	30.9
	50	11.9	8.6	34.5	11.7	9.0	34.0	11.5	9.4	33.6	11.2	9.7	32.7	10.9	10.0	32.0
	42	15.3	13.0	43.5	14.9	13.4	42.4	14.5	13.8	41.7	14.1	14.2	40.5	13.6	14.6	39.3
	44	15.9	13.1	45.0	15.5	13.5	44.0	**15.0**	**14.0**	**43.4**	14.7	14.3	41.1	14.2	14.9	40.7
PCW015T	45	16.2	13.2	45.8	15.8	13.6	44.8	15.4	14.0	44.1	15.0	14.5	42.6	14.5	15.0	41.2
	46	16.5	13.3	46.3	16.1	13.7	45.6	15.7	14.1	44.8	15.3	14.6	43.4	14.8	15.1	42.0
	48	17.1	13.4	47.3	16.7	13.8	46.6	16.3	14.2	46.2	15.8	14.7	44.7	15.4	15.3	43.6
	50	17.6	13.6	48.1	17.3	14.0	47.8	16.9	14.4	47.2	16.4	14.9	45.7	15.9	15.5	44.7
	42	19.2	15.8	52.9	18.6	16.3	52.3	18.0	16.8	51.6	17.4	17.3	49.8	16.7	17.8	48.5
	44	19.9	16.0	54.9	19.3	16.5	53.7	**18.6**	**17.0**	**52.6**	18.0	17.5	50.7	17.4	18.0	49.6
PCW020T	45	20.3	16.1	56.1	19.7	16.6	54.9	19.0	17.1	53.6	18.4	17.6	51.9	17.7	18.1	50.6
	46	20.6	16.1	56.5	20.0	16.7	55.3	19.3	17.2	54.1	18.7	17.7	52.6	18.0	18.2	51.7
	48	21.3	16.3	57.4	20.7	16.8	56.8	20.0	17.3	56.2	19.4	17.8	54.8	18.7	18.3	53.9
	50	22.0	16.4	58.9	21.3	16.9	58.6	20.6	17.4	58.4	20.0	17.9	57.0	19.3	18.5	56.2

Solution

$$\text{Capacity} = 28.1 \text{ tons} \times 12,000 \ \frac{\text{BTU/hr}}{\text{ton}}$$

$$= 337,200 \text{ BTU/hr}$$

$$\text{Power input} = 25.1 \text{ KW} \times 3410 \ \frac{\text{BTU/hr}}{\text{KW}}$$

$$= 85,590 \text{ BTU/hr}$$

Using the equation for the COP,

$$\text{COP} = \frac{\text{refrigeration capacity}}{\text{equivalent power input}}$$

$$= \frac{337,200 \text{ BTU/hr}}{85,590 \text{ BTU/hr}}$$

$$= 3.94$$

The COP found could be compared with values obtained for other possible selections, to see if an improved performance is possible without sacrificing other benefits.

TABLE 13.2 (Continued)

Model	Lvg. Chilled Water Temp. °F	Condenser Entering Water Temp. °F														
		75°			80°			85°			90°			95°		
		Cap. Tons	KW	Cond. GPM	Cap. Tons	KW	Cond. GPM	Cap. Tons	KW	Cond. GPM	Cap. Tons	KW	Cond. GPM	Cap. Tons	KW	Cond. GPM
	42	24.1	19.7	67.9	23.0	20.3	66.5	22.9	20.9	65.1	22.3	21.4	63.1	21.6	22.0	61.7
	44	25.0	19.8	70.7	24.4	20.5	69.3	**23.7**	**21.0**	**68.2**	23.0	21.7	66.0	22.4	22.3	64.9
PCW025T	45	25.5	19.9	71.8	24.8	20.6	70.5	24.1	21.2	69.1	23.4	21.8	67.2	22.8	22.4	65.9
	46	25.9	20.0	72.8	25.2	20.7	71.7	24.5	21.5	70.4	23.8	21.9	68.5	23.1	22.6	66.9
	48	26.7	20.2	75.1	26.0	20.9	73.8	25.3	21.7	72.5	24.7	22.2	70.6	24.0	22.9	69.1
	50	27.5	20.4	78.6	26.8	21.1	77.1	26.1	21.9	75.2	25.5	22.5	73.1	24.8	23.2	71.6
	42	28.3	23.2	80.0	27.5	23.9	79.6	26.6	24.6	79.2	25.9	25.3	77.0	25.1	26.0	75.5
	44	29.3	23.4	83.2	28.4	24.2	82.6	**27.6**	**25.0**	**82.5**	26.7	25.7	79.8	26.0	26.4	78.3
PCW030T	45	29.8	23.6	85.3	29.0	24.3	84.4	28.1	25.1	84.0	27.3	25.9	81.6	26.5	26.6	79.9
	46	30.3	23.7	86.3	29.7	24.5	85.2	28.5	25.2	85.5	27.8	26.1	83.1	26.9	26.9	81.5
	48	31.2	23.9	88.2	30.5	24.8	87.9	29.6	25.6	87.6	28.7	26.5	85.2	27.8	27.3	83.6
	50	32.1	24.2	88.7	31.3	25.1	90.2	30.6	26.0	88.5	29.7	26.9	86.1	28.7	27.8	85.8
	42	41.6	35.0	116.1	40.0	36.0	113.8	38.9	37.2	111.3	37.5	38.4	109.3	36.4	39.6	108.4
	44	43.0	35.4	118.8	41.7	36.5	117.1	**40.4**	**37.6**	**115.4**	39.0	38.9	112.8	38.0	40.2	112.0
PCW040T	45	43.7	35.7	120.9	42.4	36.8	119.2	41.1	37.9	117.4	39.7	39.2	115.2	38.6	40.4	114.1
	46	44.4	36.0	123.1	43.1	37.1	121.1	41.8	38.0	118.8	40.4	39.4	116.6	39.2	40.6	115.5
	48	45.9	36.6	127.0	44.6	37.8	124.8	43.3	39.0	122.5	41.9	40.2	120.1	40.6	41.3	118.8
	50	47.3	37.1	131.1	46.0	38.4	128.9	44.8	39.6	126.3	43.4	40.8	123.8	42.0	42.0	122.6

*Boldface type indicates ARI rating condition.

Notes:

1. Ratings are based on 10°F chilled water temperature range. Ratings are applicable for 6° to 14° range.

2. Condenser water flow rate data are based on tower water with a 10° rise. See Selection Procedure (page 6) for note regarding city water application.

3. Ratings are based on .0005 fouling factor in the chiller and condenser. For .001 condenser fouling factor, multiply capacity shown in ratings by .98 and kW by 1.03. For other fouling factor ratings, consult factory.

4. 50 Hz. units are full capacity except PCW040T which is 5/6 capacity.

5. Direct interpolation for conditions between ratings is permissible. Do not extrapolate.

(Courtesy of Dunham-Bush, Inc.)

13.18 INSTALLATION OF REFRIGERATION CHILLERS

The procedures for installing specific refrigeration equipment are furnished by the manufacturer. It is not the intent here to either repeat or supersede such instructions. Each manufacturer and each piece of equipment has individual features that require detailed installation instructions. These instructions are often very lengthy. If they were repeated here, they would soon be forgotten by the student who does not regularly carry out these procedures. Therefore, we will discuss here some general points of installation practice that apply to most situations.

1. Check machine for damage or refrigerant leaks (leak detecting devices and their use are described in refrigeration service manuals).
2. Locate chillers and condensers with removable tubes to provide adequate clearance on one end to allow removal of the tube bundle and on the other for removal of water box heads.
3. Allow clearance on all sides of equipment for comfortable maintenance (3 or 4 ft minimum).
4. Allow adequate clearance in front of control panels for operation and good visibility.
5. Provide vibration isolation supports under compressors and prime movers. Rubber, cork, and springs are some types available. Consult the manufacturer for the proper choice.
6. Install anchor bolts in floor or base and anchor machine. If there is any doubt about whether the floor is adequately strong for the machine or whether a special base is needed, a structural engineer must be consulted.
7. Make water, electrical, and control connections so as not to block access to the machine.

13.19 COOLING TOWERS

Operation

When water cooled condensers are used in the refrigeration plant, a steady supply of cooling water must be made available. For reasons already explained, natural sources of water are usually limited. In this case, we must arrange to cool the heated water after it passes through the condenser, and then return it to the condenser.

The *cooling tower* is the equipment that accomplishes this. It transfers heat from the condenser water to the atmospheric air (Figure 13.15). Most of the heat transfer is accomplished by the evaporation of a small percentage of the condensing water into the atmosphere. The heat required for evaporation is taken from the bulk of condenser water, thus cooling it. Water from the condenser is pumped to the top of the cooling tower and sprayed down into the tower. The tower has internal baffles called *fill*, which break up the water into finer droplets when the water splashes onto the fill. This improves the heat transfer. The cooled water collects in a basin and is then recirculated to the condenser.

In addition to the water lost due to the evaporative cooling, there are two other causes of water loss. *Drift* loss results from wind carrying water away with the air. *Blowdown* loss results from draining off and discarding a small portion of the water from the basin. This must be done at regular intervals in order to prevent a continual accumulation of minerals that would otherwise occur from

Figure 13.15
Induced draft cooling tower. (Courtesy: The Marley Cooling Tower Co.)

the evaporation and drift losses. The losses require provision for *makeup* water. This is done by providing a makeup water supply to the basin, controlled by a float valve level.

Types and Construction

The *atmospheric tower* is a type of tower where the air circulation results from air being warmed in the tower and thereby rising from natural convection. The amount of air that will circulate from this effect is quite limited, and atmospheric towers are not often used today. *Mechanical draft* towers use fans to create a high air flow rate. The *induced draft* fan type has the fan located at the tower out-

let, whereas the *forced draft* fan type blows the air through (Figure 13.16).

When the air and water move in opposite directions, the tower is called a *counterflow* type. When the air and water move at right angles to each other, the tower is called a *crossflow* type. Figure 13.17 show this difference. There is not necessarily an operating advantage in practice of one type over another. However, sometimes a crossflow tower will be lower in height (although bigger in length or width) than a counterflow tower for the same capacity. Lower height may be preferable when installed on a roof.

The tower *siding* may be wood, galvanized steel, or plastic. The structural framework may be

Figure 13.16
Forced and induced draft fan arrangements for cooling tower.

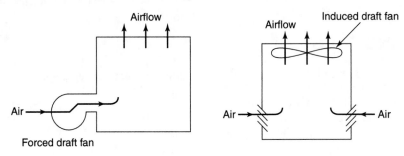

Figure 13.17
Counterflow versus crossflow of air and water in cooling tower.

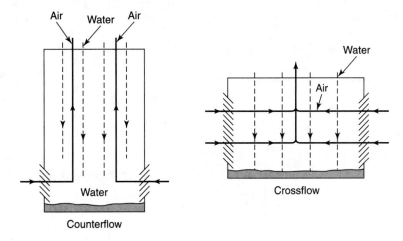

wood or steel. Redwood is ideal because water will not cause its deterioration. The *fill* may be wood, metal, or plastic. A row of baffles called *eliminators* are provided near the tower outlet to catch and prevent excessive loss of water droplets.

The capacity of a cooling tower depends on the rate of water evaporation. This rate decreases with higher water vapor content (humidity) in the ambient air. Therefore, the higher the ambient wet bulb temperature, the less the capacity of the tower.

Absorption Refrigeration System

Absorption refrigeration machines are often used for large air conditioning systems. The absence of a compressor usually has the advantages of less vibration, noise, and weight than with a vapor compression machine.

13.20 PRINCIPLES

The absorption system uses the principle that some gases will be absorbed by certain other substances. There are many pairs of substances that have this affinity for one another. We are all aware of how table salt absorbs water vapor from the air, thus making it difficult to pour. Another combination is *lithium bromide* (LiBr) and water; lithium bromide will absorb large quantities of water. Because this pair is used in many absorption systems, we will refer to them in our explanation.

Consider a tank partially filled with a *concentrated* liquid solution of lithium bromide (concentrated means that it contains very little water) as shown in Figure 13.18. The space above the liquid is evacuated of any gas, as much as possible, leaving a very low pressure. Water is then sprayed into the tank. Because of the low pressure, some of the water will evaporate, requiring heat to do so. A coil circulating water is located under the evaporating sprays. This water furnishes the heat needed for the evaporating spray, and is thereby chilled. The temperature at which the spray water evaporates will depend on the pressure in the tank, according to the saturation pressure-temperature relations of water.

Example 13.4
The pressure maintained in the evaporator of a LiBr–water absorption refrigeration machine is 0.147 psia. What is the refrigerant evaporating temperature?

Solution
Water is the refrigerant. From Table A.3, the evaporating (saturation) temperature of water at 0.147 psia is 45 F.

The lithium bromide absorbs the water as both solutions make contact. The lithium bromide eventually absorbs all of the water it can hold, however, and no longer is effective. The water vapor quantity will build up in the tank, raising its pressure, and therefore increasing its evaporating temperature above useful refrigeration temperatures.

Figure 13.18
Diagram illustrating refrigeration by absorption.

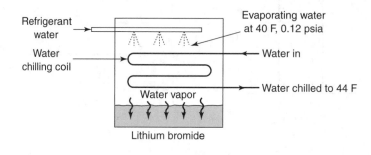

In order to have a practical absorption refrigeration system that will operate continuously, the diluted solution of lithium bromide must be reconcentrated and used again. We will now explain how the actual system functions, as shown in Figure 13.19. Typical operating temperatures and pressures are indicated on the diagram.

The *evaporator* operation is as described previously. Spray water (the refrigerant) evaporates in a tank where the pressure is very low, thus extracting heat from water circulating in a coil. The water that is chilled in the coil is distributed to air conditioning equipment as required. The spray water does not all evaporate, so the liquid water is recirculated by the *refrigerant pump.*

To prevent the pressure from building up in the evaporator, the water vapor must be absorbed by lithium bromide. A concentrated solution is stored in a tank called the *absorber.* This solution is sprayed into the absorber and recirculated by the

absorber pump. The lithium bromide absorbs and draws water vapor from the evaporator space and a low pressure is maintained there. However, the solution gradually becomes too diluted to absorb enough water. To solve this problem, diluted solution is pumped to the *concentrator* (also called *generator*) by a *concentrator pump.* Here it is heated to a temperature that will evaporate some of the water, which has a lower boiling point than the lithium bromide. The reconcentrated solution is then returned to the absorber. Steam, hot water, or a gas flame is used as a source of heat in the concentrator.

The water vapor from the concentrator flows to the *condenser,* where it is condensed to a liquid by giving up heat to water from a cooling tower or natural body of water. The condensed water is then returned to the evaporator, completing the cycle.

Two refinements to this cycle, which improve the system's efficiency and are shown in Figure

Figure 13.19

Flow diagram of lithium bromide–water absorption refrigeration system.

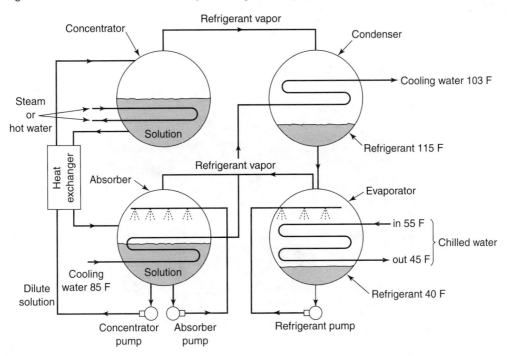

13.19, require explanation. The absorption process generates heat that would raise the temperature of the absorbing solution, making it less effective. The heat is removed by cooling water, which is circulated through a coil in the absorber. The same cooling water is then used in the condenser, as shown.

The second refinement to the cycle is the inclusion of a heat exchanger between the absorber and concentrator. The solution from the absorber is preheated by hot solution returning from the concentrator, thereby saving some of the heat needed in the concentrator.

Another pair of fluids often used in absorption systems is ammonia and water. In this case, water is the absorbent and ammonia is the refrigerant. Because of its volatility, some water boils off with the ammonia in the generator of the aqua-ammonia absorption system. This requires additional equipment (a *rectifier*) to separate the ammonia from the water. Another disadvantage of this system is that it operates at much higher pressures in the generator (about 300 psia), compared with about 30 psia for the LiBr system.

Small capacity lithium bromide–water absorption units (3–25 tons) with direct-fired generators are also available. They are popular in areas where natural gas is plentiful and inexpensive.

13.21 CONSTRUCTION AND PERFORMANCE

Absorption refrigeration machines are not actually constructed with four separate vessels as was shown in Figure 13.19. (The sketch was made that way for clarity). To economize on construction costs, the four parts are built into two or even one shell, as shown in Figure 13.20.

The machine is completely factory assembled, including *evaporator, absorber, concentrator, condenser,* and *solution pumps,* interconnecting piping, and electric controls. The pumps are hermetic to prevent any leaks into the system. The machine

Figure 13.20
Absorption refrigeration machine. (Courtesy: Machinery & Systems Division, Carrier Corp., Syracuse, N.Y.)

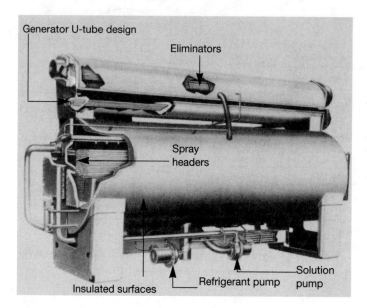

is carefully evacuated in the factory of air down to an extremely low pressure.

The heat required in the concentrator is furnished either by low pressure steam, hot water, or a gas flame. Concentrator temperatures around 240 F result in peak efficiency. A steam use rate of 18–19 lb of steam per ton of refrigeration is typical at this temperature. Selection procedures for an absorption chiller are similar to those for reciprocating or centrifugal chillers, and therefore will not be discussed in further detail. Heat source temperatures, load, chilled water, and condensing water temperatures are the factors required to select the proper machine from manufacturers' tables.

The coefficient of performance (COP) of absorption machines is much lower than systems using mechanical compression refrigeration. A COP of 0.65 is typical for large absorption equipment. This corresponds to a heat input rate of about 18,000 BTU/hr per ton of refrigeration. A large compressor-driven water chiller may have a COP of 3.5 or higher. That is, it uses only one-fifth of the energy of an absorption machine. This would seem to be an unacceptable waste of energy, but the COP does not show the whole situation. The energy used to drive the compressor is usually electricity generated by a thermal electric utility. Only about one-third of the heat from the fuel in the power plant is converted to electric energy (this is a limitation imposed by the Second Law of Thermodynamics, explained in Chapter 15.) Thus, the energy use advantage of the compression refrigeration system is greatly reduced. Further factors make the absorption machine desirable under certain circumstances. The low pressure steam used for the energy source is frequently otherwise wasted heat from a process or from a utility company. The total energy balance between the absorption or compression machine may then be equal or may even be in favor of the former.

The COP of the absorption system may be improved considerably if a two-stage generator (concentrator) is used. Heat from the vapor coming from the first generator is used to provide further vaporization of liquid from the absorber. A two-stage machine available from Japan raises the COP to about 1.0, a 50% improvement in energy efficiency.

13.22 SPECIAL APPLICATIONS

The absorption machine has considerable promise for refrigeration in conjunction with solar energy as a heat source. In this arrangement, hot water is heated by solar collection panels and then used in the concentrator as the heat source. The energy input rate required is high at the water temperatures solar heaters usually can produce (170–200 F), but because there is no depletable fuel used, this is not important.

A popular and efficient combination of refrigeration sources for air conditioning is the centrifugal–absorption combination. In this arrangement, a high pressure steam turbine is used to drive a centrifugal refrigeration machine; and the low pressure exhaust steam from the turbine is then used as a heat source in an absorption machine. The overall steam rate per ton by using the steam twice can be very attractive, about 13 lb/ton, as compared with 16 lb/ton for the turbine-driven machine alone.

It should be noted that the cooling tower required for an absorption machine will be considerably larger than that needed for a vapor compression cycle machine, due to the larger quantity of heat that must be rejected from the absorber and condenser combined.

The choice of whether to use an absorption or vapor compression machine (or combination) for an installation is largely a matter of economics, which is a function of relative fuel costs.

Since the absorption machine does not use ozone-depleting refrigerants, this is an additional attractive feature, as opposed to vapor compression machines.

13.23 CAPACITY CONTROL

Two modern methods of modulating refrigeration capacity to meet load demands are used with absorption machines. Both of them use a controller that senses and maintains a constant leaving chilled water temperature. With *heat source control,* the controller will operate a valve that controls the steam or hot water flow to the concentrator, thereby changing the machine capacity. *Solution modulation* control uses a controller that mixes the absorber solution to vary its concentration, which affects the machine capacity. Safety controls are provided with the machine and will not be discussed.

13.24　CRYSTALLIZATION

This is an important phenomenon that needs to be understood, especially by the operating engineer. If the lithium bromide solution becomes too concentrated, it changes from a liquid to a solid (crystal) form. The maximum concentration possible decreases as the solution temperature decreases. Therefore, if a solution is already near its maximum concentration and if its temperature is then lowered, it will *crystallize* (solidify).

This is a serious problem in absorption systems, because if it occurs, the crystallized LiBr blocks the piping and the machine stops working. There are three factors that can result in a drop in temperature of the solution:

1. Power failure
2. Condensing water temperature too low
3. Air leakage into the system

We will not explain here how each can cause this effect. Information can be found in manufacturers' manuals. If crystallization occurs, it is necessary to heat the piping where the blockage has occurred. Automatic methods of doing this can be provided by the manufacturer.

13.25　INSTALLATION

Detailed instructions for installation of absorption refrigeration machines are provided by each manufacturer. Some general procedures will be discussed here.

1. Investigate possible installation on an upper floor or penthouse. The light weight and lack of serious noise and vibration make this a feasible alternative to a basement. If the boiler is also located in the same space, this eliminates the need for a boiler stack and much of the piping; both are major cost items in high-rise buildings.
2. Provide rubber isolation pads under the machine.
3. Allow ample clearance for tube removal and for service access as needed on all sides.
4. Install external piping (to boiler, condenser) and electrical and control connections so as not to block access to the machine.

The Heat Pump

13.26　PRINCIPLES

The *heat pump* is a refrigeration system that can be used for both cooling and heating. To those not familiar with refrigeration cycles, the heat pump appears to be a mysterious device that operates on some unusual principle; it is not. The heat pump is usually a vapor compression refrigeration machine, which is basically no different in operation or components from that described previously. (An absorption machine can also be used as a heat pump, but this is unusual.)

Normally the purpose of a refrigeration machine is to absorb heat (in the evaporator) from a cooling load. The heat that is rejected in the condenser is thrown away to the atmosphere or a body of water. There is no reason why this heat could not be used to satisfy a heating load. When this is done, the machine is a heat "pump." The refrigeration effect, which is still occurring, may or may not be utilized, depending on need.

Figure 13.21 shows how a heat pump performs in both summer and winter. In summer, with the refrigerant flowing in the direction shown, the room coil serves as the evaporator and the room air is

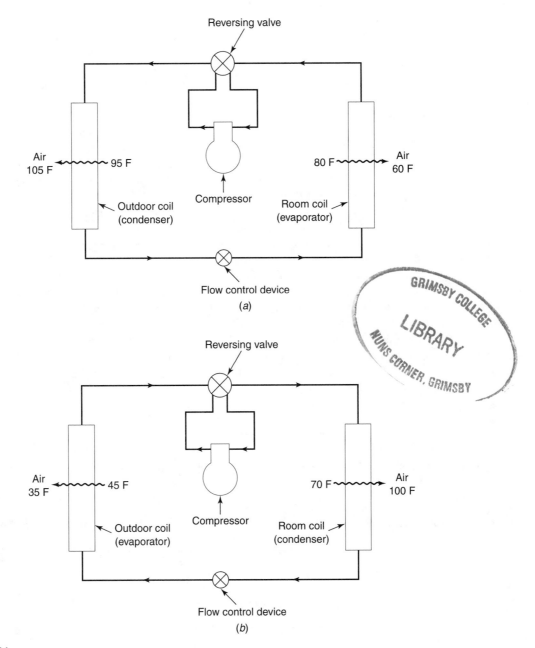

Figure 13.21
The heat pump cycle. (*a*) Summer cycle (cooling). (*b*) Winter cycle (heating).

cooled to produce summer cooling. In winter, the direction of refrigerant flow is *reversed* after leaving the compressor, so that the room coil serves as the condenser and the outdoor coil as the evaporator. The room air passing over the room coil therefore receives the heat rejected in the condenser. The heat pump now acts as a heating unit. The heat absorbed in the outdoor coil is the refrigeration effect, but of course it serves no useful purpose in this case.

The reversal of refrigerant flow to switch between heating–cooling is accomplished with a *reversing valve*. This has four ports, two of which are open at any one time to allow flow in the direction chosen. The heat pump is sometimes called a "reverse cycle" air conditioner. This is a misleading name. The cycle is the vapor compression refrigeration cycle. In the explanation given, only the refrigerant flow direction and function of the coils are reversed.

Heat pumps are often supplied as unitary equipment, with all the components assembled as a package by the manufacturer, including the air handling unit. Another arrangement, used more on larger equipment, without reversing refrigerant flow, is to reverse airflow. The evaporator and condenser coils function the same in winter and summer. However, the duct arrangement is made so that room air is circulated to the evaporator coil in summer and to the condenser in winter. The opposite is done for outside air. Therefore, in winter the room air, passing over the condenser coil, is heated.

One clear advantage of a heat pump is that it can provide heating or cooling from one machine, without any great modifications. In many cases, this means that it would have a lower first cost than using separate heating and cooling equipment.

Another advantage that is not apparent without further investigation is that it may have a lower operating cost than separate conventional heating and cooling systems, especially when electric resistance heating would be otherwise used. Although this has always caused considerable interest in the heat pump, high energy costs and shortages have created even more intense interest.

13.27   ENERGY EFFICIENCY

Consideration of how the Energy Balance principle (Chapter 3) applies to the heat pump will show how it may provide heating with a relatively small expenditure of energy. Referring to Figure 13.22, the total energy into the system equals the total energy out:

$$Q_c = Q_p + Q_e$$

where

Q_c = heat rejected from condenser

Q_e = heat absorbed in evaporator

Q_p = heat equivalent of compressor power input

Figure 13.22
Energy use in the heat pump.

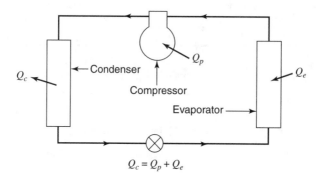

$$Q_c = Q_p + Q_e$$

The significance of this equation is that the useful heating Q_c is greater than the energy needed to drive the compressor Q_p by the amount Q_e, which does not require any energy expenditure. Contrast this with any direct heating system, either electrical or by burning a fuel to generate steam or hot water. In these cases, of course, the energy expended is at least equal to the useful heating. A relative measure of the performance of the heat pump is the heating coefficient of performance, defined as

$$COP_h = \frac{Q_c}{Q_p}$$

$$= \frac{\text{heat rejected from condenser}}{\text{heat equivalent of compressor power}}$$

Note that the coefficient of performance of a heat pump does not have the same meaning as when the unit is used for refrigeration. The heat pump COP_h is useful in illustrating the advantage of heating by using electrical energy to drive a heat pump compressor rather than using the electricity directly in resistance heaters.

For packaged heat pumps that include evaporator and condenser fans, the COP is often defined to include these auxiliary power inputs. This lowers the value of the COP slightly, of course.

13.28 SELECTION OF HEAT PUMPS— THE BALANCE POINT

As the evaporating temperature decreases, the heating capacity (the heat rejected from condenser) of a heat pump also decreases, because less heat is absorbed in the evaporator. Since the evaporator is the outdoor coil, a drop in outdoor temperature causes a decrease in the heating capacity.

For typical heating–cooling load requirements, if the heat pump is sized to handle the maximum cooling load, its heating capacity will be inadequate below outdoor temperatures often encountered in many climates. For residential applications, an outside temperature of about 30 F is a typical temperature at which the heating capacity of the unit will

just match the load. This is called the *balance point.*

At temperatures below the balance point, supplementary heating must be furnished. This is often accomplished by using one or more electric resistance heaters. As the outdoor temperature decreases, the amount of supplementary heat required increases, since both the load is increasing and the heat pump capacity is decreasing. Controls are arranged to activate the resistance heaters in steps as the outdoor temperature drops.

The supplementary heater should always be located downstream of the indoor coil in the ductwork so that the heat does not affect the condensing temperature.

Ratings of some small split system packaged heat pumps suitable for residential or similar applications are shown in Table 13.3 on page 356. The ratings in the table are based on a space temperature of 70 F for heating and 80 F DB, 67 F WB for cooling, a standard ARI test condition.

When selecting a heat pump, standard practice is to select a unit that will satisfy the design cooling load. The heating capacity of the unit is then plotted at different outdoor temperatures, and a line is drawn showing these heating capacities. Another line is drawn showing the building heating load at different outdoor temperatures (this is a straight line). The intersection of these two lines is the balance point. The supplementary heating is then sized to provide the extra heating capacity below the balance point, if any is required. Example 13.5 illustrates the selection of a heat pump and the determination of the balance point.

Example 13.5 _____

Select a heat pump for a residence in Memphis, Tennessee. The design cooling load is 44,000 BTU/hr and the design heating load is 41,000 BTU/hr. The inside design temperatures are 70 F in winter and 80 F in summer. What is the balance point? What is the size of the required supplementary resistance heaters?

Solution

Using Table A.9, the summer and winter outdoor design temperatures are 95 F and 18 F.

TABLE 13.3   RESIDENTIAL-TYPE PACKAGED HEAT PUMP RATINGS

performance
data

Fedders Flexhermetic Heat Pump Condensing Units are matched to Fedders Evaporator Blower Package Units for complete split system applications. Select performance requirements and matched components below.

Flex. Heat Pump Model No.	Matching Evap. Blower Model No.	Outside Temp. db°F	Capacities BTU/hr.		Total Watts Cool	PSIG Pressures		Outside Temp db°F Heat	Capacity BTU/hr. Heat	Total Watts Heat	PSIG Pressures	
			Cool	Sens.		Suction	Dischg.				Suction	Dischg.
CFH024A3	CFB024A2	115	20,000	14,500	4,450	80	388	60	31,000	4,000	68	310
		105	21,500	15,000	4,200	77	351	50	27,000	3,800	59	280
		*95	23,000	16,500	4,000	74	313	*45	25,000	3,600	52	263
		85	24,000	17,000	3,800	69.5	276	40	23,000	3,500	50	250
		75	26,000	18,000	3,550	65	235	30	19,500	3,300	40	225
								*20	16,000	3,150	31.5	200
								10	12,500	2,950	21	165
								0	9,000	2,800	11	135
CFH030A3	CFB036A2	115	24,000	18,000	4,800	82	387	60	35,000	4,000	74	380
		105	26,000	19,000	4,650	79	350	50	31,000	3,750	60	320
		*95	27,000	20,000	4,300	76	312	*45	29,000	3,600	53	287
		85	30,000	21,000	4,100	71.5	274	40	27,000	3,500	46	260
		75	32,000	22,000	3,750	67	234	30	23,000	3,300	37	220
								*20	18,500	3,150	30	189
								10	14,500	2,950	24	170
								0	10,000	2,800	20	159

UPFLOW

Flex. Heat Pump Model No.	Matching Evap. Blower Model No.	Outside Temp. db°F	Cool	Sens.	Total Watts Cool	Suction	Dischg.	Outside Temp db°F Heat	Capacity BTU/hr. Heat	Total Watts Heat	Suction	Dischg.
CFH036D7A	CFB036A2	115	30,000	21,000	6,250	76	400	60	41,000	5,300	63	409
		105	32,900	23,000	5,850	73	360	50	38,000	5,100	54	367
		*95	35,000	24,500	5,500	70	320	*45	35,000	4,900	50	350
		85	36,800	25,750	5,150	66	285	40	32,000	4,700	45	330
		75	38,200	26,750	4,700	62	240	30	26,000	4,300	37	295
								*20	21,000	3,900	28	260
								10	15,000	3,500	20	225
								0	10,000	3,100	10	185
CFH042D7A	CFB048A2	115	36,000	25,200	6,620	81	370	60	48,000	5,400	65	302
		105	39,000	27,300	6,320	78	335	50	44,900	5,100	55	260
		*95	42,000	29,400	6,000	75	330	*45	42,000	4,900	50	250
		85	44,000	30,800	5,650	71	265	40	37,000	4,100	45	238
		75	46,000	32,200	5,220	66	225	30	31,000	3,600	37	216
								*20	25,000	3,200	30	200
								10	22,000	3,000	22	182
								0	20,000	2,800	17	170
CFH048D3A	CFB048A2	115	39,000	27,300	7,950	79	415	60	52,000	5,900	62	318
		105	42,000	29,400	7,500	76	375	50	48,000	5,600	52	290
		*95	45,000	31,500	7,100	73	335	*45	45,000	5,500	47	280
		85	47,000	32,900	6,700	68	295	40	41,000	5,300	42	262
		75	49,000	34,300	6,150	64	250	30	34,000	5,000	33	240
								*20	28,000	4,700	25	220
								10	24,000	4,500	18	210
								0	20,000	4,300	12	204
CFH048D8A	CFB048A2	115	39,000	27,300	7,900	79	380	60	52,000	5,900	62	318
		105	42,000	29,400	7,450	76	350	50	48,000	5,600	52	290
		*95	45,000	31,500	7,100	73	310	*45	45,000	5,500	47	280
		85	47,000	32,900	6,650	68	270	40	41,000	5,300	42	262
		75	49,000	34,300	6,100	64	230	30	34,000	5,000	33	240
								*20	28,000	4,700	25	220
								10	24,000	4,500	18	210
								0	20,000	4,300	12	204
CFH060D3A	CFB060A2	115	48,000	31,500	9,600	79	390	60	66,000	7,600	60	358
		105	53,000	37,100	9,100	76	348	50	62,000	7,200	51	320
		*95	56,000	39,200	8,600	73	310	*45	57,000	6,900	45	300
		85	59,000	41,300	8,100	68	270	40	51,000	6,500	42	282
		75	61,000	42,700	7,500	64	230	30	42,000	5,800	34	250
								*20	34,000	5,200	25	225
								10	27,000	4,500	18	204
								0	21,000	3,900	10	188
CFH060D8A	CFB060A2	115	48,000	31,500	8,700	79	400	60	66,000	7,200	60	358
		105	53,000	37,100	8,650	76	360	50	62,000	6,800	51	320
		*95	56,000	39,200	8,200	73	322	*45	57,000	6,500	45	300
		85	59,000	41,300	7,750	68	284	40	51,000	6,100	42	282
		75	61,000	42,700	7,200	64	240	30	42,000	5,400	34	250
								*20	34,000	4,800	25	225
								10	27,000	4,100	18	204
								0	21,000	3,500	10	188

* A.R.I. test conditions: Inside temp. 80° F db; 67° F wb (cooling), 70° F db (heating)
Reprinted with the permission of Fedders Corp.

Using Table 13.3, a Model CFH048 unit is selected based on the design cooling load. The capacity at 95 F is 45,000 BTU/hr.

Refer to Figure 13.23. Using the columns of heating capacity versus outside heating temperatures, a curve is drawn through these points.

A straight line is drawn from the design heating load point, at 41,000 BTU/hr and 18 F, to 0 BTU/hr (no load) and 70 F. This line represents the building heating load at different outdoor temperatures.

The intersection of these two lines is about 28 F. This is the balance point. The heat pump heating capacity just matches the required building load at an outdoor temperature of 28 F.

The heat pump heating capacity at the balance point is 33,000 BTU/hr. The design heating load is 44,000 BTU/hr. Supplementary heaters are required below 28 F. At the design temperature of 18 F, the supplementary heat required is

$$\text{Supplementary heat} = \text{heating load}$$
$$- \text{heating capacity}$$
$$= 41,000 - 27,000$$
$$= 14,000 - \text{BTU/hr}$$
$$14,000 \text{ BTU/hr} \times \frac{1 \text{ KW}}{3410 \text{ BTU/hr}} = 4.1 \text{ KW}$$

Figure 13.23
Determination of heat pump balance point for Example 13.5.

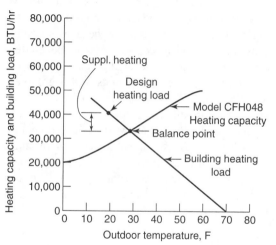

It is possible to reduce or eliminate supplementary heating by oversizing the heat pump in the cooling cycle. The larger unit will also provide more heating, as can be seen in Table 13.3. This is usually undesirable, however. The oversized unit will cycle too often in the cooling cycle, resulting in uncomfortable conditions, and a shortened life for the compressor and controls. For these reasons, heat pumps should always be sized for the cooling load, rather than the heating load.

Example 13.6 illustrates the savings in energy and operating cost by using a heat pump instead of electric resistance heating.

Example 13.6

For the heat pump used in Example 13.5, compare the amount of power saved at an outdoor temperature of 30 F by using a heat pump instead of electrical resistance heating. What is the heating COP of the heat pump at this temperature, compared to resistance heaters?

Solution
From Table 13.3, the heating capacity of the Model CFH048 heat pump at 30 F is 34,000 BTU/hr, requiring 5000 W of power for the compressor.

If electrical resistance heating is used, the equivalent to 34,000 BTU/hr is

$$\text{Resistance heating} = 34,000 \text{ BTU/hr}$$
$$\times \frac{1 \text{ W}}{3.41 \text{ BTU/hr}}$$
$$= 9970 \text{ W}$$

The savings is $9970 - 5000 = 4970$ W.

The COP_h for the heat pump at 30 F is (see Section 13.26)

$$COP_h = \frac{Q_c}{Q_p} = \frac{34,000 \text{ BTU/hr}}{5000 \text{ W} \times \dfrac{3.41 \text{ BTU/hr}}{1 \text{ W}}} = 1.99$$

The COP of electric resistance heating = 1.0. That is, the amount of heat output is the same as the amount of electric energy input (see Equation 15.11).

The relationship of the two COP_hs is

$$\frac{\text{Heat pump } COP_h}{\text{Elect. resist. } COP_h} = \frac{1.99}{1.0} \approx 2:1$$

To express this in words, at 30 F the heat pump requires about one-half the power input that electrical resistance heating does to produce the same heat output.

At higher outdoor temperatures, the heat pump will perform even better (be more energy efficient). However, at lower outdoor temperatures, it will not perform as well.

Typical actual heat pump COP_h values, when used for heating, range from 1.5–3.0. (This applies to unitary air-to-air heat pumps when operating at an outside air temperature of 47 F.) This means that the heat pump is producing 1.5–3.0 times more heat output for the same energy input than by using electrical resistance heating.

13.29   SOLAR ENERGY–HEAT PUMP APPLICATION

An effective use of the heat pump is in combination with solar energy. The coefficient of performance of the heat pump decreases with a decrease in evaporator temperature (see Chapter 15). With the conventional application of the heat pump, this temperature is lower than ambient air temperature, which results in a low COP in winter. However, a solar energy collector can be used to supply water at a much higher temperature than normally available in winter for the evaporator, say 75–100 F. This moderate temperature can be achieved with a relatively inexpensive collector, and will result in a low heat pump energy use. This arrangement is called the *solar assisted heat pump,* described in Chapter 18.

13.30   REFRIGERANTS

The refrigerants that are most widely used in compressors are in a chemical group called either *fluo-*

rinated hydrocarbons or *halocarbons.* These refrigerants have been used since the 1930s because of their excellent characteristics. They have good physical properties for performance—temperatures, pressure, oil mixing feature, heat transfer, specific heat, etc. They are nontoxic, stable, and inexpensive. It is not our intention here to investigate those matters in great detail. That is more properly left to refrigeration texts and manuals. We do wish to discuss, however, important issues about them that affect the practice of work in the HVAC industry.

All of the halocarbon refrigerants can be divided into three subgroups, according to their constituents.

1. *Chlorofluorocarbons (CFCs).* These are composed of chlorine, fluorine, and carbon atoms. Some in this group are CFC-11, CFC-12, and CFC-114. (The more familiar identification is R-11, R-12, and R-114.)
2. *Hydrochlorofluorocarbons (HCFCs).* These are composed of hydrogen, chlorine, fluorine, and carbon atoms. Some in this group are HCFC-22 (R-22) and HCFC-123 (R-123).
3. *Hydrofluorocarbons (HFCs).* These are composed of hydrogen, fluorine, and carbon atoms. Some in this group are HFC-134a (R-134a) and HFC-125 (R-125).

There are also mixtures of the above substances that are used as refrigerants. These fall into two classes: *azeotropes* and *blends* (zeotropes).

Azeotropes are mixtures that behave as a single substance. For instance, all parts of the mixture evaporate and condense at the same conditions. Two frequently used refrigerants in this group are R-500, a CFC/HFC mixture, and R-502, an HCFC/CFC mixture.

Zeotropes or *blends* are mixtures that do not always behave as a single substance. For instance, they may not evaporate or condense at a constant temperature (called *temperature glide*). This can complicate their use, especially in operating and servicing procedures.

13.31 OZONE DEPLETION

Despite their excellent properties for use as refrigerants, an extremely serious environmental problem exists with those halocarbons that contain chlorine, especially the CFCs. It has been found that they cause depletion of the ozone layer in the stratosphere.

Ozone (O_3) is an oxygen (O_2) molecule with an added oxygen atom. The ozone layer blocks out much of the harmful ultraviolet (UV) radiation from the sun. The ozone layer has been progressively depleting. One chlorine atom can destroy 100,000 ozone molecules.

Effects of a decreased ozone layer over Earth include

1. An increase in skin cancer (melanoma). This is one of the most deadly forms of cancer.
2. An increase in cataracts.
3. Reduction in immunity against disease.
4. Harmful effects on crops, timber, and marine life.

The relative ability of a substance to deplete the ozone layer is called its *ozone depletion potential (ODP)*. CFC-11 and CFC-12 have the highest (worst) value, ODP = 1.0. Table 13.4 lists some of the refrigerants and their ODP values. Note that HCFCs have a relatively low ODP, and HFCs do not cause any ozone depletion (ODP = 0).

As a response to this problem, the major industrialized nations have agreed to control the use and manufacture of CFCs and HCFCs.

TABLE 13.4 OZONE DEPLETION POTENTIAL (ODP) OF REFRIGERANTS

Refrigerants	ODP
CFC-11	1.0
CFC-12	1.0
HCFC-22	0.05
CFC-113	0.8
CFC-114	1.0
HCFC-123	0.02
HFC-134A	0.0
R-500	0.74

The production and importation of all CFCs in the United States has ceased as of December 31, 1995. All CFC use after this date must come from recovery operations. A gradual reduction of production and use of HCFCs is scheduled to result in their phase-out by the year 2030 in the United States. Some other countries have scheduled an earlier elimination, and the United States may revise its schedule.

These mandated changes have led to a search for viable temporary and permanent solutions, which include

1. Alternate refrigerants.
2. Use of refrigeration systems other than vapor compression.
3. Conservation of existing refrigerants in use.

Other than the vapor compression system, the lithium bromide absorption system is a realistic substitute in some cases, especially in medium to large air conditioning systems that would otherwise use centrifugal refrigeration compressors with CFC-11 or HCFC-22. Of course initial and operating costs would also play a factor in making a decision.

Alternate Halocarbon Refrigerants

The search for and selection of alternate non-ozone-depleting halocarbon refrigerants (HFCs) involves some difficult choices. There are generally no "drop-in" substitutes. That is, the alternate refrigerant cannot simply be placed in existing compressors. The problems may include:

1. Refrigeration compressor capacity, power requirements, and pressures may be unsatisfactory.
2. Expansion valves and desiccants may not function properly.
3. Mineral-oil-based lubricants (presently used with CFCs) cannot be used with some new refrigerants. However, polyol ester and alkylbenzene oils may be satisfactory substitutes.
4. Some new refrigerants may cause deterioration of rubber seals and hoses. In such cases, different elastic materials must be used.

5. Some new refrigerants may be less safe—more toxic and more flammable.

These restraints have led to various solutions, some of which are summarized here:

A. CFC-12 (R-12). This refrigerant is presently used largely in automotive air conditioners and household refrigerators. HFC-134a is a permanent substitute (ODI = 0). Manufacturers are now offering compressors that use this refrigerant.

B. CFC-11 (R-11). This refrigerant is used in centrifugal compressors. An interim substitute is HCFC-123, until this group is phased out. Toxicity can be a concern with HCFC-123. Research is underway for a permanent HFC replacement.

C. HCFC-22 (R-22). This refrigerant is widely used in window units, residential air conditioning, and commercial air conditioning and refrigeration. Its use will continue for some time, decreasing gradually. Possible substitutes are the HFC mixtures R-407c and R-410a. In large centrifugal and screw compressors, HFC-134a is an alternative.

Research is undergoing on numerous other new refrigerants.

Other Existing Refrigerants

New consideration is being given to previously used refrigerants, such as ammonia, carbon dioxide, and propane. These do not cause ozone depletion, but each has well known undesirable characteristics. Ammonia can be toxic and flammable, for instance.

Propane is already being used in some new household refrigerators in Europe. Apparently it is not yet being seriously considered in the United States, however, because of possible safety problems.

13.32 REFRIGERANT VENTING AND REUSE

The same concern about ozone depletion has led to regulations in the use of both CFCs and HCFCs, covered under amendments to the U.S. Clean Air Act. The specific regulations have been developed by and are enforced by the U.S. Environmental Protection Agency (EPA). A brief description of some major features of the regulations follows. They apply to R-11, R-12, R-22, and other CFCs, and HCFCs in use now.

1. Refrigerants may not knowingly be vented (released) to the atmosphere.
2. Requirements for recovery, recycling, and reclaiming refrigerants during service operations or when disposing of equipment are established.
3. Refrigeration technicians that service or dispose of equipment must be certified. This entails taking an EPA approved Certification Test.
4. Venting of refrigerants and other violations is punishable by fines up to $25,000 per violation day.

Recovery, Recycling, and Reclaiming (RRR)

There are regulations concerning the procedures and equipment involved in these practices.

Recovery is the removal of refrigerant from a system and storage in a container. All refrigerant must be recovered before opening the system.

Recycling is the cleaning of the refrigerant by removing oil, noncondensable gases, and passing the refrigerant through filter dryers to reduce moisture, acidity, and particulates.

Reclaiming is a complex cleaning process that restores the refrigerant to its original factory purity. The refrigerant must be tested to meet this standard.

Which of the three "Rs" is acceptable in each situation depends on a number of factors that are explained in the EPA regulations. The regulations also specify the equipment and procedures involved. There are a number of publications available with this information for the interested student.

13.33 GLOBAL WARMING POTENTIAL

In addition to the ozone depletion effect on the environment, halocarbon refrigerants have a global warming effect. Earth's atmosphere is being warmed due to the increase of certain gases that are products of industrial activities. These gases trap solar heat. Global warming may cause serious changes in the environment. Large land areas near sea level may flood, and agriculture will be affected.

The greatest global warming effect is from carbon dioxide (CO_2), because of the amount produced; CO_2 is the product of all fuel combustion. The ability of a substance to contribute to global warming is measured by the Global Warming Potential (GWP). Some halocarbon refrigerants have a very high GWP. For HCFC-22, the GWP = 100. For CO_2, the GWP = 1.0. For this reason, there is concern about these refrigerants. At the time of this writing, restrictions on emissions are still being considered by the industrialized nations.

13.34 WATER TREATMENT

Water used in condensers, water chillers, and boilers requires proper chemical treatment. Minerals that exist naturally in water can precipitate as solids and form *scale* that deposits on surfaces, reducing heat transfer. The water can have an acidic character that will cause *corrosion* of metals. *Biological growths* can occur that may cause deterioration of wood or coat surfaces and reduce heat transfer. This is a common problem in cooling towers, where the water is exposed to the atmosphere. The most bizarre example of contamination is "Legionnaires' disease," a bacteria that apparently has been traced in some cases to stagnant water in cooling tower basins. Water treatment in chilled water systems is usually a minor problem, because those systems are closed, but it should not be neglected. A firm that specializes in water treatment should be called in to set up a treatment plan when planning a large air conditioning system.

13.35 ENERGY CONSERVATION IN REFRIGERATION

Some methods to consider for conserving energy with refrigeration systems are:

1. Use refrigeration compressors that reduce power requirements as load decreases. For reciprocating compressors, this would involve use of cylinder unloaders or speed control. For centrifugal compressors, this would involve use of speed control or inlet guide vanes.
2. Select and operate equipment with highest evaporating (or chilled water) temperature and lowest condensing temperature consistent with maintaining satisfactory space conditions and satisfactory equipment performance.
3. Use condenser heat for heating needs by recovering heat (Chapter 15).
4. Use multiple equipment on larger projects so that each operates close to full load more often.
5. Use some form of total energy system (such as combined steam turbine centrifugal-absorption machines).

Review Questions

1. Describe with a sketch the vapor compression refrigeration system.

2. What are the four types of positive displacement compressors?

3. Explain the difference between an open and hermetic compressor.

4. Describe two methods of controlling capacity of reciprocating compressors and of centrifugal compressors.

5. What are the three types of condensers and their features?

6. Describe three types of refrigerant flow devices.

7. What are the causes of water loss in a cooling tower?

8. Describe with a sketch the lithium bromide–water absorption refrigeration system.

9. Describe with a sketch the air-to-air heat pump.

10. Explain what is meant by the term *balance point* of a heat pump.

11. Describe a heat pump defrost cycle.

Problems

13.1 Select an air-cooled condensing unit to handle a load of 15 tons of refrigeration for a bowling alley in Richmond, Virginia. Evaporating temperature is 45 F. Find the COP of the unit.

13.2 Select a package water chiller for a load of 21 tons. Chilled water is cooled from 55 to 44 F. Condenser water enters at 90 F and leaves at 100 F. Chiller and condenser fouling factors are 0.0005. Determine the COP of the unit.

13.3 Find the capacity, KW required, and COP of the unit selected in Problem 13.2 if the condenser fouling factor is 0.001.

13.4 Select a heat pump for a home in Los Angeles, California, for a design cooling load of 46,000 BTU/hr and design heating load of 42,000 BTU/hr. Determine the heating COP.

13.5 Determine the design cooling capacity and COP of the heat pump selected in Problem 13.4.

13.6 A heat pump is to be selected for summer and winter air conditioning of a home in Charleston, South Carolina. The design heating load is 32,000 BTU/hr and the design cooling load is 40,000 BTU/hr. Select a heat pump adequate to handle the summer load. What size electric booster heater, if any, is required in winter?

CHAPTER 14
Automatic Controls

The automatic controls for an HVAC system can be compared in importance and function with the brain and nervous system of a human. Without these, the body, regardless of how physically healthy it is, would be a lifeless mass. The HVAC controls must be designed and installed to fit the system and must function properly. If not, the air conditioning system will not produce satisfactory conditions.

OBJECTIVES

After studying this chapter, you will be able to:

1. Explain the purposes of automatic controls.
2. Identify and describe the elements of control systems with the aid of a diagram.
3. Explain closed- and open-loop controls.
4. Explain the types of control action.
5. Describe how electric, electronic, DDC, and pneumatic controls function.
6. Explain the types of valve flow characteristics and damper arrangements.
7. Sketch control diagrams and describe the operation of some basic control systems.

14.1 UNDERSTANDING AUTOMATIC CONTROLS

If asked to name the part of the HVAC system that they found most difficult to understand and work with, probably most designers, contractors, and service people would list the automatic control system. It is true that modern controls are often somewhat involved and that some of the devices used are complex. The major reason few people know how to deal with controls, however, is that they do not understand the basic principles and how to apply them. This is because most information focuses on hardware and how it is connected. This by itself often will not enable the service technician to determine the cause of a malfunctioning system.

Nor will it enable the designer to plan the controls to suit the type of HVAC system. In this chapter, we will emphasize the principles of control systems. These principles can be applied to electric, electronic, DDC, or pneumatic controls with equal ease. Although hardware will be discussed, we will not "lose sight of the woods for the trees."

14.2 PURPOSES OF CONTROLS

The controls can serve four different functions:

Maintain Design Conditions

Controls maintain *design conditions* (temperature, humidity) in the space. The heating and cooling capacity of the HVAC system is selected at the design load conditions. Whenever the load (heat gain or loss) is less than the design value, the system capacity is too large. If it produces its full output, the spaces will be overheated or overcooled. The controls must regulate the heating or cooling output of the system to match that of the load. The load varies mainly from changes in outdoor temperature, solar radiation, occupancy, and lights being switched on and off. Although controlling and maintaining conditions in the space are the primary function of the automatic controls, they serve other functions, as follows.

Reduce Human Labor Needed

Controls reduce the amount of human labor needed to operate and maintain the system, thus reducing labor costs and the chances of errors. For example, the controls may be designed to open an outside air damper to provide fresh air, rather than have operating personnel do this manually.

Minimize Energy Use and Costs

Controls minimize energy use and costs. One of the most important considerations in planning and operating a control system is based on its ability to minimize the use of energy at all times. For example, the controls may automatically change the amount of outside air introduced to the building so that free cooling is obtained from this air when suitable. Many of the energy conserving control applications are discussed in Chapter 15, but some of these features will be discussed here.

Controls that serve the purposes already described—maintaining space conditions, reducing labor, or conserving energy—are called *operating controls*. Often the same controls are providing all of these functions. Some controls serve a different purpose.

Keep Equipment Operation at Safe Levels

Controls keep operation of equipment at safe levels, thus preventing damage to property or injury to people. This type of control is called a *safety control*. It usually functions as a limiting device, to limit the values of temperature, pressure, or similar variables in the equipment. Safety controls have been discussed to some extent as part of the coverage of equipment in other chapters. Although some further references to them will be made here, our emphasis will be on the operating controls. Safety controls are of the utmost importance, however, and are part of the overall automatic control system.

14.3 THE CONTROL SYSTEM

One way to study a control system is to see it as a collection of many control devices. This leads to little understanding; a better approach is to recognize that every control system has similar elements. This is true regardless of how large and complex the system is, or whether it is pneumatic, electric, or electronic.

Every control system has the following elements:

1. A *controlled variable*. This is a condition that is to be controlled, such as temperature, humidity, or pressure.
2. A *controller*. This is a device that senses a signal from a change in the controlled variable and then transmits an action to a controlled device to correct the changed condition.

Thermostats, humidistats, and pressure-stats are examples of controllers. Note that a controller has two functions—to sense a signal and to transmit an action based on the signal.

3. A *source of power.* This provides the power to transmit the action from the controller to the controlled device. Two examples are compressed air (pneumatic) and electrical power sources.

4. A *controlled device.* This is a device which, when receiving the action from the controller, regulates the flow or other effect of a control agent. Examples of controlled devices are a valve, damper, or the motor driving an oil burner, pump, fan, or refrigeration compressor.

5. A *control agent.* This is the medium regulated by the controlled device. Examples are water flowing through a valve, air through a damper, or the electric current of a motor.

6. A *process plant.* The regulation of the control agent changes the output or operation of equipment called the process plant. Examples are a cooling or heating coil, oil burner, fan, pump, or compressor. The change in performance of the process plant changes the condition of the controlled variable, thus completing the desired action.

The sequence of action for any control can be shown by a *functional block control diagram* (Figure 14.1 below and Figure 14.2 on page 366).

An example of a simple control system will help identify these terms and how the control sequence functions.

Example 14.1

A student is studying in a room that has a hot water heating convector with a manual valve. The outdoor temperature drops very suddenly. Sketch a functional block control system diagram, identify the elements, and explain the sequence of control action.

Solution
The block control diagram is shown in Figure 14.3 on page 366.

1. The controlled variable is the room air temperature. A change (drop) in this temperature results from the sudden increase in heat loss to the outdoors.

2. The temperature controller (thermostat) is the student. His body senses the change in the controlled variable. He then transmits an action—turns the valve handle.

3. The energy source for transmitting the action is the human muscle power.

4. The controlled device is the manual valve.

5. The control agent (medium) is the hot water. Opening the valve increases water flow rate through it.

6. The process plant is the convector. Increased water flow rate results in more heat output from the convector to the room, causing a correction in the controlled variable condition—the room air temperature rises.

Figure 14.1
Functional block diagram for a closed-loop (feedback) control system.

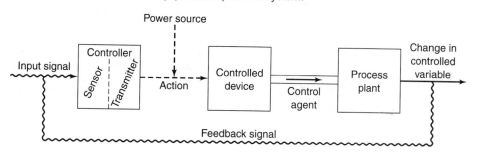

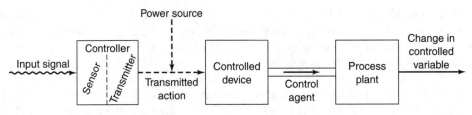

Figure 14.2
Functional block diagram for an open-loop (feed forward) control system.

Example 14.1 was of a manual control system, which is usually not very reliable or accurate. The same events occur in an automatic control system. For example, a thermostat senses the changed room air temperature and transmits an action, for instance, closing an electric circuit. An electric power source then opens an automatic, electrically operated valve on the convector.

14.4   CLOSED-LOOP (FEEDBACK) AND OPEN-LOOP CONTROL SYSTEMS

In Example 14.1, no mention was made of what might occur after the valve was opened. The room temperature might rise to an uncomfortably high level. It might happen quickly if the valve is opened too much, or gradually, if the outdoor temperature increases. Students could resolve this by closing the valve. However, they might have little time left for studying. We will now discuss how

this problem is resolved, which leads to an important concept in controls.

In the manual control system described, the controller (student) sensed the results of the corrective action (the rise in room temperature) and when this went beyond a satisfactory condition, the student took an opposite corrective action. This is an example of feedback.

Feedback is the transmission of information about the results of an action back to the sensor.

The result of the information being fed back is that the room air temperature changes were continually being corrected by the student. This is an example of a *closed-loop* control system. The functional control diagram for a real closed-loop system would be as shown in Figure 14.1 for the example of a room thermostat controlling the convector.

Note that the system will respond continuously to the feedback signal, always correcting the value of the controlled variable. This is the essence of most automatic controls. The sensor in the thermostat reacts to any change in the room temperature

Figure 14.3
Functional block control diagram for Example 14.1.

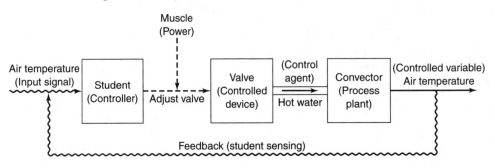

(increase or decrease) and sets in motion the corrective action.

A control system without feedback and its effect is called an *open-loop* (or feed forward) control system (Figure 14.2). In the example of the student who acted as a controller, if he had decided to open the valve because he expected colder weather, not because he felt cold, this would be a case of an open loop, because there is no feedback causing the action. The weakness of this type of control is apparent. The student does not know whether he is opening the valve a proper amount, and does not provide any corrective action unless he returns to using his sensory feeling (feedback). Open-loop controls are used in automatic control systems in certain instances, however, as will be explained later.

14.5 ENERGY SOURCES

Automatic control systems can be classified according to the source of energy they use:

Electric/Electronic

Electric energy is used to actuate the controlled devices. The controller regulates the amount of energy transmitted to the controlled device. If the controller has sensing and transmitting elements that are electronic, the system is called *electric/electronic*, or *electronic* for convenience. Otherwise the system is called *electric*. This is the only basic difference between electric and electronic control systems. The power is always transmitted electrically.

Electric control systems are often used on small installations because they are inexpensive, simple, and easy to install.

Line voltage control systems use electricity at the voltage from the power supply, usually 110 volts. *Low voltage* control systems transform the power supply to low voltages (usually 24 volts) for control use. The choice depends on cost, convenience, and safety.

Pneumatic

Compressed air is used as the source of energy to actuate the controlled devices. The controller regu-

lates the air pressure transmitted to the controlled device.

Pneumatic control systems are often used in large installations. Air compressors are required, and copper or plastic tubing is used to transmit the air. Pneumatic systems are popular because the controlled devices easily lend themselves to modulating action (Section 14.7), and they are simple.

Self-Powered

No external source of energy is used. Power to actuate the controlled devices comes from the medium being controlled. This is usually accomplished by an enclosed fluid that will change pressure in response to a temperature change. A common example is the thermal expansion valve (TEV) refrigerant flow control. Fluid in a bulb changes pressure in response to the temperature it senses. The pressure actuates the valve (Chapter 13). Self-powered controls are practical in certain applications, but are generally not used for the whole control system.

Combinations of electric, electronic, pneumatic, and self-powered controls may be used in one control system when desirable.

14.6 COMPONENT CONTROL DIAGRAM

Although the functional block control diagram is helpful in understanding the operation of the control loop, it is also useful to prepare a *component control diagram* showing the connections between components of the control system and HVAC system. Figure 14.4 on page 368 is an example of the simple control system described earlier, where a valve to a convector is controlled by a room thermostat.

A dashed line is used to represent the control action, which may come from one of a number of energy sources. The component control diagram is drawn the same regardless of type of energy source, however; this simplifies reading and understanding the diagram. An actual diagram showing all the wiring for an electric system would look much more complicated. Although it is necessary

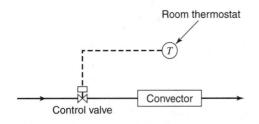

Figure 14.4
Component control diagram. Room thermostat controls position of valve.

to have detailed diagrams for each installation, they do not aid in understanding, but on the contrary, make it more difficult. For that reason, we will use component control diagrams instead of detailed wiring or compressed air piping diagrams in our explanations.

14.7 TYPES OF CONTROL ACTION

There are different types of action that the controller can impart to the controlled device, classified as follows:

Two-Position Action

This is also commonly called "on-off" action. For instance, if the controlled device is a motor, it may be started or stopped (on-off) by the two-position controller. A control valve that moves only to a fully open or closed position is another example of two-position action.

Differential is a term of importance in two-position control action. It refers to the range of controlled variable values at which action takes place. In the system, there are two differentials:

1. *Controller differential* is the range set on the control device of the variable values at which it transmits action to the controlled device. For example, if a thermostat is set to move to one position at 70 F and the other position at 72 F, it is said to have a differential of 2 F.
2. *Operating differential* is the range that actually occurs in the value of the controlled vari-

able. This differential will often be greater than the controller differential setting because there is a lag in response of the controlled device and medium. For example, when a thermostat causes a hot water heating valve to close, the convector keeps heating the room for a short time due to the hot water still in the unit.

Example 14.2 _____
A heating thermostat is set to start an oil burner on a furnace at a room temperature of 70 F. The thermostat has a 2 F differential setting. The response lag is 1 F in either direction. Between what values does the room temperature vary? What is the operating differential?

Solution
Figure 14.5 illustrates the solution. The thermostat high position (off) setting is $70 + 2 = 72$ F. With a lag of 1 F, however, the room temperature will rise to $72 + 1 = 73$ F. Similarly, the room temperature will fall to $70 - 1 = 69$ F. The operating differential is therefore $73 - 69 = 4$ F.

_____

If the controller differential is set too small, the heating or cooling equipment may cycle on and off too rapidly, a situation called *cycling* or *hunting*.

Timed Two-Position Control

If the operating differential is too great, it may cause uncomfortable conditions. This can be reduced by building *anticipation* into the controller. For example, a small heater may be contained in the thermostat. As soon as there is a signal calling for heating, the heater warms the thermostat faster than the room air would. As a result, the thermostat reaches its high setting earlier than it would otherwise, shutting off the controlled device sooner and reducing overheating. In effect, the operating differential has been reduced.

Floating Action

In floating action, the controlled device is still operated by a two-position type controller. The controlled device is constructed so that it moves

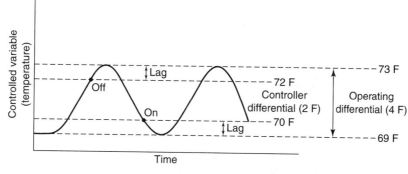

Figure 14.5
Two-position control action, as in Example 14.2.

gradually between full open and closed. The signal from the power source moves the operating part of the controlled device in one direction. There is a neutral zone (also called *dead zone*) in which no signal is transmitted, leaving the controlled device "floating" in an intermediate position until a new signal is received.

Proportional Action

In proportional action, the strength of the signal from the controller varies in proportion to the amount of change in the controlled variable. The controlled device in turn moves proportionally to the signal strength, taking a fixed intermediate position at a point relative to the change in the variable. This type of action can provide much finer response to load changes than the two position types described previously, because the response is proportional to the needs, not an all-or-nothing response. Proportional-type controllers and control devices are both required. For example, a modulating hot water valve would partially open or close at a position corresponding to the strength of a signal calling for increased or decreased heating.

There are some important terms used in proportional control that need to be defined.

The *set point* is the desired value of the controlled variable at which the controller is set to maintain.

The *control point* is the actual value of the controlled variable which the controller is maintaining at any given time.

Offset is the difference between the set point and the control point. It is also called *drift* or *deviation*.

Throttling range is the amount of change in the controlled variable required to move the controlled device from one extreme limit of travel to the other (full open to full closed). The term *proportional band* of a controller also means its throttling range.

The relationships among these terms are shown graphically in Figure 14.6.

The *sensitivity* of a controller is the relationship between changes in value of the control energy and the controlled variable. For example, a pneumatic

Figure 14.6
Proportional action control.

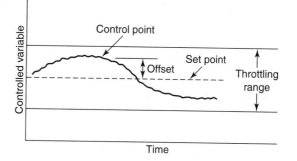

controller might have a sensitivity of 1 psi per degree F. This means that a change of 1 F in the controlled variable will change the transmitted supply pressure to the controlled device by 1 psi. The sensitivity of most controllers can be adjusted in the field to provide best control. Generally speaking, the sensitivity should be set to the maximum possible that does not cause hunting—large and continuous changes in the controlled variable.

Proportional Plus Reset (PI) Action

This type of control combines proportional action with a *reset* feature. When an offset occurs, the control point is changed automatically back toward the set point. That is, the amount of offset is reduced. The reset is accomplished by using floating action with proportional action. With proportional plus reset control, the desirability of control action proportional to the load changes is achieved without the disadvantage of large offset. Figure 14.7 shows how the controlled variable behaves with proportional plus reset; it should be compared with the previous diagrams.

Proportional plus reset action is also called *proportional-integral* (PI) action.

Proportional-Integral-Derivative (PID) Action

This type of control has the same features as PI action plus one more feature. The rate at which the

Figure 14.7
Proportional plus reset action (PI) control.

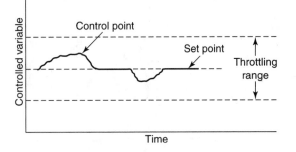

control point is moved back to the set point is part of the control action. The effect of this is that the time during which there is offset is shortened. PID control is sometimes used in room thermostats. Shortening the amount of time of offset can reduce the amount of overheating or overcooling, thus saving energy.

Stability and Hunting

It may seem from the discussions that PI or PID control action is always the most desirable. This is definitely not so. First, the proper type of control action depends on the job to be accomplished. For instance, for starting or stopping equipment, a two-position control is the only suitable one.

An important feature of a control system is its ability to maintain the control variable at a reasonably steady value. This is called *stability*. Under certain conditions, reset may cause instability, as will be explained now.

The speed of response of a reset type device is usually not very fast, due to its type of construction. Consider what happens when the controlled variable is changing very rapidly, perhaps due to sudden and frequent load changes (e.g., continual opening and closing of outside doors with a thermostat located in the room). The controller will be signaling rapidly for control action. Unfortunately, if PI control action is used, it may not be able to respond quickly enough. The controlled variable will swing widely in value and the control system will become unstable. These wide and rapid swings are called hunting. PI control is not desirable in HVAC systems where the controlled variable changes rapidly.

An example of a good application of PI control action is the chilled water temperature controller on a large water chiller. Close control of chilled water temperature (small offset) is desirable, because this results in good control of space temperature and humidity. Therefore, reset action is desirable, but a small offset could cause hunting. However, the air conditioning load usually changes slowly in a large building. Furthermore, the large mass of chilled water also reduces the rate of change of water temperature. Therefore, proportional plus reset action can be used without causing instability.

14.8   CONTROLLERS

As mentioned previously, the controller serves two functions: to *sense* the controlled variable signal and to *transmit* an action to the controlled device as a result of the signal. The variables most often requiring control in HVAC systems are temperature, humidity, pressure, and flow.

Temperature controllers are also called *thermostats*. Numerous types of sensing elements for thermostats are available.

A *bimetal element sensor* is made of two attached strips of different metals. The metals change lengths at different rates when their temperature changes, forcing a bending of the element (Figure 14.8). The bimetal strip may be straight or arranged in other shapes (see Figure 14.11 on page 372). The bimetallic sensing element is used often in *room thermostats*.

Another type of sensor uses a bulb filled with a fluid (Figure 14.9). Changes in temperature cause the fluid pressure to change, and this pressure acts to move a diaphragm or bellows.

This sensor is usually inserted in a duct or pipe. When the sensing bulb is attached directly to the control element, it is called an *immersion thermostat*. When a long capillary tube connects the sen-

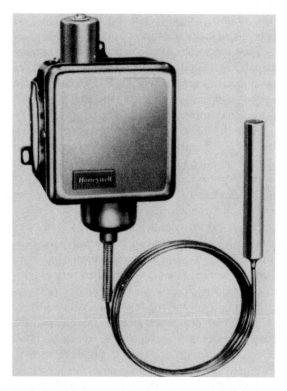

Figure 14.9
Remote thermostat with fluid-filled bulb-type sensor.
(Courtesy: Honeywell, Inc.)

sor to the control element, it is called a *remote thermostat*. This allows location of the control in a more convenient, accessible place than might otherwise be possible.

Another type of temperature sensor is called a *resistance element*. This is a thin wire whose electrical resistance changes with temperature. It is applicable to both room type and remote thermostats.

Humidity controllers are also called *humidistats*. One type of humidity sensing element uses two different materials attached together that absorb water vapor at different rates, thus bending or moving, much like a bimetal temperature sensor.

Pressure controllers are also called *pressurestats*. The sensing element is often an open tube connected directly to the fluid where pressure is to

Figure 14.8
Bimetal temperature sensor—bends with temperature change.

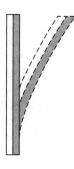

be controlled. The fluid pressure may act on a diaphragm or bellows, or a mechanical-type linkage.

Flow controllers often use pressure as a sensing signal. The velocity of the fluid where flow is to be controlled is converted to a static pressure by a sensing element such as pitot tube, and this signal is used to control flow.

After the signal is sensed by the controller sensing element, it must be transmitted by another part of the controller. Often the signal is also amplified in order to be strong enough to operate the controlled device. The transmitting element may be electric, electronic, or pneumatic.

An electric transmitter may consist simply of two electric contacts that are connected to the controlled device, as shown in the thermostat in Figure 14.10. When the bimetal element bends from temperature change, it closes or opens an electrical circuit that operates the controlled device.

An enclosed *mercury switch* is often used instead of open electric contacts. A glass tube filled with liquid mercury has two electrodes inserted in it (Figure 14.11). The sensing mechanism tips the tube so that the mercury either completes or breaks the electrical circuit through the electrodes. The mercury switch has the advantage over open contacts of being enclosed, and therefore is not subject to dirt, dust, or moisture, which could increase the resistance of the electrical contacts.

Figure 14.10
Thermostat with open electric contacts to transmit signal.

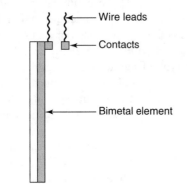

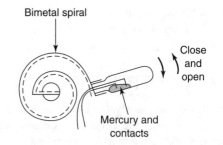

Figure 14.11
Thermostat with closed electric contacts in mercury-filled tube—sensor is spiral-shaped bimetal element.

Bimetal elements usually move slowly, resulting in a slow closing and opening of the electrical contacts. This may cause pitting of the contacts, increasing electrical resistance so that the signal is not transmitted properly. It may also result in bounce or chatter of the contacts, which may result in damage to electrical equipment in the circuit, since the circuit opens and closes many more times than normal.

This problem is resolved by causing *snap action* of the contacts, where the contacts open and close quickly. There are various means of achieving this. One way is to use a magnet that pulls the contacts quickly. A mercury switch also acts relatively fast and is therefore considered snap action.

One type of electric/electronic controller uses a resistance sensing element. The transmitting element is called a *bridge,* which is an electric circuit arranged to deliver a voltage proportional to the signal. This voltage is very small and is therefore amplified afterward. This type of device is suitable for proportional control.

A *relay* is an auxiliary device that is often used with controllers and in other parts of a control circuit. An electrical relay is a device that closes or opens one electrical circuit when a signal is received from another electric circuit. It may be used with controllers when the signal circuit is at a low voltage and the controlled device is to be operated with a high voltage. One type of relay uses a *solenoid* (Figure 14.12). A coil in the low voltage circuit acts as a magnet when electrically energized.

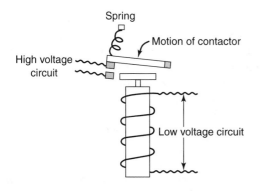

Figure 14.12
Solenoid-type electrical relay.

The magnetized iron core pulls a contact armature, closing contacts in the high voltage circuit.

Solid state relays use semiconductors to transmit the signal from the control circuit to the operating circuit. They have the advantages of no moving parts, compactness, increased reliability, and rapid action. Solid state devices have become very popular for many applications in control systems.

With pneumatic controls, a pneumatic transmitting element adjusts the air pressure that is supplied to the controlled device. In one type, the signal from the sensor moves a flapper that covers the opening to a branch of the tube carrying the control air (Figure 14.13). As the flapper moves away from the opening, some air bleeds out and the pressure in the main line decreases. This reduces the pressure transmitted to the controlled device, causing a changed action. The opposite happens when the signal moves the flap toward the opening. This arrangement is called a *bleed*-type controller. That

is, some of the control air is bled off from the control circuit. A *nonbleed*-type arrangement is also often used. It has the advantage of not using as much compressed air.

Pneumatic controllers have the desirable feature of being inherently proportional-type devices. The amount of air pressure, which varies with the flapper position, varies the position of the controlled device.

Special Purpose Thermostats

A *limiting thermostat* has a built-in maximum or minimum setting of the set point. For example, it may be constructed so that the maximum heating set point is 74 F. If the occupant sets the control temperature at 80 F, it will still control at 74 F. The obvious use of this is to conserve energy.

A *day–night thermostat* is actually two thermostats in one, with two different set points. At night or on weekends, the control temperature is set back to conserve energy. It is usually controlled by a time clock.

A *summer–winter thermostat* is a dual-temperature thermostat like a day–night type. It may be controlled manually or by the outdoor temperature.

A *master–submaster* thermostat arrangement is where one (master) thermostat controls and changes the set point of another (submaster) thermostat. This might be used to have an outdoor thermostat reset the control point of a thermostat controlling the hot water temperature in a heating system. This control function is called *reset control* (see Section 14.12).

Figure 14.13
Operation of pneumatic thermostat (bleed-type).

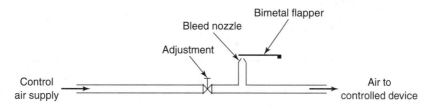

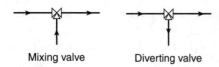

Figure 14.14
Three-way valves.

A *dead band* thermostat has a wide differential band (e.g., 8–10 F) within which the thermostat does not call for heating or cooling. This may result in significant energy savings in some applications.

14.9 CONTROLLED DEVICES

Valves, dampers, relays, and motors are examples of controlled devices in HVAC systems.

Control valves may be either two- or three-way devices. Three-way valves are either of the *mixing* or *diverting* type (Figure 14.14). A mixing valve has two inlets and one outlet. A diverting valve has one inlet and two outlets.

Two-way valves are used to vary flow rate to the heating or cooling equipment by throttling. Mixing and diverting valves can also be used to vary flow rates through the unit, as shown in Figure 14.15, while still maintaining the same total flow rate.

Mixing and diverting valves can also be used to control capacity by varying water temperature instead of quantity (Figure 14.16). In this application, supply water from the boiler and return water are mixed to provide water at the desired temperature.

The capacity of a heating or cooling coil can be changed either by varying the water flow rate or the temperature. However, the output does not change as much with flow rate variation as it does with water temperature. For this reason, water temperature control is often preferred. On the other hand, flow rate control with a two-way valve is usually less expensive, and thus is often used on room terminal units.

Valves have four different characteristics concerning how the flow varies with valve stroke; this depends on the shape of the valve opening. Valves are classified into three groups: *quick opening, linear,* and *equal percentage.* The difference in performance is shown in Figure 14.17. The equal percentage valve is usually best for automatic control of water flow rate in coils because more variation in flow rate can be achieved for a given movement of valve stroke than can be achieved with the other types. This results in better modulation of heating or cooling capacity, since considerable throttling of the water flow rate is required to reduce capacity. When a smaller range of throttling capacity is required, a linear flow valve is adequate.

A quick opening valve is used when almost full flow must occur even with a small change in the controlled variable. It is used, for example, with an outside air preheat coil. When heat is called for, the valve should open wide to prevent freezing of the steam in the coil.

Two-position electrically operated valves use a solenoid to move the valve stem to an open or closed position in response to the signal (Figure 14.18).

Modulating electric valves use a motor as a valve operator that moves the valve stem gradually in response to the signal. The operator for pneumatic valves is either a diaphragm, bellows, or

Figure 14.15
Use of three-way valves to control flow rate.

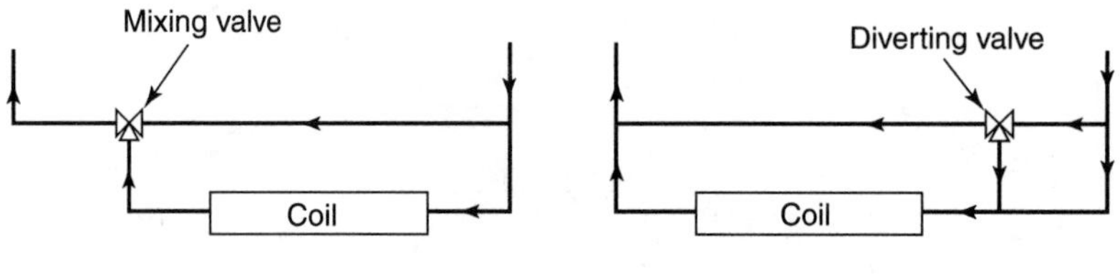

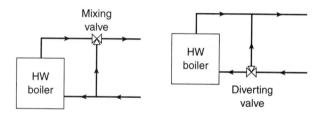

Figure 14.16
Use of three-way valves to control supply water temperature.

piston that responds to pressure and moves the valve stem (Figure 14.19 on page 376).

Automatic dampers are used as controlled devices for varying airflow, mixing air, or for bypassing (diverting) air. As with valves, the purpose is usually to vary heating or cooling capacity of the equipment. Except for very small sizes, they are of multiblade construction. Two arrangements are available; *parallel* or *opposed* blade (Figure 14.20 on page 376). The opposed blade arrangement will give better modulation of air flow rate. Parallel blade dampers should be used only for two-position (open-closed) control.

Figure 14.17
Flow characteristics of control valves.

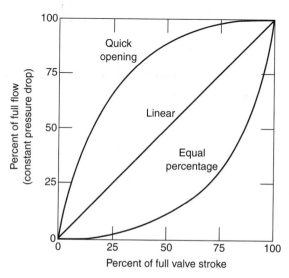

Electric motors are used for modulating dampers in electric control systems, and pistons or diaphragms are used as damper operators in pneumatic systems.

There are other auxiliary devices used in control systems that will not be discussed here. Although they are of practical importance, a description of them will not add to our understanding of control principles.

14.10 CHOICE OF CONTROL SYSTEMS

There are countless choices and arrangements of controls for HVAC systems. This is why our approach in this chapter up until now has not been merely to describe control systems, but instead has focused on principles. We will now look at some examples of how controls are used.

There is a choice of from where to control the HVAC system. Control can be provided at the heating/cooling source, the pump or fan, or terminal units. For example, the burner or compressor can be started-stopped or modulated. Control can also

Figure 14.18
Two-position solenoid electric valve.

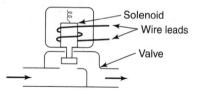

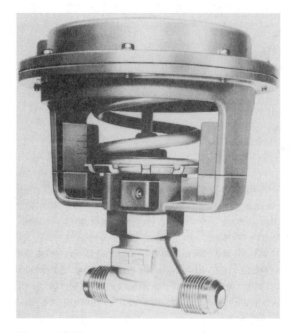

Figure 14.19
Modulating pneumatic valve. (**Courtesy: Honeywell, Inc.**)

be provided by varying air or water flow rates using dampers or valves. On many systems, a combination of these is used.

In addition to selecting which part of the HVAC system is to be controlled, there are choices as to what controlled variable to use for control—the space, the medium, or outdoor air. Of course, the space air is the final temperature being controlled, and it would therefore seem obvious that control should be here, using a room thermostat. On many systems, however, additional control is provided from thermostats sensing outdoor air or the cooling/heating fluid medium.

It is the desire to provide better control and to conserve energy that often determines the choices made. Examples will be described later.

Safety controls are also required and are part of the control system. The student should note that systems have safety controls that are not fully discussed here.

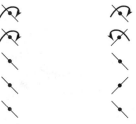

Parallel   Opposed

Figure 14.20
Multiblade damper arrangements.

14.11 CONTROL FROM SPACE TEMPERATURE

Control of Burner or Compressor

A simple control arrangement is to have a room thermostat control an oil or gas burner or a refrigeration compressor. In Figure 14.21, a room thermostat T starts and stops a refrigeration compressor motor M of an air conditioning unit. This is the type of control used on a window unit, with the room thermostat mounted on the unit. Larger systems usually have more complex controls, to provide better control and to conserve energy.

A similar arrangement for a hot water boiler is shown in Figure 14.22. The room thermostat T starts and stops the oil burner motor in response to room air temperature. In a gas-fired boiler, the thermostat opens or closes a valve in the gas supply line. Not shown, but always required, is a high limit thermostat, a safety control that shuts off the burner when the water rises above a set temperature.

Figure 14.21
Space control of refrigeration compressor motor.

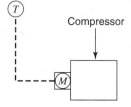

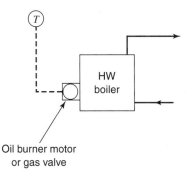

Figure 14.22
Space control of HW boiler burner motor or gas valve.

A warm air furnace would use a similar control.

Control of Flow Rate Through Valves

A room thermostat may be used to vary the flow of hot water, chilled water, or refrigerant to a terminal unit or coil in a duct (Figure 14.23) using automatic valves, thus varying the heating or cooling output.

Control of Volume Dampers

The room thermostat varies the supply air quantity by controlling a modulating damper *D*, as seen in Figure 14.24 on page 378. This control is used in variable air volume (VAV) systems.

Control of Mixing Dampers

A room thermostat varies the proportions of hot and cold air from two ducts (Figure 14.25 on page 378). Two sets of dampers move together so that one closes as the other opens on call from the thermostat. This arrangement is used in both dual duct and multizone systems (Chapter 12).

Control of Face and Bypass Dampers

The quantity of air flowing over the cooling coil or bypassed around the coil is varied by the opposing motion of the two dampers (Figure 14.26 on page 379).

14.12 CONTROL FROM OUTDOOR AIR

Although the outdoor air temperature rather than space temperature could be used to control space temperature, this is seldom done, because it does not provide feedback, as explained previously. However, it is used in combination for certain purposes that will be explained.

Control of Outside and Return Air Proportions

Controls on larger systems are often used to vary amounts of outside air, from a minimum fresh air requirement to all outside air. This is done so that the outside air can be used for cooling when suitable. Control may be provided from a mixed air

Figure 14.23
Space control of water flow rate through terminal unit or coil in duct.

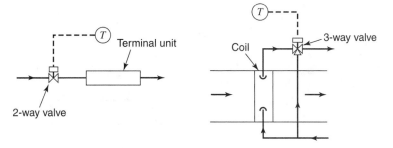

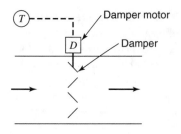

Figure 14.24
Space control of air flow rate through damper.

thermostat that adjusts the outside and return air dampers to provide cool outside air when required, as seen in Figure 14.27.

The minimum outside air damper is open during the coldest weather. When outdoor air temperature rises, mixed air thermostat T_1 gradually opens maximum outside air dampers and closes return air dampers to provide outside air in the range of 50–60 F. Thermostat T_1 operates through a high limit thermostat T_2. When the outdoor air temperature rises to a level at which it has no cooling effect (near room temperature), thermostat T_2 takes over and closes the maximum outside air damper.

An arrangement that will offer even better energy conservation uses an *enthalpy controller* (Figure 14.28 on page 380). The controller senses wet bulb temperature and therefore enthalpy of the outside and return airstreams and sets the air proportion so that outside air is used for cooling whenever its enthalpy, not its temperature, is lower than that of the return air. There are days when the humidity, and therefore enthalpy, of the outside air may be low enough so that it is useful for cooling, even though its temperature does not indicate this.

Whether using temperature or enthalpy sensing, the outside air (OA) and return air (RA) dampers are modulated to provide cooling from the outside air whenever it is suitable. Controlling these dampers saves operating the refrigeration equipment and also prevents the introduction of excess outside air at high temperatures, both conserving energy. For these reasons, this system is called *economizer* control.

Outdoor Temperature Reset

A control arrangement that is sometimes included as part of the control system is to have an outdoor thermostat *reset* (change) the temperature at which a variable is controlled. For example, it may reset the water temperature in a boiler (Figure 14.29 on page 380). An immersion thermostat T_2 controls the boiler water temperature at its set point through the burner motor. As the outdoor temperature rises,

Figure 14.25
Space control of mixing dampers for multizone and dual duct systems. (a) Multizone unit. (b) Dual duct and mixing box.

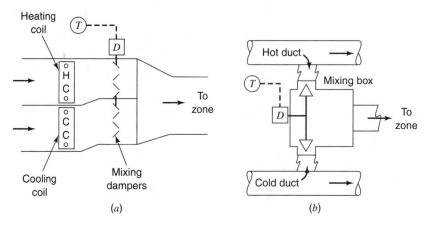

(a)

(b)

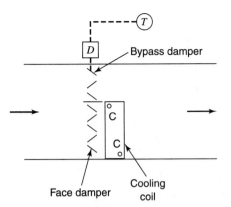

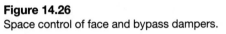

Figure 14.26
Space control of face and bypass dampers.

overheating. Further space control might be furnished through variable volume or other means. In these examples, T_1 is the *master* thermostat and T_2 is the *submaster* thermostat.

14.13 CONTROL FROM HEATING/COOLING MEDIUM

It is often advantageous to control equipment from controllers sensing the conditions in the heating or cooling medium. In the examples described previously, the immersion and duct thermostats are controlled from water and supply air temperatures. Another example is shown in Figure 14.31 on page 381.

A water chiller has a refrigeration compressor whose capacity is controlled by an immersion thermostat in the chilled water supply line. The thermostat modulates the compressor capacity to maintain a constant chilled water temperature. This is often done on large chilled water HVAC systems, with separate space control of water flow rate. In this application, medium control is useful because it is desirable to keep the chilled water temperature at a constant value in order to ensure proper dehumidification. Another reason for medium control is that faster response may be achieved by controlling the supply air or water temperature. The methods by which compressor capacity is modulated are discussed in Chapter 13.

outdoor thermostat T_1 resets the control point of T_2 lower. In this way, the hot water supply temperature is inversely proportional to the outside temperature, and overheating in mild weather is reduced. This of course also results in energy conservation. The system usually also includes additional control from space temperature, such as water flow rate.

The duct heating system in Figure 14.30 on page 381 operates in a similar manner. Duct thermostat T_1 controls the air supply temperature to the space through the automatic valve. Outdoor thermostat T_1 resets the control point of T_2 to reduce

Figure 14.27
Outdoor temperature control of outside and return air dampers for energy conservation.

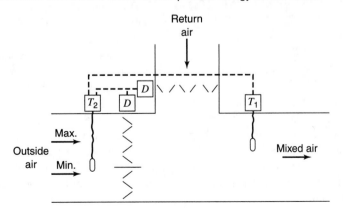

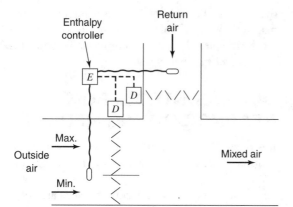

Figure 14.28
Enthalpy control of outside and return air dampers for energy conservation.

14.14 HUMIDITY CONTROL

For humidification in heating systems, a space humidistat controls a humidifier. Steam or water spray humidifiers located in the ductwork are used. The cooling coils are often used for both cooling and dehumidifying in a cooling system. In this case, some form of reheating after cooling is used. One arrangement is shown in Figure 14.32.

The cooling coil is controlled by the room thermostat as long as room humidity is below the humidistat setting. When the humidity rises above the control point, the humidistat takes control of the cooling coil, calling for cooling. If the room temperature becomes too low, the thermostat operates the reheat coil. The psychrometrics of this process are explained in Chapter 7.

14.15 COMPLETE CONTROL SYSTEMS

One of the individual temperature control arrangements described previously may serve as the complete control system in a simple heating or cooling system. Often, however, combinations of space, outdoor, and medium control are used. This may be done because controls of temperature, humidity, and ventilation are all required, or to provide closer

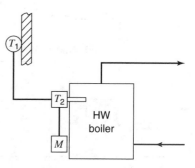

Figure 14.29
Outdoor reset of water temperature.

control. It may also be done to conserve energy. Two examples of possible control system arrangements will be described.

A hot water heating control system with individually controlled rooms or zones is shown in Figure 14.33 on page 382. The controls operate as follows:

1. The immersion (medium) thermostat T_1 controls the hot water supply temperature through the burner operation.
2. The outdoor thermostat T_2 resets the control point of thermostat T_1 as the outdoor temperature varies.
3. Room thermostats T_3 control the terminal unit valves to maintain desired space temperatures.
4. The outdoor thermostat shuts off the pump when the outdoor temperature rises to a value that requires no building heating.

This control system provides good temperature control and also conserves energy. By reducing water supply temperature on mild winter days, overheating of rooms is avoided, providing greater comfort and less energy use. By stopping the pump automatically when no heating is needed, further energy is conserved.

An example of a control system for a single zone year-round air conditioning system is shown in Figure 14.34 on page 382. The HVAC system provides summer and winter space temperature control and ventilation, but no humidity control. It operates as follows:

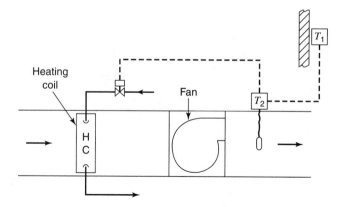

Figure 14.30
Outdoor reset of supply air temperature.

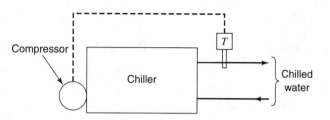

Figure 14.31
Control of chilled water temperature.

1. The enthalpy controller (or a temperature controller) positions the return air and maximum outside air dampers so that maximum free cooling is achieved during the cooling season. During the heating season, minimum outside air is used.

2. The air discharge thermostat T_1 controls the cooling coil (summer) or heating coil (winter) to regulate discharge air temperature.

Figure 14.32
Space temperature and humidity control of cooling and heating coils.

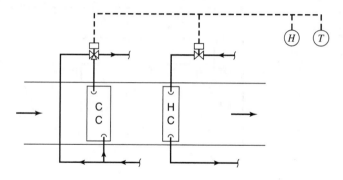

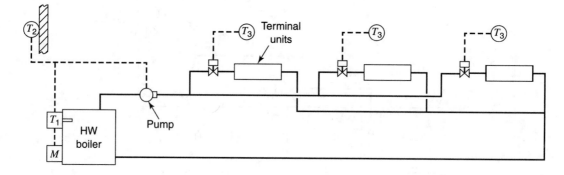

Figure 14.33
Hot water heating control system example.

3. The room thermostat T_2 acts as a master controller to the submaster $T_1.$ In response to a change in room temperature, the room thermostat resets the control point of the discharge thermostat.

Using the room thermostat to control the discharge thermostat provides faster response of the system to changes in room temperature. The use of enthalpy control to conserve energy has been explained previously. A summer–winter room thermostat might be used to control at two different room temperatures.

These examples are given to illustrate how controls can be combined. There are hundreds of other arrangements. In each case, however, they are composed of the basic elements and, with a little patience, the student can analyze their operations.

Direct Digital Control (DDC)

As we have learned, the conventional control arrangement is to have each controller sense a signal and then send an action directly to a controlled device. The use of digital microcomputers in HVAC control systems has changed this control sequence. The scheme using computers is called *direct digital control (DDC)*.

We will outline the basic operation and a few advantages of DDC, avoiding the complex terminology and structure of DDC systems. In DDC, conventional controllers are not used. A sensor

Figure 14.34
Year-round air conditioning control system example.

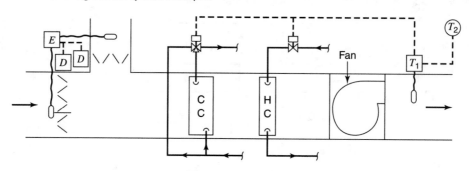

transmits the sensed signal (e.g., temperature) to the computer. (The signal from the sensor is usually conditioned by intermediate devices so that it is in a form that the computer can understand.) The computer then sends out a signal to operate the controlled device (e.g., a valve) as needed. This signal is also usually conditioned so that it is suitable for the controlled device. This sequence is complex even in our simplified explanation, but achieves some important advantages.

Control changes (e.g., set points) can be made at one central point (the computer) instead of having to be done at each controller. This results in improved conditions and reduced operation and maintenance work.

Energy conservation strategies are easily handled in the computer program, rather than relying on many pieces of hardware and their connections that can get out of calibration and break down.

Review Questions

1. Explain the purposes of automatic control.

2. Sketch a functional block control diagram for an open loop and for a closed loop system. Label and describe basic elements.

3. What are the two most common energy sources for control systems?

4. With the aid of a sketch, describe two-position control action. Explain controller differential and operating differential and why they are different.

5. With the aid of a sketch, describe proportional control action. Explain the terms *set point, control point, offset,* and *throttling range.*

6. With the aid of a sketch, describe proportional plus reset control action.

7. Describe three types of thermostat sensors.

8. With the aid of a sketch, describe a mixing valve and a diverting valve.

9. Sketch the piping connections of a mixing valve and a diverting valve to control flow rate through a coil.

10. With the aid of a sketch, describe the three types of flow characteristics of control valves. What are the applications of each type?

11. With the aid of a sketch, describe the two types of multiblade damper arrangements. What are the proper applications of each type?

12. With the aid of a sketch, describe the control of a boiler burner from space temperature.

13. With the aid of a sketch, describe the control of a multizone unit mixing dampers from space temperature.

14. With the aid of a sketch, describe temperature economizer control of outside and return air dampers.

15. With the aid of a sketch, describe outdoor rest control of HW boiler supply temperature.

16. With the aid of a sketch, describe space temperature and humidity control of a cooling and heating coil.

17. With the aid of a sketch, describe a hot water heating system control with individual room temperature control and outdoor reset of supply water temperature.

18. With the aid of a sketch, describe a year-round single zone air conditioning system that controls space temperature with economizer control for energy conservation.

19. Explain the term *snap action,* its purpose, and two ways of accomplishing it.

20. Explain what a dead band thermostat is. What is its purpose?

Problems

14.1 A warm air heating system has a room thermostat that controls the furnace oil burner motor. Draw a functional block control diagram and identify the elements.

14.2 A three-way mixing valve on a water chiller is controlled by an immersion thermostat to maintain constant chilled water temperature. Draw a functional block control diagram and identify the elements.

14.3   A room thermostat controls a two-way modulating valve to a hot water heating coil in a central air conditioning unit. A duct thermostat in the supply airstream maintains a minimum discharge air temperature. Draw a component control diagram.

14.4   An air conditioning system has a DX cooling coil with face and bypass dampers. A room thermostat opens a two-position solenoid valve at minimum load and then modulates the face and bypass dampers. Draw a component control diagram.

14.5   A cooling thermostat is set to control flow through a fan coil unit at a room temperature of 78 F. The thermostat has a 2 F differential setting. The response lag is 2 F. Between what values does the room temperature vary? What is the operating differential?

CHAPTER 15
Energy Utilization and Conservation

In the past, little attention was usually given to conserving the energy used by HVAC systems because of the relatively low cost of fuel. Sharply rising fuel prices and concerns of shortages have changed this situation. Costs for energy used in building operations have become such a significant expense that it is necessary that they be kept to a minimum level. This requires a thorough analysis of energy use and conservation in HVAC design, installation, and operation.

In this chapter, we will explain procedures for analyzing energy use, energy conservation in design, and efficiency in operation. Some of the energy recovery equipment that has become popular will be described. The use of computers as a tool in analyzing these problems will be discussed.

Some of the energy material covered in other chapters may be repeated here. This is done intentionally so that all of the information is presented together. On the other hand, our presentation can cover only a small part of this subject. The scope of techniques for energy conservation is vast, and new ideas are constantly developing. It should also be noted that energy use and conservation in HVAC systems are closely related to energy use in lighting and other building systems. Our discussion of these subjects will also be limited.

The intention of this chapter is to give the student an idea of how to approach the problem of energy conservation in an organized manner, to indicate what factors should be considered, and how to find this information.

OBJECTIVES

After studying this chapter, you will be able to:

1. Calculate thermal energy conservation values as specified in energy codes.

2. Determine the efficiency, COP, or energy efficiency ratio (EER) of energy conversion equipment.

3. Determine the seasonal heating requirements and fuel costs for a building using the degree day method.

4. Describe energy recovery equipment used in air conditioning.

5. Suggest energy conservation procedures in construction, design, installation, and maintenance.

6. Describe uses of computers in the HVAC field.

15.1 ENERGY STANDARDS AND CODES

It must be recognized that use of energy studies and conservation techniques is no longer optional on the part of the HVAC designer, contractor, or operating personnel. Standards have been developed that have already been adopted as part of almost all state building codes. No one can successfully practice in the HVAC industry without a reasonable knowledge of this subject. *ASHRAE Standard 90.1, Energy Efficient Design of New Buildings* is widely used by HVAC designers as a basis for building energy use standards, as well as a basis for many state building energy codes.

The data and procedures used here will be selected and adapted from energy codes of actual states. However, they will intentionally not be identical to any particular code. Also, bear in mind that each locality regularly revises its codes according to new developments. The purpose of our presentation is to show the student the approaches used in sound HVAC energy efficient design. For real applications, the student must follow the actual energy code for his/her state.

There are three methods that are permitted in order to meet specified HVAC energy standards/codes.

A. *Component performance* or *prescriptive method*. In this procedure, the maximum permissible overall thermal performance values U_o of the building components are prescribed. The actual values are calculated by formulas to confirm compliance with the prescribed criteria.

B. *System performance method*. In this procedure, the whole building envelope is considered a system whose thermal performance must meet the prescribed standards.

C. *Energy cost budget method (ECB)*. In this procedure, the proposed building energy use and cost is determined and compared to the prescribed values for compliance.

The prescriptive method is the simplest but offers the least flexibility in varying proposed construction features to meet the standard. It lends itself to manual or computerized solutions. Because of the voluminous database involved in the other two methods, a computerized approach is recommended. Software is available from some of the sources listed in the Bibliography.

We will describe the type of calculation procedures involved in the prescriptive method. The data we present is only a small selection from a typical code, and is modified for learning value.

Heating Design Requirements

The overall thermal performance U_o for a component such as a wall, roof/ceiling or floor, is defined as

$$U_o = \frac{U_w \times A_w + U_g \times A_g + U_d \times A_d}{A_o} \quad (15.1)$$

where

U_o = overall thermal performance of exterior wall, roof/ceiling or floor, BTU/hr-ft²-F

A_o = total area of exterior wall, roof/ceiling or floor, ft²

U_w = U-value of opaque portion or exterior wall, roof/ceiling or floor

A_w = area of opaque wall, roof/ceiling or floor

U_g = U-value of glass

A_g = area of glass

U_d = U-value of door

A_d = area of door

The U_o-values are calculated for a proposed construction and then compared to the allowable values. Table 15.1 is a simplified version of such values. Actual codes and standards specify different

TABLE 15.1   MAXIMUM ALLOWABLE OVERALL THERMAL PERFORMANCE U_o-VALUES FOR BUILDING ENVELOPE COMPONENTS, BTU/HR-FT²-F

Envelope Component	Maximum Thermal Value
Walls	$U_o = 0.20$
Roofs and floors	$U_o = 0.05$

U_o-values based on climate, building use, and other factors. Conceptually they are similar, however.

The following example illustrates the use of the component performance method.

Example 15.1

A building has the following specifications:

$$A_w = 4000 \text{ ft}^2, \; U_w = 0.15 \text{ BTU/hr-ft}^2\text{-F}$$
$$A_g = 1000 \text{ ft}^2, \; U_g = 0.69$$
$$A_d = 100 \text{ ft}^2, \; U_d = 0.40$$
$$\text{Degree days} = 6000$$

Determine the overall wall U_o, and compare it to the suggested standard.

Solution

Equation 15.1 will be used:

$$U_{ow} = \frac{0.15 \times 4000 + 0.69 \times 2000 + 0.40 \times 100}{6100}$$

$$= 0.33 \text{ BTU/hr-ft}^2\text{-F}$$

From Table 15.1, maximum allowable

$$U_{ow} = 0.28$$

The wall in Example 15.1 does not meet the standard specified, and construction presumably would not be permitted. One solution would be to change the architectural design to reduce the amount of glass, or to change materials to reduce U-values, or both.

This approach may restrict the architect and engineer unnecessarily. In such a case, the performance or energy cost budget methods might be used. In the performance method, codes present combinations of window and opaque areas, U-factors, etc., that

satisfy the criteria. The designer can then choose from these offerings.

The energy cost budget method sets a maximum value on energy consumption in a building. For example, a total design (maximum) energy use standard of 35 BTU/hr per square foot of floor area for new residences and 70 BTU/hr for new commercial buildings has been suggested as a reasonable standard for many climates. This standard usually includes energy for HVAC, lighting, and other building operations. There are many existing commercial buildings operating that have a design consumption of 150 BTU/hr per square foot, or higher, far exceeding these standards.

Another variation on energy use standards is to specify the maximum amount that should be used for a whole year; for example, 55,000 BTU/hr per square foot per year has been suggested for residences in some climates.

A limitation of relying solely on an energy budget type standard is the difficulty in predicting the energy consumption of a building. If a building is designed to meet a certain standard, and the owner finds it does not, the resolution of the conflict may not be clear. Is the fault with the architect, engineer, or operating procedures?

Cooling Design Requirements

For summer HVAC energy use, a factor called the OTTV (overall thermal transfer value) has been developed and adopted in many codes. It is defined as follows for walls:

$$\text{OTTV}_w = (U_w \times A_w \times \text{TD}_w + A_g \times \text{SF} \times \text{SC}$$
$$+ U_g \times A_g \times \text{TD})/A \qquad (15.2)$$

where

OTTV_w = overall thermal transfer value, BTU/hr-ft²

U_w, U_g = U-values for exposed opaque wall, glass, BTU/hr-ft²-F

A_w, A_g = area of exposed opaque wall, glass, ft²

SF = solar factor, BTU/hr-ft² (Figure 15.1 on page 388)

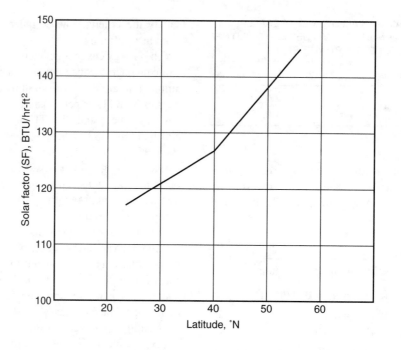

Figure 15.1
Solar factor (SF) values for use in Equation 15.2.
(Adapted from the California Building Energy Efficiency Standards, 1988.)

SC = shade coefficient (Table 6.6)

TD_w = wall temperature difference, F (Table 15.2).

TD = design inside–outside temperature difference, F

A = total exposed area of walls and glass, ft^2

Figure 15.1 lists the SF values for use with Equation 15.2.

Maximum recommended allowable values of the OTTV$_w$ for nonresidential buildings are shown in Figure 15.2.

A similar approach is used to calculate the OTTV$_r$ for roofs (not shown here).

TABLE 15.2   VALUES OF TD$_W$ FOR USE WITH EQUATION 15.2

Wall weight, lb/ft^2	0–25	26–40	41–70	Over 70
TD$_W$, F	44	37	30	23

The following example illustrates the application of the component performance method.

Example 15.2

A building has the following specifications:

$A_w = 8000$ ft^2, $U_w = 0.20$ BTU/hr-ft^2-F

$A_g = 3000$ ft^2, $U_g = 0.65$

Design TD = 15 F. Location = 36° N latitude

Wall weight = 35 lb/ft^2. SC = 0.61

Determine the wall OTTV value, and compare it with the maximum allowable.

Solution

Equation 15.2 will be used. From Table 15.2, TD$_w$ = 37 F. From Figure 15.1, SF = 125.

$$OTTV_w = (0.20 \times 8000 \times 37 + 3000 \times 125$$
$$\times 0.61 + 0.65 \times 3000 \times 15)/11,000$$
$$= 28.9 \text{ BTU/hr-ft}^2$$

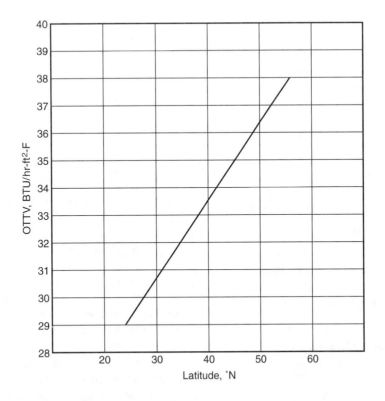

Figure 15.2
Recommended overall thermal transfer values (OTTV) for walls, nonresidential buildings.
(Adapted from the California Building Energy Efficiency Standards, 1988.)

From Figure 15.2, the maximum $OTTV_w = 32.4$. The building meets the standards.

The system performance and energy cost budget methods are also applicable to cooling design requirements.

Infiltration A typical code infiltration standard is shown in Table 3.4.

Design Conditions Typical code indoor design conditions of temperature, humidity, and ventilation were shown in Tables 1.1 and 6.17. Outdoor design conditions are as shown in Table A.9.

There are further energy code standards that apply to use of energy-conserving HVAC controls, minimum energy efficiency performance standards of equipment, and other related factors. Since these subjects are discussed in appropriate locations in this book, and the codes are quite specific, we will not consider this further here.

The energy codes also include prescriptions for lighting and service hot water heating.

15.2 SOURCES OF ENERGY

The energy sources that are available for HVAC system operation can be classified in two groups—*depletable* (nonrenewable) and *nondepletable* (renewable). Depletable energy is stored in fossil fuels: coal, oil, and gas. These sources are quite limited in supply in relation to the rate that they are being used, and therefore will be depleted within a relatively short time (estimates range from 20–100 years for gas and oil to 1000 years for coal). Nondepletable energy sources have an expected long life in terms of human existence, so that they can be considered everlasting. Examples are the sun's radiation, nuclear energy, wind, ocean currents and tides, and geothermal energy (hot water or steam from deep underground). The nondepletable sources are often spread out or diffuse and therefore are very difficult to utilize. However, two principles should always be considered when selecting energy sources for HVAC systems:

1. It is always desirable to minimize the use of energy derived from depletable sources. These sources will eventually be depleted and they are expensive.
2. It is desirable to use energy from nondepletable sources. Of course this energy should still be used efficiently. Even though the energy is sometimes free of cost, the cost of equipment to utilize it increases as more energy is used.

15.3 PRINCIPLES OF ENERGY UTILIZATION

To be able to evaluate possible energy conservation techniques, it is important to understand some principles of energy use and availability. These principles are largely based on the First and Second Laws of Thermodynamics, discussed in Chapter 2, which should be reviewed. The implications of these laws for energy conservation will be emphasized here.

The stored energy in a fossil fuel is released through combustion, in the form of heat. The heat may be used directly if needed or the heat energy may be converted into work or power if needed. (Power is the rate of doing work.)

When the energy in the fuel is released, it is theoretically possible to use all of the equivalent heat released, as seen in Figure 15.3. However, some of the heat energy released is always wasted. It would be too expensive to build equipment that would have negligible losses. Figure 15.4 is an example of a more realistic situation.

For each 100 BTU of energy released from the fuel in the boiler, about 20 BTUs are lost in the

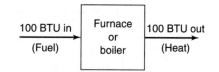

Figure 15.3
Maximum energy available from furnace or boiler.

form of the hot combustion gases in the flue. (There are other small losses not shown in the example, such as radiation of heat from the surface of the equipment.) Some of the major efforts in energy conservation are directed to recovering this wasted heat.

Fuel is often burned to heat air, water, or steam for a direct use such as heating of buildings. Another major use, however, is to generate heat that is then converted into power. An example is the generation of electrical power, as shown in Figure 15.5.

Combustion of fuel in the boiler heats steam that is used to drive a steam turbine. The heat in the steam has been converted into mechanical power in the turbine. The turbine drives an electrical generator, creating electrical power. In this case, however, there is a very severe limit on how much of the heat energy can be converted into work, according to the Second Law of Thermodynamics. Under the best practical circumstances, only about 30 BTU of the 80 BTU delivered to the turbine can be converted into work. The remainder (50 BTU) must be rejected in the form of heat. Of course, this heat may still have a use.

Figure 15.4
Typical actual energy available from furnace or boiler.

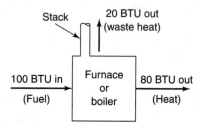

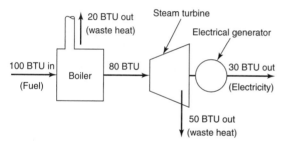

Figure 15.5
Typical actual energy available from power plant.

These facts lead to further recommendations concerning energy conservation.

1. It is generally unnecessary and often wasteful to convert heat into power and then back into heat again. An example of this is generating electricity at a power plant, as above, and then delivering it to buildings, where through electric resistance heaters, it is converted back to heat. As seen in the examples, only 30 BTUs of the original 100 BTUs in the fuel is used, whereas 80 BTUs would be available if the heat was used directly. There are sometimes other advantages to using electricity directly for heating (convenience, control), but the energy conservation aspects should always be considered.
2. If heat is converted into power, the remainder of the heat that cannot be converted into power should be used for heating and not wasted, if practical. This is the basis for total energy systems, to be discussed later.

Some other effects related to energy conservation, also explained in Chapter 2, are summarized here:

3. Friction causes loss of useful energy and should be minimized, especially in pipe and duct flow.
4. Rapid expansion of fluids causes loss of useful energy and should be minimized. An example where this may occur is in pressure reducing devices.

5. Mixing of fluids at different temperatures causes loss of useful energy. This occurs in dual duct and three-pipe systems and should be minimized.

15.4 MEASURING ENERGY UTILIZATION IN POWER-PRODUCING EQUIPMENT (EFFICIENCY)

In order to decide whether or not equipment is utilizing energy efficiently, there must be standards for measuring its performance. Some of these will be discussed here. Figure 15.6 shows a schematic arrangement of a power-producing device such as a turbine or engine which receives heat and converts some of the heat into power.

Heat (Q_1) flows from the heat source (usually a fuel) at a high temperature to the engine which converts some of this heat to the power (P) which it produces. The remainder of the heat (Q_2) is rejected to a heat sink at a lower temperature. This is usually water or the atmosphere. The values of Q_1, Q_2, and P are related to each other, according to the Energy Equation, as

$$P = Q_1 - Q_2 \qquad (15.3)$$

The efficiency (E_p) of an engine in percent is defined as

$$E_p = \frac{\text{power output}}{\text{equivalent heat input}} \times 100 = \frac{P}{Q_1} \times 100$$
$$(15.4)$$

Figure 15.6
Energy flow for an engine.

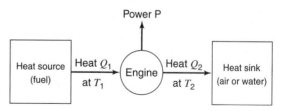

The power and heat must be expressed in the same units.

Equation 15.4 is a useful measure of energy conservation because we always wish to get the maximum conversion of heat Q_1 to power P, and as small an amount of waste heat Q_2 as possible.

Example 15.3

A Diesel engine-generator is being used to generate electricity in a hospital. The operating engineer wants to know how efficiently it is operating. Over a period of one hour, the engineer measures 5 gallons of fuel oil consumed and 50 KWH of electricity produced. The fuel oil has a heating value of 140,000 BTU/gal. What is the efficiency of the engine-generator?

Solution

Using Equation 15.5, after converting terms into the same units,

$$Q_1 = 5\ \frac{gal}{hr} \times 140,000\ \frac{BTU}{gal}\ \frac{1KWH}{3410\ BTU}$$

$$= 205\ KWH$$

$$E_p = \frac{P}{Q_1} \times 100 = \frac{50\ KWH}{205\ KWH} \times 100$$

$$= 24.4\%$$

The engine in Example 15.3 is converting 24% of the energy in the fuel into electrical power. The remainder is converted into heat which usually is wasted to heat sinks, in this case to both the combustion gases going out the stack and the engine cooling water. Much of this wasted energy could be recovered if heating is needed. A heat exchanger might be used to heat water from the hot combustion gases. This is an example of *heat recovery equipment*.

Example 15.4

An energy consultant called in to investigate using the heat wasted in the engine in Example 15.3 estimates that a heat exchanger will recover 50% of the wasted heat. How much heat can be made available?

Solution

The wasted heat, as seen from Figure 15.6, is Q_2. Using Equation 15.3, solving for Q_2,

$$Q_2 = Q_1 - P = 205 - 50 = 155\ KW$$

$$= 155\ KW \times 3410\ \frac{BTU/hr}{KW}$$

$$= 528,550\ BTU/hr$$

of which 50% can be recovered, or 264,300 BTU/hr.

The definition of efficiency given previously is correct in relation to the engine efficiency, but in HVAC applications heat, when used, is just as useful a form of energy as power. An overall efficiency could therefore be defined when heat is also used, as

$$E_0 = \frac{\text{useful energy}}{\text{energy input}} \times 100 \qquad (15.5)$$

The same units must be used for both energy terms.

Example 15.5

If the recommendations of the energy consultant in Example 15.5 are followed, what is the overall efficiency of energy use?

Solution

The electrical energy and the recovered heat are both useful:

$$\text{Useful energy} = P + 0.5\ Q_2$$

$$= 50\ KW + 264,300$$

$$\times \frac{1\ KW}{3410\ BTU\ hr}$$

$$= 127.5\ KW$$

$$E_0 = \frac{127.5\ KW}{205\ KW} \times 100 = 62.2\%$$

Notice the tremendous energy conservation achieved here—more than doubling the useful energy of the total energy consumed, as compared to before.

Example 15.6

If the heat recovery device in Example 15.5 were not installed, the building management would have to purchase steam for the heat needed, at a cost of $6.00 per thousand pounds. The amount of recovered heat is needed for 1500 hours per year. How much will be saved on the energy bill if the device is installed?

Solution

Using a figure of 1000 BTU/lb for the latent heat given up by the steam, the savings are

$$\text{Savings} = 264{,}300 \; \frac{\text{BTU}}{\text{hr}} \times \frac{1 \text{ lb}}{1000} \text{BTU}$$

$$\times \frac{\$6.00}{1000 \text{ lb}} \times 1500 \text{ hr} = \$2380$$

Although the actual efficiency of an engine expresses how much energy is useful, it does not show what the best possible efficiency could be. There is a limiting maximum value of the efficiency of any engine that converts heat into power, as can be shown from the Second Law of Thermodynamics. This is expressed by the following equation:

$$E_m = \frac{T_1 - T_2}{T_1} \times 100 \tag{15.6}$$

where

E_m = maximum efficiency of an engine

T_1 = temperature at which heat is received from heat source

T_2 = temperature at which heat is rejected to heat sink

In this equation, temperatures must be expressed in absolute units.

Example 15.7

The combustion gases driving an engine are at 2000 F and they are exhausted to the stack at 600 F. What is the maximum efficiency the engine could have?

Solution

Using Equation 15.6 with the temperatures expressed in Rankine (absolute units),

$$T_1 = 2000 \text{ F} + 460 = 2460 \text{ R}$$

$$T_2 = 600 \text{ F} + 460 = 1060 \text{ R}$$

$$E_m = \frac{T_1 - T_2}{T_2} \times 100 = \frac{1400}{2460} \times 100 = 57\%$$

Example 15.7 shows that the maximum possible efficiency of an engine is always much less than 100%. This is important to realize, because in some cases, it may not be worthwhile to seek very small possible improvements in efficiency.

15.5 MEASURING ENERGY CONSERVATION IN COOLING EQUIPMENT—THE COP AND EER

Refrigeration and air conditioning equipment consumes rather than produces power—the opposite of engines. Efficiency, as described in the previous section, is a measure of the relative power output of an engine, and therefore is not meaningful when applied to power consuming equipment. The *Coefficient of Performance* (COP) is a useful measure devised to measure and compare performance of air conditioning and refrigeration equipment.

Figure 15.7 shows the energy flows in a refrigeration machine. Heat (Q_1) flows from the heat source at a low temperature, cooling it. This is the useful refrigeration effect. Power (P) is required in a compressor. Heat (Q_2) flows to the heat sink. According to the Energy Equation,

$$Q_2 = Q_1 + P \tag{15.7}$$

Figure 15.7
Energy flow for a refrigeration machine.

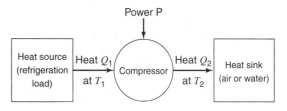

The COP is defined as

$$COP = \frac{\text{useful cooling capacity}}{\text{equivalent power input}} = \frac{Q_1}{P} \qquad (15.8)$$

Both terms must be expressed in the same energy units. The COP is a good measure of energy conservation because we always want to get the maximum amount of cooling Q_1 with the minimum value of power input P.

Example 15.8

What is the COP of a model PCW-030T (Table 13.2) refrigeration water chiller when producing chilled water at 44 F and with entering condensing water temperature of 85 F?

Solution
From Table 13.2, the capacity is listed as 27.6 tons with 25.0 KW required motor power. Using Equation 15.8, after converting to the same units for both terms,

$$Q_1 = 27.6 \text{ tons} \times \frac{12,000 \text{ BTU/hr}}{1 \text{ ton}}$$

$$\times \frac{1 \text{ KW}}{3410 \text{ BTU/hr}}$$

$$= 96.8 \text{ KW}$$

$$COP = \frac{Q_1}{P} = \frac{96.8 \text{ KW}}{25.0 \text{ KW}} = 3.9$$

Another measure of energy conservation quite similar to the COP is the *Energy Efficiency Ratio* (EER). It is defined as

$$EER = \frac{\text{useful cooling capacity in BTU/hr}}{\text{power input in watts}} = \frac{Q_1}{P}$$

$$(15.9)$$

The EER has exactly the same two terms as the COP, only they are expressed in different units, as noted. The EER has been developed because it is easier for the consumer to use and understand. It has already become a legal requirement for manufacturers to label the EER values for certain types of air conditioning equipment. These EER values are measured at a standard set of temperature conditions, so that equipment is comparable.

Example 15.9

A consumer shopping for a window air conditioning unit sees a 9200 BTU/hr capacity unit in a store, with a labeled EER rating of 6.2. Another unit does not have an EER label, but a sales brochure lists the cooling capacity at 9200 BTU/hr and the power input at 1200 watts, at standard conditions. Which unit will be the most efficient in energy use?

Solution
Using Equation 15.9 to compute the EER of the second unit,

$$EER = \frac{9200 \text{ BTU/hr}}{1200 \text{ W}} = 7.7$$

The second unit has a significantly better energy use efficiency. The power consumption of the first unit at full load would be

$$P = \frac{Q_1}{EER} = \frac{9200 \text{ BTU/hr}}{6.2 \text{ BTU/hr/W}} = 1480 \text{ W}$$

and would therefore be less desirable. Of course, the consumer should find out why the second unit does not have the required label!

Any compressor-driven cooling device has a maximum possible COP, similar to the maximum efficiency of an engine. This is expressed by the following equation:

$$COP_m = \frac{T_1}{T_2 - T_1} \qquad (15.10)$$

where

COP_m = maximum possible COP for a cooling device

T_1 = temperature at which heat is absorbed from cooling load

T_2 = temperature at which heat is rejected to heat sink

The temperatures must be in absolute units in the equation.

Example 15.10

A refrigerating unit operates at an evaporating temperature of 40 F and a condensing temperature of 100 F. What is the maximum coefficient of performance it could have?

Solution

Using Equation 15.10, with temperatures in Rankine units,

$$COP_m = \frac{T_1}{T_2 - T_1} = \frac{500}{60} = 8.3$$

15.6 MEASURING ENERGY CONSERVATION IN THE HEAT PUMP

When refrigeration equipment is used as a heat pump for heating (Chapter 13), a modified form of the COP is used as a measure of efficient energy use, as follows:

$$COP_h = \frac{\text{useful heat output}}{\text{heat equivalent of power input}} = \frac{Q_2}{P}$$

$$(15.11)$$

The maximum possible COP for a heat pump when used for heating is expressed by Equation 15.12:

$$COP_{hm} = \frac{T_2}{T_2 - T_1} \qquad (15.12)$$

where T_2 and T_1 have the same meaning as in Equation 15.10 and are, as before, in absolute units.

Example 15.11

A contractor is offered a heat pump that is claimed to have a heating capacity of 800,000 BTU/hr with a compressor requiring 25 KW, while operating between 30 F and 100 F. The price is right, so the contractor buys it. Was this a good decision?

Solution

The claimed COP should be compared with the maximum possible COP:

$$COP_h = \frac{800,000 \text{ BTU/hr}}{25 \text{ KW} \times 3410 \text{ BTU/hr/KW}} = 9.4$$

$$COP_{hm} = \frac{T_2}{T_2 - T_1} = \frac{560}{70} = 8.0$$

The claimed performance is better than the maximum possible under the conditions, so the contractor is being misled. Actually any real COP would be much lower than 8.0 due to energy losses from friction and other causes.

15.7 MEASURING ENERGY CONSERVATION IN HEATING EQUIPMENT

For equipment that converts energy directly to heat (boilers, furnaces), efficiency (E) defined as a measure of energy conservation as follows:

$$E = \frac{\text{useful heat output}}{\text{equivalent heat input}} \times 100 \qquad (15.13)$$

As discussed previously, unlike engines and refrigeration equipment, there is no theoretical limit below 100% on efficiency in heating equipment. Practically, however, efficiency never approaches 100%, except for electric resistance heating. There are always some energy losses, and a reasonable balance must be made between equipment cost and efficiency. The term *heating Coefficient of Performance* (COP_h), similar to that for the heat pump, is also used to express efficiency of heating equipment.

Example 15.12

A packaged gas-fired steam boiler has a catalog rated output of 2 million BTU/hr and a listed fuel consumption of 2510 ft^3/hr of natural gas with a heating value of 1000 BTU/ft^3. What is the boiler full load efficiency?

Solution

The heat input of the boiler will be calculated, and then Equation 15.13 will be used:

$$\text{Input} = 2510 \text{ ft}^3 \text{ hr} \times 1000 \text{ BTU/lb}$$

$$= 2,510,000 \text{ BTU/hr}$$

$$E = \frac{2,000,000}{2,510,000} \times 100 = 79.7\%$$

Equipment Performance Standards

Minimum energy efficiency requirements have been developed for HVAC equipment. Table 15.3 is an example of this type of standard. The requirements are for seasonal performance. AFUE is the seasonal fuel utilization efficiency (Chapter 4) of heating equipment. SEER is the seasonal energy efficiency ratio of cooling equipment. HSPF is the heating seasonal performance factor of a heat pump, defined as the heat output in BTU/hr divided by the energy input in watts.

15.8   MEASURING ENERGY CONSERVATION IN PUMPS AND FANS

Efficiency of energy use in pumps and fans is defined as

$$E = \frac{\text{power output}}{\text{power input}} \times 100 \qquad (15.14)$$

Both terms must be in the same units. If the power input is taken as the power at the fan or pump shaft, it is called brake horsepower (BHP), and the efficiency is that of the pump or fan alone. If the power input is that required at the motor, the combined efficiency of the motor and pump or fan is found. The power output of a pump was given in Equation 11.1.

TABLE 15.3   HVAC EQUIPMENT MINIMUM PERFORMANCE[1]

Furnaces		
Gas-fired	70% AFUE	(78% AFUE)
Oil-fired	75% AFUE	(78% AFUE)
Boilers		
Gas-fired	70% AFUE	(80% AFUE)
Oil-fired	75% AFUE	(80% AFUE)
Central Air Conditioners		
Split system	9.5 SEER	(10.0 SEER)
Single package	9.5 SEER	(9.7 SEER[2])
Heat Pumps		
Split system	8.5 SEER	(10.0 SEER)
	2.6 COP @ 47°F db	(6.8 HSPF)
Single package	8.5 SEER	(9.7 SEER[2])
	2.6 COP @ 47°F db	(6.6 HSPF[2])

[1] Except as provided in footnote 2, performance requirements in parentheses applicable to equipment manufactured after January 1, 1992.
[2] Applicable to single package systems manufactured after January 1, 1993.
Reprinted from the New York State Energy Conservation Construction Code, 1991.

Example 15.13
An operating engineer wants to determine the efficiency of a condenser water pump and motor. The pump flow rate is 260 GPM, with a total head of 40 ft. The electrical meters show the pump motor is drawing 4.2 KW. What is the pump–motor efficiency?

Solution
Finding the power output (Equation 11.1),

$$\text{WHP} = \frac{260 \times 40}{3960} = 2.63 \text{ HP}$$

$$= 2.63 \text{ HP} \times 0.746 \text{ KW/HP} = 1.96 \text{ KW}$$

Using Equation 15.14,

$$E = \frac{1.96 \text{ KW}}{4.2 \text{ KW}} \times 100 = 46.7\%$$

After determining the energy use efficiency of equipment by one of the above methods, specific recommendations for equipment improvement can be made, if needed. Although many of these recommendations have been discussed previously with each type of equipment, they will be referred to again later in this chapter. Before any steps are taken, however, an energy analysis of the whole system should be made.

15.9 MEASURING ENERGY USE IN EXISTING BUILDING HVAC SYSTEMS

Before individual energy conservation steps can be decided upon, an analysis of the energy consumption of the system should be carried out. In existing buildings, this is best done by an actual hourly measurement of energy used throughout the system for the whole year. This is often not fully possible because of the lack of suitable instrumentation. It is recommended that instruments be added where practical and that this procedure be carried out.

The most complete procedure in analyzing energy use in existing building includes measuring energy consumption of each piece of equipment, flow rates, temperatures, and pressures; noting types of systems, controls, and their operations. Other interrelated system operations would also be analyzed (e.g., lighting). The physical conditions of the building would be noted. The results will suggest what conservation steps should be implemented.

The amount of detail to which the analysis is carried out largely depends on how much conservation gain is expected and the cost of the analysis itself. It is not possible to describe one single energy analysis procedure, because each case will be different. A list of some energy conserving measures that may result from studies will be listed later.

If it is not possible to measure the actual energy consumption in an existing building, a reasonably accurate estimate can be made by an energy simulation of the system operation, as is done in the design stages of new buildings. This will be explained in the next section.

If the cost of a detailed energy analysis is prohibitive, a "walk-through" survey and recommendations by an experienced energy consultant may suffice, but are not usually as effective.

15.10 MEASURING ENERGY USE IN NEW BUILDING HVAC SYSTEMS

A thorough energy conservation study requires determination of how much energy a building uses. In new building design, this requires predicting energy consumption. The most complete analysis would consist of the following steps, in order:

1. Calculation of building heating and cooling loads on an hourly basis for the whole year. A complete heat balance is carried out, taking into account heat storage, internal gains, and changing climatic conditions.
2. Simulation of the system performance. This involves describing the performance or each component of the system by equations, and then solving these equations using the values of loads calculated above. Load requirements of boilers and refrigeration equipment are

determined. Duct and piping flow rates and pressure drops are calculated, from which pump and fan power requirements are determined. Seasonal changes are accounted for, such as proportions of outside air used.

3. The equipment loads are transformed into energy requirements by utilizing efficiency characteristics of the equipment. Using unit fuel costs, the total energy costs can then be calculated.

These steps may be done for every hour of the year or perhaps for a selected number of hours and then summed up for the year.

To utilize it as an energy conservation tool, the analysis is carried out for different building construction features to see which uses less energy. For example, the type and amount of glass may be changed, or varied solar shading devices may be considered. Different types of HVAC systems may also be compared to see which is most efficient.

These procedures usually require the use of computers. The calculations are far too lengthy and complex to attempt them manually. Many computer programs are available that will handle these tasks satisfactorily.

A detailed energy conservation analysis is required in most large buildings because of new code design standards and also because of client demand. However, for initial approximate energy estimates, simpler procedures are available. These may be adequate for a final energy analysis for small buildings, particularly residences.

15.11   THE DEGREE DAY METHOD

The degree day method is a well-established single calculation procedure for estimating energy requirements for heating. It can be done for the whole year or on a monthly basis. It is based on the assumption that a building will be in a heat balance with its surroundings at a certain average daily outside temperature, called the *balance temperature.* In the past, this temperature was assumed to be 65 F. Over a one-day period at this average tempera-

ture, solar and internal heat gains produce heat storage effects that result in no heat requirements, yet a comfortable temperature is maintained.

It is assumed that the heating load is proportional to the temperature difference between 65 F and the average daily winter outdoor temperature. This leads to the *degree day* concept. For example, if the average daily temperature is 50 F on a given day, that day has 15 degree days (65 F–50 F). An average 35 F day would have 30 degree days. Having twice the number of degree days, the heating load would be expected to be twice as much on the 35 F day as on the 50 F day.

The degree day is therefore based on weather data, and can be calculated from known average daily temperatures, as above. This data can be summed up for a month or a whole heating season. Tables 18.4 and A.6 list expected degree days for a number of cities.

The degree day can be used directly to estimate the heating energy requirements of a building after the design heating load is determined.

Equation 15.15, resulting from the degree day concept, will give the heating requirements over a time period desired:

$$Q_0 = \frac{Q}{TD} \times D \times 24 \qquad (15.15)$$

where

Q_0 = heat energy consumed over period considered, BTU

Q = design heating load, BTU/hr

TD = design temperature difference, F

D = number of degree days per period

Example 15.14 _____

A building has a design heating load of 86,000 BTU/hr. The design temperature difference is 70 F. On a day with an average temperature of 30 F, how much heat is required for the whole day?

Solution

The number of degree days on this day is $D = 65 - 30 = 35$. Using Equation 15.15,

$$Q_0 = \frac{86{,}000 \text{ BTU/hr}}{70 \text{ F}} \times 35 \text{ F} \times 24 \text{ hr}$$

$$= 1{,}030{,}000 \text{ BTU}$$

Example 15.14 is used to explain how the degree day is used, but the procedure needs to be modified in practical use. In the first place, recent investigation has found that, in modern homes, the assumption of 65 F outdoor temperature as that at which no heat loss occurs is often too high, because of increased internal heat gains and better insulation. This is accounted for by a correction factor C_D (Figure 15.8). The values shown are typical but may vary considerably due to factors such as type of construction and use.

A correction factor E, which is essentially the Annual Fuel Utilization Efficiency, is used to account for equipment seasonal performance effects. Values of E may range from 0.5 to 0.9 for the heating equipment, depending on factors such as the steady-state efficiency, energy conservation devices, and operating procedures (see Chapter 4).

For electric resistance heating, a value of 1.0 for E should be used, because all of the input energy is converted to useful heat.

The inclusion of the two correction factors results in a *modified degree day* method and equation for calculating the seasonal energy requirements, as follows:

$$Q_0 = \frac{Q}{\text{TD}} \times D \times 24 \times C_D \times \frac{1}{E} \qquad (15.16)$$

where

Q_0, Q, TD, D are as in Equation 15.15, and

C_D = correction factor for degree days

E = correction factor for heating equipment seasonal performance

The design heating load Q_0 should include an allowance for duct or pipe heat losses to unconditioned space (see Chapter 3) and additional infiltration allowance from stack effect, if significant. The *ASHRAE Systems Handbook* has further discussion on this and related subjects.

Figure 15.8
Correction factor C_D for degree days. (Reprinted with permission from the *1980 Systems, ASHRAE Handbook & Product Directory*.)

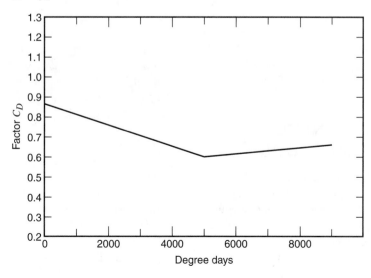

Example 15.15 _____
A residence located in Milwaukee, Wisconsin has a design heating load of 62,000 BTU/hr. Inside design temperature is 70 F. Estimate the annual energy consumption.

Solution
From Table A.6, $D = 7863$ degree days, TD = 70 − (−7) = 77 F. From Figure 15.8, $C_D = 0.65$. Assuming $E = 0.55$, using Equation 15.16,

$$Q_0 = \frac{62,000}{77} \times 7640 \times 24 \times 0.65 \times \frac{1}{0.55}$$

$$= 174,500,000 \text{ BTU/yr}$$

Using the heating value of the fuel and its unit cost, the estimated annual energy cost can be found.

Example 15.16 _____
Gas with a heating value of 1000 BTU/ft^3 and a cost of $0.60 per 100 ft^3 is used to heat the house in the previous example. What is the expected average yearly heating cost?

Solution

$$\text{Cost} = 174,500,000 \text{ BTU} \times \frac{1 \text{ ft}^3}{1000 \text{ BTU}}$$

$$\times \frac{\$0.60}{100 \text{ ft}^3}$$

$$= \$1047$$

The reader should compare the estimated heat required in Example 15.15 with that found if the unmodified degree day equation is used.

15.12 OTHER ENERGY MEASURING METHODS

A simple method for estimating cooling load energy requirements is based on the *equivalent full load hours* of operation of summer cooling equipment. This is equal to the total number of summer hours that equipment would operate at full load to produce the same amount of seasonal cooling as

the actual number. The equivalent full load hours are found from experience in each city.

Example 15.17 _____
A building has design cooling load of 210 tons. Equivalent full load hours for the summer are 800. The refrigeration and air conditioning equipment requires 1.5 KW per ton. Electricity costs $0.07/KWH. What is the energy cost for a summer season?

Solution

Energy = 800 hr × 210 tons × 1.5 KW/ton

$\quad\quad$ = 252,000 KWH

Cost = 252,000 KWH × $0.07/KWH

$\quad\quad$ = $17,640

Equivalent full load data can often be obtained from utility companies. This method is very approximate and is recommended only for preliminary energy conservation studies.

The degree day method can also be used for estimating seasonal cooling load energy requirements. Cooling degree day values are available from weather data. An equation similar to 15.15 is used to determine the cooling energy for the cooling season.

The degree day methods do not fully account for seasonal variations in equipment efficiency, the balance point, and similar factors. The *bin method* of estimating energy use attempts to do this. The outdoor temperatures are divided into a number of intervals (bins) from highest to lowest. In each bin, the energy consumed is calculated, using the modified degree day method. The sum is then the annual energy used.

As mentioned before, energy estimating methods that do an hourly analysis, including the simulation of equipment performance, provide the most accurate results; this type of method may be required to be used in meeting energy codes and standards. The most widely known is DOE 2.2, developed by the U.S. Department of Energy. Software is available for implementing this analysis.

15.13 AIR-TO-AIR HEAT RECOVERY

One of the most beneficial ways of conserving energy is to recover heat that is generated during building operations which would otherwise be wasted. After an energy analysis is carried out, the major energy losses are known and can be considered for recovery. A common source of useful waste heat is the exhaust air from HVAC systems. This air can be used to preheat outdoor ventilation air in the winter using air-to-air heat exchangers. In the summer, the reverse heat transfer can be accomplished, reducing cooling requirements. In some applications, there are other similar sources of waste heat (e.g., exhaust air from an industrial oven). Figure 15.9 shows the operation of an air-to-air heat recovery device. Some types of this equipment will now be described.

The Heat Wheel

This air-to-air heat exchanger has a rotating wheel that is packed loosely with wire mesh or similar material (Figure 15.10).

One side of the wheel is ducted to the exhaust air and the other side is ducted to the outdoor air. The warm exhaust air (in winter) flows through the packing material, heating it. As the wheel slowly rotates, the heated packing is exposed to the cold airstream, which picks up this heat. If there is a significant amount of latent heat to be exchanged, a moisture-absorbing material may be used.

A small proportion of exhaust air is always carried over by the wheel to the fresh airstream. If cross-contamination is a factor, as in some medical facilities, the contaminant must be removed. Condensation or freezing of moisture may also be a problem.

The heat wheel is a simple and effective device. It can recover up to 80% of the available heat in a typical application.

Example 15.18

A heating system uses 20,000 CFM of ventilation and exhaust air. The exhaust air temperature is 72 F and the average intake air temperature is 35 F. The system operates 1000 hours a season. The building owner is considering saving energy by using a heat wheel to heat the intake air with the exhaust air. A manufacturer offers a wheel that will heat the intake air to 61 F. Fuel oil with a heating value of 140,000 BTU/gal, and costing $0.90 a gallon, is used to heat the building. How much energy, oil, and money would be saved if the heat exchanger were installed?

Solution

The energy saved is found from the sensible heat equation for preheating the intake air.

$$Q = 1.1 \times CFM \times TC$$
$$Q = 1.1 \times 20{,}000 \times (61 - 35)$$
$$= 572{,}000 \text{ BTU/hr}$$
$$= 572{,}000 \text{ BTU/hr} \times 1000 \text{ hr}$$
$$= 572{,}000{,}000 \text{ BTU}$$

Figure 15.9
Air-to-air heat recovery device typical operation.

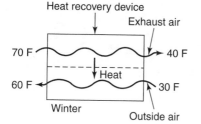

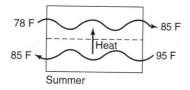

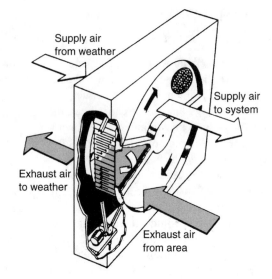

Figure 15.10
Heat wheel. (Courtesy: Cargocaire Engineering Corp.)

The fuel oil saved is

$$572,000,000 \text{ BTU} \times 1 \text{ gal}/140,000 \text{ BTU}$$
$$= 4086 \text{ gal}$$

The cost saving is

$$4086 \text{ gal} \times \$0.90/\text{gal} = \$3677$$

A small amount of energy is required by the fan to move the air through the wheel. This should be accounted for to find the net savings.

The Plate-Type Heat Exchanger

A heat exchanger composed of fixed plates separating the two streams of air is shown in Figure 15.11.

Because the airstreams are completely separated, no contamination can occur. The proportion of heat recovered is usually 40–60%.

The Run-Around Coil

This consists of two conventional air conditioning coils piped together in series (Figure 15.12).

Water or an antifreeze solution is circulated between the coils by a pump. The liquid picks up the heat from the warm airstream and gives up heat to the cold stream. An advantage of the run-around system is that the airstreams can be located far from each other. Heat recovery efficiency ranges from 40–60%.

The Heat Pipe

This device is a closed tube containing a porous capillary wick around the inside surface and a working fluid (Figure 15.13).

Figure 15.11
Plate-type heat exchanger.

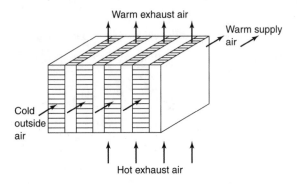

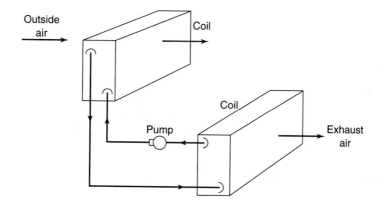

Figure 15.12
Run-around coil.

The warm airstream flows over one end of the tube, and the heat evaporates the liquid. The vapor flows because of its pressure to the other end of the tube. The cold airstream flowing over this end of the tube absorbs heat from the fluid, which condenses. The liquid is absorbed in the wick and migrates through it to the warm end of the tube, completing its circuit.

Heat pipes can be stacked in banks according to the capacity needed.

15.14 REFRIGERATION CYCLE HEAT RECOVERY

Most large buildings require cooling in interior zones and heating in exterior zones at the same time during some periods of the year. The heat rejected to the condenser in the refrigeration plant, otherwise wasted, may be a source for part of this heating requirement.

When water chillers are used, a special *double-bundle condenser* may be used to recover this heat (Figure 15.14 on page 404).

One tube bundle in the condenser is connected to the cooling tower and is used in the conventional manner when no heating is required. Another tube bundle is connected to the perimeter hot water heating system. The condenser water is directed through this circuit when hot water is required. A separate tube bundle is used because the water circulated to a cooling tower may become quite dirty and would soon foul the closed hydronic system.

The *heat pump* of course is designed to use the condenser heat for heating. Because the theory of heat pump operation has been described in Chapter 13, it will not be repeated here. Simultaneous heating and cooling can be achieved by utilizing both the rejected condenser heat and evaporator cooling at the same time.

Another method of conserving energy is to use the refrigeration cycle for *free cooling*. When the outdoor air wet bulb temperature is low, the refrigeration compressor is not operated, but the cooling tower and condensing water pumping system are

Figure 15.13
Heat pipe.

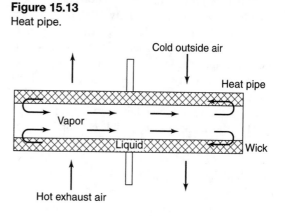

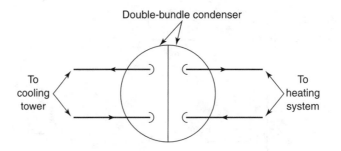

Figure 15.14
Double-bundle condenser for recovery of refrigeration system heat.

operated. The water from the cooling tower is cooled by evaporative cooling (Chapter 7) to a temperature suitable for air conditioning. This water is too dirty to circulate through the chilled water hydronic system, however, so a heat exchanger must be included to transfer heat from the chilled water system, which is operated in its usual manner. The free cooling operation is effective in the fall and spring when wet bulb temperatures are low and the refrigeration load is light. There are other variations on this arrangement that will not be discussed here.

15.15   THERMAL STORAGE

The building heating or cooling equipment can be used to heat or cool a medium which is then used to supplement or replace the heating or cooling effect of the equipment at a later time. This procedure is called *thermal storage.*

With *cool storage,* the refrigeration equipment is used to produce a large quantity of a cold medium stored in tanks such as cold water, brine, or ice (during off-peak times). When needed, this stored medium is then used to supplement the refrigeration equipment. For instance, the stored cold medium can be used in a heat exchanger to provide chilled water for the air handling units' cooling coils.

There are a number of possible economic incentives that make cool thermal storage attractive. Most of these incentives arise from the fact that, in most buildings, air conditioning peak loads occur only for brief periods (perhaps 5% of the time), and average loads are often a fraction of peak loads.

Among the economic reasons for considering use of cool thermal storage are:

1. Smaller refrigeration equipment. The higher loads are handled by the smaller refrigeration equipment supplemented by drawing down the stored cooling medium effect.
2. Smaller air handling equipment and ducts. In the design of cool storage systems, the storage medium is often cooled to a temperature much lower than the chilled water would be cooled to with conventional systems. Subsequently, the chilled water for the air conditioning cooling coils can be lower, say 35 F instead of 45 F. This in turn makes it possible to cool the supply air to say 50 F instead of 60 F. As a result, a much smaller flow rate of supply air can be used to achieve the same cooling effect, reducing the size of ducts, fans, coils, etc.
3. Energy cost savings. Most utility companies offer a lower electrical demand and/or use rate at night and on weekends.
4. Energy use savings. The refrigeration equipment may require less total energy because it is operating closer to full load more of the time, where its efficiency may be higher.

In addition, at night the condensing temperature will be lower, which also increases the efficiency of the refrigeration equipment.

If the refrigeration unit cools the storage medium to a temperature lower than the conventional chiller arrangement, this of course will decrease the efficiency of the equipment. In this case, an economic analysis must be made of the trade-off with reducing air-side equipment size and energy use.

In any case, a thorough economic analysis is required of the various options and conditions to determine the proportion of load to be handled by the storage system.

Heat thermal storage can also be considered for projects. The stored heat may be used for supplemental heating or for heating service hot water.

Sources of heat may be refrigeration equipment condenser water or solar energy. The storage medium may be water or other liquids, or solids such as rocks. The building mass itself may serve as a passive storage element (see Chapter 18).

15.16 LIGHT HEAT RECOVERY

A considerable amount of heat is generated by the high level lighting systems in modern buildings. There are numerous ways of recovering this heat. Room return air can be circulated around special water-cooled lighting fixtures that are used to provide hot water. In addition to furnishing heat for any required uses, removing part of the lighting heat from the room also reduces the cooling load, thus achieving a double energy conservation.

15.17 TOTAL ENERGY SYSTEMS

A project which generates all of its energy needs (power and heat) at the site and which utilizes the maximum part of this energy that is practically available is called a *total energy system* (TES). The term *cogeneration* of power and heat is also used. There are many types and combinations of these systems. Electrical power may be generated from a gas turbine, reciprocating engine, or steam turbine, any of which drive electrical generators. This equipment always rejects heat.

The aim of a total energy system is to utilize the waste heat rejected from one piece of equip-

ment for heating or to operate other equipment. The decision about whether or not to use a TES relies primarily on energy cost and conservation. When electricity is generated at a utility, about 65% of the energy in the fuel is rejected as heat. The argument for a TES is that if the power is generated at the buildings, it is convenient and perhaps less expensive to utilize the waste heat for required heating needs, rather than purchasing both heating fuel and power. (Of course there are some cases where power generating systems have been planned which heat and distribute hot water or steam long distances.) Detailed cost analyses are necessary before making a decision to use a TES. Sometimes only part of a building's energy needs are selected for total energy use. Two examples will illustrate some of the many possible combinations of total energy systems. In Figure 15.15 on page 406, a reciprocating internal combustion engine is used to drive a generator for electric power and light. About 70% of input energy from the fuel oil goes to heat in the combustion gases, engine cooling water, and lubricating oil. Instead of wasting this energy, most of it can be used for building heating needs.

The stack gases, cooling water, and lubricating oil are circulated through heat exchangers to produce hot water. In this way, about 80%, rather than 30%, of the energy input can be used.

An example of a partial total energy system is shown in Figure 15.16 on page 406. A high pressure steam boiler is used in winter for the building heating system, using steam or hot water from a heat exchanger. In summer, the steam is used to drive a steam turbine–centrifugal compressor refrigeration machine combination. Low pressure steam exhausting from the turbine is used to operate an absorption refrigeration machine. Further refinements could be use of condenser water from the refrigeration machines for heating needs (service hot water), as well as extracting heat from the boiler combustion gases.

In this case, only the building heating and cooling needs are integrated into a total energy system. Electricity would be purchased and used in a conventional manner.

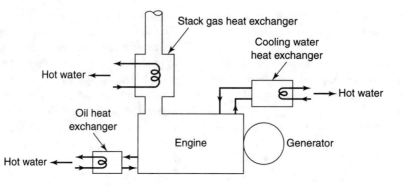

Figure 15.15
Heat recovery from an engine-generator for hot water.

15.18   ENERGY CONSERVATION METHODS

We have discussed procedures for measuring the efficiency of energy use and have described specific devices for recovering waste heat. Our next task is to determine how to achieve more energy conservation, if it is required. We will consider separately the areas of HVAC system design, installation, and operation. In each case, a few specific suggestions will be given. These are not intended as complete checklists, which would be far too lengthy and are not appropriate here.

For convenience and clarity, energy conservation in design will be divided into four categories: building construction, design criteria, system design, and controls.

Methods of achieving conservation in design usually are considered in the planning of new buildings. Some of the methods may be applicable to existing buildings through *retrofitting*—changes to the existing system. Whether or not a specific method is practical in existing buildings depends on the nature of each case. In some cases the decision is obvious (an existing building could not be turned around to reduce solar heat gain).

Figure 15.16
Use of steam waste heat for cooling.

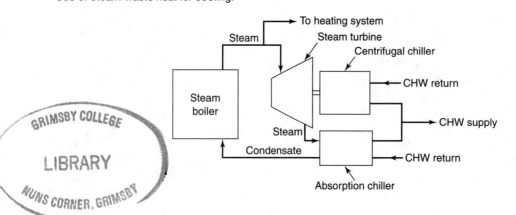

15.19 BUILDING CONSTRUCTION

Reduced HVAC energy consumption by building construction methods usually is a direct result of minimizing heat gains and losses. Some suggestions follow:

1. Use exterior wall and roof materials with high thermal resistance. An *R*-20 value is not unreasonable for roofs, considering present energy costs.
2. Avoid excessive use of exterior glass (which has a low *R*-value and high solar heat gain). An exception to this is a residence where there is a great solar intensity during most of the winter and a moderate temperature. Some homes are designed with a large southern facing glass expanse to utilize the resulting winter solar radiation heat.
3. Plan the site for reduced gains or losses. Trees or other objects may reduce solar heat gains and infiltration.
4. Orient the building to minimize solar heat gain in summer (and maximize it in winter).
5. Use internal shading devices and external shading over hangs or side reveals.
6. Use double or even triple window glazing in severe winter climates.
7. Avoid excessive lighting requirements.
8. Use window sash that has a tight seal at the frame. Consider possible use of nonoperable windows. However, this prevents use of natural ventilation in case the mechanical system fails.

Example 15.19 _____

A building is being designed that will have 20,000 ft² of roof area. The winter design temperature difference is 70 F. Degree days are 6500. The architect asks the consulting engineer to approximate how much gas will be saved in the winter if a roof with an insulation value of *R*-20 is substituted for a roof with an *R*-5 value. Gas at 1000 BTU/ft³ is used for heating.

Solution

The design heat losses for the two types of roof are calculated, and the difference found:

$$\text{for } R\text{-5 roof } U = \frac{1}{R} = \frac{1}{5} = 0.20 \text{ BTU/hr-ft}^2\text{-F}$$

$$R\text{-20 roof } U = \frac{1}{20} = 0.05 \text{ BTU/hr-ft}^2\text{-F}$$

R-5:

$$Q = 0.20 \times 20,000 \times 70 = 280,000 \text{ BTU/hr}$$

R-20:

$$Q = 0.05 \times 20,000 \times 70 = 70,000$$
$$\text{Difference} = \overline{210,000} \text{ BTU/hr}$$

To find the annual savings, the degree days are used. For a rough approximation, we will use the unmodified degree day method:

$$\text{BTU saved} = \frac{210,000}{70} \times 6500 \times 24$$
$$= 468,000,000 \text{ BTU/yr}$$

$$\text{Gas saved} = \frac{468,000,000 \text{ BTU/hr}}{1000 \text{ BTU/ft}^3}$$
$$= 468,000 \text{ ft}^3/\text{yr}$$

15.20 DESIGN CRITERIA

HVAC system design values used in the past have often resulted in systems that consume excessive amounts of energy. Some recommended energy conserving design factors are listed here. Many of these are based on ASHRAE Standards.

1. Use a 68–72 F maximum winter indoor design temperature (75 F and over has often been used in the past), except for special occupancy.
2. Use the 97½% winter outdoor design temperature (Table A.6). As an example of the savings here, this is 15 F for New York City, instead of 0 F, the recommended value found in practically all old tables. This does not reduce the heat loss directly, of course, because it is proportional to

the actual temperature difference. However, it will result in less oversized equipment, and therefore lower energy consumption.

3. Use a 78 F summer indoor design temperature (75 F or lower has often been used in the past).

4. Use the 2½% summer outdoor design temperature. For Cleveland, Ohio, this is 88 F DB. Many old tables list a temperature of 100 F!

5. Use the summer outdoor wet bulb temperature coinciding with the DB, not the maximum, which usually occurs at a different time. For Dallas, Texas, this would result in a 75 F WB instead of 78 F.

6. Use a design window infiltration rate based on latest standards and technology.

7. Use recommended design ventilation rates consistent with codes.

8. Use cooling load calculation procedures and data that account for the building thermal storage.

9. Use correct procedures and data for calculating duct and piping friction losses, accounting for duct pressure regain if significant.

15.21   SYSTEM DESIGN

The type of HVAC system and equipment affects the consumption of energy. Some suggestions that should be considered are:

1. Examine the possible use of a total energy system compared with conventional systems.

2. Recover waste heat in exhaust air by use of heat recovery devices to heat or cool ventilation air.

3. Use refrigeration condenser water for heating.

4. Recover waste heat from hot combustion gases with a heat exchanger.

5. Compare the energy consumption of a heat pump with separate systems for heating and cooling.

6. Examine the feasibility of a solar energy collector system for heating service hot water or even for space heating or cooling.

7. Use a variable air volume (VAV) type air conditioning system, if suitable. This is generally the most energy efficient system, because the reduction in air quantity directly lowers power use.

8. Avoid terminal reheat systems, which cool and then reheat air, unless the reheat comes from waste energy.

9. Dual duct and multizone systems or any other types that mix hot and cold air (or water) are inherently energy wasteful, although steps can be taken to minimize energy losses.

10. Choose the highest chilled water (evaporating) temperature and the lowest condensing temperature that results in satisfactory space conditions. This results in minimum refrigeration machine power consumption.

11. Choose the largest satisfactory chilled water and hot water temperature range, thereby reducing water flow rates and pumping power.

12. Choose lowest reasonable friction loss in ducts and piping, reducing fan and pumping power.

13. Select pumps and fans near the most efficient operating point.

14. Use multiple equipment so that at part loads some equipment can be shut down. The equipment operating then will be closer to full load, at which it is often more efficient.

15. For equipment that is to be operated at part load, speed reduction is usually the most efficient means of control.

16. Try to avoid using preheat coils for outside air. This often results in heating and recooling, a ridiculous waste of energy. Instead, design the intake plenum so that the outside and return air mix thoroughly.

17. Consider reducing CHW and HW pump flow rates when load reduces.

15.22   CONTROLS

Considerable energy conservation can be achieved through selection of proper automatic controls. Some of the suggestions have already been discussed in this chapter because they involve both controls and other parts of the system. Further suggestions are:

1. Use enthalpy control for the supply air. Enthalpy control devices sense and measure the total enthalpy of outside and return air, and adjust the dampers to provide the air mixture proportion that will provide the most economical natural cooling, thus reducing or eliminating need for refrigeration at times.
2. Use automatic time switching to start and stop equipment according to need.
3. Use night and weekend automatic temperature setback for unoccupied spaces.
4. Investigate the possible use of a central computerized control system designed to provide the most efficient operation at all times.

The energy-saving features of control systems are discussed further in Chapter 14.

15.23 INSTALLATION

The system must be installed so that it does not waste energy. Some suggestions are:

1. Install the system as designed, unless some apparent energy wasteful decision has been made; if so, verify before a change is made.
2. Provide airtight duct joints, test for leakage, and seal if necessary.
3. Avoid obstructions in ducts.
4. Make gradual transitions in ducts.
5. Use wide radius turns in ducts and use turning vanes if necessary.
6. Provide ample insulation on ducts, piping, and equipment.
7. Use dampers that have tight closing features.
8. Install air vents at proper locations in water and steam distribution systems.
9. Locate thermostats so that they do not sense an abnormal condition. For example, a cooling thermostat exposed to solar radiation might cause the cooling system to operate unnecessarily.
10. Use low pressure drop valves in water lines, consistent with function. For example, butterfly valves have a lower pressure loss than globe valves.

11. Install ducts and piping with shortest lengths of run and fewest changes of direction.
12. Test and balance the complete HVAC system in accordance with design requirements.

15.24 OPERATION AND MAINTENANCE

No energy conservation design procedures will be successful without follow-up of system operation to achieve and maintain minimum energy consumption. Some suggestions are:

1. Retest and rebalance the system completely at regular scheduled intervals during its lifetime. Often the initial balancing procedures are all a system ever receives, and it gradually loses efficiency. Think of how inefficiently a car would perform if it never had a tune up!
2. Shut down any equipment when not needed. The use of the space must be examined before this decision is made. For example, even though a space is unoccupied, possible freeze ups must be prevented.
3. Set back temperatures when spaces are not occupied. In winter, the temperature can often be set back to 55 F.
4. Start equipment shutdown or temperature setback shortly before occupants leave.
5. Start equipment as late as possible before occupants arrive, consistent with achieving comfort.
6. Utilize natural precooling at night from outdoor air (enthalpy control may do this automatically).
7. Close outdoor air dampers near the beginning and end of operation each day when equipment is providing heating or cooling.
8. Turn off unnecessary lighting.
9. Check that solar shading devices are used properly.
10. Check and replace or clean air filters when resistance reaches design conditions.
11. Clean all heat transfer surfaces regularly (coils, tubes).

12. Set chilled water temperatures (high) and condensing water temperatures (low), consistent with comfort and design values. Condensing water temperatures must not be so low that the refrigeration equipment malfunctions.
13. Adjust burners and draft on furnaces and boilers periodically to ensure complete combustion and minimum excess air.
14. Arrange for reliable water treatment services from a specialist for boiler and condensing water systems. Dirt in the system will reduce heat transfer as well as harm equipment.
15. Check operation of air vents regularly.
16. Limit the maximum electric power demand of the system (utility companies usually charge extra for high demands). This is accomplished by starting each piece of equipment in sequence, with ample time in between. In addition, some equipment (e.g., electrically driven refrigeration compressors) can be furnished with devices that limit their power demand.
17. Check ductwork regularly for leaks that may develop.
18. Check outside windows and doors regularly for abnormal cracks.
19. Perform routine maintenance regularly (lubrication, belt tension, inspection for damage or breakage).
20. Check that objects (books, clothing, furniture) are not obstructing airflow through room terminal units.
21. Check that air distribution outlets have not been obstructed or tampered with.
22. Check that room thermostats or humidistats have not been readjusted by unauthorized personnel. Provide locks if necessary.
23. Set thermostats at 68–70 F in winter and 78–80 F in summer, except for special occupancies or uses.

15.25 COMPUTERS IN HVAC SYSTEMS

Computers have a growing application in the design, installation, and operation of HVAC and other building systems. The computer can improve energy conservation efforts, reduce construction and operating costs, and result in a better quality system.

In HVAC system design, the computer can be used for the following tasks:

1. Load calculations. The computer can perform heating and cooling load calculations more accurately and at a much faster rate than can be done manually. It enables the designer to consider calculating the loads for varied types of building construction to find which is the minimum.
2. Simulation of system performance and energy estimation (as described previously).
3. Calculation of duct and piping sizes.
4. Selection of equipment. Manufacturers often have computer programs for this task.
5. Preparing HVAC drawings (CAD).

A computer can be used by the mechanical contractor to assist installation in the following ways:

1. To maintain an inventory of supplies and equipment so that shortages do not develop.
2. To determine costs of the installation as it proceeds.
3. To plan the installation for the shortest construction time and lowest cost. This would involve determining when equipment would be shipped to the job, when it should be installed, how many workers to use, and similar decisions a contractor must make. The computer can do this more accurately than can be done manually, thus resulting in an installation that costs less and is completed more quickly.

Computers and related equipment can be used in building HVAC operations to sense and control conditions. Some examples are:

1. Space conditions throughout the building (room temperatures, humidity).
2. System conditions (pressures, flow rates).
3. Equipment conditions (power, speed).

The information may be transmitted to a central control console, shown visually on screens, and may be recorded automatically. For situations that may become dangerous, a visual or audible alarm signal may be included. A central console is shown in Figure 15.17.

If it is necessary to adjust any unsatisfactory conditions, this can be done manually, or a more advanced computerized control system can be installed to send signals to control devices (valves, dampers) to adjust conditions automatically. This more advanced function of computers will undoubtedly grow in application because it provides the best energy conservation.

Design Software

Programs for carrying out HVAC design tasks were once usually written by the user, using a suitable language. Today, however, commercially available and public domain software is available for virtually every major HVAC task. The appendix lists some software sources and uses.

Figure 15.18 on pages 412–413 is an example of the output from a cooling load program based on the CLTD method developed for educational purposes.

Computer-aided design and drafting (CAD, or CADD) is also now extensively used. HVAC-oriented software is available for these tasks. There are many advantages to preparing contract (engi-

neering) drawings and shop drawings using CAD instead of manual drafting. Drawing changes are simplified. Coordination and interferences with other trades is improved. Some programs provide takeoff features. That is, a list of all piping and ductwork can be made. An example of the product of a CAD program is shown in Figure 15.19 on page 414.

Problems

15.1 A gas turbine with an efficiency of 26% is used to drive a centrifugal refrigeration compressor. The turbine uses 32,000 ft^3/hr of natural gas. What HP is available to drive the compressor?

15.2 A heat recovery exchanger is used to recover 30% of the wasted heat from the gas turbine in Problem 15.1 to heat hot water from 50 F to 120 F. How many GPM of water can be heated if all the recovered waste heat is used?

15.3 Determine the overall efficiency of the gas turbine heat recovery exchanger used in Problems 1 and 2.

15.4 A gas turbine burns fuel at a temperature of 1500 F. The stack gases are at 500 F. What is the maximum efficiency that could be achieved?

15.5 A refrigeration machine with a cooling capacity of 34 tons of refrigeration has a COP of 3.6. What is the power input to the compressor?

15.6 A refrigeration unit operates at an evaporating temperature of 38 F and a condensing temperature of 95 F. What is the maximum COP it could have at these conditions?

15.7 A wise consumer wants to check if the window unit she purchased is operating as efficiently as claimed. The unit is supposed to have EER of 7.8 and a capacity of 9500 BTU/hr at design conditions. A portable watt-meter is clamped on the outlet cord and it reads 1560 watts. Does the unit meet its claims?

Figure 15.17
Central console for controlling HVAC system.
(Courtesy: Honeywell, Inc.)

COIL COOLING LOAD CALCULATION FORM

Project ___EC430#1___   Location ___BROOKLYN, NY___

Coil Number ___101___   Drawing Number ___HVAC-1___   Date ___01/04/94___

Design Conditions: Latitude ___40 N___   Month ___JULY___

Outdoor DB __89__ °F,   WB __73__ °F,   RH __53__ %,   W __97__ gr/lb,   Daily Range __17__ °F

Average Temp __80.5__ °F

Indoor DB __78__ °F,   RH __50__ %,   WB __65__ °F,   W __71__ gr/lb

VENTILATION AIR FLOW AND SENSIBLE COOLING LOAD				
Solar Time	14	15	16	17
Room No. 101 Vent Air CFM	1200	1200	1200	1200
Room No. Vent Air CFM				
Room No. Vent Air CFM				
Room No. Vent Air CFM				
Room No. Vent Air CFM				
∑ Room Vent Air CFM	1200	1200	1200	1200
Outdoor Air DB °F	89	89	89	89
Vent Air Sensible CLD	14520	14520	14520	14520

TOTAL SENSIBLE CLD	DT SA FAN CLD% = 10	BT SA FAN CLD% = 0	RA FAN CLD% = 0	
Room No. 101 Sensible CLD	228895	237783	243536	245905
Room No. Sensible CLD				
Room No. Sensible CLD				
Room No. Sensible CLD				
Room No. Sensible CLD				
∑ Room Sensible CLDs	228895	237783	243536	245905
Draw Through SA Fan CLD	22890	23778	24354	24591
Total Coil Room Sens CLD	251785	261561	267890	270496
Blow Through SA Fan CLD	0	0	0	0
Return Air Fan CLD	0	0	0	0
Room No. 101 Ret Air CLD	15829	16444	16853	17263
Room No. Ret Air CLD				
Room No. Ret Air CLD				
Room No. Ret Air CLD				
Room No. Ret Air CLD				
RA Fan CLD + ∑ RA CLD	15829	16444	16853	17263
Vent Air Sensible CLD	14520	14520	14520	14520
Total Coil Sensible CLD	282133	292525	299263	302279

LATENT COOLING LOAD		Outdoor Air Spec Hum = 97 gr/lb	Indoor Air Spec Hum = 71 gr/lb		
Solar Time		14	15	16	17
Room No. 101 Latent CLD		18282	18282	18282	18282
Room No. Latent CLD					
Room No. Latent CLD					
Room No. Latent CLD					
Room No. Latent CLD					
Σ Room Latent CLDs		18282	18282	18282	18282
Σ Room Vent Air CFM		1200	1200	1200	1200
Vent Air Latent CLD		21216	21216	21216	21216
Total Coil Latent CLD		39498	39498	39498	39498

TOTAL COIL CLDs AND SUPPLY AIR FLOW RATE	Sup Air DB = 55 °F		Room DB = 78 °F	
Total Coil Room Sens CLD	251785	261561	267890	270496
Supply Air CFM	9952	10338	10589	10691
Σ Room Latent CLDs	18282	18282	18282	18282
Coil Room SHR	0.932	0.935	0.936	0.937
Total Coil Sensible CLD	282133	292525	299263	302277
Total Coil Latent CLD	39498	39498	39498	39498
Total Coil CLD in BTU/hr	321631	332021	338761	341771
Total Coil CLD in MBH	321.63	332.02	338.76	341.78
Total Coil CLD in Tons	26.80	27.67	28.14	28.48
Coil Total SHR	0.877	0.881	0.883	0.884

Figure 15.18
Example of computer program output of cooling load calculations. (**Courtesy:** Prof. A. C. Finger, New York City Technical College/CUNY.)

15.8 A building being planned will have a design heating load of 840,000 BTU/hr. A decision is to be made as to which of the following heating systems shall be used:

A. No. 2 oil-fired boiler, 78% efficiency.

B. Heat pump, $COP_h = 3.2$.

C. Electric resistance heaters.

On the basis of minimum energy use, which system should be selected?

15.9 In Problem 15.8, if fuel oil is $0.85/gallon and electricity is $0.07/KWH, which system would be chosen for minimum operating cost?

15.10 A pump with a mechanical efficiency of 68% delivers 87 GPM of water at a total head of 31 ft w. It is driven by a motor with a 92% efficiency. Electricity costs $0.06/KWH. The pump operates 10 hours a day, 50 weeks a year. What is the expected yearly electric billing cost?

15.11 A house in Boston, Massachusetts has a design heating load of 71,000 BTU/hr. The

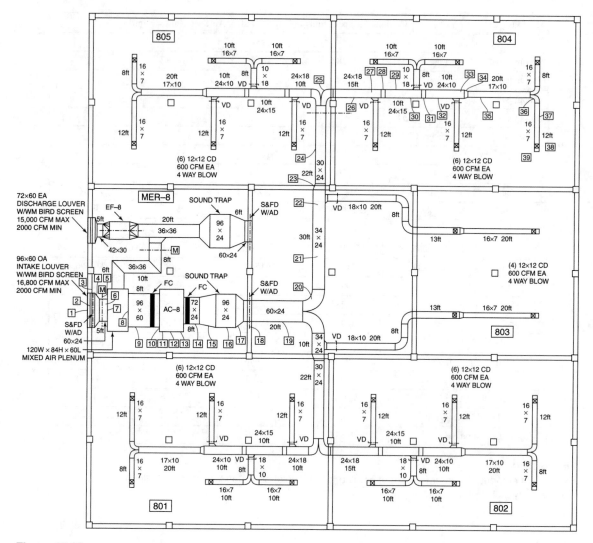

Figure 15.19

Example of a computer-aided drawing (CAD) of an air conditioning system. (**Courtesy: Professor A. C. Finger, New York City Technical College/CUNY.**)

inside design temperature is 72 F. Using the modified degree day method, determine the expected yearly energy consumption if No. 2 oil at $0.85/gal is used as fuel. What would the yearly heating cost be?

15.12 Compare the expected yearly energy consumption found in Problem 15.11 with that of the unmodified degree day method.

15.13 The heating system in a building utilizes 16,000 CFM of ventilation and exhaust air. The inside temperature is 70 F and the average outdoor winter temperature is 32 F. The heating system operates 1600 hours in winter. A heat recovery device with an 80% efficiency is being considered to recover heat in the exhaust air. The building uses No. 6

fuel oil at \$0.95/gal. What would be the yearly cost savings if the heat recovery device were installed?

15.14 The following two choices are being considered for the construction of a building's walls and windows:

A. $A_w = 7200$ ft^2,

$U_w = 0.14$ BTU/hr-ft^2-F

$A_g = 1600$ ft^2, $U_g = 1.10$

B. $A_w = 4100$ ft^2, $U_w = 0.10$

$A_g = 4700$ ft^2, $U_g = 0.62$

Determine which design has the better overall wall thermal transmittance.

15.15 A building in Albany, New York, has a roof constructed of built-up roofing, 4 in. l.w. concrete, and a ½ in. acoustic tile coiling. Does the roof meet the minimum heat transfer coefficient suggested in this book? If not, what would be the R-value of insulation required to meet the standard?

15.16 A building has the following specifications:

$A_w = 3400$ ft^2, $U_w = 0.24$ BTU/hr-ft^2-F

$A_g = 1200$ ft^2, $U_g = 1.06$

Location: Tampa, Florida. Inside design temperature = 78 F.

Wall weight = 45 lb/ft^2. SC = 0.56

Determine the OTTV of the wall and compare it with the maximum allowable suggested in this book.

15.17 Calculate the overall heat transfer coefficient for your home, apartment house, and school, and compare it with the standard in your state.

15.18 Calculate the OTTV for your home, apartment house, and school, and compare it with the standard in your state.

15.19 Calculate the design energy budget in BTU/hr-ft^2 for heating and cooling your home, apartment house, and school, and compare it with the standards suggested in this book (or other standards).

15.20 An energy budget code requires buildings to use a maximum of 75 BTU/hr-ft^2 of energy from depletable sources from all uses in summer. A 35,000 ft^2 building has a design air conditioning load of 95 tons. The air conditioning system requires 1.3 KW of power per ton. Lighting and miscellaneous power sources use 140 KW of power. Determine if this building meets the code.

CHAPTER 16
Instrumentation, Testing, and Balancing

The design and installation of an air conditioning system may be carried out properly, but if it is not adjusted and balanced to meet design conditions, it will not perform satisfactorily. The subject of this chapter is the instrumentation and procedures used in balancing systems.

The complexity of modern air conditioning systems may make the balancing process quite involved. The student and even some experienced personnel frequently are unaware of the difficulties and requirements of balancing. In the past, it was often possible to get by with making a few adjustments and checking if people were comfortable. This is no longer satisfactory on large systems. Organized procedures are required to balance a system so that it will result in comfort in all seasons.

Furthermore, the increased need for minimizing energy waste also requires correct balancing techniques. An improperly balanced system will almost certainly use excess energy.

The testing and balancing process is often carried out by the contractor upon completion of the installation, but for large systems, a whole new profession has grown, with organizations and specialists who do only this work. This has happened not only because of the complexity of the task, but because it is advantageous that an independent organization verifies that the system is operating correctly.

Above all, skill in balancing requires a sound understanding of air conditioning principles, as described in this book or elsewhere.

OBJECTIVES

After studying this chapter, you will be able to:

1. Choose appropriate temperature, pressure, and velocity measuring instruments for balancing HVAC systems.

2. Carry out duct traverse calculations.
3. List the data to be collected for balancing systems.
4. List the procedures for balancing air and water flow systems.

16.1 DEFINITIONS

The phrase *testing, adjusting,* and *balancing* (TAB) has become popular as a description of what is commonly called "balancing." *Testing* is the process of operating and checking the performance of equipment. *Balancing* is the process of proportioning the correct flow of air and water throughout the system, through mains, branches, equipment, and terminal units. *Adjusting* is the process of regulating and setting variables so that a balanced system is achieved. The variables may be speeds, temperatures, pressures, flow rates, and so forth, and are regulated by adjusting some device that controls a variable. For example, a valve may be regulated to adjust flow rates in the process of balancing. Hereafter we will use the abbreviation *TAB* to refer to the complete operation.

16.2 INSTRUMENTATION

The success of TAB depends on adequate *instrumentation*. These instruments are used for measuring temperature, pressure, velocity, flow rate, speed, heat flow, and electrical energy. We will discuss types of instruments and their applications. Of course, some of these instruments are also used in routine operating and maintenance work.

Instruments have varied degrees of accuracy. For TAB work, an accuracy within 5% of the true value is more than adequate and is usually available without exorbitant instrument costs. Instruments should always be *calibrated* before each use. That is, the accuracy of their readings should be checked against those of an instrument or procedure known to be more accurate, and then adjusted. This should be done over the whole range of values, not just at one point. Some instruments have a means of calibration adjustment to correct their settings. With others, a *calibration curve* can be made and then the readings corrected by using the curve.

Example 16.1 _____

Figure 16.1 is a calibration curve made for a certain pressure gage by comparing its readings with

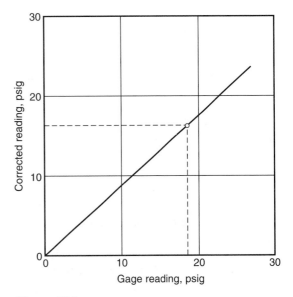

Figure 16.1
Calibration curve for pressure gage in Example 16.1.

that of a manometer. During balancing, a reading of 18 psig on the gage is taken at some station. What is the correct pressure?

Solution
Using Figure 16.1, at a reading of 18 psig on the gage, reading the corrected value, it is 16 psig.

_____

It is important to not increase errors by reading instruments incorrectly. The eye of the reader should be directly in line with both the scale and pointer. If the eye is to the side, above or below, a serious error can be made.

16.3 TEMPERATURE

Temperature measuring instruments can all be called *thermometers*, although one type, the *liquid-in-glass* thermometer, is most commonly known. All temperature measuring devices utilize, as a means of developing a temperature scale, the fact that properties of materials change with changes in their temperature.

The *liquid-in-glass*, or *stem* thermometer utilizes the fact that liquids change volume with change in temperature. Mercury is the liquid most frequently used. A bulb at one end of the thermometer is inserted in the fluid. When reading temperatures of liquids in pipes, a permanent *well* should be installed at points where readings are required, because obviously the pipe cannot be cut into to take readings. The well is an insert that can be filled with oil, into which the thermometer is placed. Thermometers with an accuracy of ±0.5 F (± 1 C) are suitable for HVAC work.

Liquid-in-glass thermometers can be affected significantly by radiant heat, causing erroneous readings. When there is radiant energy nearby, they can be protected by a suitable shield.

The *resistance* thermometer utilizes the fact that the electrical resistance of a metal wire changes with temperature. The resistance is measured by an electrical device whose indicator reads temperature directly.

When two wires of different metals have their ends joined together, it is found that they will generate a very small voltage. This arrangement is called a *thermocouple*. The voltage generated varies with temperature, and this provides a means for constructing a temperature measuring device. The wires are connected to a *potentiometer*, an electrical device that measures the voltage and whose scale reads temperature directly (Figure 16.2).

The thermocouple and resistance thermometers have the great advantage of allowing temperature measurements at remote points. Furthermore, a large number of points can be connected to one readout instrument by simple switching. They are often used to provide the system operating engineer with temperature readings from hundreds of locations at one central console.

A *dial-type* thermometer with a metal stem (Figure 16.3) has a *bimetallic* element in the stem that moves with temperature change, actuating the dial. They are very convenient for quick field checks, but have limited accuracy. One type, the *contact* thermometer, can be placed directly on a pipe to read the water temperature inside, although the reading is only approximate.

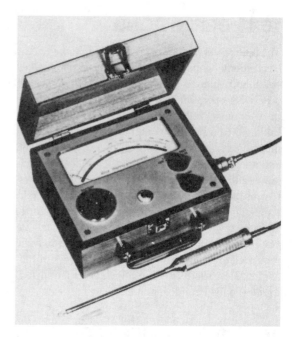

Figure 16.2
Temperature measuring instrument using thermocouple and potentiometer. (Courtesy: Alnor Instrument Co.)

Figure 16.3
Dial-type thermometer with bimetal sensing stem. (Courtesy: Airserco Manufacturing Co.)

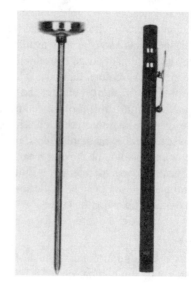

16.4 PRESSURE

Manometers are simple and accurate pressure measuring instruments, which utilize the pressure that will lift a column of liquid. Water and mercury are the two most common liquids used. The *U-tube manometer* is the simplest arrangement (Figure 16.4). One end of the manometer is connected to the location where pressure is to be measured, and the other is usually left open to the atmosphere. The reading on the manometer is therefore *gage pressure*—the difference between the pressure being measured and the atmospheric pressure. Sometimes the difference between the pressures at two locations in a pipe or duct is to be measured. In this case, the ends of the manometer are connected to the two locations.

For measuring smaller pressures, water is used in a manometer. Mercury is used for measuring larger pressures.

Figure 16.4
U-tube manometer. **(Courtesy: Dwyer Instruments, Inc.)**

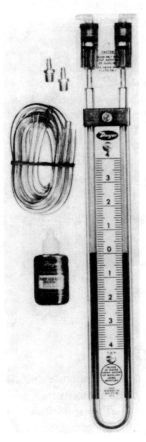

Example 16.2
A TAB technician is going to use a U-tube manometer to measure a pressure difference. It is known that the pressure difference is in the range of 3–4 psi. Should the technician use a mercury or water manometer?

Solution
Assuming water is used, the conversion factor (Table A.2) to change to feet of water is

$$p = 4 \text{ psi} \times \frac{2.3 \text{ ft w.}}{1 \text{ psi}} = 9.2 \text{ ft w.}$$

If mercury is used,

$$p = 4 \text{ psi} \times \frac{2.04 \text{ in. Hg}}{1 \text{ psi}} = 8.2 \text{ in. Hg}$$

This means that a manometer more than 9 ft in height would be needed to read these pressures if water were used, a very clumsy arrangement! Because mercury has a much greater density than water, it should be used.

Example 16.3
A technician takes a mercury U-tube manometer reading of 9 in. Hg. One end of the manometer is attached to the location where pressure is to be measured, the other is open to the atmosphere. What is the pressure at the location, in psig?

Solution
Using the proper conversion factor from Table A.2,

$$p = 9 \text{ in. Hg} \times \frac{1 \text{ psi}}{2.04 \text{ in. Hg}} = 4.4 \text{ psig}$$

A mercury *barometer* is a single vertical tube manometer that is closed at the top, evacuated, and filled with mercury. The bottom end has an open pool of mercury which is exposed to atmospheric pressure. Therefore, the mercury barometer is used to read atmospheric pressure.

For reading small differences in pressure more accurately, an *inclined manometer*, also called a *draft gage*, is used. By inclining the manometer, the vertical distance can be divided more finely, allowing more accurate readings. The inclined manometer is used often to measure small pressures in air ducts.

The *Bourdon tube* pressure gage is the most commonly used type for installation in pipe lines and on vessels. A hollow metal curved tube changes its shape with pressure. By means of a linkage to a pointer, the pressure is read on a scale on the face of the gage. This type of device reads gage pressures (Figure 16.5). A *compound gage* is used to read pressures above or below atmospheric pressure.

The Bourdon tube pressure gage is rugged, reliable, and relatively inexpensive. It is not suitable for reading very small pressure differences and gets out of correct calibration easily.

A popular dial-type pressure gage which has a diaphragm that moves with pressure is shown in Figure 16.6. There is no mechanical linkage to the pointer—the transmission of pressure is by use of a magnetic field. It can read pressure differences to accuracies of 0.01 in. w. and is relatively inexpensive. It is quite popular for measuring pressure differences in ducts and across air filters.

16.5   VELOCITY

For very approximate air velocity studies, smoke can be used. The smoke, generated by a mechanical device or candle, is injected into the airstream and observed. More practical use of smoke is to examine the air distribution patterns in rooms and to find air leaks from ducts.

Figure 16.5
Pressure gage, Bourdon tube type. (Courtesy: Weksler Instruments Corp.)

Figure 16.6
Pressure gage, magnetic field actuated.
(Courtesy: Dwyer Instruments, Inc.)

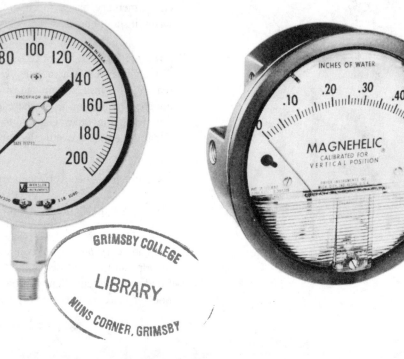

Anemometers and the *Pitot tube* are most frequently used for measuring air velocities. The Pitot tube is used together with manometers to determine velocity. It consists of two concentric tubes (Figure 16.7). The inner tube has an opening at the tip which is pointed directly at the oncoming airstream and therefore reads *total pressure*. The outer tube has holes in the circumference that are perpendicular to the airstream, and therefore *static pressure* is transmitted through this tube.

By connecting the total pressure outlet to one leg of a manometer and the static pressure outlet to the other leg, as shown in Figure 16.8, the difference between these pressures is read on the manometer. The difference between total and static pressure is called *velocity pressure*.

As was explained in Chapter 8, the velocity pressure is related to velocity for air by Equation 8.7:

$$V = 4000 \sqrt{H_v}$$

where

V = air velocity, ft/min

H_v = velocity pressure, in. w.

Using the equation above, a table relating velocity (V) to velocity pressure (H_v) can be constructed

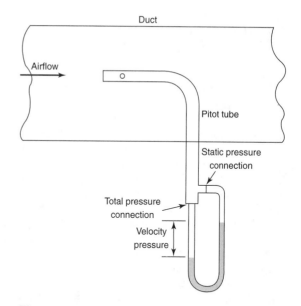

Figure 16.8
Use of Pitot tube to measure velocity pressure in a duct.

(Table 16.1 on page 422). This table is convenient to use in field testing and balancing work when measuring velocities and flow rates, as an example in the next section illustrates.

The *rotating vane* anemometer has a propeller that spins as the air flows past it (Figure 16.9 on page 422). The dial reads total distance in linear feet of air. A stopwatch is used with the device so that the velocity in feet per minute can be found.

The *deflecting vane* anemometer (Figure 16.10 on page 422) has a vane in a case with an opening. As air flows by, it deflects the vane. There is a linkage to a pointer and scale where the air velocity is read.

The *hot wire* anemometer utilizes the fact that the electrical resistance of a wire changes with temperature. The temperature of the wire in turn depends on the cooling effect of the air velocity flowing past it. The sensor is connected to an electrical readout device which reads air velocity directly. It is a very sensitive instrument that is used to measure low velocities, such as drafts in occupied spaces.

Figure 16.7
Pitot tube. (Courtesy: Dwyer Instruments, Inc.)

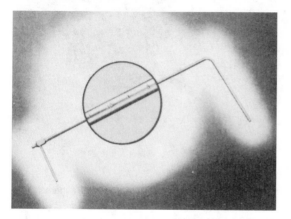

TABLE 16.1   VELOCITY, V (FPM) VS. VELOCITY PRESSURE, H_V (IN. W. G.) FOR STANDARD AIR (.075 lbs / ft³)

H_v	V	H_v	V	H_v	V
.01	400	.31	2230	.61	3127
.02	566	.32	2260	.62	3153
.03	693	.33	2301	.63	3179
.04	801	.34	2335	.64	3204
.05	895	.35	2369	.65	3229
.06	981	.36	2403	.66	3254
.07	1060	.37	2436	.67	3279
.08	1133	.38	2469	.68	3303
.09	1201	.39	2501	.69	3327
.10	1266	.40	2533	.70	3351
.11	1328	.41	2563	.71	3375
.12	1387	.42	2595	.72	3398
.13	1444	.43	2626	.73	3422
.14	1498	.44	2656	.74	3445
.15	1551	.45	2687	.75	3468
.16	1602	.46	2716	.76	3491
.17	1651	.47	2746	.77	3514
.18	1699	.48	2775	.78	3537
.19	1746	.49	2804	.79	3560
.20	1791	.50	2832	.80	3582
.21	1835	.51	2850	.81	3604
.22	1879	.52	2888	.82	3625
.23	1921	.53	2916	.83	3657
.24	1962	.54	2943	.84	3669
.25	2003	.55	2970	.85	3690
.26	2042	.56	2997	.86	3709
.27	2081	.57	3024	.87	3729
.28	2119	.58	3050	.88	3758
.29	2157	.59	3076	.89	3779
.30	2193	.60	3102	.90	3800

Figure 16.9
Rotating vane anemometer. (Courtesy: Davis Instruments Manufacturing Co.)

16.6   FLOW RATES

The Pitot tube and types of anemometers described earlier can be used to measure air flow rate by applying the continuity Equation 8.1:

$$\text{VFR} = A \times V$$

where

VFR = volume flow rate of fluid

A = cross-sectional area of duct

V = velocity of fluid

Figure 16.10
Deflecting vane anemometer. (Courtesy: Bacharach Instrument Co.)

The Pitot tube is often used to find air flow rate in a duct. The velocity at any section usually varies across the duct, and therefore an average velocity must be found. This is done by placing the tube at a number of different locations (called *traverses*) measuring the velocity pressures, calculating velocities, and averaging them. For rectangular ducts, readings at 16 or more locations are taken (Figure 16.11). When using a Pitot tube, care must be taken that the probe points directly into the airstream.

Example 16.4

Figure 16.12 is a cross section of a duct, listing the velocity pressure traverse readings taken with a Pitot tube. Readings are in in. w.g. What is the air flow rate in CFM through the duct?

Solution
Using Table 16.1, the velocity at each traverse reading corresponding to the velocity pressure are recorded, as follows, and the average velocity is then calculated:

Hv	V
.05	895
.07	1060
.06	981
.05	895
.05	895
.09	1201
.08	1133
.06	981
.06	981
.10	1266
.10	1266
.05	895
.04	801
.06	981
.06	981
.04	801

$$V = \frac{16,013}{16}$$
$$= 1000 \text{ FPM}$$

The flow rate in CFM is now found:

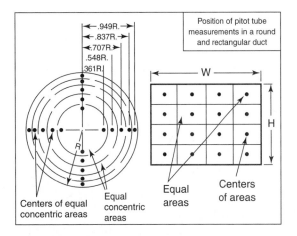

Figure 16.11
Positions of Pitot tube traverse measurements for round and rectangular ducts.

$$A = 24 \times \frac{12}{144} = 2 \text{ ft}^2$$

$$\text{VFR} = A \times V = 2 \text{ ft}^2 \times 1000 \text{ ft/min}$$
$$= 2000 \text{ CFM}$$

The rotating vane and deflecting vane anemometers are used mainly for measuring flow rates through air diffusers and grilles. However, a modified version of the continuity equation must be used. The outlet area of the air distribution device differs somewhat from the measured area because the air does not flow through it evenly. The outlet manufacturer will furnish this information. Furthermore, each type of anemometer will read slightly different velocities. The continuity equation is revised when using an anemometer to read

$$\text{VFR} = A_k \times V_k$$

Figure 16.12
Duct traverse Pitot tube readings for Example 16.4.

24 in.			
.05	.07	.06	.05
.05	.09	.08	.06
.06	.10	.10	.05
.04	.06	.06	.04

12 in.

where

VFR = volume flow rate, CFM

A_k = effective air distribution outlet area, ft^2

V_k = average air velocity through outlet, ft/min

Water flow rates are often measured by using the relationship between velocity and pressure loss in piping. Two devices used are the *orifice plate* (Figure 16.13) and the *Venturi tube*. The pressure loss, which varies with the velocity, is measured with a manometer. Equations are developed for each device so that the velocity and flow rate can be calculated.

The orifice plate and Venturi tube must be installed in a straight length of pipe with no turns or obstructions for a considerable distance both upstream and downstream as specified by the supplier or by ASME, to prevent erroneous readings.

The system pump can be used as an instrument for approximate measure of system flow rate. By measuring the head across the pump and referring to the pump characteristic curves, the flow can be found. Similarly, any other device (e.g., chiller, valve) whose pressure-flow characteristics are known, can be used in the same way.

There are other types of liquid flow meters, but they will not be discussed here.

16.7   HEAT FLOW

Instruments have been developed that read heat flow through a surface such as a building wall or pipe. This device utilizes the heat conduction equation (Chapter 3). By measuring temperatures on both sides of the wall and with a known thermal resistance, it determines the heat flow. It can also be used to measure thermal resistance of a material.

A relatively new method of measuring heat flow is to use *infrared photography*. Temperature variations show different shades of brightness on infrared photos (Figure 16.14). This can be used to locate sources of high or low heat flow. It has become popular as an aid in energy conservation by pinpointing high heat losses through building surfaces.

16.8   HUMIDITY

The *sling psychrometer* (Figure 16.15) is the instrument used most often to find the humidity of air in HVAC work. It consists of two liquid-in-glass thermometers, one with a cloth wick wrapped around the sensing bulb. The wick is soaked in water and the apparatus is spun rapidly. The wetted stem thermometer will read the air *wet bulb* temperature and

Figure 16.13
Venturi tube and orifice plate for measuring flow rate in pipes.

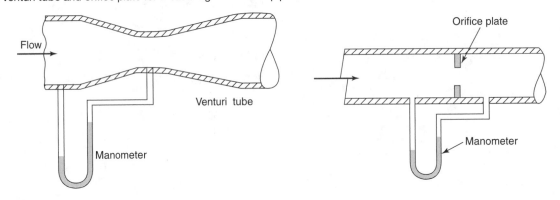

Figure 16.14
Infrared photograph locating excessive heat losses.
(Courtesy: Thermo Test Division, Fuel Savers, Inc.)

the other thermometer reads the *dry bulb* temperature. The moisture content of the air can then be found by using the psychrometric chart (Chapter 7). There are also instruments called *hygrometers* that read humidity directly. They usually have a material that changes shape with change in humidity, and use a linkage to a pointer on a dial face.

16.9 EQUIPMENT SPEED

Measuring speeds of rotating equipment (fans, pumps, compressors) is often required in TAB work. A chronometric *tachometer* has a shaft with a tip that is placed on the countersunk end of the

Figure 16.15
Sling psychrometer. (Courtesy: Airserco Manufacturing Co.)

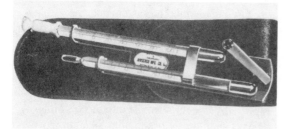

rotating shaft of the equipment whose speed is to be measured. The dial reads RPM directly. Other types are the electric and vibrating reed tachometers.

The *stroboscope* is a very useful speed measuring device used when physical contact with the rotating equipment is difficult. It has a flashing light whose frequency is adjustable. The light is pointed toward any rotating part of the equipment. When the frequency of flashing is equal to the speed of rotation (or any multiple), the equipment appears to stand still.

16.10 ELECTRICAL ENERGY

The measurement of current, voltage, and electric power use of motors is required in TAB work. *Ammeters, voltmeters,* and *wattmeters* are sometimes permanently wired into the system to measure these characteristics. For field TAB use, however, the portable *clamp-on* combination voltmeter and ammeter is very convenient (Figure 16.16 on page 426). The U-shaped jaw can be opened and clipped around an electric wire. The magnetic field created by the current flow operates the instrument.

16.11 TESTING AND BALANCING

Organized procedures for TAB are necessary to avoid an endless, wasted process. If the work is done blindly by simply going around from one outlet to another, good balancing will never be achieved, particularly on larger systems. The procedures for air and water systems will be discussed separately. The work can be divided into two stages in each case: preparatory steps and the main balancing procedure. Air system balancing is usually more complex and lengthy than water system balancing.

16.12 PREPARATION FOR AIR SYSTEM BALANCING

Before starting the system balancing routine, it is advisable to carry out a series of preliminary

Figure 16.16
Clamp-on volt amp ohmmeter. **(Courtesy: Amprobe Instrument.)**

checks and steps. Otherwise time will be wasted later in carrying out these useful tasks. Some of this work may be the responsibility of the mechanical contractor, and some the responsibility of the TAB technician. This varies from job to job. In any case, all of it should be done. The following steps are recommended:

1. Obtain ductwork drawings that can be used to indicate air flow rates and velocities through all ducts and air distribution devices. If either the engineering drawings or shop drawings are suitable, use them. Otherwise prepare simple single line drawings. List all design velocities and flow rates on the drawings. Show the locations of all dampers.

2. Obtain all equipment data from the design specifications and from manufacturers. This includes any performance information required for testing and balancing. The following list is helpful but is not intended to cover every case:

General. Indoor and outdoor design conditions (winter and summer).

Fans. Type, size, CFM, static pressure, BHP, RPM, motor data. Obtain performance curves if possible.

Filters. Type, CFM, pressure loss, effective area (A_k), effective velocity (V_k) for balancing device to be used, air outlet temperature.

Coils. CFM, air pressure loss, face area, air temperatures and humidity, capacity (both sensible and latent for cooling coils).

Dampers. Types, pressure loss, CFM.

Mixing Boxes, VAV Units, Induction Units. CFM, pressure loss, any special information furnished by manufacturer.

3. Prepare standard TAB report forms for recording data. An apparatus report form and air outlet report form are recommended.

4. Select and obtain the instrumentation most suitable for each task.

5. Decide where all measurements will be taken and check the installation to see if access is available. Arrange for suitable access with the contractor if necessary. For example, access doors in ducts may have to be provided.

6. Check the system to see if all dampers and controls are in correct positions for balancing.

7. Schedule times for the TAB work. One period in summer and one in winter are usually necessary (for a year-round system), and preferably close to outdoor design conditions.

8. Start and operate the system and check that all equipment is performing satisfactorily. This would usually be done under the supervision of the mechanical contractor.

9. It is advisable on large systems, especially high-pressure systems, to check for major leaks by injecting smoke into the system. Any leaks should be closed with sealant or tape.

If the above preliminary procedures are carried out, considerable time and expense can be saved during balancing.

16.13 THE AIR SYSTEM BALANCING PROCESS

1. Check that the preparatory steps have been completed.
2. Measure fan speeds and adjust to design values.
3. Measure total system CFM by one or more of the following methods:
   A. Pitot tube traverse in the main duct.
   B. Anemometer readings across coils in the air handling unit. Readings across the filters or dampers are not acceptable—they will be inaccurate.
   C. Static pressure reading across the fan and reference to the fan performance curve. If this method is used, the static pressure reading must be corrected for the system effect (Chapter 10).

Although any of the above methods can be used, the Pitot tube traverse is more accurate. If the CFM is within ±10% of design, proceed to the next step. If not, first check if there is some major problem such as closed dampers, malfunctioning equipment, or incorrect design, and correct the problem. If everything is satisfactory, adjust fan speed to bring CFM to within ±10% of design. Most large fans are furnished with adjustable speed drives.

4. Measure and adjust CFM to within ±10% of design in major branches using branch dampers. Use a Pitot tube traverse.
5. Measure and adjust flow at air outlets as follows:
   A. Start with outlets farthest from fan, working backward.
   B. If there are a number of outlets on one branch, adjust total CFM to branch to approximately design value.
   C. Measure and adjust flow at each outlet to ±10% of design. Remember to use

effective area and effective velocity readings. A deflecting vane-type anemometer should be used. Rotating vane types sometimes do not give satisfactory readings.
   D. Repeat the process at each outlet a second time. After the first set of adjustments, it will often be found that some of the outlets do not read correctly. Repeat a third time, if necessary. Accuracy should be at least within ±10% of design. Some installations require ±5%. Record all readings.

6. Check room air distribution for drafts or dead spots. Make minor adjustments if needed. (This may be the contractor's responsibility.)
7. Check and record all performance data on fans, filters, and other equipment listed previously.
8. Measure air DB and WB temperatures before and after coils, preferably at full load. If not at full load, the manufacturer may furnish information so that the predicted full load performance can be determined.
9. Carry out a similar procedure for the return air system. This should be done at the same time as the supply system, not afterward.
10. Send in the bill for your fee!

Some TAB technicians prefer to start at the first outlet on a branch, rather than the last. For small installations, some of the steps may be unnecessary. For instance, if the air handling equipment is packaged, much of the testing of individual components is not practical and perhaps not necessary. A straightforward adjustment of airflow at each outlet is the main balancing task.

16.14 PREPARATION FOR WATER SYSTEM BALANCING

1. Obtain or prepare a piping flow diagram showing all equipment. Record flow rates and temperatures on the diagram.
2. Obtain all equipment data from the design specifications and from the manufacturers. This

includes any performance information required for testing and balancing. The following list is helpful but is not intended to cover every case:

Pumps. Type, size, GPM, head, BHP, RPM, motor data. Obtain performance curves.

Water Coils. Physical characteristics (rows, circuiting), capacity (sensible and latent), GPM, water pressure loss.

Chillers. Capacity, GPM, water pressure loss, motor data.

Boilers. Capacity, water or steam temperatures, flow rates.

Terminal Units. Type, size, capacity (sensible and latent), GPM, pressure loss, water temperatures, motor data (if any).

Cooling Tower. Type, size, capacity, air DB and WB temperatures, water temperatures, and flow rates.

3. Prepare report forms for recording data.
4. Select and obtain the instrumentation most suitable for each TAB task. Calibrate all instruments.
5. Decide where all measurements will be taken and check the installation to see if access is available.
6. Check the system to see if all valves and controls are in correct positions for balancing.
7. Schedule times for the TAB work. One period in summer and one in winter are usually necessary (for a year-round system), and preferably close to outdoor design conditions.
8. Start and operate the system and check that all equipment is performing satisfactorily.

16.15   THE WATER SYSTEM BALANCING PROCESS

1. Check that the preparatory steps (Section 16.14) have been completed.
2. Check pump speed with design. Pumps do not usually have adjustable speed drives, so it is unlikely there will be any significant difference between actual and design speeds.

3. Gradually close the pump discharge balancing valve, recording suction and discharge head and motor amps and volts. Do this for a number of settings from valve full open to full closed. Use this information to plot an actual pump performance curve. Correct the pump head for any significant differences in velocity heads entering and leaving (Chapter 8). Use one pressure gage with alternate connections to the suction and discharge to avoid errors between gages. Compare the results with the manufacturer's data. *Note*: Do not operate the pump at shut-off (with the valve completely closed) for any significant length of time, as it may overheat.
4. Adjust system flow to about 110% of design GPM, according to pump curve.
5. If the decision has been made to balance flow rates in mains and branches manually, adjust manual balancing valves and read flow rates on the instruments installed (orifice plates or Venturi tubes) until flows are approximately correct.

    This detailed procedure is sometimes not carried out on water systems, because adequate balancing can be accomplished at the equipment. Furthermore automatic flow control valves are often used at the equipment.
6. Check and balance flow rates through chiller and large coils. This can be done by using instruments or reading pressure loss through coil and using manufacturer's curves of pressure loss versus flow rate. Adjust balancing valves to within ±10% of design GPM.
7. Check and balance flow rates to terminal units also using pressure loss versus flow rate data to within ±10% of design GPM.

    An alternate method of checking design flow is to measure water temperature in and out of units. This is not as accurate, however, and is recommended only as an additional check.
8. Repeat the balancing process until no change is found.
9. Measure pump head and motor data and check flow rate.

10. Measure and adjust water flow to cooling tower, using pump curve data or instruments. Check performance of cooling tower by measuring water flow rate and temperatures, air DB and WB temperatures in and out. Calculate capacity. If not done on a design day, the manufacturer can furnish data on predicting full load performance.

11. Carry out performance tests of boilers, chillers, and cooling towers. Details will not be explained here because this is usually done on large systems with the aid of and in the presence of the manufacturer's field engineer. These are called *witness tests*. Measurements taken are similar to those described for other equipment.

On small systems, many of the above steps are combined or are not necessary.

16.16   ENERGY CONSERVATION

A special listing of energy conservation methods in TAB work would be repetitious, in view of our description of balancing procedures. Obviously an unbalanced system can be highly wasteful in pump and fan power usage, and in the excess energy required if it is overheating or overcooling. It is shocking to see how much effort and expense are put into the design and installation of some HVAC systems, whereas the TAB procedures are done inadequately.

Even when a system is thoroughly and properly balanced, it will inevitably become unbalanced with time. Piping will become rough, settings on dampers and valves may be changed by vibrations or unauthorized tampering, and equipment may get dirty or wear out. Included in all maintenance schedules should be consideration of a thorough rebalancing procedure every few years.

16.17   SOUND MEASUREMENT

Overall sound levels are measured with a sound level meter. Acceptable room noise levels are often specified on the dBA weighted scale (Chapter 10), which most instruments are designed to read.

If sound level readings are required at individual frequencies, a meter called a *sound analyzer* is used. This is useful when trying to discover sources of noise. For example, a centrifugal compressor may produce high-pitched sounds, which would be picked up when the analyzer was set to high frequencies.

Review Questions

1. Explain the terms *testing, adjusting,* and *balancing.*

2. List four types of temperature measuring instruments and how they function.

3. Explain the difference between a U-tube and an inclined manometer.

4. Describe three types of anemometers.

5. Sketch and describe a Pitot tube connection to measure velocity.

6. List the equipment data to be obtained for air system balancing.

7. List the equipment data to be obtained for water system balancing.

Problems

16.1   A mercury manometer is going to be used to test the performance of a pump whose maximum head is 28 ft w. What is the minimum height the manometer should have?

16.2   A rotating vane anemometer is used to measure the air flow rate exiting from a 20 in. × 10 in. duct. The average reading on the instrument after 30 sec is 800 linear ft. Determine the flow rate in CFM.

16.3   A Pitot tube traverse is used to find the air flow rate in a duct. Figure 16.17 on page 430 indicates the velocity pressure readings at each point in in. w. Find the air flow rate in CFM.

16.4   A contact thermometer is used to make an approximate check of the performance of a

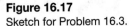

|←————— 38" —————→|

0.28	0.30	0.30	0.27
0.27	0.32	0.33	0.29
0.27	0.34	0.32	0.28
0.27	0.31	0.31	0.26

26"

Figure 16.17
Sketch for Problem 16.3.

hot water convector. The convector has a rated output of 12,000 BTU/hr. It is circulating 4 GPM of water entering at 218 F. What should the thermometer read if it is placed against the outlet pipe?

16.5   List all the locations and probable causes of excess heat loss indicated by the infrared photograph shown in Figure 16.14.

16.6   In testing the discharge head produced by a pump, the contractor uses the gage whose calibration curve is shown in Figure 16.1. If the required head is 46 ft w., what should be the reading on the gage?

16.7   An absorption refrigeration machine being tested is supposed to have a pressure in the absorber of 0.12 psia. A mercury manometer is used to check the pressure. What should be the reading on the manometer in mm of Hg vacuum?

16.8   A 12 in. air diffuser is to be balanced to a flow rate of 700 CFM. The effective outlet area is 0.57 ft². What should be the average velocity reading on an anemometer?

CHAPTER 17
Planning and Designing the HVAC System

In this chapter, we will explain the steps followed in the planning and design of two HVAC projects. Many of the individual items of knowledge studied previously will be utilized. Understanding these procedures will not only be useful to those interested primarily in designing HVAC systems; it will also increase the knowledge of the contractor, operator, and service engineer as to how the system functions, thereby enabling them to perform their work more efficiently. Practical problems that occur on an actual job will be faced.

OBJECTIVES

After studying this chapter, you will be able to:

1. Design a small hydronic heating system.
2. Design a small air conditioning system.

17.1 PROCEDURES FOR DESIGNING A HYDRONIC SYSTEM

The steps in planning and designing hydronic systems are outlined as follows (the numbers following each item are the principal chapters in which the subject is covered). In addition, recommendations on energy conservation (Chapter 15) should always be considered for each step.

1. Calculate the heating/cooling loads (3, 6).
2. Select terminal unit types and location (5).
3. Choose the type of system piping arrangement. Plan the distribution to the terminal units (5).
4. Determine the water flow rates and temperatures throughout the system (5).
5. Select the sizes of terminal units (5).
6. Determine pipe sizes (8).
7. Plan the piping layout in the building and locate valves (5, 9, 17).
8. Select the pump (11).

9. Select the boiler/chiller (4, 13).
10. Select the compression tank (11).
11. Provide and locate accessories required for proper operation and maintenance: air vents, drains, unions, expansion devices, anchors, supports, insulation (9, 11).
12. Select the control system (14).
13. Prepare final plans and specifications (17).

Although these steps are generally carried out in the order shown, the nature of the project some-times requires a slight change in sequence. On smaller projects, steps are often combined.

We will describe each of these steps in some de-tail, as they apply to the following project.

PROJECT I. *Design a hot water heating system for the residence whose floor plans are shown in Figure 17.1. The building is located in Des Moines, Iowa, with construction as follows:*

Figure 17.1
Floor plan of residence for Project I.

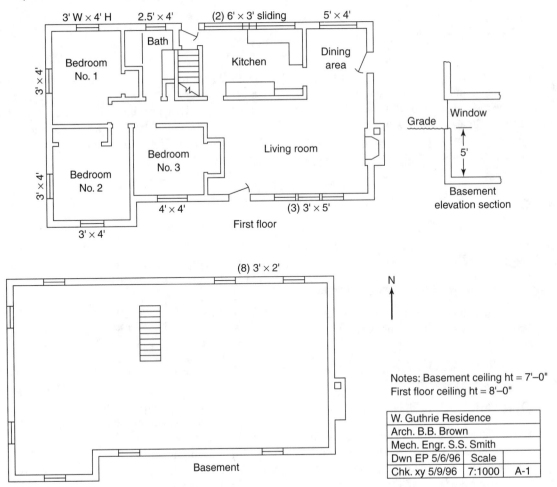

Notes: Basement ceiling ht = 7'–0"
First floor ceiling ht = 8'–0"

W. Guthrie Residence		
Arch. B.B. Brown		
Mech. Engr. S.S. Smith		
Dwn EP 5/6/96	Scale	
Chk. xy 5/9/96	7:1000	A-1

FIRST FLOOR

Walls: *frame with 4 in. brick veneer, sheathing, R-7 insulation, inside finish. (U = 0.09 BTU/hr-ft²-F).*

Roof: *wood frame pitched roof on rafters, R-11 insulation, finished ceiling U = 0.07 BTU/hr-ft²-F).*

Windows: *insulating (double) glass with ¼ in. air space. Double-hung vinyl sash. Infiltration rate = 0.5 CFM/ft. Window dimensions give width first and height last.*

Doors: *1 in. wood, 7 ft H × 3 ft W. Infiltration rate = 1.0 CFM/ft².*

BASEMENT

Walls: *8 in. concrete block below grade.*

Windows: *single glass, vinyl casement type. Infiltration rate = 0.8 CFM/ft. (Note that this is a high rate. These windows probably need better weatherstripping.)*

It should be noted that we have arbitrarily chosen a hot water heating system for the project. A warm air system might also have been chosen. The choice depends on costs, type of construction, convenience, and individual tastes—items we will not discuss further.

17.2 CALCULATING THE HEATING LOAD

The steps outlined in Chapter 3 will be followed here as they apply to the project. The data is recorded on a room heating load calculation form (Figure 17.2 on pages 434–435).

1. The recommended outdoor design temperature is selected from Table A.9, using the 97.5% level. An indoor design temperature of 72 F is chosen as a comfortable yet energy conserving design value (thermostat temperature may be set lower).
2. The room dimensions are taken from the architectural plans. Dimensions are converted to decimals for calculations. For example, in Bedroom No. 1, the north wall length 9 ft 6 in. is recorded as 9.5 ft. It is more precise to take dimensions from center lines of partitions and walls, rather than inside dimensions. Closet areas are included as part of a room when of significant size. Very small corners or offsets may be neglected. Small hallways usually may be neglected. The decision in each case depends on whether the heat transfer values would be significant.
3. The wall areas are calculated from room plan dimensions and height. For example, for Bedroom No. 1, the exposed wall length is 9.5 + 12 = 21.5 ft. The gross exposed wall area is therefore 21.5 × 8 = 172 ft². Allowing for the window openings, the net wall area is 172 − 24 = 148 ft².

   The stairs are included as part of the kitchen area. Some designers might neglect this small area.

   The Dining Area is calculated as if it were a separate room because it will be heated by an individual terminal unit.
4. The heat transfer coefficients *U* are found in Tables A.6 to A.8.
5. Using the conduction heat transfer equation (3.8), the rate of heat transfer through each exposed material surface is calculated. If there is no temperature difference across the material, then there is no heat transfer.

   The basement is open and it is assumed that it will be at room temperature from heat given off by equipment and piping. The ground temperature is 50 F.

   The total heat transfer from each room is listed in the appropriate subtotal row. Note that all calculated values are rounded off to three places of accuracy.
6. Length of cracks are found from the plans.
7. The air infiltration rates are as specified (this information is obtained from the window manufacturer).
8. Infiltration heat losses are calculated from Equation 3.10. For corner rooms, the infiltration on that side with the largest value is used.

Room Heating Load Calculations

P. 1 of 2 PP.

Project W. Guthrie Residence — Location Des Moines, IA — Indoor DB 72 F

Engrs. Enersave Assoc. — Calc. by EP 8-9-96 — Chk. by KM 8-12-96 — Outdoor DB −5 F

Room	Bedroom No. 1				
Plan Size	9.5' × 12' + 2' × 3'				
Heat Transfer	*U* ×	*A* ×	TD =		BTU/hr
Walls	.09	148	77		1030
Windows	.51	24	77		940
Doors					
Roof/ceiling	.07	120	77		650
Floor					
Partition					
Heat Transfer Loss					2620
Infiltration	1.1 ×	(CFM) *A* ×	*B* ×	TC =	
Window	1.1	.50	17	77	720
Door	1.1				
Infiltration Heat Loss					720
Room Heating Load					3340

Room	Bedroom No. 2				
Plan Size	11' × 12.5'				
Heat Transfer	*U* ×	*A* ×	TD =		BTU/hr
Walls	.09	164	77		1140
Windows	.51	24	77		940
Doors					
Roof/ceiling	.07	138	77		740
Floor					
Partition					
Heat Transfer Loss					2820
Infiltration	1.1 ×	(CFM) *A* ×	*B* ×	TC =	
Window	1.1	.50	17	77	720
Door	1.1				
Infiltration Heat Loss					720
Room Heating Load					3540

Room	Bedroom No. 3				
Plan Size	9.5' × 9.5' + 2' × 4'				
Heat Transfer	*U* ×	*A* ×	TD =		BTU/hr
Walls	.09	60	77		420
Windows	.51	16	77		630
Doors					
Roof/ceiling	.07	98	77		530
Floor					
Partition					
Heat Transfer Loss					1580
Infiltration	1.1 ×	(CFM) *A* ×	*B* ×	TC =	
Window	1.1	.50	20	77	850
Door	1.1				
Infiltration Heat Loss					850
Room Heating Load					2430

Room	Bathroom				
Plan Size	7' × 9' + 4' × 2'				
Heat Transfer	*U* ×	*A* ×	TD =		BTU/hr
Walls	.09	46	77		320
Windows	.51	10	77		390
Doors					
Roof/ceiling	.07	55	77		300
Floor					
Partition					
Heat Transfer Loss					1010
Infiltration	1.1 ×	(CFM) *A* ×	*B* ×	TC =	
Window	1.1	.50	15.5	77	660
Door	1.1				
Infiltration Heat Loss					660
Room Heating Load					1670

Room	Kitchen				
Plan Size	17' × 9'				
Heat Transfer	*U* ×	*A* ×	TD =		BTU/hr
Walls	.09	82	77		570
Windows	.51	36	77		1410
Doors	.64	21	77		1030
Roof/ceiling	.07	153	77		820
Floor					
Partition					
Heat Transfer Loss					3830
Infiltration	1.1 ×	(CFM) *A* ×	*B* ×	TC =	
Window	1.1	.50	33	77	1400
Door	1.1	1.0	21	77	1780
Infiltration Heat Loss					3180
Room Heating Load					7010

Room	Living Room				
Plan Size	22' × 13'				
Heat Transfer	*U* ×	*A* ×	TD =		BTU/hr
Walls	.09	217	77		1500
Windows	.51	45	77		1770
Doors	.64	21	77		1030
Roof/ceiling	.07	286	77		1540
Floor					
Partition					
Heat Transfer Loss					5840
Infiltration	1.1 ×	(CFM) *A* ×	*B* ×	TC =	
Window	1.1	.50	57	77	2410
Door	1.1	1.0	21	77	1780
Infiltration Heat Loss					4190
Room Heating Load					10,030

Infiltration CFM	Column A	Column B
Windows	CFM per ft	Crack length, ft
Doors	CFM per ft^2	Area, ft^2

Figure 17.2
Room heating load calculations for Project I.

Room Heating Load Calculations

Project __W. Guthrie Res.__ Location _____ Indoor DB _____ F

Engrs. _____ Calc. by _____ Chk. by _____ Outdoor DB _____ F

Room	Dining Area					Basement								
Plan Size	9' × 9'					42' × 22' + 11' × 3'								
Heat Transfer	U	×	A	×	TD	=	BTU/hr	U	×	A	×	TD	=	BTU/hr	U	×	A	×	TD	=	BTU/hr
Walls	.09	106	77	730	.09	670	27	1630				
					.08	220	77	1360				
Windows	.51	20	77	790	.98	48	77	3620				
Doors	.64	21	77	1030								
Roof/ceiling	.07	81	77	440								
Floor					.04	957	27	1030				
Partition												
	Heat Transfer Loss			2990				7640				
Infiltration	(CFM) $1.1 \times$ A × B × TC =				(CFM) $1.1 \times$ A × B × TC =				(CFM) $1.1 \times$ A × B × TC =			
Window	1.1				1.1	.80	30	77	2030	1.1			
Door	1.1	1.0	21	77	1780	1.1				1.1			
	Infiltration Heat Loss			1780				2030				
	Room Heating Load			4770				9670				

Room														
Plan Size														
Heat Transfer	U	×	A	×	TD	=	BTU/hr	U	×	A	×	TD	=	BTU/hr	U	×	A	×	TD	=	BTU/hr
Walls												
Windows												
Doors												
Roof/ceiling												
Floor												
Partition												
	Heat Transfer Loss											
Infiltration	(CFM) $1.1 \times$ A × B × TC =				(CFM) $1.1 \times$ A × B × TC =				(CFM) $1.1 \times$ A × B × TC =			
Window	1.1				1.1				1.1			
Door	1.1				1.1				1.1			
	Infiltration Heat Loss											
	Room Heating Load											

Infiltration CFM	Column A	Column B
Windows	CFM per ft	Crack length, ft
Doors	CFM per ft²	Area, ft²

Figure 17.2
(Continued)

9. The room heat transfer losses and infiltration losses are added to find the room total heat loss (load).
10. Building heating load. Figure 17.3 shows the calculation of the building heating load. The areas and infiltration rates are found directly from building plans, rather than by adding up individual room areas in Figure 17.2.

The estimated infiltration is found by choosing one-half the total infiltration of all sides. The calculations are shown in Figure 17.3.

Based on knowledge of building conditions and operations, additions are made to the calculated building net load. In this example, we have assumed that the piping heat loss is included in the basement load and that night setback of temperatures requires 40% excess capacity for pickup (see Chapters 3 and 4).

The final total is the required boiler gross output. A suitable boiler should be selected using this value.

17.3 TYPE AND LOCATION OF TERMINAL UNITS

Baseboard radiation will be used, because it is inexpensive, occupies little space, and is attractive—features that are desirable in a residence. Terminal units will be located under windows to prevent cold downdrafts. When there is more than one window in a room, units will be located under each, if possible. The basement will not be heated with terminal units, because it will be "unfinished." Heat from the boiler and pipe surface will be adequate.

17.4 PIPING SYSTEM ARRANGEMENT

For a small, modest home, either a series loop system or a one-pipe main, with the advantage of regulating flow to each unit, would be used. The advantages of a two-pipe arrangement do not gain enough benefits in such a small installation to justify the higher cost.

We will use a split series loop piping arrangement. Actually, the temperature drop in a single series loop will probably not be excessive on such a small project, but the split loop is chosen only to make a slightly more complicated design procedure, providing more instructional value for the student. (It is suggested that the student also design other systems for the additional experience.)

The connection of piping to units is now determined. The boiler should be located near the chimney in the basement, so that the vent connection is short. The building shape lends itself to splitting the loop naturally on the north and south sides, starting with Bedrooms 1 and 2.

A schematic piping sketch showing this arrangement is now made (Figure 17.4 on page 438). Note that there are terminal units under both windows in Bedrooms 1 and 2 for increased comfort. In preparing this diagram, it is often helpful to use transparent sketch paper placed over the architectural plans. The designer then sketches the locations of terminal units and connecting piping. These sketches are also used to aid in preparation of the finished drawings.

17.5 FLOW RATES AND TEMPERATURES

The procedures described in Chapter 5 will be followed. Let us try a system water temperature drop of 14 F. The required system flow rate to handle the building heating load (Equation 5.2) is

$$\text{Flow rate} = \frac{Q}{500 \times \text{TC}} = \frac{40{,}170}{500 \times 14}$$
$$= 5.7 \text{ GPM (use 6 GPM)}$$

Because there are two loops, the flow splits between them. We could either determine the quantity to be delivered to each, or we could assume each has exactly a 14 F temperature drop and then calculate the required flow rate in each. Because the two parts of the building have about the same load, it is satisfactory to assume 3 GPM each. If the loads were very different, using the same flow rate

Building Heating Load Calculations

Project: W. Guthrie Residence
Location: Des Moines, IA
Engineers: Enersave Assoc.

Calc. by: EP 9-8-96
Chk. by: KM 9-10-96

	DB, F	W, gr/lb
Indoor	72	
Outdoor	–5	
Diff.	77	

Heat Transfer	Q =	U	×	A	×	TD	=	BTU/hr
Roof		.07		957		77		5160
Walls (1st floor)		.09		1072		77		7430
(base, below grade)		.08		670		27		1450
(base, above grade)		.09		220		77		1520
Windows (1st floor)		.51		175		77		6870
(basement)		.98		48		77		3620
Doors		.64		63		77		3100
Floor (basement)		.04		957		27		1030

	BTU/hr
Heat Transfer Subtotal	30,180
Infiltration $Q_s = 1.1 \times$ ___118___ CFM × ___77___ TC =	9990
Building Net Load	40,170
Ventilation $Q_s = 1.1 \times$ _____ CFM × _____ TC =	
$Q_L = 0.68 \times$ _____ CFM × _____ gr/lb =	
Duct Heat Loss _____ %	
Duct Heat Leakage _____ %	
Piping and Pickup Allowance ___40___ %	16,070
Service HW Load	
Boiler or Furnace Gross Load	56,240

Calculation of Infiltration CFM

	N	E	S	W
1st floor windows	44		47	17
doors	21	21	21	
base, windows	24		24	16
Subtotals	89	21	92	33

½ Total CFM = (89 + 21 + 92 + 33) × ½ = 118 CFM

Figure 17.3
Building heating load calculations for Project I.

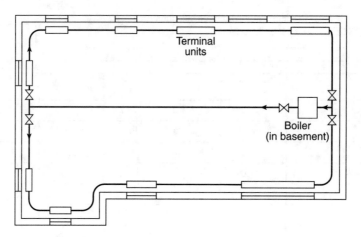

Figure 17.4
Schematic piping arrangement for Project I.

in each would result in very unequal temperature drops in each loop. This might occasionally cause problems in selection of a unit.

A supply temperature is now chosen in the recommended range. We will choose 210 F, which is a somewhat arbitrary decision. Higher temperatures might reduce the length of radiation slightly, but it is only a small part of the cost of the installation. Furthermore, the higher the temperature, the more severe a burn might occur from contact with exposed pipe. Even lower temperatures might be preferred by some designers for residential systems.

17.6   SELECTION OF TERMINAL UNITS

The size of each terminal unit is selected with the aid of the manufacturer's rating tables. The water flow rate and average temperature for the unit must be known. A flow rate of 3 GPM has already been chosen.

The average temperature for each unit could be determined by calculating the temperature drop through each unit, using Equation 5.2. In this system, however, it will be accurate enough to use the average water temperature for the whole system as if it were that for each unit. This is because the ca-

pacity of the baseboard does not vary much with small temperature differences. If the temperature changes were large, this would not be allowable, nor would it be allowable for a type of unit whose capacity varies greatly with water temperature. (An important case is chilled water terminal units.) The system average water temperature is

$$T_{ave} = 210 - 14/2 = 203 \text{ F}$$

The nearest lower listing in Table 5.1 is 200 F, so this value will be used in selecting each unit. A ¾ in. pipe size will be used. The capacity at these conditions is 740 BTU/hr per foot of length. For Bedroom 1, the length of baseboard needed to produce the heat output required is therefore

$$\text{Length} = \frac{3340 \text{ BTU/hr}}{740 \text{ BTU/hr per ft}} = 4.5 \text{ ft (use 5 ft)}$$

For this particular room, the radiation is to be installed under both windows, so two units, each 2½ ft long, or perhaps one 3 ft and the other 2 ft long, would be used. The selection of the units for other rooms is carried out in a similar manner. The results are shown in Table 17.1.

Sometimes the required length of a unit is greater than the outside wall length available. One solution is to use adjacent inside walls. Of course,

TABLE 17.1 SELECTION OF TERMINAL UNITS FOR PROJECT I

Room	Unit	BTU/hr	GPM	t_{ave}	BTU/hr per ft	L, ft	Type
Bedroom 1	A	1670	3	200	740	3	Base
Bedroom 1	B	1670	3	200	740	2	Base
Bath	C	1670	3	200	740	3	Base
Kitchen	D	7010	3	200		4	Convector
Dining	E	4770	3	200	740	7	Base
Living	F	10,030	3	200	740	14	Base
Bedroom 3	G	2430	3	200	740	4	Base
Bedroom 2	H	1770	3	200	740	3	Base
Bedroom 2	I	1770	3	200	740	2	Base

if there are cabinets or other obstructions, this prevents use of a wall. In such cases, a convector or other type of unit with high capacity may solve the problem, as in the kitchen.

17.7 PIPE SIZING

The procedure described in Chapter 8 will be used to find the pipe diameters and system pressure loss.

1. The piping system sketch is shown in Figure 17.4.
2. The flow rate in each line has been determined and is shown on the sketch.
3, 4. Copper tubing will be used because it is easy to work with. Furthermore, the baseboard radiation is copper tube. A trial size of the main is made. Using Figure 8.15, either a 1 in. or 1¼ in. Type L will have a friction loss within the recommended range with a flow rate of 6 GPM (2.8 and 1.0 ft of water per 100 ft, respectively). The 1 in. diameter will be chosen for a small initial cost saving. However, operating cost will be slightly higher.
5. Using the chosen friction loss rate as a guide (equal friction method), the two loop pipe sizes will be chosen. The flow rate is 3 GPM in each. From Figure 8.15, it is seen that a ¾ in. diameter provides the friction loss rate closest to that desired, 3.0 ft of water per 100 ft. The velocity is about 2 FPS, an acceptable value.

We have used the suggested equal friction method for sizing the pipe. Actually in such a simple system the procedure does not have to be followed rigorously. Either a 1 in. or 1¼ in. pipe could have been used for the larger line and a ¾ in. or ½ in. pipe for the sub-loops, because balancing will be a simple operation.

17.8 PIPING OR DUCT LAYOUT

The physical layout of the piping, including all changes in direction, valves, and other items in the lines, must be known in order to find the system pressure loss. This information is then used to determine the required pump head. Although the final drawings show the detailed piping layout, it suffices to make freehand sketches to scale for the purpose here. Often the simple sketch made previously is used, with some imagination necessary to visualize the changes in direction.

To plan the piping layout, place transparent sketch paper over each floor plan and sketch the desired location of piping and equipment to be installed on that floor. Some recommended good practice guidelines for locating piping (which also apply to ductwork), are:

1. Piping and ducts are usually run parallel or perpendicular to building walls.
2. Use the minimum number of changes in direction consistent with item 1.

3. The location should minimize the difficulty of installation. For example, avoid locating piping in cramped quarters where the pipe fitter cannot work easily.

4. The location should provide for ease of operation. For instance, a pipeline containing a valve that is frequently used and yet is difficult to reach would be poor planning. Piping that obstructs equipment control panels is another example of a poor layout.

5. The layout should provide for ease of maintenance. Piping and ducts should not be located where they interfere with access to equipment requiring servicing or replacement.

6. The layout should not interfere with normal use of the space. For this reason, horizontal pipes are usually run overhead and close to walls. They cannot be located where they block or displace furniture or openings.

7. The location should not physically interfere with or disturb proposed installations of other trades (electrical and plumbing). As an example, ductwork running directly underneath a lighting fixture would block some light. HVAC planners must coordinate their layouts with the other trades to prevent possible interferences. This requires studying their plans.

8. Avoid penetration of structural members (columns or beams). If this seems unavoidable, permission from the structural engineer must be obtained.

9. The layout should not create any safety hazards. For example, a valve located above a transformer would be poor practice, in case of a valve leak.

10. The layout should conform with aesthetic requirements of the architecture. An obvious example is that piping and ducts must be concealed above hung ceilings or behind walls in many types of spaces.

11. The layout must meet all applicable Code requirements.

In addition to the general recommendations in this list, each project has its own peculiarities that affect the choice of piping and duct location.

The piping layout developed for this project is shown in Figure 17.5. The student should check the plans to see if it follows good practice. Lines are run in the basement overhead. Note that where obstructions are encountered on the occupied floor, the piping drops down to the basement.

Gate valves shall be located on both sides of any equipment to be serviced (see Figure 4.9). Globe valves or other flow regulating valves have been located in each loop so that flow can be balanced to the design values.

17.9 PUMP SELECTION

The first step in selecting the pump is to calculate the system pressure loss. From inspection of the piping layout, it is clear that the circuit with the greatest pressure loss is the loop on the north side of the building. Not only is the total length of piping longer, but there are more changes in direction. The lengths are measured, the fittings are counted, and equivalent lengths are determined. Pressure losses in each section are calculated, then added. The results are recorded in Table 17.2 on page 442. Students should go through the procedure to see if they obtain the same results. Be sure to include all vertical distances and elbows. When some information is not given, an estimate is often accurate enough. For example, a vertical distance of 2 ft from the horizontal lines in the basement to the terminal units is assumed.

The system friction loss is 8.7 ft w. Because the piping is a closed loop, there is no additional pump pressure needed to lift the water, and therefore the required pump head is also 8.7 ft w. An in-line pump will be used for its low cost and small capacity needed. From Figure 11.4, a No. 102 pump is adequate, developing about 11 ft w. head at 6 GPM, using a ⅛ HP motor.

17.10 BOILER SELECTION

A boiler will be selected from Figure 4.21. It is assumed that the fuel is natural gas. On an actual

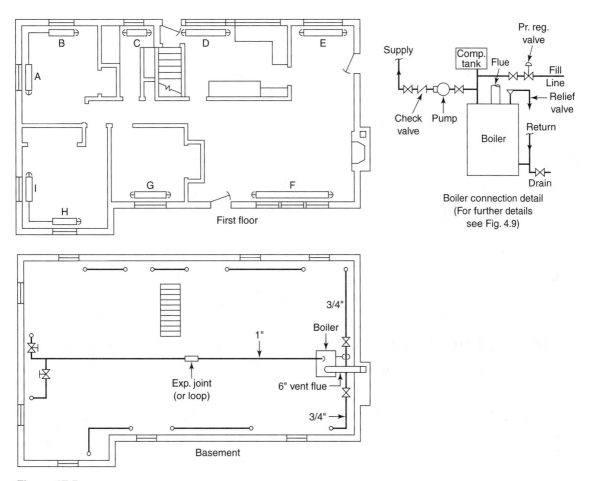

Figure 17.5
HVAC plans for Project I.

project, the choice of fuel is made on the basis of cost, convenience, and availability. The required boiler gross output (DOE capacity) is 56,240 BTU/hr. The boiler selected to satisfy the load is

Model no. GG75HED

Gross (DOE) output = 64,000 BTU/hr

Input = 75,000 BTU/hr

Flue diameter = 5 in.

AFUE = 84.78%

Note that in order to obtain the boiler gross capacity, we have used the piping and pickup losses known from job conditions in the load calculations. If no specific knowledge of these losses was available, we might have used the industry standard losses (Chapter 4).

The boiler steady state efficiency can also be determined:

$$E = \frac{64,000}{75,000} \times 100 = 85.3\%$$

TABLE 17.2   PRESSURE DROP CALCULATIONS FOR PROJECT I

Section	Item	D in.	GPM	V FPS	EL ft	No.	TEL, ft	H_f ft w./ 100 ft	H_f ft w.
Main	Pipe	1	6	2.3	43		43	2.8	
	Boiler				11		11		
	Gate				1.0	3	3		
	Check				42	2	84		
	El				2.6	7	18		
	Tee				2.6	1	3		
					Subtotal		162	$\times 2.8 / 100 =$	4.5
Loop A	Pipe	¾	3	2.0	78		78	3.0	
	Globe				22	1	22		
	El				2.0	18	36		
	Tee				4.0	1	4		
					Subtotal		140	$\times 3.0 / 100 =$	4.2
								Total $H_f =$	8.7

17.11 COMPRESSION TANK

The size of the compression tank is found by using Equation 11.7. The volume of water in the system must first be determined, using Table 9.2. Lengths are approximated from building plans:

$$1 \text{ in. tube, } 43 \text{ ft} \times 0.044$$
$$\text{gal/ft} = 1.9 \text{ gal}$$
$$¾ \text{ in. tube, } 160 \text{ ft} \times$$
$$0.025 \text{ gal/ft} = 40$$
$$\text{Boiler} = 12$$
$$\text{System volume} = \overline{54 \text{ gal}}$$

The value for the boiler water volume is furnished by the manufacturer.

The terms in Equation 11.7 are

$t = 210 \text{ F}$

$H_a = 14.7 \text{ psi} \times 2.3 \text{ ft w./1 psi} = 34 \text{ ft w.}$

$H_t = 5 \times 2.3 + 4 + 34 = 50 \text{ ft w. abs}$

$H_O = 45 \text{ psia} \times 2.3 = 104 \text{ ft w. abs}$

In these calculations, a fill pressure of 5 psi was used, and an estimated system height of 4 ft above the tank elevation used to find H_t. The boiler relief valve setting is 30 psig. Substituting in the equation,

$$V = \frac{[0.00041(210) - 0.0466]54}{34/50 - 34/104} = 6.0 \text{ gal}$$

(Practically for such a small system, a pre-selected diaphragm-type expansion tank would be furnished.)

17.12 ACCESSORIES

The location and types of valves to be used have already been discussed. Air vents will be installed at high points. This would include points where the horizontal piping from the radiation drops down to the basement. A drain connection with valve will be installed at the return to the boiler. Connections to the boiler and pump will be made with unions so that they can be removed easily. An expansion loop will be provided in the long 1 in. line. The ¾ in. lines have enough natural offsets to permit expansion. All piping in the basement will be insulated. Proper support hangers will be furnished for all horizontal piping.

17.13 CONTROLS

The application calls for a simple, inexpensive control system. A room thermostat controls the pump operation in response to a call for heat. A high limit immersion thermostat controls the gas burner to maintain a set hot water temperature. A control diagram is shown in Figure 17.6. Safety controls would be provided in addition to these operating controls.

The room thermostat is located on an inner wall of the living room. Although direct control of temperature is from this room only, the smallness of the house should result in limited temperature variations in other rooms. A dual thermostat will be used to set back night temperatures and conserve energy.

17.14 PLANS AND SPECIFICATIONS

The final step of the project is to prepare finished *plan* and *specifications* of the heating system. The drawings would be similar to those in Figure 17.5, made to a scale convenient for easy reading. A *plan view* is made for each floor, and *details* of connections to equipment are shown separately.

The specifications will include descriptions of materials and equipment. Some excerpts from a typical specification will be given to familiarize the student with how they are written. Much of the information is prepared with the aid of the manufacturer's equipment specifications.

BOILER *Furnish and install a low pressure hot water heating boiler of the gas-fired, packaged, cast-iron type, designed and constructed in accordance with an approved by ASME, I-B-R, and AGA Standards. It shall have a DOE rating of at least 64,000 BTU/hr and an AFUE of no less than 84%.*

The boiler shall include all of the following: pre-assembled heat exchanger with built-in air eliminator, base; flue collector; gas burners; gas orifices and manifold assembly; combination gas valve including manual shut-off, pressure regulator, pilot adj., and automatic pilot-thermocouple safety; hi limit control; altitude; pressure and temperature gauge; pressure relief valve (ASME); draft hood; draft hood spill switch; rollout safety switch; pre-assembled insulated semi-extended jacket (extended as shown); and automatic vent damper.

Optional equipment for the boiler includes: room thermostat; millivolt (self-energized) controls; combination gas valve; combination limit controls and millivolt thermostat; air package consisting of diaphragm expansion tank, fill and pressure reducing valve, and automatic air vent; combustible floor kit; and intermittent pilot ignition system.

The boiler shall be as manufactured by ____ or approved equal.

BASEBOARD RADIATION *Furnish and install baseboard heating elements where shown on the plans, of the capacity specified. They shall be of ¾ in. nominal copper tubing with aluminum fins. Furnish complete enclosures for wall-to-wall covering, including front and back panels, end caps, corner, and trim. All components shall have shell white baked-on finish. Furnish support brackets, expansion cradles, and dampers. Baseboard shall be as manufactured by ____ or approved equal.*

Figure 17.6
Control diagram for Project I.

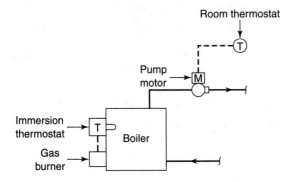

Similar specifications would be included for all other items in the system. The specifications must also include legal and contractual statements, covering subjects such as guarantees and liabilities.

17.15 ENERGY USE AND CONSERVATION

The recommendations for energy conservation have been applied in planning the system. This includes the design temperatures chosen, the R-values of construction materials, the use of storm sash, and windows with small infiltration rates. The control system includes night setback of temperature. The pump is operated intermittently rather than continuously.

An estimation of fuel used can be made using the modified degree day method. Applying Equation 15.16,

$$Q_O = \frac{Q}{TD} \times D \times 24 \times C_D \times \frac{1}{E}$$

$$= \frac{40,170}{77} \times 6588 \times 24 \times 0.62 \times \frac{1}{0.83}$$

$$= 61,650,000 \text{ BTU}$$

Using natural gas with a heating value of 1000 BTU ft^3, the fuel quantity used is

$$61,650,000/1000 = 61,650 \text{ ft}^3 \text{ per year}$$

17.16 PROCEDURES FOR DESIGNING AN ALL-AIR SYSTEM

The steps in planning and designing all-air systems are outlined as follows (numbers in parentheses are chapter references):

1. Calculate the heating/cooling loads (3, 6).
2. Determine the supply air conditions (7).
3. Choose the type of system (12).
4. Plan the equipment and duct locations (17).
5. Determine duct sizes (8).
6. Determine sizes of air distribution devices (10).
7. Select equipment (5, 10, 12, 13).
8. Provide and locate accessories required for proper operation and maintenance (9, 16).
9. Select the control system (14).
10. Prepare final plans and specifications (17).

These steps are generally carried out in the order shown, but occasionally some change in the sequence is necessary, particularly when information required to complete one step is dependent on data found in a later step. Sometimes steps can be combined. The recommendations on energy conservation should always be considered for each step.

We will describe each of these steps in some detail as they apply to the following project.

PROJECT II. Design an all-air summer air conditioning system for the department store whose floor plans are shown in Figure 17.7. The building is located in Fort Worth, Texas, with construction and conditions as follows:
Walls: 4 in. face brick, 8 in concrete block, gypsum wallboard finish. Dark colored.
Roof: 4 in. lightweight concrete deck, suspended ceiling.
Partitions: 4 in. cinder block.
Building construction: mediumweight.
Glass: single clear plate glass aluminum frame. No shading.
Lighting: 3.0 W/ft^2, including ballast factor.
Occupancy: 500 people.
Equipment: 16 motors ¼ HP each, operating 50% of time (vending machines and similar devices).
Door infiltration: 1 CFM/ft^2. Doors are glass. Only sales area will be air conditioned. Floors are not carpeted. Store closes at 6 PM standard time.

17.17 CALCULATING THE COOLING LOAD

The steps outlined in Chapter 6 will be followed here as they apply to the project. The data are

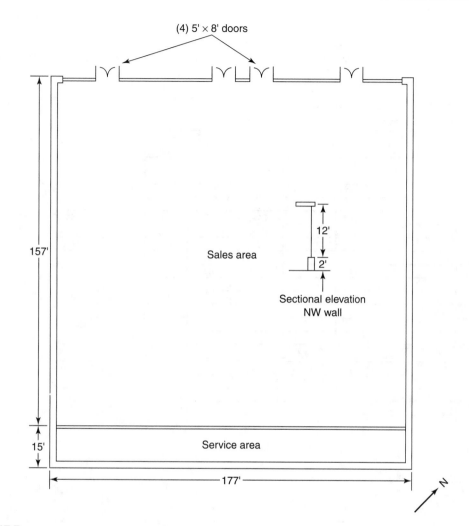

Figure 17.7
Floor plan of building for Project II.

recorded on the commercial cooling load calculations form (Figure 17.8 on page 446).

Step 1. The recommended outdoor design conditions are selected from Table A.9, at the 2.5% level, including the mean coincident wet bulb. An indoor design condition of 78 F DB and 50% RH is chosen as comfortable yet energy conserving. Perhaps even 80 F could have been used, because occupancy is for short periods of time.

Step 2–3. Dimensions are taken from the architectural plan. Areas are calculated and recorded. The NW wall has a 12 ft height of glass the full length of the building. Doors are glass.

Step 4. Heat transfer coefficients are found from Tables 6.1, 6.3, or A.4–A.8.

Steps 5–7. The peak load time must be determined. Because there is only one air conditioned space, there is no problem of calculating peak loads for both individual rooms and for the

COMMERCIAL COOLING LOAD CALCULATIONS

Project __Bargain Dept. Store__   Bldg./Room __Building, peak__

Location __Ft. Worth, TX__   Engrs. __Miller & Miroze__   Calc. by __EP 9/6/96__   Chk. by __PC 9/8/96__

Design Conditions		DB F	WB F	RH %	W' gr/lb	
	Outdoor	99	74		86	Daily Range __22__ F Ave. __88__ F
	Room	78		50	71	Day __June 21__ Time __5 PM (ST)__

Lat. __32° N__

Conduction	Dir.	Color	U	A, ft²		CLTD, F		SCL BTU/hr
				Gross	Net	Table	Corr.	
Glass			1.01		2124	13	16	34,320
Wall Group D	SW	D	.22		2198	27	28	13,540
	NE	D	.22		2198	24	29	14,020
	NW	D	.22		354	18	23	1790
Roof/ceiling			D	.13	27,790	65	70	252,890
Floor								
Partition			.40		2478		12	11,890
Door								

Solar	Dir.	Sh.	SHGF	A	SC	CLF	
	NW	no	176	2124	.64	.88	
Glass							210,540

Lights __83,370__ W × 3.41 × __—__ BF × __1__ CLF   284,290

Lights _____ W × 3.41 × _____ BF × _____ CLF

People __250__ SHG × __500__ n × __1__ CLF   125,000

    __200__ LHG × __500__ n

Equipment __1000 × 16 × 0.5__   8000

Equipment _____

Infiltration 1.1 × _____ CFM × _____ TC

    0.68 × _____ CFM × _____ gr/lb

	LCL BTU/hr
	100,000

Subtotal   956,280   100,000

SA duct gain __5%__   47,810

SA duct leakage __5%__   47,810

_____   5000   Total CL

SA fan gain (draw through) __5%__   47,810   BTU/hr

Room/Building Cooling Load   1,099,710   150,000   1,249,710

SA fan gain (blow through) _____

Ventilation 1.1 × __7500__ CFM × __19__ TC   156,750

    0.68 × __7500__ CFM × __15__ gr/lb   76,500

RA duct gain _____

RA fan gain __2.5%__   23,910

Pump gain _____

Cooling Coil Load   1,280,370   226,500   1,506,870

Refrigeration Load   1,506,870

Figure 17.8
Cooling load calculations for Project II.

whole building. From Table 6.6, at 32°N latitude, the peak NW glass solar gain occurs in June (176 BTU/hr). Because the CLTD for the roof will also be maximum at this time of year (Table 6.1), this is the peak month. (If peak solar gain was in the fall, we would have to calculate the solar and conduction gains at that time to see if the total is a peak. This might be true if the glass faced south.)

The peak time of day for the load must now be determined. The CLF value from Table 6.8 shows that the glass solar load peaks at 5 PM (the store closes at 6 PM). From Table 6.1, the roof CLTD also peaks at 5 PM. Since the glass solar load and roof conduction are far larger than the other external heat gains, we can safely say that 5 PM is the peak time. Those items that are not read directly from the table and that may need clarification will be explained.

The CLTD values are corrected for month, latitude, color of surface, inside temperature, and average daily temperature.

The spaces adjacent to the air-conditioned area are assumed to be at 88 F (halfway between inside and outside temperatures). If exhaust air from the air-conditioned area is used to ventilate these spaces, this might be a reasonable assumption, otherwise a higher temperature might result.

Lighting
The lighting total is

$$30 \text{ W/ft}^2 \times 27{,}790 \text{ ft}^2 = 83{,}370 \text{ W}$$

The ballast factor BF has already been included in the value for the lighting intensity. All the lights are presumed turned on. The cooling system will be shut down as soon as the store closes, so CLF = 1.0.

People
The number of people (500) presumably was determined by the building owner. If this information is not available, Table 6.17 can be used as an estimate. The CLF is 1.0, because the cooling system is shut down when the store closes. Table 6.3 is used to find the heat gains.

Step 8. The items comprising the heat gains can now be calculated and are recorded in Figure 17.8. The values are found from the following tables:

Item	Table
Design conditions	A.9, 1.1
SHGF (Sensible Heat Gain Factor)	6.6
SC (Shading Coefficient)	6.7
CLF (Cooling Load Factor), solar	6.8, 6.9
CLTD (Cooling Load Temperature Difference), glass	6.5
CLTD, wall	6.2, 6.3
CLTD, roof	6.1
U	6.1, 6.3, A.4–A.8
CLF, people	6.14
Ventilation	6.17
People gain	6.13
Motor gain	6.16

The student should confirm all values listed in Figure 17.8.

Equipment
The equipment (motors) estimate has been furnished by the building owner. The resulting heat output, assuming 50% operation, is

$$1000 \text{ BTU/hr} \times 16 \times 0.5 = 8000 \text{ BTU/hr}$$

Infiltration
The ventilation system fans will pressurize the building, preventing significant infiltration.

Duct Heat Gains
The ductwork will be located above the hung ceiling and therefore the heat gain to it must be included. An estimate of 5% of the sensible load is taken.

There will be no significant heat gain to the return air ducts, because their lengths will be very short.

The heat equivalent of the duct air leakage must be determined, because the ducts are outside the conditioned space. An estimate of 5% of the sensible and latent loads will be used.

Fan Heat Gain

The temperature rise and heat gain from the fans must be determined. It is assumed that the supply fan SP = 2.0 in. w.g. A draw-through fan arrangement will be used; therefore, the heat gain will become part of the sensible load. From Chapter 6, the value is 5%.

Step 9. The ventilation load can now be determined. The number of occupants at peak time is estimated at 500. Table 6.17 recommends 15 CFM per person for this application.

The outside air temperature used for calculating the ventilation sensible load is corrected for the time of day, as shown in Table 6.18.

The sensible, latent, and total cooling loads are found by adding room gains to ventilation and return air gains. (Any blow-through fan and pump gains would also be added.) There will be no diver-

sity of loads, because all lights are on at all times, and the people occupancy occurs at the same time.

Step 10. The required design supply air conditions can now be determined. The solution will be worked out with the aid of a psychrometric chart (Figure 17.9), as explained in Chapter 7.

The supply air CFM is determined. A trial value of the coil leaving air temperature is chosen, 58 F. The required CFM is

$$CFM = \frac{RSCL}{1.1 \times TC} = \frac{1,099,710}{1.1 \times (78 - 58)}$$

$$= 49,990 \text{ CFM (say 50,000 CFM)}$$

The RSHR line is drawn. The RSHR is

$$RSHR = \frac{RSCL}{RTCL} = \frac{1,099,710}{1,249,710} = 0.88$$

Figure 17.9
Psychrometric chart analysis of air conditioning processes for Project II.

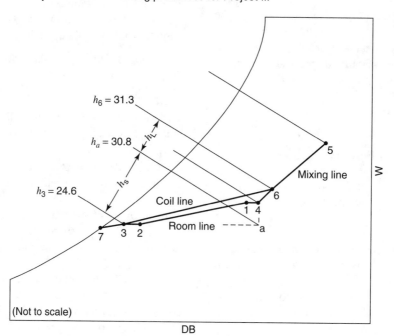

Since the supply air is cooled 20 F below the room temperature, the temperature rise from both the supply air fan and supply air duct heat gains (5% each) is 0.10×20 F = 2 F.

This establishes the supply air temperature to the room at $58 + 2 = 60$ F. The point (2) is located on the RSHR line. A horizontal line is shown from point 2 to 58 F, establishing the supply air condition off the coil, point (3).

The temperature rise due to the return air fan heat (2.5%) is 0.025×20 F = 0.5 F. The return air temperature to the cooling coil is therefore $78 + 0.5 = 78.5$ F, point (4).

The mixing line is now drawn between points (4) and (5), the outside air condition. The mixed air condition, point (6), is located on the mixing line, based on the proportion of outside air to total air quantity.

The coil process line is then drawn, from point (6) to (3). The coil process line extended to the saturation line establishes point (7), the coil effective surface temperature (EST).

The ratio of the lengths (6) − (3) to (6) − (7) is the coil contact factor, CF = 0.92. The coil bypass factor is then BF = 1 − CF = 0.08.

The cooling coil load can be calculated from the psychrometric chart, using the enthalpy values shown in Figure 17.9.

$$
\begin{aligned}
\text{CTCL} &= 4.5 \times \text{CFM}_{sa} \, (h_6 - h_3) \\
&= 4.5 \times 50{,}000 \, (31.3 - 24.5) \\
&= 1{,}530{,}000 \text{ BTU/hr} \\
\text{CSCL} &= 4.5 \times \text{CFM}_{sa} \, (h_a - h_3) \\
&= 4.5 \times 50{,}000 \, (30.2 - 24.5) \\
&= 1{,}282{,}500 \text{ BTU/hr} \\
\text{CLCL} &= 4.5 \times \text{CFM} \, (h_6 - h_a) \\
&= 4.5 \times 50{,}000 \, (31.3 - 30.2) \\
&= 247{,}500 \text{ BTU/hr}
\end{aligned}
$$

These figures check closely with the cooling load calculation values, which would be used to specify the cooling coil requirements. If a coil is available, the trial value of supply air temperature is satisfactory. If not, a new trial value would be selected.

17.18 TYPE OF SYSTEM

An all-air single zone system will be used, with DX refrigerant coils, because there is only one space, and the equipment can all be located in one place, with air distributed directly to the space from there.

17.19 EQUIPMENT AND DUCT LOCATIONS

The suggestions in Section 17.8 on planning the installation should be referred to as a guideline. The shape and size of the building suggest a few rows of ducts running the length or width of the building, above the suspended ceiling. The refrigeration and air handling equipment will be located on the end of the building. Roof-mounted units will be used because of their low cost and convenience for a one-story building. This also saves usable floor space in the building. Four duct mains will be used, with four air diffusers in each. This seems to provide a reasonably simple duct arrangement, but with enough diffusers so that good air distribution will result.

A separate air handling unit will be used for each main. It is possible that two units could be used for the building, but this would leave the system with only 50% of capacity if one were out of service. Furthermore, 50 tons is about the limit at which many roof-mounted units are manufactured. A sketch of the arrangement is shown in Figure 17.10 on page 450.

17.20 DUCT SIZES

Because the total supply air quantity is 50,000 CFM, there is 12,500 CFM for each of the four air handling units, if the air is distributed equally among them. In this particular application, consideration might be given to supplying a greater proportion of air to the side of the building receiving the solar load through the glass. On the other hand, the load tends to equalize throughout the space by internal heat transfers. We will assume

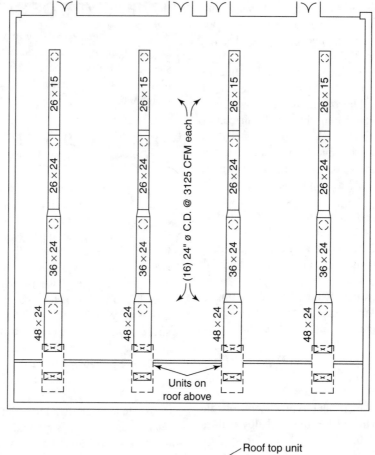

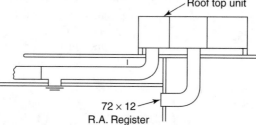

Elevation view of connections to roof units

Figure 17.10
HVAC plan for Project II.

an equal air distribution is satisfactory, taking note that on an actual installation, this should be studied further before a decision is made. Normally one checks similar existing installations for their experiences.

The air quantities in each section of duct can now be determined. Because there are 16 air diffusers, each supplies 3125 CFM. The second section of duct has 12,500 − 3125 = 9375 CFM. The remaining sections are found in the same manner.

Low pressure, galvanized steel rectangular ductwork will be used. The ducts will be sized by the equal friction method. A friction loss rate of 0.1 in. w. per 100 ft will be used. Using Figure 8.21, for the first duct section at 12,500 CFM, the velocity is 1800 FPM. This is a reasonable maximum velocity for noise levels in retail stores (Table 8.6). The equivalent round diameter is listed as 35 in.

The available depth above the hung ceiling for the duct is determined from architectural and structural drawings. Assume that the maximum duct depth allowable is 24 in. Using Figure 8.23, a 48 in. × 24 in. rectangular duct is approximately equivalent to a 36 in. round duct. The second duct section has an equivalent 33 in. diameter. A 40 in. × 24 in. size is selected. Only one dimension of the duct is changed each time, if possible, to make a simpler transition fitting. In addition, the duct shape is kept as square as possible. This minimizes the aspect ratio, reducing the amount of sheet metal and insulation used. The other duct sections are sized in the same manner. The results are shown as follows, and are also indicated on Figure 17.10:

Section	CFM	V, FPM	Friction Loss in. w./ 100 ft	Eq. D, in.	Duct Size, in.
AB	12,500	1800	0.1	36	48 × 24
BC	9375	1650	0.1	32	36 × 24
CD	6250	1500	0.1	27	26 × 24
DE	3125	1250	0.1	21	26 × 15

17.21 AIR DISTRIBUTION DEVICES

Each diffuser is covering a radius of approximately 20 ft equally in all directions. The type listed in Table 10.4 will be used. A 24 in. diffuser is the best selection. It has approximately a 20 ft throw radius with an NC-38 noise level at 3125 CFM. Both the NC level and velocity are suitable for retail stores. A 20 in. diffuser has a suitable radius of diffusion,

but an NC-level of about 47, which is above that recommended.

Return air grilles will be located in the ceiling directly under the air handling units on one end of the building. There will be a total of 40,000 CFM of return air. The remainder of the supply will be used to ventilate the service areas in the rear of the building (this system is not shown). The return grilles will be located well above the occupied zone. Therefore, a reasonably high face velocity can be used without resulting in excessive noise. Each of the returns handles 10,000 CFM. Grilles with a face area of 12 ft^2 will be used.

17.22 EQUIPMENT

Packaged roof-mounted units will be used. They will consist of compressors, evaporator coil, air-cooled condenser, supply air fan, dampers, and cabinet. Permanent, cleanable, panel-type filters will be used. A gas-fired furnace section can also be furnished if the unit is to be used for heating. A detailed specification will be described later.

The required fan pressure characteristics must be determined. Information on pressure drops through each component is obtained from the manufacturer. This data, plus the friction loss through the ducts, are summarized in Table 17.3 on page 452.

The required supply fan total pressure is 1.79 in. w.g. In the calculations, a radius-type elbow was assumed. No significant pressure losses occur in the gradual transition pieces. The values for pressure losses in the takeoff and diffuser were obtained from the manufacturers. Note that a filter pressure drop of 0.40 in. w.g. has been used. This allows for dirt build-up, because the clean filter pressure drop is much less. If the latter figure were used, the fan would be undersized. The operating engineer must be sure to change filters at 0.4 in. w.g., not greater.

A return air fan will be included. A short section of ductwork from the return air grilles is required. The fan pressure requirements would be determined in a manner similar to that for the supply fan.

TABLE 17.3   SUMMARY OF FAN PRESSURE CHARACTERISTICS

Item	CFM	Duct Size	V, FPM	Friction Loss in. w./ 100 ft	L, ft	Loss coeff, C	V.P. in. w.	Pr. loss in w.
UNIT								
Inlet	12,500							0.32
Filter								0.40
Coil								0.41
Outlet								0.25
DUCT								
AB	12,500	48 × 24	1800	0.1	20			0.02
Elbow						0.20	0.20	0.04
BC	9375	36 × 24			40			0.04
CD	6250	26 × 24			40			0.04
DE	3125	26 × 15			40			0.04
Takeoff								0.08
Diffuser								0.15
						Total system pressure loss =		1.79

The selection of the roof-mounted units would be based on the design sensible latent and total loads and conditions. Because there are four units, all of the same size, the requirements for each would be

$$Q_s = \frac{1,280,370}{4} = 320,090 \text{ BTU/hr}$$

$$Q_l = \frac{226,500}{4} = 56,625$$

$$Q_t \qquad = 376,715 \text{ BTU/hr}$$

$$= 31.4 \text{ tons}$$

Entering air = 81.3 F DB, 66.7 F WB

Leaving air = 58 F DB, 57 F WB

CFM = 12,500

A selection from a manufacturer's catalog would be made with this data.

17.23   ACCESSORIES

Fire dampers and smoke detectors might be required in the ductwork, as determined by Codes.

Control grids and balancing dampers will be provided in the collar to each diffuser. Return, exhaust, and outside dampers will be furnished. All supply ductwork will be insulated with vapor barrier covered insulation for an overall *R*-8.

17.24   AUTOMATIC CONTROL SYSTEM

The controls in roof-mounted package units usually come in one of a few fixed choices as a "programmed" arrangement. That is, the controls are completely wired and built into the equipment, and only certain combinations are available. The cooling system controls will be specified as follows:

1. An economizer control will be furnished. When the outdoor air temperature is between 60 F and 78 F, a mixed air thermostat shall modulate outside and return air dampers to provide free cooling when called for. When the outdoor air temperature reaches the high limit setting, an outdoor thermostat will position the dampers to minimum outside air. This

type of control is similar to that described and shown in Chapter 14.

2. On call for further cooling, a two-stage room thermostat shall operate the first stage of the compressor. This may be accomplished either by using a compressor furnished with one step of cylinder unloading, or by furnishing a unit with two compressors. On call for further cooling, the second stage of compressor cooling shall operate. The opposite sequence of events will occur when the room temperature is satisfied.

The refrigeration compressor will have a pump down cycle (Chapter 13). A solenoid valve to the evaporator coil will open and close as called for by the room thermostat. When the solenoid valve closes, the compressor will continue to operate until the suction pressure reaches a set value. At this point, a pressurestat will stop the compressor. The compressor will have a normal complement of safety controls—high and low pressure cutouts and oil pressure failure control. The condenser will have a pressure control that maintains adequate head pressure when the outdoor air temperature is low. This will function by cycling the condenser fans. The room thermostats will be mounted on interior columns at shoulder height, away from drafts or heat sources. The control system will be low voltage electric.

17.25   PLANS AND SPECIFICATIONS

The floor plans and any necessary details are prepared, using the previous sketches as a guide. The architectural and structural plans and those of the other trades are used to ensure that there are no interferences. The plans would be similar to that shown in Figure 17.10, drawn to scale.

The specifications would be of a form similar to those described for Project I. For example, the specifications for the roof-mounted unit might read as follows:

Furnish and install where shown on the plans four (4) roof-mounted air conditioning units complete with compressors, evaporator, fans, air-cooled condenser, cabinet, and accessories. Each unit shall have a total cooling capacity of 376,715 BTU/hr and a sensible cooling capacity of 320,090 BTU/hr with 12,500 CFM of air entering at 81.3 F DB and 66.7 F WB with 99 F DB air entering the condenser.

The supply fan shall have a 1.80 in. w.g. total pressure with 7.5 BHP. The fan motor shall be 7½ HP, 220 V., 3 phase, 60 hz. The pressure drop through the unit components shall not exceed 1.4 in. w.g.

The cabinet shall be of 20-gauge galvanized steel panel construction with an acrylic paint and completely weatherized, using stainless steel fasteners. Access panels shall provide easy access to all components. Panels shall have weatherproof gaskets and chrome-plated handles. The cabinet shall be lined inside with 1 in. of glass fiber insulation and neoprene coated.

The unit shall have two semihermetic compressors (or one compressor with automatic one-stage unloaders), complete with oil pump, suction and discharge service valves, oil level sight glass, suction gas filter, and crankcase heater.

The evaporator coil shall be of copper tube and aluminum fin with the number of rows required to handle the sensible and total cooling capacities specified. The refrigeration equipment shall be completely piped and charged, with thermal expansion valve, distributor, sight glass, and drier, using copper tubing.

The air-cooled condenser shall be complete with coils, fans, and motors. Coils shall be copper tube with aluminum fins.

All electrical components shall be weatherproof, in accordance with UL (Underwriter's Laboratory) requirements. The unit shall have a return air section complete with centrifugal fan and fresh air, return air, and exhaust dampers. Furnish a roof mounting curb. The units shall be completely tested at the factory. The units shall be as manufactured by ____ or approved equal.

17.26   ENERGY CONSERVATION

The recommended 2.5% outdoor design dry bulb temperature and coincident mean wet bulb have been used. Cooling load calculation procedures that account for energy storage have been used, thus reducing the design load considerably. (The load calculated by one of the older methods, not accounting for storage, is about 20% greater!) This will result in smaller equipment that will operate closer to full load, and therefore more efficiently.

The roof load is fairly high, and perhaps more insulation should be used. The glass solar load is quite high. Solar heat absorbing glass or internal shading is probably unacceptable to the owners, however, as they want potential customers to be attracted by the visibility of the interior. An outside shading overhang might be considered, but it is possible that it would have to be too long to be effective; this should be determined. Door infiltration could be significant. Revolving doors or a vestibule would reduce infiltration greatly, but this might slow down customer ingress.

The control system has been designed for outside air free cooling, using temperature sensing. Enthalpy sensing would provide even more energy conservation, but is perhaps excessive in cost for the number of hours it would be effective. The weather data could be checked to determine this.

One major energy conservation device that probably should be considered is an air-to-air heat exchanger between ventilation and exhaust air, especially if the system is also used for winter heating. A calculation will show the value of summer operation. Assuming that a heat recovery wheel with 80% efficiency is used, if it recovers both sensible and latent heat, the energy saved at design load is

$$Q = 4.5 \times 7500 \times 0.80 \, (37.7 - 30.2)$$
$$= 303,500 \text{ BTU/hr} = 16.9 \text{ tons}$$

This is not only a savings in energy consumption, but the size of all the cooling equipment could be reduced by 16.9 tons. Assuming an equivalent full load cooling season of 1500 hours, and 1.4 KW per ton required by the compressor and fans, and with an electric power cost of $0.08/KWH, the annual summer savings would be:

$$16.9 \text{ tons} \times 1.4 \text{ KW/ton} \times 1500 \text{ hr}$$
$$\times \$0.08/\text{KWH} = \$2840$$

This is a substantial and continual saving in both energy use and operating cost.

Problems

17.1   Plan and design a series loop hydronic heating system for the residence shown in Figure 17.1, for construction in your town. Use construction materials and conditions suitable for the climate, as recommended by the energy standards in this book or in your state.

17.2   Design a reverse return heating system for the residence shown in Figure 17.1, as recommended in Problem 17.1.

17.3   Design a split series loop (or reverse return) heating system for the residence described in Problem 3.20, to be built in your town, using materials and conditions recommended by energy conservation standards.

17.4   Plan and design an all-air summer air conditioning system for the factory described in Problem 3.21, to be constructed in your town, using materials and conditions recommended for energy conservation.

17.5   Plan and design an all-air summer air conditioning system for the store described in Example 6.17, to be constructed in your town, using materials and conditions recommended for energy conservation.

17.6   Carry out the designs recommended in Problems 17.1–17.5 for other locations.

CHAPTER 18
Solar Heating and Cooling Systems

Solar energy systems can be used to provide domestic hot water and hot water or warm air for air conditioning, in place of boilers or furnaces. The advantage, of course, is that there is no depletable fuel consumed and thus no cost for the heat energy used. The disadvantages have been the equipment cost and problems in collecting this energy. However, sharply increasing energy costs have stimulated the development of practical solar heating systems. Solar heating of hot water is already quite common in Japan, Israel, and Australia.

The growth of solar energy for space heating of buildings is also developing rapidly in many countries and is practical where fuel costs are relatively high.

We will discuss basic principles and features of solar heating equipment and systems here. A relatively simple yet practical method of system design and collector sizing will be explained.

A solar heating system is composed of three essential parts: a collector, a storage system, and a distribution and terminal system.

OBJECTIVES

After studying this chapter, you will be able to:

1. Explain the construction of a flat plate solar collector.
2. Sketch basic solar hot water and space heating systems.
3. Calculate the required size of a solar heat collector and storage tank.
4. Perform an economic payback analysis for solar heating.

18.1 SOLAR COLLECTORS

The function of the *collector* is to receive and capture as much of the radiant energy from the sun as is practical. There are three types of collectors: concentrating, flat plate, and evacuated tube.

The *flat plate* collector has a receiving surface that is flat in shape. The collector is stationary and therefore receives the maximum radiation only when it faces the sun directly. At all other times, it "sees" the sun at an angle, and its effective surface area is smaller, resulting in collection of less energy.

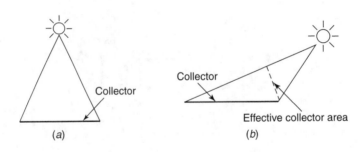

Figure 18.1
Effect of the angle between the collector and sun on solar radiation received. (a) Collector directly facing sun—all surface area effective. (b) Collector not directly facing sun—effective area smaller than actual surface area.

Figure 18.1 demonstrates this. The effective surface is equal only to the projected area facing the sun.

A typical flat plate collector construction for heating water or other liquids is shown in Figure 18.2. The *transparent cover* (either one or two plates) admits solar radiation. Both glass and plastics have been used. The *absorber plate* collects the radiant energy. It is constructed of a flat panel and tubing through which the water is circulated as it is heated. The plate and tubing may be steel, aluminum, plastic, or copper. It is often coated black to increase energy absorption. The cover plates reduce the heat that would be reradiated out from the hot panel. The solar radiation received has a short wavelength and passes through the transparent cover plate. The absorber plate is heated by the radiation it absorbs and in turn heats the water. Some of the energy received is reradiated outward by the absorber plate. This radiation, however, has a longer wavelength, which does not pass outward

Figure 18.2
Typical flat plate collector for heating liquids.

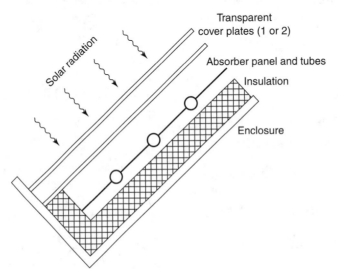

through the cover plate. The cover also reduces the amount of convected heat lost to the surrounding air. Insulation is provided on the other sides to reduce heat losses there.

A flat plate collector construction for heating air is shown in Figure 18.3. Air is circulated through the space behind the absorber plate and is heated directly. In the example shown, the plate has fins to increase the heat transfer surface.

Concentrating or focusing collectors have a concave surface that concentrates the rays received. The result is a collector that can produce a much higher temperature than the flat plate collector. Concentrating collectors are often designed to move and track the sun, thus they always receive maximum radiation. The flat plate collector, although less efficient, has been favored due to its usually lower cost, but new and improved types of concentrating collectors are being developed.

The *evacuated tube* collector has a tube-within-a-tube construction. The outer tube is transparent and transmits the solar radiation. The inner tube contains the circulating liquid that is heated. The annular space between the tube is evacuated of air. This greatly reduces the convection heat losses from the collector, similar to how a thermos bottle functions. The efficiency of the evacuated tube collector is quite high, as much as double that of the flat plate collector. It can also produce water at a very high temperatures (over 300 F). With reduced production costs, it should add greatly to the use of solar heating and cooling systems.

18.2 STORAGE AND DISTRIBUTION SYSTEMS

The collector does not collect heat at night or on overcast days. Therefore, it is common to store any excess heat collected when operating to be used when the sun is not shining. Two types of storage systems are in common use. In one, hot water is heated in a large storage tank and stored until needed (see Figure 18.6). In another arrangement, hot air from the collector is passed through a bin filled with pebbles (see Figure 18.7). The heated pebbles serve as a heat storage medium. When heat is needed, air is circulated through the hot pebble bed and then delivered to the building.

A third type of storage uses the latent heat of fusion to store heat. Hot water from the solar collector is circulated to a coil surrounded by a substance

Figure 18.3
Typical flat plate collector for heating air.

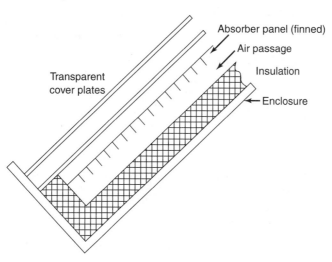

Transparent cover plates

Absorber panel (finned)

Air passage

Insulation

Enclosure

that melts at a temperature convenient for heating. A separate coil that circulates water used for building heating is also surrounded by this substance. The melted substance gives up its latent heat to this water, solidifying again. This method is not yet in common use, but has good potential.

In most climates, it is too expensive to provide a storage system large enough to deliver the building heating requirements for long periods of cloudy weather, so a conventional auxiliary heat source, either a hot water boiler or warm air furnace, is used as a back-up.

The distribution system is similar to those in conventional fuel-fired heating systems circulating either hot water or warm air to rooms.

18.3   TYPES OF SOLAR HEATING SYSTEMS

The simplest use of a solar heat source is to heat domestic hot water. Figure 18.4 shows one arrangement. The loop circulating the heated water from the collector passes through the heat exchanger tank, heating the water that is used in the building. An auxiliary electric heater may also be used. In the example shown, the collector loop water flows due to the natural density difference created by the temperature change (*thermosyphon* effect). A pump may be used to increase the flow rate, particularly in larger systems.

Although the collector system could be drained of water when necessary in climates subject to freezing, it is more common to use an antifreeze solution such as ethylene glycol to avoid this problem. When the system is used to heat domestic hot water, plumbing codes may require a *double wall heat exchanger* (Figure 18.5) to prevent contamination of the potable water in case there is a break in the collector heating coil in the tank.

A solar heating system using a liquid collector loop and hydronic heating system is shown in Figure 18.6. The heated water or antifreeze solution from the collector heats the water in the storage tank, which is circulated to the room terminal units. A hot water boiler is used to supplement the heat from the solar collector. The liquid collector and hot water storage system could also be used with a heating coil in a duct, if the building is heated with air.

A solar heating system using an air collector, pebble bed storage, and warm air heating system is

Figure 18.4
Solar domestic hot water heater with natural (thermosyphon) circulation.

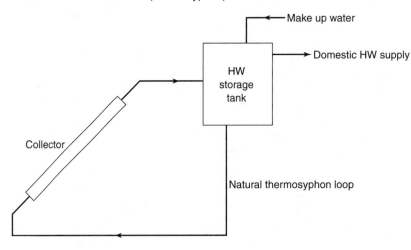

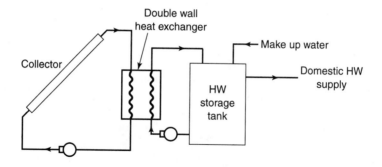

Figure 18.5
Double wall heat exchanger to prevent contamination of potable water.

shown in Figure 18.7 on page 460. The heated air from the collector is delivered directly to the building heating system when required. When the collector capacity exceeds building heating needs, the air can be diverted through the pebble bed to charge the storage system. A warm air furnace is used to supplement the heat from the solar collector.

Solar air systems as compared with water systems have the same disadvantages that occur in conventional systems—equipment and ducts are larger due to the lower heat-carrying capacity of air. On the other hand, leakage, corrosion, and freezing problems do not exist.

The heated water or air from a solar heating system can be used as the low temperature heat source for a conventional heat pump operating in the winter cycle (Chapter 13), instead of using outside air. This increases the low side temperature of the heat pump considerably, thereby increasing its coefficient of performance and reducing energy con-

sumption significantly. This is a very popular use of a solar heating system, because it can then also be used for direct heating when the heating load is sufficiently low, and supplemented or replaced by the heat pump when necessary.

18.4 SOLAR COOLING SYSTEMS

Although solar energy systems produce only heat, any cooling system that receives its operating energy from a heat source can be driven by using the solar system heat. One method is to use the solar heated hot water to operate an absorption refrigeration system. Because the efficiency of flat plate solar collectors decreases greatly as the required temperature increases, the cost of the solar system may be quite high to produce the water temperatures required to operate absorption machines (about 190 F).

Figure 18.6
Solar heating system with liquid collector and hydronic distribution.

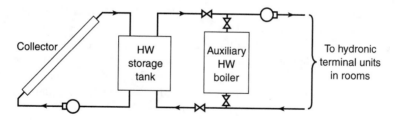

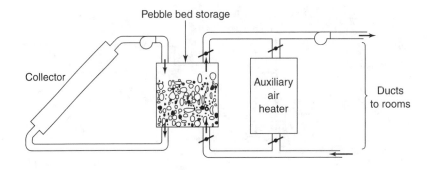

Figure 18.7
Solar heating system with air collector and warm air distribution.

A solar heating system can be used as a heat source of a Rankine power cycle turbine which in turn is used to drive a conventional refrigeration compressor. The Rankine cycle is the same as that in a steam turbine. When used with a solar heating system, however, a fluid is used that will vaporize at a temperature available from the solar collector. The cooling system for the building shown in the frontispiece illustration uses this method.

18.5   SOLAR RADIATION ENERGY

In order to determine the capacity of solar heating equipment, the quantity of solar energy that is received by the collector must be found. The solar radiation received just outside Earth's atmosphere is approximately 429 BTU/hr-ft^2 on a surface normal to (faces directly) the sun. It is about 14 BTU/hr-ft^2 higher in winter and lower in summer, due to the slightly elliptical orbit of Earth around the sun. The solar energy received at Earth's surface, called *insolation* (I), is much less than the above value, due to the absorbing and scattering effect of the atmosphere. That part of the solar radiation which reaches Earth's surface and is not affected by the atmosphere is called *direct* radiation. It ranges from 0 to about 350 BTU/hr-ft^2 on a surface normal to the sun, depending on the length of path through the atmosphere and its condition. A por-

tion of the scattered radiation also reaches Earth's surface. This is called *diffuse* radiation. Radiation may also be *reflected* back from the ground onto surfaces.

In order to design solar energy collectors, it is necessary to find that part of the energy which is received on surfaces that are not normal to the sun. Collectors are usually fixed in position and therefore cannot collect the full insolation on a surface that is always normal to the sun's rays. There is an hourly variation in the angle between the collector surface and the sun due to the rotation of Earth. There is also a daily variation because, in winter, Earth's axis is tilted further from the sun than in summer. This results in the sun being low on the horizon in winter and high in summer. These variations also change according to the distance from the equator (latitude).

Figure 18.8 illustrates how the tilt of Earth's axis relative to the sun (called *declination*) affects the amount of solar radiation received. When Earth is tilted away from the sun, as in winter, the length of path of the radiation through the atmosphere is longer, resulting in less radiation reaching Earth's surface, despite the fact that Earth is slightly closer to the sun in the winter.

The change in declination results in the sun appearing higher in the sky in the summer than in the winter. This affects the desirable angle that a collector should be tilted at in order to maximize the radiation received. Some general recommendations

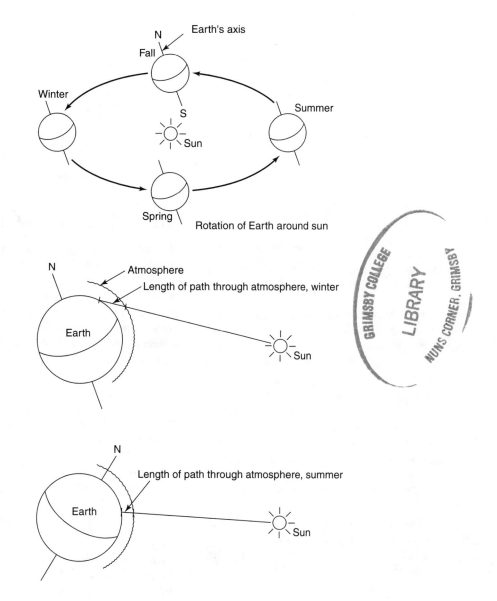

Figure 18.8
Effect of tilt of Earth's axis on solar radiation received.

will be given later. The elevation of the sun above the horizon is called its *altitude* and its horizontal position relative to south is called its *azimuth*. Both are expressed in degrees. Figure 18.9 on page 462 illustrates this.

18.6 INSOLATION TABLES

The solar insolation values for clear days have been determined and reported in tables. Table 18.1 on pages 463–466 is an example, shown for latitudes

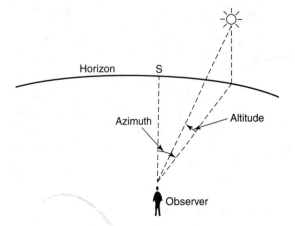

Figure 18.9
Illustration of the altitude and azimuth of the sun.

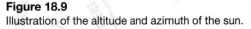

of 28°N to 48°N. *Solar time* is used; this is the time as read on a sun dial facing true south. Equations are available for converting solar time to local time, if necessary. The table shows insolation values monthly and hourly, for flat surfaces—those normal to the sun, horizontal (flat on the ground), and tilted at various elevations above the horizontal. In all cases, the surfaces are facing south. The insolation values listed are the total of direct and diffuse radiation except for the values listed for a normal surface, which includes only direct insolation. Interpolation between values can be used when accuracy requires it.

It is possible for radiation to be reflected from the ground onto a tilted surface. Reflected radiation is not included in the table, because it varies with local conditions.

Example 18.1

Find the solar insolation at 10 AM and 12 N solar time on January 21 in Phoenix, Arizona, for surfaces tilted as follows: horizontal, 32°, 42°, and 90°, facing south.

Solution
Phoenix is at 33°N latitude, so the 32°N latitude table may be used without significant error. Reading from the table the results are:

Time	Insolation in BTU/hr-ft²-F at Tilt Angles Listed			
	horiz.	32°	42°	90°
10 AM	167	256	269	221
12 N	209	308	321	253

The insolation tables also show the total daily insolation for each tilt angle, which is simply the sum of the hourly values for a day.

18.7 CLEARNESS FACTOR

The natural clearness of the atmosphere is not the same at all locations. It varies with the natural amount of water vapor and dust in the air. For example, the clearness in the southeastern part of the United States is less than in the Rocky Mountain area. In order to allow for this, corrections to the insolation values may be required. The values shown in Table 18.1 are based on a *clearness factor* (CF) of 1.0. Figure 18.10 on page 467 is a map that shows clearness factors (also called clearness numbers) for the continental United States. Note that the values may differ for summer and winter in some localities.

Example 18.2

Find the total daily insolation for a clear day on March 21 in Denver, Colorado, for a south facing surface tilted at a 50° elevation above the horizontal.

Solution
Denver is at 40°N latitude. From Table 18.1, the total daily insolation on March 21 for a surface at a 50° tilt is 2284 BTU/ft². From Figure 18.10, however, it is noted that CF = 1.05 for that part of Colorado. Correcting the value from the table,

$$I = 2284 \times 1.05 = 2398 \text{ BTU/ft}^2 \text{ per day}$$

The clearness factor does not account for the effects of industrial air pollution. When smog is present, it may reduce insolation. The solar heating system designer must determine this from personal knowledge of the area.

TABLE 18.1 CLEAR DAY INSOLATION VALUES

SOLAR POSITION AND INSOLATION, 24°N LATITUDE

Date	Solar Time		Solar Position		BTUH/Sq. Ft. Total Insolation on Surfaces		South Facing Surface Angle With Horiz.				
	AM	PM	ALT	AZM	Normal	Horiz.	14	24	34	54	90
Jan 21	7	5	4.8	65.6	71	10	17	21	25	28	31
	8	4	16.9	58.3	239	83	110	126	137	145	127
	9	3	27.9	48.8	288	151	188	207	221	228	176
	10	2	37.2	36.1	308	204	246	268	282	287	207
	11	1	43.6	19.6	317	237	283	306	319	324	226
	12		46.0	0.0	320	249	296	319	332	336	232
	Surface Daily Totals				2766	1622	1984	2174	2300	2360	1766
Feb 21	7	5	9.3	74.6	158	35	44	49	53	56	46
	8	4	22.3	67.2	263	116	135	145	150	151	102
	9	3	34.4	57.6	298	187	213	225	230	228	141
	10	2	45.1	44.2	314	241	273	286	291	287	168
	11	1	53.0	25.0	321	276	310	324	328	323	185
	12		56.0	0.0	324	288	323	337	341	335	191
	Surface Daily Totals				3036	1998	2276	2396	2446	2424	1476
Mar 21	7	5	13.7	83.8	194	60	63	64	62	59	27
	8	4	27.2	76.8	267	141	150	152	149	142	64
	9	3	40.2	67.9	295	212	226	229	225	214	95
	10	2	52.3	54.8	309	266	285	288	283	270	120
	11	1	61.9	33.4	315	300	322	326	320	305	135
	12		66.0	0.0	317	312	334	339	333	317	140
	Surface Daily Totals				3078	2270	2428	2456	2412	2298	1022
Apr 21	6	6	4.7	100.6	40	7	5	4	4	3	2
	7	5	18.3	94.9	203	83	77	70	62	51	10
	8	4	32.0	89.0	256	160	157	149	137	122	16
	9	3	45.6	81.9	280	227	227	220	206	186	41
	10	2	59.0	71.8	292	278	282	275	259	237	61
	11	1	71.1	51.6	298	310	316	309	293	269	74
	12		77.6	0.0	299	321	328	321	305	280	79
	Surface Daily Totals				3036	2454	2458	2374	2228	2016	488
May 21	6	6	8.0	108.4	86	22	15	10	9	3	5
	7	5	21.2	103.2	203	98	85	73	59	44	12
	8	4	34.6	98.5	248	171	159	145	127	106	15
	9	3	48.3	93.6	269	233	224	210	190	165	16
	10	2	62.0	87.7	280	281	275	261	239	211	22
	11	1	75.5	76.9	286	311	307	293	270	240	34
	12		86.0	0.0	288	322	317	304	281	250	37
	Surface Daily Totals				3032	2556	2447	2286	2072	1800	246
Jun 21	6	6	9.3	111.6	97	29	20	12	12	11	7
	7	5	22.3	106.8	201	103	87	73	58	41	13
	8	4	35.5	102.6	242	173	158	142	122	99	16
	9	3	49.0	98.7	263	234	221	204	182	155	18
	10	2	62.6	95.0	274	280	269	253	229	199	18
	11	1	76.3	90.8	279	309	300	283	259	227	19
	12		89.4	0.0	281	319	310	294	269	236	22
	Surface Daily Totals				2994	2574	2422	2230	1992	1700	204
Jul 21	6	6	8.2	109.0	81	23	16	11	10	9	6
	7	5	21.4	103.8	195	98	85	73	59	44	13
	8	4	34.8	99.2	239	169	157	143	125	104	16
	9	3	48.4	94.5	261	231	221	207	187	161	18
	10	2	62.1	89.0	272	278	270	256	235	206	21
	11	1	75.7	79.2	278	307	302	287	265	235	32
	12		86.6	0.0	280	317	312	298	275	245	36
	Surface Daily Totals				2932	2526	2412	2250	2036	1766	246
Aug 21	6	6	5.0	101.3	35	7	5	4	4	4	2
	7	5	18.5	95.6	186	82	76	69	60	50	11
	8	4	32.2	89.7	241	158	154	146	134	118	16
	9	3	45.9	82.9	265	223	222	214	200	181	39
	10	2	59.3	73.0	278	273	275	268	252	230	58
	11	1	71.6	53.2	284	304	309	301	285	261	71
	12		78.3	0.0	286	315	320	313	296	272	75
	Surface Daily Totals				2864	2408	2402	2316	2168	1958	470
Sep 21	7	5	13.7	83.8	173	57	60	60	59	56	26
	8	4	27.2	76.8	248	136	144	146	143	136	62
	9	3	40.2	67.9	278	205	218	221	217	206	93
	10	2	52.3	54.8	292	258	275	278	273	261	116
	11	1	61.9	33.4	299	291	311	315	309	295	131
	12		66.0	0.0	301	302	323	327	321	306	136
	Surface Daily Totals				2878	2194	2342	2366	2322	2212	992
Oct 21	7	5	9.1	74.1	138	32	40	45	48	50	42
	8	4	22.0	66.7	247	111	129	139	144	145	99
	9	3	34.1	57.1	284	180	206	217	223	221	138
	10	2	44.7	43.8	301	234	265	277	282	279	165
	11	1	52.5	24.7	309	268	301	315	319	314	182
	12		55.5	0.0	311	279	314	328	332	327	188
	Surface Daily Totals				2868	1928	2198	2314	2364	2346	1442
Nov 21	7	5	4.9	65.8	67	10	16	20	24	27	29
	8	4	17.0	58.4	232	82	108	123	135	142	124
	9	3	28.0	48.9	282	150	186	205	217	224	172
	10	2	37.3	36.3	303	203	244	265	278	283	204
	11	1	43.8	19.7	312	236	280	302	316	320	222
	12		46.2	0.0	315	247	293	315	328	332	228
	Surface Daily Totals				2706	1610	1962	2146	2268	2324	1730
Dec 21	7	5	3.2	62.6	30	3	7	9	11	12	14
	8	4	14.9	55.3	225	71	99	116	129	139	130
	9	3	25.5	46.0	281	137	176	198	214	223	184
	10	2	34.3	33.7	304	189	234	258	275	283	217
	11	1	40.4	18.2	314	221	270	295	312	320	236
	12		42.6	0.0	317	232	282	308	325	332	243
	Surface Daily Totals				2624	1474	1852	2058	2204	2286	1808

TABLE 18.1 (Continued)

SOLAR POSITION AND INSOLATION, 32°N LATITUDE

Date	Solar Time AM	Solar Time PM	Solar Position ALT	Solar Position AZM	Normal	Horiz.	22	32	42	52	90
Jan 21	7	5	1.4	65.2	1	0	0	0	0	1	1
	8	4	12.5	56.5	203	56	93	106	116	123	115
	9	3	22.5	46.0	269	118	175	193	206	212	181
	10	2	30.6	33.1	295	167	235	256	269	274	221
	11	1	36.1	17.5	306	198	273	295	308	312	245
	12		38.0	0.0	310	209	285	308	321	324	253
	Surface Daily Totals				2458	1288	1839	2008	2118	2166	1779
Feb 21	7	5	7.1	73.5	121	22	34	37	40	42	38
	8	4	19.0	64.4	247	95	127	136	140	141	108
	9	3	29.9	53.4	288	161	206	217	222	220	158
	10	2	39.1	39.4	306	212	266	278	283	279	193
	11	1	45.6	21.4	315	244	304	317	321	315	234
	12		48.0	0.0	317	255	316	330	334	328	222
	Surface Daily Totals				2872	1724	2188	2300	2345	2322	1644
Mar 21	7	5	12.7	81.9	185	54	60	60	59	56	32
	8	4	25.1	73.0	260	129	146	147	144	137	78
	9	3	36.8	62.1	290	194	222	224	220	209	119
	10	2	47.3	47.5	304	245	280	283	278	265	150
	11	1	55.0	26.8	311	277	317	321	315	300	170
	12		58.0	0.0	313	287	329	333	327	312	177
	Surface Daily Totals				3012	2084	2378	2403	2358	2246	1276
Apr 21	6	6	6.1	99.9	66	14	9	6	6	5	3
	7	5	18.8	92.2	206	86	78	71	62	51	10
	8	4	31.5	84.0	255	158	156	148	136	120	35
	9	3	43.9	74.2	278	220	225	217	203	183	68
	10	2	55.7	60.3	290	267	279	272	256	234	95
	11	1	65.4	37.5	295	297	313	306	290	265	112
	12		69.6	0.0	297	307	325	318	301	276	118
	Surface Daily Totals				3076	2390	2444	2356	2206	1994	764
May 21	6	6	10.4	107.2	119	36	21	13	13	12	7
	7	5	22.8	100.1	211	107	88	75	60	44	13
	8	4	35.4	92.9	250	175	159	145	127	105	15
	9	3	48.1	84.7	269	233	223	209	188	163	33
	10	2	60.6	73.3	280	277	273	259	237	208	56
	11	1	72.0	51.9	285	305	305	290	268	237	72
	12		78.0	0.0	286	315	315	301	278	247	77
	Surface Daily Totals				3112	2582	2454	2284	2064	1788	469
Jun 21	6	6	12.2	110.2	131	45	26	16	15	14	9
	7	5	24.3	103.4	210	115	91	76	59	41	14
	8	4	36.9	96.8	245	180	159	143	122	99	16
	9	3	49.6	89.4	264	236	221	204	181	153	19
	10	2	62.2	79.7	274	279	268	251	227	197	41
	11	1	74.2	60.9	279	306	299	282	257	224	56
	12		81.5	0.0	280	315	309	292	267	234	60
	Surface Daily Totals				3084	2634	2436	2234	1990	1690	370

Date	Solar Time AM	Solar Time PM	Solar Position ALT	Solar Position AZM	Normal	Horiz.	22	32	42	52	90
Jul 21	6	6	10.7	107.7	113	37	22	14	13	12	8
	7	5	23.1	100.6	203	107	87	75	60	44	14
	8	4	35.7	93.6	241	174	158	143	125	104	16
	9	3	48.4	85.5	261	231	220	205	185	159	31
	10	2	60.9	74.3	271	274	269	254	232	204	54
	11	1	72.4	53.3	277	302	300	285	262	232	69
	12		78.6	0.0	279	311	310	296	273	242	74
	Surface Daily Totals				3012	2558	2422	2250	2030	1754	458
Aug 21	6	6	6.5	100.5	59	14	9	7	6	6	4
	7	5	19.1	92.8	190	85	77	69	60	50	12
	8	4	31.8	84.7	240	156	152	144	132	116	33
	9	3	44.3	75.0	263	216	220	212	197	178	65
	10	2	56.1	61.3	276	262	272	264	249	226	91
	11	1	66.0	38.4	282	292	305	298	281	257	107
	12		70.3	0.0	284	302	317	309	292	268	113
	Surface Daily Totals				2902	2352	2388	2296	2144	1934	736
Sep 21	7	5	12.7	81.9	163	51	56	56	55	52	30
	8	4	25.1	73.0	240	124	140	141	138	131	75
	9	3	36.8	62.1	272	188	213	215	211	201	114
	10	2	47.3	47.5	287	237	270	273	268	255	145
	11	1	55.0	26.8	294	268	306	309	303	289	164
	12		58.0	0.0	296	278	318	321	315	300	171
	Surface Daily Totals				2808	2014	2288	2308	2264	2154	1226
Oct 21	7	5	6.8	73.1	99	19	29	32	34	36	32
	8	4	18.7	64.0	229	90	120	128	133	134	104
	9	3	29.5	53.0	273	155	198	208	213	212	153
	10	2	38.7	39.1	293	204	257	269	273	270	188
	11	1	45.1	21.1	302	236	294	307	311	306	209
	12		47.5	0.0	304	247	306	320	324	318	217
	Surface Daily Totals				2696	1654	2100	2208	2252	2232	1588
Nov 21	7	5	1.5	65.4	2	0	0	0	1	1	1
	8	4	12.7	56.6	196	55	91	104	113	119	111
	9	3	22.6	46.1	263	118	173	190	202	208	176
	10	2	30.8	33.2	289	166	233	252	265	270	217
	11	1	36.2	17.6	301	197	270	291	303	307	241
	12		38.2	0.0	304	207	282	304	316	320	249
	Surface Daily Totals				2406	1280	1816	1980	2084	2130	1742
Dec 21	8	4	10.3	53.8	176	41	77	90	101	108	107
	9	3	19.8	43.6	257	102	161	180	195	204	183
	10	2	27.6	31.2	288	150	221	244	259	267	226
	11	1	32.7	16.4	301	180	258	282	298	305	251
	12		34.6	0.0	304	190	271	295	311	318	259
	Surface Daily Totals				2348	1136	1704	1888	2016	2086	1794

464

TABLE 18.1 (Continued)

SOLAR POSITION AND INSOLATION, 40°N LATITUDE

Date	Solar Time AM	Solar Time PM	Solar Position ALT	Solar Position AZM	BTUH/Sq. Ft. Total Insolation on Surfaces Normal	Horiz.	South Facing Surface Angle With Horiz. 30	40	50	60	90
Jan 21	8	4	8.1	55.3	142	28	65	74	81	85	84
	9	3	16.8	44.0	239	83	155	171	182	187	171
	10	2	23.8	30.9	274	127	218	237	249	254	223
	11	1	28.4	16.0	289	154	257	277	290	293	253
	12		30.0	0.0	294	164	270	291	303	306	263
	Surface Daily Totals				2182	948	1660	1810	1906	1944	1726
Feb 21	7	5	4.8	72.7	69	10	19	21	23	24	22
	8	4	15.4	62.2	224	73	114	122	126	127	107
	9	3	25.0	50.2	274	132	195	205	209	208	167
	10	2	32.8	35.9	295	178	256	267	271	267	210
	11	1	38.1	18.9	305	206	293	306	310	304	236
	12		40.0	0.0	308	216	306	319	323	317	245
	Surface Daily Totals				2640	1414	2060	2162	2202	2176	1730
Mar 21	7	5	11.4	80.2	171	46	55	55	54	51	35
	8	4	22.5	69.6	250	114	140	141	138	131	89
	9	3	32.8	57.3	282	173	215	217	213	202	138
	10	2	41.6	41.9	297	218	273	276	271	258	176
	11	1	47.7	22.6	305	247	310	313	307	293	200
	12		50.0	0.0	307	257	322	326	320	305	208
	Surface Daily Totals				2916	1852	2308	2330	2284	2174	1484
Apr 21	6	6	7.4	98.9	89	20	11	8	7	7	4
	7	5	18.9	89.5	206	87	77	70	61	50	12
	8	4	30.3	79.3	252	152	153	145	133	117	53
	9	3	41.3	67.2	274	207	221	213	199	179	93
	10	2	51.2	51.4	286	250	275	267	252	229	126
	11	1	58.7	29.2	292	277	308	301	285	260	147
	12		61.6	0.0	293	287	320	313	296	271	154
	Surface Daily Totals				3092	2274	2412	2320	2168	1956	1022
May 21	5	7	1.3	114.7	1	0	0	0	0	0	0
	6	6	12.7	105.6	144	49	25	15	14	13	9
	7	5	24.0	96.6	216	114	89	76	60	44	13
	8	4	35.4	87.2	250	175	158	144	125	104	25
	9	3	46.8	76.0	267	227	221	206	186	160	60
	10	2	57.5	60.9	277	267	270	255	233	205	89
	11	1	66.2	37.1	283	293	301	287	264	234	108
	12		70.0	0.0	284	301	312	297	274	243	114
	Surface Daily Totals				3160	2552	2442	2264	2040	1760	724
Jun 21	5	7	4.2	117.3	22	4	3	3	3	2	1
	6	6	14.8	108.4	155	60	30	18	17	16	10
	7	5	26.0	99.7	216	123	92	77	59	41	14
	8	4	37.4	90.7	246	182	159	142	121	97	16
	9	3	48.8	80.2	263	233	219	202	179	151	47
	10	2	59.8	65.8	272	272	266	248	224	194	74
	11	1	69.2	41.9	277	296	296	278	253	221	92
	12		73.5	0.0	279	304	306	289	263	230	98
	Surface Daily Totals				3180	2648	2434	2224	1974	1670	610

Date	Solar Time AM	Solar Time PM	Solar Position ALT	Solar Position AZM	BTUH/Sq. Ft. Total Insolation on Surfaces Normal	Horiz.	South Facing Surface Angle With Horiz. 30	40	50	60	90
Jul 21	5	7	2.3	115.2	2	0	0	0	0	0	0
	6	6	13.1	106.1	138	50	26	17	15	14	9
	7	5	24.3	97.2	208	114	89	75	60	44	14
	8	4	35.8	87.8	241	174	157	142	124	102	24
	9	3	47.2	76.7	259	225	218	203	182	157	58
	10	2	57.9	61.7	269	265	266	251	229	200	86
	11	1	66.7	37.9	275	290	296	281	258	228	104
	12		70.6	0.0	276	298	307	292	269	238	111
	Surface Daily Totals				3062	2534	2409	2230	2006	1728	702
Aug 21	6	6	7.9	99.5	81	21	12	9	8	7	5
	7	5	19.3	90.0	191	87	76	69	60	49	12
	8	4	30.7	79.9	237	150	150	141	129	113	50
	9	3	41.8	67.9	260	205	216	207	193	173	89
	10	2	51.7	52.1	272	246	267	259	244	221	120
	11	1	59.3	29.7	278	273	300	292	276	252	140
	12		62.3	0.0	280	282	311	303	287	262	147
	Surface Daily Totals				2916	2244	2354	2258	2104	1894	978
Sep 21	7	5	11.4	80.2	149	43	51	51	49	47	32
	8	4	22.5	69.6	230	109	133	134	131	124	84
	9	3	32.8	57.3	263	167	206	208	203	193	132
	10	2	41.6	41.9	280	211	262	265	260	247	168
	11	1	47.7	22.6	287	239	298	301	295	281	192
	12		50.0	0.0	290	249	310	313	307	292	200
	Surface Daily Totals				2708	1788	2210	2228	2182	2074	1416
Oct 21	7	5	4.5	72.3	48	7	14	15	17	17	16
	8	4	15.0	61.9	204	68	106	113	117	118	100
	9	3	24.5	49.8	257	126	185	195	200	198	160
	10	2	32.4	35.6	280	170	245	257	261	257	203
	11	1	37.5	18.7	291	199	283	295	299	294	229
	12		39.5	0.0	294	208	295	308	312	306	238
	Surface Daily Totals				2454	1348	1962	2060	2098	2074	1654
Nov 21	8	4	8.2	55.4	136	28	63	72	78	82	81
	9	3	17.0	44.1	232	82	152	167	178	183	167
	10	2	24.0	31.0	268	126	215	233	245	249	219
	11	1	28.6	16.1	283	153	254	273	285	288	248
	12		30.2	0.0	288	163	267	287	298	301	258
	Surface Daily Totals				2128	942	1636	1778	1870	1908	1686
Dec 21	8	4	5.5	53.0	89	14	39	45	50	54	56
	9	3	14.0	41.9	217	65	135	152	164	171	163
	10	2	20.7	29.4	261	107	200	221	235	242	221
	11	1	25.0	15.2	280	134	239	262	276	283	252
	12		26.6	0.0	285	143	253	275	290	296	263
	Surface Daily Totals				1978	782	1480	1634	1740	1796	1646

TABLE 18.1 (Continued)

SOLAR POSITION AND INSOLATION, 48°N LATITUDE

Date	Solar Time AM	Solar Time PM	Solar Position ALT	Solar Position AZM	Normal	Horiz.	38	48	58	68	90
Jan 21	8	4	3.5	54.6	37	4	17	19	21	22	22
	9	3	11.0	42.6	185	46	120	132	140	145	139
	10	2	16.9	29.4	239	83	190	206	216	220	206
	11	1	20.7	15.1	261	107	231	249	260	263	243
	12		22.0	0.0	267	115	245	264	275	278	255
	Surface Daily Totals				1710	596	1360	1478	1550	1578	1478
Feb 21	7	5	2.4	72.2	12	1	3	4	4	4	4
	8	4	11.6	60.5	188	49	95	102	105	106	96
	9	3	19.7	47.7	251	100	178	187	191	190	167
	10	2	26.2	33.3	278	139	240	251	255	251	217
	11	1	30.5	17.2	290	165	278	290	294	288	247
	12		32.0	0.0	293	173	291	304	307	301	258
	Surface Daily Totals				2330	1080	1880	1972	2024	1978	1720
Mar 21	7	5	10.0	78.7	153	37	49	49	47	45	35
	8	4	19.5	66.8	236	96	131	132	129	122	96
	9	3	28.2	53.4	270	147	205	207	203	193	152
	10	2	35.4	37.8	287	187	263	266	261	248	195
	11	1	40.3	19.8	295	212	300	303	297	283	223
	12		42.0	0.0	298	220	312	315	309	294	232
	Surface Daily Totals				2780	1578	2208	2228	2182	2074	1632
Apr 21	6	6	8.6	97.8	108	27	13	9	8	7	5
	7	5	18.6	86.7	205	85	76	69	59	48	21
	8	4	28.5	74.9	247	142	149	141	129	113	69
	9	3	37.8	61.2	268	191	216	208	194	174	115
	10	2	45.8	44.6	280	228	268	260	245	223	152
	11	1	51.5	24.0	286	252	301	294	278	254	177
	12		53.6	0.0	288	260	313	305	289	264	185
	Surface Daily Totals				3076	2106	2358	2266	2114	1902	1262
May 21	5	7	5.2	114.3	41	9	4	4	4	3	2
	6	6	14.7	103.7	162	61	27	16	15	13	10
	7	5	24.6	93.0	219	118	89	75	60	43	13
	8	4	34.7	81.6	248	171	156	142	123	101	45
	9	3	44.3	68.3	264	217	217	202	182	156	86
	10	2	53.0	51.3	274	252	265	251	229	200	120
	11	1	59.5	28.6	279	274	296	281	258	228	141
	12		62.0	0.0	280	281	306	292	269	238	149
	Surface Daily Totals				3254	2482	2418	2234	2010	1728	982
Jun 21	5	7	7.9	116.5	77	21	9	9	8	7	5
	6	6	17.2	106.2	172	74	33	19	19	16	12
	7	5	27.0	95.8	220	129	93	77	59	39	15
	8	4	37.1	84.6	246	181	157	140	119	95	35
	9	3	46.9	71.6	261	225	216	198	175	147	74
	10	2	55.8	54.8	269	259	262	244	220	189	105
	11	1	62.7	31.2	274	280	291	273	248	216	126
	12		65.5	0.0	275	287	301	283	258	225	133
	Surface Daily Totals				3312	2626	2420	2204	1950	1644	874
Jul 21	5	7	5.7	114.7	43	10	5	5	5	4	3
	6	6	15.2	104.1	156	62	28	18	16	15	11
	7	5	25.1	93.5	211	118	89	75	59	42	14
	8	4	35.1	82.1	240	171	154	140	121	99	43
	9	3	44.8	68.8	256	215	214	199	178	153	83
	10	2	53.5	51.9	266	250	261	246	224	195	116
	11	1	60.1	29.0	271	272	291	276	253	223	137
	12		62.6	0.0	272	279	301	286	263	232	144
	Surface Daily Totals				3158	2474	2386	2200	1974	1694	956
Aug 21	6	6	9.1	98.3	99	28	14	10	9	8	6
	7	5	19.1	87.2	190	85	75	67	58	47	20
	8	4	29.0	75.4	232	141	145	137	125	109	65
	9	3	38.4	61.8	254	189	210	201	187	168	110
	10	2	46.4	45.1	266	225	260	252	237	214	146
	11	1	52.2	24.3	272	248	293	285	268	244	169
	12		54.3	0.0	274	256	304	296	279	255	177
	Surface Daily Totals				2898	2086	2300	2200	2046	1836	1208
Sep 21	7	5	10.0	78.7	131	35	44	44	43	40	31
	8	4	19.5	66.8	215	92	124	124	121	115	90
	9	3	28.2	53.4	251	142	196	197	193	183	143
	10	2	35.4	37.8	269	181	251	254	248	236	185
	11	1	40.3	19.8	278	205	287	289	284	269	212
	12		42.0	0.0	280	213	299	302	296	281	221
	Surface Daily Totals				2568	1522	2102	2118	2070	1966	1546
Oct 21	7	5	2.0	71.9	4	0	1	1	1	1	1
	8	4	11.2	60.2	165	44	86	91	95	95	87
	9	3	19.3	47.4	233	94	167	176	180	178	157
	10	2	25.7	33.1	262	133	228	239	242	239	207
	11	1	30.0	17.1	274	157	266	277	281	276	237
	12		31.5	0.0	278	166	279	291	294	288	247
	Surface Daily Totals				2154	1022	1774	1860	1890	1866	1626
Nov 21	8	4	3.6	54.7	36	5	17	19	21	22	22
	9	3	11.2	42.7	179	46	117	129	137	141	135
	10	2	17.1	29.5	233	83	186	202	212	215	201
	11	1	20.9	15.1	255	107	227	245	255	258	238
	12		22.2	0.0	261	115	241	259	270	272	250
	Surface Daily Totals				1666	596	1336	1448	1518	1544	1442
Dec 21	9	3	8.0	40.9	140	27	87	98	105	110	109
	10	2	13.6	28.2	214	63	164	180	192	197	190
	11	1	17.3	14.4	242	86	207	226	239	244	231
	12		18.6	0.0	250	94	222	241	254	260	244
	Surface Daily Totals				1444	446	1136	1250	1326	1364	1304

Reprinted with permission from the 1974 ASHRAE Transactions, Vol. 80, Part II.

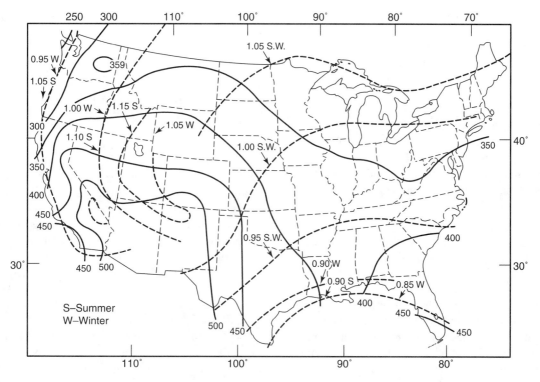

Figure 18.10
Clearness factors for the continental United States. (*Note:* Dashed lines show clearness factors; solid lines are average annual insolation in Langleys/day [Langleys /day × 2.69 = BTU/hr]).

18.8 ORIENTATION AND TILT ANGLES

In order to maximize the total daily amount of radiation collected, a flat plate collector should be *oriented south.* This is why Table 18.1 lists insolation only at this orientation. If, for some reason, a collector is not oriented south, corrections can be made to the insolation values from data available in the literature. However, for an orientation as much as 15° away from south, the loss in radiation is negligible for practical purposes.

There may be special circumstances where it is desirable to orient the collector other than south. Some local climates have a regular pattern of fog or clouds all morning or afternoon, for example. In that case, an orientation slightly southeast or southwest might be better. Interference from nearby objects, of course, might dictate the collector orientation.

The optimum *tilt angles* above the horizontal for collecting the maximum daily total radiation have been determined. Because the elevation of the sun changes seasonally, the angle depends on the use of the solar system. Recommended values are:

Service	Optimum Tilt Angle
Heating only	Latitude angle + 10°
Heating and cooling	Latitude angle
Cooling only	Latitude angle − 10°

Small variations in the above tilts will not affect the results significantly. Some authorities recommend a tilt of latitude + 15° for heating.

18.9　SUNSHINE HOURS

Table 18.1 lists insolation for clear days, without clouds. In order to determine the economic feasibility of installing a solar heating system, the amount of solar energy available monthly and seasonally must be found. Therefore, an estimate must be made of the amount of time the sun is shining. Table 18.2 lists the mean percentage of sunshine hours expected for various locations, on a monthly and annual basis. The total monthly values found from Table 18.1 must be corrected for this.

Example 18.3 _____

Determine the monthly solar insolation in March for the surface described in Example 18.2.

Solution

From Example 18.2, the daily insolation is 2398 BTU/ft² per day on a clear day. From Table 18.2, the mean percentage of sunshine hours in March in Denver is 65. Applying this factor, the monthly solar insolation in March is

$$2398 \text{ BTU/ft}^2 \text{ per day} \times 31 \text{ days} \times 0.65$$
$$= 48,320 \text{ BTU/ft}^2$$

_____

A condition that is not accounted for in this correction for cloudy periods is the effect of very short periods of sunshine. On days when the sun appears and disappears often, there may not be enough sustained direct radiation to bring the collector up to any significant operating temperature. Data on this effect are not yet readily available. However, the procedures described here are relatively conservative, thus balancing this effect.

The method we have used does not allow for diffuse radiation on cloudy days. But it is questionable if this radiation would usually be great enough to operate or "drive" a typical flat plate collector. Below a certain insolation value, a collector will not absorb any net amount of heat.

Under special local conditions, ground reflectance of radiation in the collector surface should be included. For example, if it is known that the area surrounding the collector will always have snow in winter, the resulting reflected radiation should be calculated. We will not discuss this subject here.

18.10　COLLECTOR PERFORMANCE

The energy actually collected by a solar collector will be considerably less than the amount received. Some of the energy will be lost back to the surroundings, because the collector is at a higher temperature. The efficiency of a collector is defined as the ratio of (heat) energy output to energy input. The efficiency depends on the physical construction of the collector. Increasing the number of cover plates raises the efficiency. The materials and configuration of the absorber plate, coating, and insulation also affect the efficiency. Manufacturers furnish specific data on their collectors.

For a given collector, the efficiency will vary with the temperature difference between the collector and the surrounding ambient air, and with the amount of insolation received. This data can be plotted as shown in Figure 18.11 on page 471.

Example 18.4 _____

Find the amount of heat collected per hour by a 500 ft² flat plate collector of the type shown in Figure 18.11, with two cover plates, at a time when the solar insolation received is 200 BTU/hr-ft² and the surrounding air temperature is 40 F. The inlet water temperature is 110 F.

Solution

Calculating the value used on the horizontal axis of Figure 18.11,

$$\frac{t_{in} - t_{air}}{I} = \frac{110 - 40}{200} = 0.35$$

From the figure, the efficiency is read on the curve for two cover plates as 0.45. The heat collected is

$$200 \text{ BTU/hr-ft}^2 \times 0.45 \times 500 \text{ ft}^2$$
$$= 45,000 \text{ BTU/hr}$$

_____

TABLE 18.2 MEAN PERCENTAGE OF POSSIBLE SUNSHINE

Mean Percentage of Possible Sunshine for Selected Locations

State and Station	Years	Jan.	Feb.	Mar.	Apr.	May	June	July	Aug.	Sept.	Oct.	Nov.	Dec.	Annual
Ala. Birmingham	56	43	49	56	63	66	67	62	65	66	67	58	44	59
Montgomery	49	51	53	61	69	73	72	66	69	69	71	64	48	64
Alaska. Anchorage	19	39	46	56	58	50	51	45	39	35	32	33	29	45
Fairbanks	20	34	50	61	68	55	53	45	35	31	28	38	29	44
Juneau	14	30	32	39	37	34	35	28	30	25	18	21	18	30
Nome	29	44	46	48	53	51	48	32	26	34	35	36	30	41
Ariz. Phoenix	64	76	79	83	88	93	94	84	84	89	88	84	77	85
Yuma	52	83	87	91	94	97	98	92	91	93	93	90	83	91
Ark. Little Rock	66	44	53	57	62	67	72	71	73	71	74	58	47	62
Calif. Eureka	49	40	44	50	53	54	56	51	46	52	48	42	39	49
Fresno	55	46	63	72	83	89	94	97	97	93	87	73	47	78
Los Angeles	63	70	69	70	67	68	69	80	81	80	76	79	72	73
Red Bluff	39	50	60	65	75	79	86	95	94	89	77	64	50	75
Sacramento	48	44	57	67	76	82	90	96	95	92	82	65	44	77
San Diego	68	68	67	68	66	60	60	67	70	70	70	76	71	68
San Francisco	64	53	57	63	69	70	75	68	63	70	70	62	54	66
Colo. Denver	64	67	67	65	63	61	69	68	68	71	71	67	65	67
Grand Junction	57	58	62	64	67	71	79	76	72	77	74	67	58	69
Conn. Hartford	48	46	55	56	54	57	60	62	60	57	55	46	46	56
D. C. Washington	66	46	53	56	57	61	64	64	62	62	61	54	47	58
Fla. Apalachicola	26	59	62	62	71	77	70	64	63	62	74	66	53	65
Jacksonville	60	58	59	66	71	71	63	62	63	58	58	61	53	62
Key West	45	68	75	78	78	76	70	69	71	65	65	69	66	71
Miami Beach	48	66	72	73	73	68	70	65	67	62	62	65	65	67
Tampa	63	63	67	71	74	75	66	61	64	64	67	67	61	68
Ga. Atlanta	65	48	53	57	69	68	68	62	63	65	67	60	47	60
Hawaii. Hilo	9	48	42	41	34	31	41	44	38	42	41	34	36	39
Honolulu	53	62	64	60	62	64	66	67	70	68	68	63	60	65
Lihue	9	48	48	48	46	51	60	58	59	67	58	51	49	54
Idaho, Boise	20	40	48	59	67	68	75	89	86	81	66	46	37	66
Pocatello	21	37	47	58	64	66	72	82	81	78	66	48	36	64
Ill. Cairo	30	46	53	59	65	71	77	82	79	75	73	56	46	65
Chicago	66	44	49	53	56	63	69	73	70	65	61	47	41	59
Springfield	59	47	51	54	58	64	69	76	72	73	64	53	45	60
Ind. Evansville	48	42	49	55	61	67	73	78	76	73	67	52	42	64
N. Mex. Albuquerque	28	70	72	72	76	79	84	76	75	81	80	79	70	76
Roswell	47	69	72	75	77	76	80	76	75	74	74	74	69	74
N.Y. Albany	63	43	51	53	53	57	62	63	61	58	54	39	38	53
Binghamton	63	31	39	41	44	50	56	54	51	47	43	29	26	44
Buffalo	49	32	41	49	51	59	67	70	67	60	51	31	28	53
Canton	43	37	47	50	48	54	61	63	61	54	45	30	31	49
New York	83	49	56	57	59	62	65	66	64	64	61	53	50	59
Syracuse	49	31	38	45	50	58	64	67	63	56	47	29	26	50
N. C. Asheville	57	48	53	56	61	58	63	62	59	62	64	59	48	58
Raleigh	61	50	56	59	64	67	65	62	62	63	64	62	52	61
N. Dak. Bismarck	65	52	58	56	57	58	61	73	69	62	59	49	48	59
Devils Lake	55	53	60	59	60	59	62	71	67	59	56	44	45	58
Fargo	39	47	55	56	58	62	63	73	69	60	57	39	46	59
Williston	43	51	59	60	63	66	66	78	75	65	60	48	48	63
Ohio. Cincinnati	44	41	46	52	56	62	69	72	68	68	60	46	39	57
Cleveland	65	29	36	45	52	61	67	71	68	62	54	32	25	50
Columbus	65	36	44	49	54	63	68	71	68	66	60	44	35	55
Okla. Oklahoma City	62	57	60	63	64	65	74	78	78	74	68	64	57	68
Oreg. Baker	46	41	49	56	61	63	67	83	81	74	62	46	37	60
Portland	69	27	34	41	49	52	55	70	65	55	42	28	23	48
Roseburg	29	24	32	40	51	57	59	79	77	68	42	28	18	51
Pa. Harrisburg	60	43	52	55	57	61	65	68	63	62	58	47	43	57
Philadelphia	66	45	56	57	58	61	62	64	61	62	61	53	49	57
Pittsburgh	63	32	39	45	50	57	62	64	61	62	54	39	30	51
R. I. Block Island	48	45	54	47	56	57	60	62	62	60	59	58	44	56
S. C. Charleston	61	58	60	65	72	73	70	66	66	67	68	68	57	66
Columbia	55	53	57	62	68	65	68	63	65	64	68	64	51	63
S. Dak. Huron	62	55	60	62	62	65	68	76	72	66	61	52	49	63
Rapid City	53	58	62	63	62	61	66	73	73	69	66	58	54	64
Tenn. Knoxville	62	42	49	53	59	64	66	64	59	64	64	53	51	57
Memphis	55	44	51	57	64	68	74	73	74	70	69	58	45	64
Nashville	63	42	47	54	60	65	69	69	68	69	65	55	42	59
Tex. Abilene	14	64	68	73	66	73	86	83	85	73	71	72	66	73
Amarillo	54	71	71	75	75	75	82	81	81	79	76	76	70	76
Austin	33	46	50	57	60	62	72	76	79	70	70	57	49	63

TABLE 18.2 (Continued)

Mean Percentage of Possible Sunshine for Selected Locations

State and Station	Years	Jan.	Feb.	Mar.	Apr.	May	June	July	Aug.	Sept.	Oct.	Nov.	Dec.	Annual
Ft. Wayne	48	38	44	51	55	62	69	74	69	64	58	41	38	57
Indianapolis	63	41	47	49	55	62	68	74	70	68	64	48	39	59
Iowa. Des Moines	66	56	56	56	59	62	66	75	70	64	64	53	48	62
Dubuque	54	48	52	52	58	60	63	73	67	61	55	44	40	57
Sioux City	52	55	58	58	59	63	67	75	72	67	65	53	50	63
Kans. Concordia	52	60	60	62	63	65	73	79	76	72	70	64	58	67
Dodge City	70	67	66	68	68	68	74	78	78	76	75	70	67	71
Wichita	46	61	63	64	64	66	73	80	77	73	69	61	59	69
Ky. Louisville	59	41	47	52	57	64	73	72	69	68	64	51	39	59
La. New Orleans	69	49	50	57	63	66	64	58	60	65	70	60	46	59
Shreveport	18	48	54	58	60	69	78	79	80	79	77	65	60	69
Maine. Eastport	58	45	51	52	52	51	53	55	57	54	50	37	40	50
Mass. Boston	67	47	56	57	56	59	62	64	63	61	58	48	48	57
Mich. Alpena	45	29	43	52	56	59	64	70	64	52	44	24	22	51
Detroit	69	34	42	48	52	58	65	69	66	61	54	35	29	53
Grand Rapids	56	26	37	48	54	60	66	72	67	58	50	31	22	49
Marquette	55	31	40	47	52	53	56	63	57	47	38	24	24	47
S. Ste. Marie	60	28	44	50	54	54	59	63	58	45	36	21	22	47
Minn. Duluth	49	47	55	60	58	58	60	68	63	53	47	36	40	55
Minneapolis	45	49	54	55	57	60	64	72	69	60	54	40	40	56
Miss. Vicksburg	66	46	50	57	64	69	73	69	72	74	71	60	45	64
Mo. Kansas City	69	55	57	59	60	64	70	76	73	70	67	59	52	65
St. Louis	68	48	49	56	59	64	68	72	68	67	65	54	55	61
Springfield	45	48	54	57	60	63	69	77	72	71	65	58	48	63
Mont. Havre	55	49	58	61	63	63	65	78	75	64	57	48	46	62
Helena	65	46	50	57	60	59	63	77	74	63	57	48	43	60
Kalispell	50	28	40	49	57	58	60	77	73	67	50	28	20	53
Nebr. Lincoln	55	57	57	59	60	60	63	69	71	67	66	59	55	64
North Platte	53	63	63	64	62	64	72	78	74	72	70	62	58	68
Nev. Ely	21	61	64	68	65	67	79	79	81	81	73	67	62	72
Las Vegas	19	74	77	78	81	85	91	84	86	92	84	83	75	82
Reno	51	59	64	69	75	77	82	90	89	86	76	68	56	76
Winnemucca	53	52	60	64	70	76	83	90	90	86	75	62	53	74
N. H. Concord	44	48	53	55	53	51	56	57	58	55	50	43	43	52
N.J. Atlantic City	62	51	57	58	59	62	65	67	66	65	54	58	52	60

Mean Percentage of Possible Sunshine for Selected Locations

State and Station	Years	Jan.	Feb.	Mar.	Apr.	May	June	July	Aug.	Sept.	Oct.	Nov.	Dec.	Annual
Brownsville	37	44	49	51	57	65	73	78	78	67	70	54	44	61
Del Rio	36	53	55	61	63	60	66	75	80	69	66	58	52	63
El Paso	53	74	77	81	85	87	87	78	78	80	82	80	73	80
Ft. Worth	33	56	57	65	66	67	75	78	78	74	70	63	58	68
Galveston	66	50	50	55	61	69	76	72	71	70	74	62	49	63
San Antonio	57	48	51	56	58	60	69	74	75	69	67	55	49	62
Utah. Salt Lake City	22	48	53	61	68	73	78	82	82	84	73	56	49	69
Vt. Burlington	54	34	43	48	47	53	59	62	59	51	43	25	24	46
Va. Norfolk	60	50	57	60	63	67	66	66	66	63	64	60	51	62
Richmond	56	49	55	59	63	67	66	65	62	63	64	58	50	61
Wash. North Head	44	28	37	42	48	48	48	50	46	48	41	31	27	41
Seattle	26	27	34	42	48	53	48	62	56	53	36	28	24	45
Spokane	62	30	41	53	63	64	64	82	79	68	53	28	22	58
Tatoosh Island	49	26	36	39	45	47	46	48	44	47	38	26	23	40
Walla Walla	44	24	35	51	63	67	72	86	84	72	59	33	20	60
Yakima	18	34	49	62	70	72	74	86	86	74	61	38	29	65
W. Va. Elkins	55	33	37	42	47	55	55	56	53	55	51	41	33	48
Parkersburg	62	30	36	42	49	56	60	63	60	60	53	37	29	48
Wis. Green Bay	57	44	51	55	56	58	64	70	65	58	52	40	40	55
Madison	59	44	49	52	53	58	64	70	66	60	56	41	38	56
Milwaukee	59	44	48	53	56	60	65	73	67	62	56	44	39	57
Wyo. Cheyenne	63	65	66	64	61	59	68	70	68	69	69	65	63	66
Lander	57	66	70	71	66	65	74	76	75	72	67	61	62	69
Sheridan	52	56	61	62	61	61	67	76	74	67	60	53	52	64
Yellowstone Park	35	39	51	55	57	56	63	73	71	65	57	45	38	56
P. R. San Juan	57	64	69	71	66	59	62	65	67	61	63	65	65	65

Based on period of record through December 1959, except in a few instances.

These charts and tabulation derived from "Normals, Means, and Extremes" table in U.S. Weather Bureau publication Local Climatological Data.

Reprinted from ASHRAE TRANSACTIONS 1974, Volume 80, PART II, by permission of the American Society of Heating, Refrigerating and Air-Conditioning Engineers, Inc.

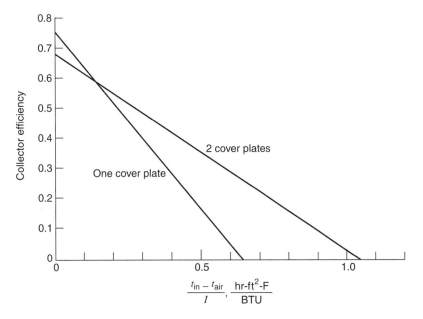

Figure 18.11
Typical flat plate collector efficiency.

18.11 SIZING THE COLLECTOR

The main task in designing a solar heating system is to determine the size of the collector to be used. The information we have developed can be used to do this in a number of ways, one of which we will describe here. In this method, the percentage of total yearly building heating load that the solar heating system shall handle is first decided on, and based on this, the collector size can be determined. Experience has shown that for many typical applications, a range of 50–75% of the total yearly load is a good range to assign to the task of the solar heating system, to result in an economic combination of owning and operating costs. The steps in sizing the collector are as follows:

1. Calculate the total solar insolation received for the year.
2. Calculate the design building heating load. Use degree day values (Chapter 15) to find the total yearly energy required.

3. Determine the efficiency of the collector to be used.
4. Based on a percentage of total energy to be handled by the solar system, determine the collector size.

Example 18.5 _____
A solar heating system to handle 65% of the yearly heating requirements is to be installed for a home in Washington, D.C. The design heating load is 41,000 BTU/hr and the indoor temperature is 70 F. The collector average efficiency is 42%. Collector orientation is south and the tilt is 50°. Determine the required collector size. Assume CF = 1.0.

Solution
The insolation received is calculated as explained previously. Table 18.3 on page 472 shows the results of heat collected.

The total yearly heating requirement, using the degree day method, is

TABLE 18.3   RESULTS OF HEAT COLLECTION

	Insolation			Collected Energy		
	$\dfrac{BTU}{day\text{-}ft^2}$	$\dfrac{\%}{Sun}$	$\dfrac{BTU}{day\text{-}ft^2}$ average	Coll. eff.	$\dfrac{BTU}{day\text{-}ft^2}$	$\dfrac{BTU}{month\text{-}ft^2}$
Sept.	2182	68	1484	.42	623	18,690
Oct.	2098	60	1259	.42	529	16,400
Nov.	1870	46	860	.42	361	10,830
Dec.	1740	39	678	.42	285	8,840
Jan.	1906	41	781	.42	328	10,170
Feb.	2202	46	1013	.42	425	11,900
Mar.	2284	52	1188	.42	499	15,470
Apr.	2168	56	1214	.42	510	15,300
May	2040	62	1265	.42	531	16,460

Collected Yearly Total $= 124{,}060$ BTU/ft^2

$$\frac{41{,}000 \text{ BTU/hr}}{(70-19)} \times 24 \text{ hr} \times 4224 \text{ degree days}$$

$$= 81{,}500{,}000 \text{ BTU}$$

The size of the solar collector to handle 65% of the heat required is

$$\text{Collector area} = \frac{81{,}500{,}000 \text{ BTU}}{124{,}060 \text{ BTU/ft}^2} \times 0.65$$

$$= 427 \text{ ft}^2$$

In the procedure above, we assume an average yearly collector efficiency. For a more precise calculation, the varying collector efficiency would be determined from manufacturer's curves for the water and ambient temperatures, and insolation values expected each month.

18.12   ECONOMIC ANALYSIS

When considering the installation of a solar heating system, either in a new building or retrofitting an existing system, an analysis must be made to determine if the fuel cost savings will offset the increased capital equipment expenditures. The first step in doing this is to simply find the annual fuel costs savings from that part of the heating requirements handled by the solar system.

Example 18.6 _____

Determine the expected yearly energy and fuel cost savings for the home in Example 18.5 if the solar heating system described is installed. Fuel oil at $0.90 a gallon is normally used, with a heating value of 145,000 BTU/gal. The auxiliary heating system has an average efficiency of 60%.

Solution

The solar system handles 65% of the yearly requirements. The net yearly depletable energy saved is therefore

$$81{,}500{,}000 \text{ BTU} \times 0.65 = 52{,}975{,}000 \text{ BTU}$$

The amount of fuel oil saved is

$$52{,}975{,}000 \text{ BTU} \times \frac{1 \text{ gal}}{145{,}000 \text{ BTU}} \times \frac{1}{0.60}$$

$$= 609 \text{ gal}$$

The yearly fuel cost saving is

$$609 \text{ gal} \times \$0.90/\text{gal} = \$548$$

_____

In order to decide whether the annual savings are justified, we must determine how long it will take to recover our initial investment. If this period is reasonable, then we are willing to make the investment, because thereafter the annual savings will in a sense be profit. There are numerous economic procedures for estimating the justification of the investment. One popular procedure is called the *present worth* method. This approach will tell us how much of an initial investment is justified if we wish to have it paid back in a given number of years. The following equations are used to determine this:

$$P = \frac{S(1 + i_e)^t - 1}{i_e(1 + i_e)^t} \quad (18.1)$$

$$i_e = \frac{1 + i}{1 + j} - 1 \quad (18.2)$$

where

P = initial investment justifiable (present worth)

S = initial yearly savings, $

t = desired payback period, years

i = annual interest rate for borrowing money, decimals

j = annual rate of fuel price increase, decimals

i_e = effective interest rate, from Equation 18.2

An important characteristic of this method is that it accounts directly for rising fuel prices.

Example 18.7
Mr. and Mrs. U. R. Wise, owners of the house in Example 18.5, want to recover their investment on a retrofit solar heating system addition in 10 years. They can borrow money at 6% under a special government energy conservation program. Fuel costs rise at an annual rate of 12%. What should be the maximum initial investment?

Solution
Applying the present worth method equations,

$$i_e = \frac{1 + 0.06}{1 + 0.12} - 1 = -0.054$$

$$P = \frac{S(1 + i_e)^t - 1}{i_e(1 + i_e)^t}$$

$$= 548 \frac{(1 - 0.054)^{10} - 1}{-0.054(1 - 0.054)^{10}} = \$7530$$

If the initial cost of adding the solar heating system is $7530 or less, the owners would be wise to install it.

18.13 STORAGE SYSTEM SIZING

Studies have been made which indicate that a storage tank size of 1.5–2.5 gallons of water per square foot of collection area is usually the optimum economic tank size, regardless of climate.

This provides, at most, about one day's heat storage.

Example 18.8
A storage tank of 2.0 gal/ft^2 of collector area is used for the solar heating system for the home in Example 18.5. How long can the system operate on storage on a typical March day?

Solution
The heat required for a day in March is first determined. To do this, the monthly degree day data is required. Table 18.4 on pages 474–475 presents this information for some cities.

First, the BTU/deg day are determined:

$$\frac{41,000 \text{ BTU/hr}}{(70 - 19)\text{F}} \times 24 \text{ hr}$$
$$= 19,294 \text{ BTU/deg day}$$

There are 626 degree days in March. The average degree days in a day are

$$\frac{626}{31} = 20.2 \text{ deg days}$$

The daily heat required is

$$19,294 \text{ BTU/deg day} \times 20.2 \text{ deg days}$$
$$= 389,740 \text{ BTU}$$

The storage tank volume is

$$2.0 \text{ gal/ft}^2 \times 427 \text{ ft}^2 = 854 \text{ gal} \times 8.3 \text{ lb/gal}$$
$$= 7088 \text{ lb}$$

TABLE 18.4 AVERAGE MONTHLY DEGREE DAYS FOR SELECTED LOCATIONS

Location	July	August	September	October	November	December	January	February	March	April	May	June	Yearly Total
Alabama, Birmingham	0	0	6	93	363	555	592	462	363	108	9	0	2551
Alaska, Anchorage	245	291	516	930	1284	1572	1631	1316	1293	897	592	315	10,864
Arizona, Tucson	0	0	0	25	231	406	471	344	242	75	6	0	1800
Arkansas, Little Rock	0	0	9	127	465	716	756	577	434	126	9	0	3219
California, San Francisco	81	78	60	143	306	462	508	395	363	279	214	126	3015
Colorado, Denver	6	9	117	428	819	1035	1132	938	887	558	288	66	6283
Connecticut, Bridgeport	0	0	66	307	615	986	1079	966	853	510	208	27	5617
Delaware, Wilmington	0	0	51	270	588	927	980	874	735	387	112	6	4930
District of Columbia, Washington	0	0	33	217	519	834	871	762	626	288	74	0	4224
Florida, Tallahassee	0	0	0	28	198	360	375	286	202	86	0	0	1485
Georgia, Atlanta	0	0	18	124	417	648	636	518	428	147	25	0	2961
Hawaii, Honolulu	0	0	0	0	0	0	0	0	0	0	0	0	0
Idaho, Boise	0	0	132	415	792	1017	1113	854	722	438	245	81	5809
Illinois, Chicago	0	12	117	381	807	1166	1265	1086	939	534	260	72	6639
Indiana, Indianapolis	0	0	90	316	723	1051	1113	949	809	432	177	39	5699
Iowa, Sioux City	0	9	108	369	867	1240	1435	1198	989	483	214	39	6951
Kansas, Wichita	0	0	33	229	618	905	1023	804	645	270	87	6	4620
Kentucky, Louisville	0	0	54	248	609	890	930	818	682	315	105	9	4660
Louisiana, Shreveport	0	0	0	47	297	477	552	426	304	81	0	0	2184
Maine, Caribou	78	115	336	682	1044	1535	1690	1470	1308	858	468	183	9767
Maryland, Baltimore	0	0	48	264	585	905	936	820	679	327	90	0	4654
Massachusetts, Boston	0	9	60	316	603	983	1088	972	846	513	208	36	5634
Michigan, Lansing	6	22	138	431	813	1163	1262	1142	1011	579	273	69	6909
Minnesota, Minneapolis	22	31	189	505	1014	1454	1631	1380	1166	621	288	81	8382
Mississippi, Jackson	0	0	0	65	315	502	546	414	310	87	0	0	2239
Missouri, Kansas City	0	0	39	220	612	905	1032	818	682	294	109	0	4711
Montana, Billings	6	15	186	487	897	1135	1296	1100	970	570	285	102	7049
Nebraska, Lincoln	0	6	75	301	726	1066	1237	1016	834	402	171	30	5864
Nevada, Las Vegas	0	0	0	78	387	617	688	487	335	111	6	0	2709

TABLE 18.4 (Continued)

Location	July	August	September	October	November	December	January	February	March	April	May	June	Yearly Total
New Hampshire, Concord	6	50	177	505	822	1240	1358	1184	1032	636	298	75	7383
New Jersey, Atlantic City	0	0	39	251	549	880	936	848	741	420	133	15	4812
New Mexico, Albuquerque	0	0	12	229	642	868	930	703	595	288	81	0	4348
New York, Syracuse	6	28	132	415	744	1153	1271	1140	1004	570	248	45	6756
North Carolina, Charlotte	0	0	6	124	438	691	691	582	481	156	22	0	3191
North Dakota, Bismarck	34	28	222	577	1083	1463	1708	1442	1203	645	329	117	8851
Ohio, Cleveland	9	25	105	384	738	1088	1159	1047	918	552	260	66	6351
Oklahoma, Stillwater	0	0	15	164	498	766	868	664	527	189	34	0	3725
Oregon, Pendleton	0	0	111	350	711	884	1017	773	617	396	205	63	5127
Pennsylvania, Pittsburgh	0	9	105	375	726	1063	1119	1002	874	480	195	39	5987
Rhode Island, Providence	0	16	96	372	660	1023	1110	988	868	534	236	51	5954
South Carolina, Charleston	0	0	0	59	282	471	487	389	291	54	0	0	2033
South Dakota, Rapid City	22	12	165	481	897	1172	1333	1145	1051	615	326	126	7345
Tennessee, Memphis	0	0	18	130	447	698	729	585	456	147	22	0	3232
Texas, Dallas	0	0	0	62	321	524	601	440	319	90	6	0	2363
Utah, Salt Lake City	0	0	81	419	849	1082	1172	910	763	459	233	84	6052
Vermont, Burlington	28	65	207	539	891	1349	1513	1333	1187	714	353	90	8269
Virginia, Norfolk	0	0	0	136	408	698	738	655	533	216	37	0	3421
Washington, Spokane	9	25	168	493	879	1082	1231	980	834	531	288	135	6655
West Virginia, Charleston	0	0	63	254	591	865	880	770	648	300	96	9	4476
Wisconsin, Milwaukee	43	47	174	471	876	1252	1376	1193	1054	642	372	135	7635
Wyoming, Casper	6	16	192	524	942	1169	1290	1084	1020	657	381	129	7410
Alberta, Calgary	109	186	402	719	1110	1389	1575	1379	1268	798	477	291	9703
British Columbia, Vancouver	81	87	219	456	657	787	862	723	676	501	310	156	5515
Manitoba, Winnipeg	38	71	322	683	1251	1757	2008	1719	1465	813	405	147	10,679
New Brunswick, Fredericton	78	68	234	592	915	1392	1541	1379	1172	753	406	141	8671
Nova Scotia, Halifax	58	51	180	457	710	1074	1213	1122	1030	742	487	237	7361
Ontario, Ottawa	25	81	222	567	936	1469	1624	1441	1231	708	341	90	8735
Quebec, Montreal	9	43	165	521	882	1392	1566	1381	1175	684	316	69	8203
Saskatchewan, Regina	78	93	360	741	1284	1711	1965	1687	1473	804	409	201	10,806

Source: *Heating, Ventilating, and Air Conditioning*, F.C. McQuiston and J. D. Parker, Copyright c 1977, John Wiley & Sons. Reprinted by permission.

We will assume the water is heated from 90–140 F in the tank. The heat available is therefore

$$Q = m \times c \times TC$$
$$= 7088 \text{ lb} \times 1 \text{ BTU/lb-F} \times 50 \text{ F}$$
$$= 354,440 \text{ BTU}$$

The tank has slightly less than one day's storage to handle the heating requirements on a March day. Extended periods of cloudiness and conditions in colder months would require use of the conventional auxiliary heat source.

18.14 APPROXIMATE SYSTEM DESIGN DATA

The U.S. Department of Energy has furnished information based on typical conditions that can be used for approximate solar heating and hot water system sizing and economic analysis. Figure 18.12 shows climatic zones in the United States. This map is used in conjunction with Tables 18.5 and 18.6 for preliminary estimates of collector and storage tank sizes and

TABLE 18.5   APPROXIMATE COLLECTOR AND STORAGE TANK SIZES REQUIRED TO PROVIDE THE HEATING AND HOT WATER NEEDS OF A 1500 SQUARE FOOT HOME

Climatic Zone	Percent of Energy Supplied by Solar	Collector Area, Square Feet	Storage Tank Capacity, Gallons
1	71	800	1500
2	72	500	750
3	66	800	1500
4	73	300	500
5	75	200	280
6	70	750	1500
7	70	500	750
8	71	200	280
9	72	600	1000
10	58	500	750
11	85	200	280
12*	85	45	80

energy savings for a 1500 ft² home and 10,000 ft² building. Note that the figures are not grossly different from those resulting from our calculations.

Figure 18.12
Climatic zones in the United States.

TABLE 18.6   APPROXIMATE COLLECTOR AND STORAGE TANK SIZES REQUIRED TO PROVIDE THE HEATING NEEDS OF A 10,000 SQUARE FOOT BUILDING LOCATED IN EACH CLIMATIC ZONE

Climatic Zone	Percent of Energy Supplied by Solar	Collector Area, Square Feet	Storage Tank Capacity, Gallons
1	74	5330	10,000
2	71	3330	5000
3	68	5330	10,000
4	73	2000	4000
5	75	1210	2000
6	72	5000	7500
7	71	3330	5000
8	74	1330	2000
9	75	4000	6000
10	60	3330	5000
11	77	1000	1500
12	*		

18.15   PASSIVE SOLAR HEATING SYSTEMS

A solar heating system that utilizes the building structure to provide some or all of the heating is called a *passive* system. An *active* system uses collectors and ducts or piping. Those that combine both are called *hybrid* systems.

In a simple passive system design, the building is constructed with a large southward facing glass window. This arrangement is called a *direct* passive system because the sun directly heats the interior materials, which then heat the room air. If the system is properly designed, heat which is stored in the walls and floor in the day will continue to heat the room at night. Heavy mass materials are desirable for this purpose. Sometimes the sides of the building not facing south are built into a hillside.

In an *indirect* passive solar heating system, the sun heats a thermal storage element (e.g., a heavy concrete wall) which in turn heats the room. An example is shown in Figure 18.13 on page 478. Behind the southward facing double glass (this is not a window!) there is a thermal storage inner wall with openings into the room at the top and bottom. See Figure 18.13(*a*). The inner wall has a large mass so that it can store considerable heat. It may be made of concrete or rocks, or it may be a tankful of water. This arrangement is often called a *Trombe* wall after one of its developers.

The low winter sun heats the inner wall which in turn heats the air between the wall and glass. The warmed air rises and enters the room through the top opening. Cool air from the room flows through the bottom opening to replace the warm air and a convection current is created. See Figure 18.13(*b*). Sometimes a fan is used to increase the flow of air. At night, the warmed wall will still heat the room, and the openings in the wall are closed to prevent circulation. See Figure 18.13(*c*). Furthermore, by providing vent openings under the roof, cool air from the north side can be circulated in summer. See Figure 18.13(*d*).

There are many other ingenious passive solar heating systems that have been developed in conjunction with the architectural design. For example, greenhouses which have been integrated as part of the building and roofs with water heat storage ponds.

Problems

18.1   Find the solar insolation on a south facing surface at 12° N, 3 PM, and the daily total on January 21 in Los Angeles, California, for a surface tilted at 32°, 42°, and normal to the sun.

18.2   What is the optimum collector tilt angle to collect the maximum daily solar insolation in November for (a) Boston, Massachusetts and (b) your town?

18.3   Find the total daily clear day insolation on February 21 for a collector at a tilt angle recommended for domestic hot water heating in a home in Savannah, Georgia.

18.4   Find the average daily insolation in January for a surface tilted at the recommended angle for heating in (a) Dallas, Texas, and (b) Detroit, Michigan.

18.5 Determine the efficiency of a flat plate collector for a solar heating system, of the type shown in Figure 18.11, with a single cover plate, at 1 PM on a clear February day in Springfield, Illinois. The outdoor temperature is 30 F and the inlet water temperature is 80 F. What is the heat output of a 500 ft² collector at this time?

18.6 If two cover plates are used on the collector in Problem 18.5, what is the heat output at the time listed?

18.7 Determine the collector size for a home in Kansas City, Missouri, with a design heating load of 45,000 BTU/hr, to handle 70% of the yearly heating requirements. The average collector efficiency is 44%.

18.8 Determine the collector size for (a) your home and (b) your school, for a solar heating system to handle 70% of the yearly heating requirements.

18.9 Determine the expected yearly fuel and cost savings for the solar heating systems in Problems 18.7 and 18.8. Average furnace efficiency is 60%. Use prevailing fuel costs.

18.10 Determine a suitable storage tank size for the solar heating systems in Problems 18.7 and 18.8. Find the amount of heat stored and compare it with that required on an average February day. The water temperature draw down in the tank is 55 F.

Figure 18.13
Passive solar heating arrangement (called "Trombe" wall).

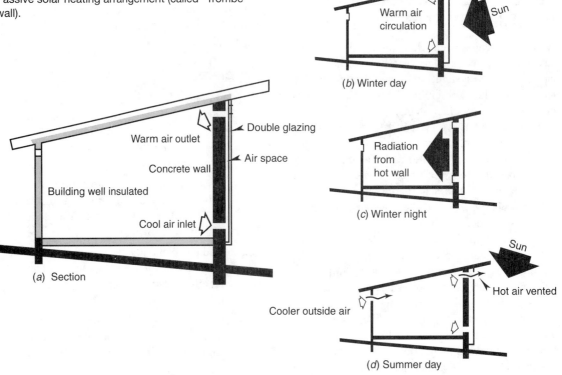

(b) Winter day

Double glazing

Warm air outlet

Air space

Concrete wall

Building well insulated

Cool air inlet

(a) Section

Radiation from hot wall

(c) Winter night

Cooler outside air

Hot air vented

(d) Summer day

18.11 Compare the preliminary estimate sizes of collector and storage tank and energy savings from Tables 18.5 and 18.6 to those found in your calculations in Problems 18.7 through 18.10.

18.12 For the building described in Problems 18.7 and 18.8, determine how much it is worth investing in a solar heating system added to the present heating system. Special loans are available at 5%. Assume a payback period of 8 years. Repeat the problem for a payback period of 15 years.

BIBLIOGRAPHY

REFERENCE MATERIALS

The following organizations provide reference material related to subjects contained in this book, as well as various codes and standards.

1. Air Conditioning and Refrigeration Institute (ARI), Arlington, VA.
2. Air Movement and Control Association, Inc. (AMCA), Arlington Heights, IL.
3. American Gas Association (AGA), Arlington, VA.
4. American Society of Heating, Refrigerating and Air-Conditioning Engineers (ASHRAE), Atlanta, GA.
5. American Society of Mechanical Engineers (ASME), New York, NY.
6. Hydronics Institute, Berkeley Heights, NJ.
7. Sheet Metal and Air Conditioning Contractors Association, Inc. (SMACNA), Vienna, VA.

SOFTWARE

The following suppliers provide software related to subjects contained in this book.

1. Elite Software, Bryan, TX 77806
   HVAC load calculations, energy analysis, psychrometric analysis, and duct and pipe sizing.
2. Horgan Vertical Applications Co., Coatesville, PA 19320
   HVAC load calculations and labor and material cost estimates.
3. The Trane Company, La Crosse, WI 54601
   HVAC load calculations, duct and piping design, and energy and economic analysis.
4. Air Conditioning Contractors of America, Washington, DC 20036
   HVAC load calculations, energy analysis, psychrometric analysis, and duct and pipe sizing.
5. Carmel Software, Cleveland, OH 44121
   HVAC load calculations (residential), psychrometric processes, and cost analysis.
6. Carrier Corporation, Software Systems Network, Syracuse, NY 13221
   HVAC load calculations, duct and piping design, energy and economic analysis, refrigerant pipe sizing, sheet metal layout, and estimating.

Note: Many manufacturers provide software for equipment selection, and have sites on the Internet via the World Wide Web to provide technical assistance.

Appendix

TABLE A.1 ABBREVIATIONS AND SYMBOLS

A	area	COP	coefficient of performance, cooling
ACH	air changes per hour	COP_h	coefficient of performance, heating
AFUE	annual fuel utilization efficiency	c	specific heat
BF	bypass factor	clo	clothing thermal resistance
BF	ballast factor	D	diameter
BHP	brake horsepower	D	degree days
BTU	British thermal unit	DB	dry bulb temperature
C	degrees Celsius	DP	dew point temperature
C	loss coefficient	d	density
C	thermal conductance	d.a.	dry air
C_D	correction factor for degree days	dB	decibel(s)
C_F	correction factor for equipment part load efficiency	E	edge factor
		E	efficiency
CF	contact factor	EER	energy efficiency ratio
CFM	ft³/min	E.L.	equivalent length
CHW	chilled water	EST	effective surface temperature
CLF	cooling load factor	ETD	equivalent temperature difference
CLTD	cooling load temperature difference	F	degrees Fahrenheit
$CLTD_c$	cooling load temperature difference, corrected	F	force

483

TABLE A.1   (Continued)

f	friction factor	Pa	Pascal
f	correction factor for ceiling ventilation	p	pressure
ft	foot/feet	p_{abs}	absolute pressure
GLF	glass load factor	p_{atm}	atmospheric pressure
GPM	gallons per minute	p_g	gage pressure
g	gravitational constant	p_{vac}	vacuum pressure
g	gram(s)	psf	lb/ft^2
gal	gallon(s)	psi	lb/in.^{2}
gr	grain(s)	psia	lb/in.^{2} absolute
H	total enthalpy	psig	lb/in.^{2} gage
H	head pressure	Q	heat; heat transfer rate
H_e	elevation head	Q_l	latent heat
H_f	friction loss	Q_s	sensible heat
H_p	pump or fan head	Q_t	total heat
H_s	static head	R	degrees Rankine
$\mathbf{H}_t$	total head	R	thermal resistance
H_v	velocity head	R	gas constant
Hg	mercury	R	recovery factor
Hz	frequency	RH	relative humidity
HP	horsepower	RLCL	room latent cooling load
HW	hot water	RSCL	room sensible cooling load
h	specific enthalpy	RSHR	room sensible heat ratio
h_f	specific enthalpy of saturated liquid	RTCL	room total cooling load
h_{fg}	latent heat of vaporization	SC	shade coefficient
h_g	specific enthalpy of saturated vapor	SEER	seasonal energy efficiency ratio
hr	hour	SHGF	solar heat gain factor
in.	inch(es)	SHR	sensible heat ratio
J	Joule	SPR	static pressure regain
K	unit length conductance	s.g.	specific gravity
K	correction factor for color of surface	sec	second(s)
K	degrees Kelvin	T	temperature, absolute
KW	kilowatt(s)	TC	temperature change
k	thermal conductivity	TD	temperature difference
kg	kilogram(s)	t	temperature
L	length	U	overall heat transfer coefficient
LF	latent factor	U_o	overall thermal performance
LM	correction factor for latitude and month	V	volume
lb	pound(s)	V	velocity
ME	mechanical efficiency	VFR	volume flow rate
m	mass	v	specific volume
m	meter(s)	W	work
min	minute(s)	W	watt(s)
N	newton	W	humidity ratio
N	speed, revolutions per minute	W'	humidity ratio, gr/lb d.a.
NC	noise criteria	WB	wet bulb temperature
NPSH	net positive suction head	WHP	water horsepower
n	number of people	w	weight
OTTV	overall thermal transfer value	w.	water
P	power	w.g.	water gage

TABLE A.2 UNIT EQUIVALENTS (CONVERSION FACTORS)

To change from one set of units to another, multiply known quantity and unit by the ratio of unit equivalents that results in the desired units.

LENGTH
U.S.: 12 in. = 1 ft = 0.333 yd
metric: 1 m = 100 cm
$\quad\quad$ = 1000 mm = 10^{-3}
$\quad$ km = 10^6 microns
U.S.—metric: 1 ft = 0.30 m
SI unit is the m

AREA
U.S.: 144 in.^{2} = 1 ft^2
U.S.—metric: 1 ft^2 = 0.093 m^2
SI unit is the m^2

VOLUME
U.S.: 1728 in.^{3} = 1 ft^3 = 7.48 gal
U.S.—metric: 1 ft^3 = 0.0283 m^3
SI unit is the m^3

MASS
U.S.: 1 lb = 16 oz = 7000 gr
metric: 1 kg = 1000 g
U.S.—metric: 2.2 lb = 1 kg
SI unit is the kg

FORCE
U.S.—metric: 1 lb = 4.45 N
SI unit is the N

VELOCITY
U.S.: 1 ft/sec = 0.68 mi/hr
SI unit is the m/sec

DENSITY
U.S.—metric: 1 lb/ft^3 = 16.0 kg/m^3
SI unit is the kg/m^3

PRESSURE
U.S.: 1 psi = 2.3 ft w. = 2.04 in. Hg
metric: 1 atm = 101,300 N/m^3
$\quad$ 1 mm Hg = 133.3 Pa
U.S.—metric: 14.7 psi = 1 atm
SI unit is the N/m^2 (Pa)

TEMPERATURE
U.S.: F = R − 460
metric: C = K − 273
U.S.—metric: F = (9 / 5) C + 32
$\quad\quad\quad\quad$ C = 5 / 9(F − 32)
SI unit is the K

ENERGY
U.S.: 1 BTU = 778 ft-lb
metric: 1 J = 1 W-sec = 0.239 cal
U.S.—metric: 1 BTU = 1055 J = 252 cal
SI unit is the J

POWER (RATE OF ENERGY)
2545 BTU/hr = 1 hp = 0.746 KW
$\quad\quad\quad\quad$ = 33,000 ft-lb/min
3410 BTU/hr = 1 KW
1 ton of refrigeration = 12,000 BTU/hr
$\quad\quad\quad\quad$ = 4.72 HP = 3.52 KW

$\quad$ *SI* unit is the W

SPECIFIC HEAT
U.S.—metric: 1 BTU/lb-F = 1 cal/gm C
$\quad\quad\quad\quad$ = 4.2 kJ/kg C

HEAT TRANSFER COEFFICIENT U
U.S.—metric: 1 BTU/hr-ft^2-F = 5.68 W/m^2 C

VOLUME FLOW RATE
U.S.—metric: 1 CFM = 1.70 m^3/hr

APPROXIMATE EQUIVALENTS FOR WATER ONLY (AT 60°F)
Density: 8.33 lb = 1 gal
$\quad\quad$ 62.4 lb = 1 ft^3
Flow rate: 1 GPM = 500 lb/hr

TABLE A.3   PROPERTIES OF SATURATED STEAM AND SATURATED WATER

		Temperature Table Specific vol, cu ft/lb		Enthalpy, BTU/lb					Pressure Table Specific vol, cu ft/lb		Enthalpy, BTU/lb		
Temp F	Abs press, psi	Sat liquid v_f	Sat vapor v_g	Sat liquid h_f	Evap h_{fg}	Sat vapor h_g	Abs press, psi	Temp, F	Sat liquid v_f	Sat vapor v_g	Sat liquid h_f	Evap h_{fg}	Sat vapor h_g
32	0.08854	0.01602	3306	0.00	1075.8	1075.8	0.50	79.58	0.01608	641.4	47.6	1048.8	1096.4
35	0.09995	0.01602	2947	3.02	1074.1	1077.1	1.0	101.74	0.01614	333.6	69.7	1036.3	1106.0
40	0.12170	0.01602	2444	8.05	1071.3	1079.3	2.0	126.08	0.01623	173.73	94.0	1022.2	1116.2
45	0.14752	0.01602	2036.4	13.06	1068.4	1081.5	3.0	141.48	0.01630	118.71	109.4	1013.2	1122.6
50	0.17811	0.01603	1703.2	18.07	1065.6	1083.7	4.0	152.97	0.01636	90.63	120.9	1006.4	1127.3
55	0.2141	0.01603	1430.7	23.07	1062.7	1085.8	5.0	162.24	0.01640	73.52	130.1	1001.0	1131.1
60	0.2563	0.01604	1206.7	28.06	1059.9	1088.0	6.0	170.06	0.01645	61.98	138.0	996.2	1134.2
65	0.3056	0.01605	1021.4	33.05	1057.1	1090.2	7.0	176.85	0.01649	53.64	144.8	992.1	1136.9
70	0.3631	0.01606	867.9	38.04	1054.3	1092.3	8.0	182.86	0.01653	47.34	150.8	988.5	1139.3
75	0.4298	0.01607	740.0	43.03	1051.5	1094.5	9.0	188.28	0.01656	42.40	156.2	985.2	1141.4
80	0.5069	0.01608	633.1	48.02	1048.6	1096.6	10	193.21	0.01659	38.42	161.2	982.1	1143.3
85	0.5959	0.01609	543.5	53.00	1045.8	1098.8	14.7	212.00	0.01672	26.80	180.0	970.4	1150.4
90	0.6982	0.01610	468.0	57.99	1042.9	1100.9	20	227.96	0.01683	20.089	196.2	960.1	1156.3
95	0.8153	0.01612	404.3	62.98	1040.1	1103.1	25	240.07	0.01692	16.303	208.5	952.1	1160.6
100	0.9492	0.01613	350.4	67.97	1037.2	1105.2	30	250.33	0.01701	13.746	218.8	945.3	1164.1
105	1.1016	0.01615	304.5	72.95	1034.3	1107.3	40	267.25	0.01715	10.498	236.0	933.7	1169.7
110	1.2748	0.01617	265.4	77.94	1031.6	1109.5	50	281.01	0.01727	8.515	250.1	924.0	1174.1
115	1.4709	0.01618	231.9	82.93	1028.7	1111.6	60	292.71	0.01738	7.175	262.1	915.5	1177.6
120	1.6924	0.01620	203.27	87.92	1025.8	1113.7	70	302.92	0.01748	6.206	272.6	907.9	1180.6
125	1.9420	0.01622	178.61	92.91	1022.9	1115.8	80	312.03	0.01757	5.472	282.0	901.1	1183.1
130	2.2225	0.01625	157.34	97.90	1020.0	1117.9	90	320.27	0.01766	4.896	290.6	894.7	1185.3
135	2.5370	0.01627	138.95	102.9	1017.0	1119.9	100	327.81	0.01774	4.432	298.4	888.8	1187.2
140	2.8886	0.01629	123.01	107.9	1014.1	1122.0	110	334.77	0.01782	4.049	305.7	883.2	1188.9
145	3.281	0.01632	109.15	112.9	1011.2	1124.1	120	341.25	0.01789	3.728	312.4	877.9	1190.4
150	3.718	0.01634	97.07	117.9	1008.2	1126.1	130	347.32	0.01796	3.455	318.8	872.9	1191.7
155	4.203	0.01637	86.52	122.9	1005.2	1128.1	140	353.02	0.01802	3.220	324.8	868.2	1193.0
160	4.741	0.01639	77.29	127.9	1002.3	1130.2	150	358.42	0.01809	3.015	330.5	863.6	1194.1
165	5.335	0.01642	69.19	132.9	999.3	1132.2	160	363.53	0.01815	2.834	335.9	859.2	1195.1
170	5.992	0.01645	62.06	137.9	996.3	1134.2	170	368.41	0.01822	2.675	341.1	854.9	1196.0
175	6.715	0.01648	55.78	142.9	993.3	1136.2	180	373.06	0.01827	2.532	346.1	850.8	1196.9
180	7.510	0.01651	50.23	147.9	990.2	1138.1	190	377.51	0.01833	2.404	350.8	846.8	1197.6
185	8.383	0.01654	45.31	152.9	987.2	1140.1	200	381.79	0.01839	2.288	355.4	843.0	1198.4
190	9.339	0.01657	40.96	157.9	984.1	1142.0	250	400.95	0.01865	1.8438	376.0	825.1	1201.1
200	11.526	0.01663	33.64	168.0	977.9	1145.9	300	417.33	0.01890	1.5433	393.8	809.0	1202.8
212	14.696	0.01672	26.80	180.0	970.4	1150.4	350	431.72	0.01913	1.3260	409.7	794.2	1203.9
220	17.186	0.01677	23.15	188.1	965.2	1153.4	400	444.59	0.0193	1.1613	424.0	780.5	1204.5
240	24.969	0.01692	16.323	208.3	952.2	1160.5	450	456.28	0.0195	1.0320	437.2	767.4	1204.6
260	35.429	0.01709	11.763	228.6	938.7	1167.3	500	467.01	0.0197	0.9278	449.4	755.0	1204.4
280	49.203	0.01726	8.645	249.1	924.7	1173.8	600	486.21	0.0201	0.7698	471.6	731.6	1203.2
300	67.013	0.01745	6.466	269.6	910.1	1179.7	700	503.10	0.0205	0.6554	491.5	709.7	1201.2
350	134.63	0.01799	3.342	321.6	870.7	1192.3	800	518.23	0.0209	0.5687	509.7	688.9	1198.6
400	247.31	0.01864	1.8633	375.0	826.0	1201.0	900	531.98	0.0212	0.5006	526.6	668.8	1195.4
450	422.6	0.0194	1.0993	430.1	774.5	1204.6	1000	544.61	0.0216	0.4456	542.4	649.4	1191.8
500	680.8	0.0204	0.6749	487.8	713.9	1201.7	1200	567.22	0.0223	0.3619	571.7	611.7	1183.4
550	1045.2	0.0218	0.4240	549.3	640.8	1190.0	1500	596.23	0.0235	0.2760	611.6	556.3	1167.9

TABLE A.4 THERMAL RESISTANCE *R* OF BUILDING AND INSULATING MATERIALS
(hr · ft²-F/BTU)

Description		Density lb/ft³	Resistance (*R*)	
			Per Inch	Per Listed Thickness
BUILDING BOARD				
Boards, Panels, Subflooring, Sheathing				
Woodboard Panel Products				
Asbestos-cement board		120	0.25	--
Asbestos-cement board	0.125 in.	120	--	0.03
Asbestos-cement board	0.25 in.	120	--	0.06
Gypsum or plaster board	0.375 in.	50	--	0.32
Gypsum or plaster board	0.5 in.	50	--	0.45
Gypsum or plaster board	0.625 in.	50	--	0.56
Plywood (Douglas Fir)		34	1.25	--
Plywood (Douglas Fir)	0.25 in.	34	--	0.31
Plywood (Douglas Fir)	0.375 in.	34	--	0.47
Plywood (Douglas Fir)	0.5 in.	34	--	0.62
Plywood (Douglas Fir)	0.625 in.	34	--	0.77
Plywood or wood panels	0.75 in.	34	--	0.93
Vegetable fiber board				
Sheathing, regular density	0.5 in.	18	--	1.32
	0.78125 in.	18	--	2.06
Sheathing, intermediate density	0.5 in.	22	--	1.22
Nail-base sheathing	0.5 in.	25	--	1.14
Shingle backer	0.375 in.	18	--	0.94
Shingle backer	0.3125 in.	18	--	0.78
Sound deadening board	0.5 in.	15	--	1.35
Tile and lay-in panels, plain or acoustic		18	2.50	--
	0.5 in.	18	--	1.25
	0.75 in.	18	--	1.89
Laminated paperboard		30	2.00	--
Homogeneous board from repulped paper		30	2.00	--
Hardboard				
Medium density		50	1.37	--
High density, service temp. service underlay		55	1.22	--
High density, std. tempered		63	1.00	--
Particleboard				
Low density		37	1.85	--
Medium density		50	1.06	--
High density		62.5	0.85	--
Underlayment	0.625 in.	40	--	0.82
Wood subfloor	0.75 in.		--	0.94
BUILDING MEMBRANE				
Vapor—permeable felt		--	--	0.06
Vapor—seal, 2 layers of mopped 15-lb felt		--	--	0.12
Vapor—seal, plastic film		--	--	Negl.
FINISH FLOORING MATERIALS				
Carpet and fibrous pad		--	--	2.08
Carpet and rubber pad		--	--	1.23
Cork tile	0.125 in.	--	---	0.28
Terrazzo	1 in.	--	--	0.08
Tile—asphalt, linoleum, vinyl, rubber		--	--	0.05
vinyl asbestos				
ceramic				
Wood, hardwood finish	0.75 in.			0.68
INSULATING MATERIALS				
Blanket and Batt				
Mineral fiber, fibrous form processed				
from rock, slag, or glass				
approx. 2–2.75 in.		0.3–2.0	--	7
approx. 3–3.5 in.		0.3–2.0	--	11
approx. 3.50–6.5 in.		0.3–2.0	--	19
approx. 6–7 in.		0.3–2.0		22
approx. 8.5 in.		0.3–2.0		30

TABLE A.4   (Continued)

Description	Density lb/ft³	Resistance (R) Per Inch	Resistance (R) Per Listed Thickness
Board and Slabs			
Cellular glass..	8.5	2.63	--
Glass fiber, organic bonded.................................	4–9	4.00	--
Expanded rubber (rigid)......................................	4.5	4.55	--
Expanded polystyrene extruded			
Cut cell surface...	1.8	4.00	--
Expanded polystyrene extruded			
Smooth skin surface......................................	2.2	5.00	--
Expanded polystyrene extruded			
Smooth skin surface......................................	3.5	5.26	--
Expanded polystyrene, molded beads...................	1.0	3.57	--
Expanded polyurethane (R-11 exp.).....................	1.5	6.25	--
(Thickness 1 in. or greater).............................	2.5		
Mineral fiber with resin binder...........................	15	3.45	--
Mineral fiberboard, wet felted			
Core or roof insulation...................................	16–17	2.94	--
Acoustical tile..	18	2.86	--
Acoustical tile..	21	2.70	--
Mineral fiberboard, wet molded			
Acoustical tile..	23	2.38	--
Wood or cane fiberboard			
Acoustical tile .... 0.5 in.	--	--	1.25
Acoustical tile .... 0.75 in.	--	--	1.89
Interior finish (plank, tile).................................	15	2.86	--
Wood shredded (cemented in preformed slabs)........	22	1.67	--
LOOSE FILL			
Cellulosic insulation (milled paper or wood pulp)......	2.3–3.2	3.13–3.70	--
Sawdust or shavings..	8.0–15.0	2.22	--
Wood fiber, softwoods......................................	2.0–3.5	3.33	--
Perlite, expanded...	5.0–8.0	2.70	--
Mineral fiber (rock, slag, or glass)			
approx. 3.75–5 in	0.6–2.0		11
approx. 6.5–8.75 in	0.6–2.0		19
approx. 7.5–10 in	0.6–2.0		22
approx. 10.25–13.75 in.	0.6–2.0		30
Vermiculite, exfoliated	7.0–8.2	2.13	--
	4.0–6.0	2.27	--
Roof Insulation			
Preformed, for use above deck			1.39
Different roof insulations are available in different			
thicknesses to provide the design C values listed.			to
Consult individual manufacturers for actual			
thickness of their material			8.33
MASONRY MATERIALS			
Concretes			
Cement mortar..	116	0.20	--
Gypsum-fiber concrete 87.5% gypsum,			
12.5% wood chips..	51	0.60	--
Lightweight aggregates including expanded	120	0.19	--
shale, clay, or slate; expanded slags;	100	0.28	--
cirders; pumice; vermiculite; also	80	0.40	--
cellular concretes	60	0.59	--
	40	0.86	--
	30	1.11	--
	20	1.43	
Perlite, expanded...	40	1.08	
	30	1.41	
	20	2.00	
Sand and gravel or stone aggregate (oven dried)	140	0.11	
Sand and gravel or stone aggregate (not dried)...........	140	0.08	
Stucco..	116	0.20	

TABLE A.4 (Continued)

Description		Density lb/ft³	Resistance (R)	
			Per Inch	Per Listed Thickness
MASONRY UNITS				
Brick, common...		120	0.20	--
Brick, face		130	0.11	--
Clay tile, hollow:				
1 cell deep.......................................	3 in.	--	--	0.80
1 cell deep.......................................	4 in.	--	--	1.11
2 cells deep	6 in.	--	--	1.52
2 cells deep	8 in.	--	--	1.85
2 cells deep	10 in.	--	--	2.22
3 cells deep	12 in.	--	--	2.50
Concrete blocks, three oval core:				
Sand and gravel aggregate	4 in.	--	--	0.71
....	8 in.	--	--	1.11
....	12 in.	--	--	1.28
Cinder aggregate	3 in.	--	--	0.86
....	4 in.	--	--	1.11
....	8 in.	--	--	1.72
....	12 in.	--	--	1.89
Lightweight aggregate	3 in.	--	--	1.27
(expanded shale, clay, slate,	4 in.	--	--	1.50
or slag; pumice)	8 in.	--	--	2.00
....	12 in.	--	--	2.27
Concrete blocks, rectangular core:				
Sand and gravel aggregate				
2 core, 8 in. 36 lb...		--	--	1.04
Same with filled cores		--	--	1.93
Lightweight aggregate (expanded shale,				
clay, slate, or slag; pumice):				
3 core, 6 in. 19 lb...		--	--	1.65
Same with filled cores		--	--	2.99
2 core, 8 in. 24 lb...		--	--	2.18
Same with filled cores		--	--	5.03
3 core, 12 in. 38 lb...		--	--	2.48
Same with filled cores		--	--	5.82
Stone, lime or sand..		--	0.08	--
Gypsum partition tile:				
3 × 12 × 30 in. solid		--	--	1.26
3 × 12 × 30 in. 4-cell		--	--	1.35
4 × 12 × 30 in. 3-cell		--	--	1.67
PLASTERING MATERIALS				
Cement plaster, sand aggregate		116	0.20	--
Sand aggregate	0.375 in.	--	--	0.08
Sand aggregate	0.75 in.	--	--	0.15
Gypsum plaster:				
Lightweight aggregate	0.5 in.	45	--	0.32
Lightweight aggregate	0.625 in.	45	--	0.39
Lightweight aggregate on metal lath	0.75 in.	--	--	0.47
Perlite aggregate		45	0.67	--
Sand aggregate		105	0.18	--
Sand aggregate..	0.5 in.	105	--	0.09
Sand aggregate..	0.625 in.	105	--	0.11
Sand aggregate on metal lath	0.75 in.	--	--	0.13
Vermiculite aggregate		45	0.59	--
ROOFING				
Asbestos-cement shingles		120	--	0.21
Asphalt roll roofing		70	--	0.15
Asphalt shingles		70	--	0.44
Built-up roofing	0.375 in.	70	--	0.33
Slate	0.5 in.	--	--	0.05
Wood shingles, plain and plastic film faced..............		--	--	0.94

TABLE A.4 (Continued)

Description	Density lb/ft³	Resistance (R) Per Inch	Resistance (R) Per Listed Thickness
SIDING MATERIALS (on flat surface)			
Shingles			
Asbestos-cement	120	--	0.21
Wood, 16 in., 7.5 exposure............................	--	--	0.87
Wood, double, 16 in., 12 in. exposure..................	--	--	1.19
Wood, plus insul. backer board, 0.3125 in	--	--	1.40
Siding			
Asbestos-cement, 0.25 in., lapped....................	--	--	0.21
Asphalt roll siding..................................	--	--	0.15
Asphalt insulting siding (0.5 in. bed.)..............	--	--	1.46
Wood drop, 1 x 8 in..................................	--	--	0.79
Wood, bevel, 0.5 x 8 in., lapped....................	--	--	0.81
Wood, bevel, 0.75 x 10 in., lapped..................	--	--	1.05
Wood, plywood, 0.375 in., lapped....................	--	--	0.59
Wood, medium density siding, 0.4375 in..............	40	0.67	
Aluminum or steel, over sheathing			
Hollow-backed	--	--	0.61
Insulating-board backed nominal			
0.375 in..	--	--	1.82
Insulating-board backed nominal			
0.375 in., foil backed............................			2.96
Architectural glass................................	--	--	0.10
WOODS			
Maple, oak, and similar hardwoods	45	0.91	--
Fir, pine, and similar softwoods	32	1.25	--
Fir, pine, and similar softwoods........... 0.75 in.	32	--	0.94
.... 1.5 in.		--	1.89
.... 2.5 in.		--	3.12
.... 3.5 in.		--	4.35

TABLE A.5 THERMAL RESISTANCE *R* OF SURFACE AIR FILMS AND AIR SPACES (hr · ft²-F/BTU)

SURFACE AIR FILMS

	Direction of Heat Flow	*R*-Value
STILL AIR (interior surfaces)		
Horizontal	Upward	0.61
Sloping–45 degree	Upward	0.62
Vertical	Horizontal	0.68
Sloping–45 degree	Downward	0.76
Horizontal	Downward	0.92
MOVING AIR (exterior surfaces)		
15 mph Wind (Winter)	Any	0.17
7.5 mph Wind (Summer)	Any	0.25

AIR SPACES

Position of Air Space	Direction of Heat Flow	Thickness of Air Space ½″	¾″	1½″	3½″
		R-Value			
Horizontal	Up	0.84	0.87	0.89	0.93
45° Slope	Up	0.90	0.94	0.91	0.96
Vertical	Horizontal	0.91	1.01	1.02	1.01
Horizontal	Down	0.92	1.02	1.14	1.21
45° Slope	Down	0.92	1.02	1.09	1.05

TABLE A.6 TYPICAL BUILDING ROOF AND WALL CONSTRUCTION CROSS-SECTIONS AND OVERALL HEAT TRANSFER COEFFICIENTS, BTU/HR-FT²-F

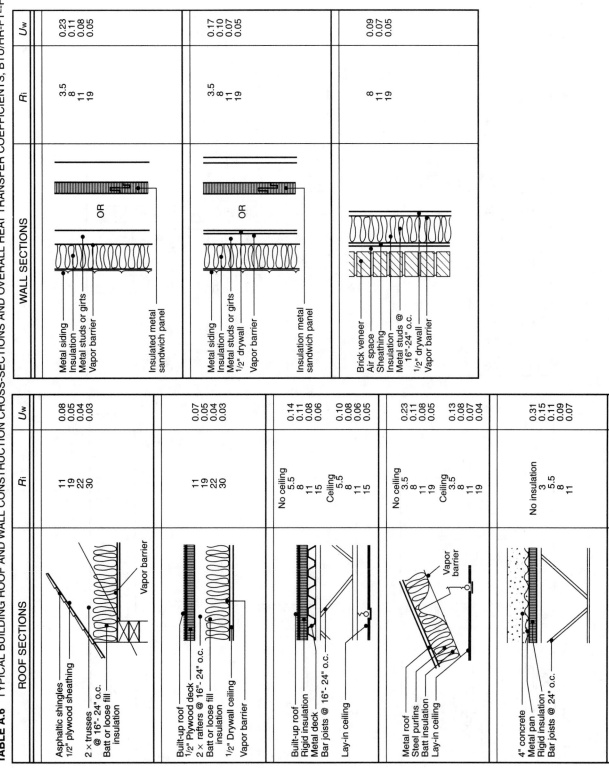

ROOF SECTIONS

	R_i	U_w
Asphaltic shingles 1/2" plywood sheathing 2 × trusses @ 16"- 24" o.c. Batt or loose fill insulation Vapor barrier	11 19 22 30	0.08 0.05 0.04 0.03
Built-up roof 1/2" Plywood deck 2 × rafters @ 16"- 24" o.c. Batt or loose fill insulation 1/2" Drywall ceiling Vapor barrier	11 19 22 30	0.07 0.05 0.04 0.03
Built-up roof Rigid insulation Metal deck Bar joists @ 16"- 24" o.c. Lay-in ceiling	No ceiling 5.5 8 11 15 Ceiling 5.5 8 11 15	0.14 0.11 0.08 0.06 0.10 0.08 0.06 0.05
Metal roof Steel purlins Batt insulation Lay-in ceiling Vapor barrier	No ceiling 3.5 8 11 19 Ceiling 3.5 8 11 19	0.23 0.11 0.08 0.05 0.13 0.08 0.07 0.04
4" concrete Metal pan Rigid insulation Bar joists @ 24" o.c.	No insulation 3 5.5 8 11	0.31 0.15 0.11 0.09 0.07

WALL SECTIONS

	R_i	U_w
Metal siding Insulation Metal studs or girts Vapor barrier OR Insulated metal sandwich panel	3.5 8 11 19	0.23 0.11 0.08 0.05
Metal siding Insulation Metal studs or girts 1/2" drywall Vapor barrier OR Insulation metal sandwich panel	3.5 8 11 19	0.17 0.10 0.07 0.05
Brick veneer Air space Sheathing Insulation Metal studs @ 16"-24" o.c. 1/2" drywall Vapor barrier	8 11 19	0.09 0.07 0.05

491

TABLE A.6 (Continued)

WALL SECTIONS	DENSITY	R_i	U_w
Face brick, Air space, 8" Conc. block, Core insulation	Conc. Block 80#/cu ft	no insulation core insulation	0.23 0.12
Face brick, Air space, Rigid insulation, 8" Conc. block, Core insulation	Conc. Block 120#/cu ft	no insulation core insulation	0.27 0.16
	Conc. Block 80#/cu ft	Core insulation and R_i 3 5.5 8 11	0.09 0.07 0.06 0.05
	Conc. Block 120#/cu ft	Core insulation and R_i 3 5.5 8 11	0.11 0.09 0.07 0.06
Face brick, Air space, 8" Conc. block, Core insulation, Insulation, 1/2" drywall, Vapor barrier / Rigid insulation, Vapor barrier	Conc. Block 80#/cu ft	Core insulation and R_i 3.5 (1 × 2 @ 16" o.c.) 5.5 8 (2 × 3 @ 16" o.c.)	0.08 0.07 0.06
	Conc. Block 120#/cu ft	Core insulation and R_i 3.5 (1 × 2 @ 16" o.c.) 5.5 8 (2 × 3 @ 16" o.c.)	0.10 0.08 0.07

WALL SECTIONS	R_i	U_w
Wood or metal siding, Sheathing, Insulation, 2 × 4 studs @ 16"-24" o.c., 1/2" drywall, Vapor barrier OR Rigid insulation	8 11 14	0.09 0.08 0.07
2 × 6 studs, same construction as above	8 11 19	0.09 0.07 0.05
Face brick, Air space, Sheathing, Insulation, 2 × 4 studs @ 16"-24" o.c., 1/2" drywall, Vapor barrier OR Rigid insulation	8 11 19	0.09 0.07 0.06
2 × 6 studs, same construction as above	8 11 19	0.08 0.07 0.05

TABLE A.6 (Continued)

WALL SECTIONS	DENSITY	W	R_i	U_w
 Concrete	80#/cu ft	6" 8" 12"	— — —	0.31 0.25 0.18
	120#/cu ft	6" 8" 12"	— — —	0.50 0.42 0.32
 Concrete Insulation Metal furring/studs @ 16" to 24" o.c. 1/2" drywall Vapor barrier OR Rigid insulation Vapor barrier	80#/cu ft	6"	3 5.5 8 11	0.15 0.11 0.09 0.07
		8"	3 5.5 8 11	0.13 0.10 0.08 0.06
		12"	3 5.5 8 11	0.11 0.09 0.07 0.06
	120#/cu ft	6"	3 5.5 8 11	0.18 0.13 0.10 0.07
		8"	3 5.5 8 11	0.17 0.12 0.09 0.07
		12"	3 5.5 8 11	0.15 0.11 0.09 0.07

WALL SECTIONS	DENSITY	W	R_i	U_w
 Concrete block	Conc. Block 80#/cu ft	6" 8" 12"	— — —	.37 .34 .29
	Conc. Block 120#/cu ft	6" 8" 12"	— — —	.47 .43 .38
 Concrete block Core insulation	Conc. Block 80#/cu ft	6" 8" 12"	— — —	.18 .14 .10
	Conc. Block 120#/cu ft	6" 8" 12"	— — —	.26 .22 .17
 Concrete block Core insulation Batt or rigid insulation 1/2" drywall Vapor barrier	Conc. Block 80#/cu ft	6"	3.5 5.5 8	.11 .09 .07
		8"	3.5 5.5 8	.09 .08 .05
		12"	3.5 5.5 8	.07 .06 .05
	Conc. Block 120#/cu ft	6"	3.5 5.5 8	.13 .10 .08
		8"	3.5 5.5 8	.12 .09 .06
		12"	3.5 5.5 8	.10 .08 .06

TABLE A.7 OVERALL HEAT TRANSFER COEFFICIENT *U* FOR BUILDING CONSTRUCTION COMPONENTS, BTU/HR-FT²-F

Construction	*U*-Value in BTU/hr-ft²-F	
	Summer	Winter
WALLS		
Frame with wood siding, sheathing, and inside finish:		
No insulation	.22	.23
R-7 insulation (2 in.–2½ in.)	.09	.09
R-11 insulation (3 in.–3½ in.)	.07	.07
Frame with 4 in. brick or stone veneer, sheathing, and inside finish:		
No insulation	.24	.24
R-7 insulation	.09	.09
R-11 insulation	.07	.07
Frame with 1 in. stucco, sheathing, and inside finish:		
No insulation	.29	.29
R-7 insulation	.10	.10
R-11 insulation	.07	.07
Masonry:		
8 in. concrete block, no finish	.49	.51
12 in. concrete block, no finish	.45	.47
Masonry (8 in. concrete block):		
Inside finish:		
furred gypsum wallboard (½ in.); no insulation	.29	.30
furred, foil-backed gypsum wallboard (½ in.); no insulation	.29	.30
1 in. polystyrene insulation board (*R*-5), and ½ in. gypsum wallboard	.13	.13
Masonry (8 in. cinder block or hollow clay tile):		
Inside finish:		
furred gypsum wallboard (½ in.); no insulation	.25	.25
furred, foil-backed gypsum wallboard (½ in.); no insulation	.17	.17
1 in. polystyrene insulation board (*R*-5), and ½ in. gypsum wallboard	.12	.12
Masonry (4 in. face brick and 8 in. cinder block or 8 in. hollow clay tile):		
Inside finish:		
furred gypsum wallboard (½ in.); no insulation	.22	.22
furred, foil-backed gypsum wallboard (½ in.); no insulation	.15	.16
1 in. polystyrene insulation board (*R*-5), and ½ in. gypsum wallboard	.12	.12
Masonry (12 in. hollow clay tile or 12 in. cinder block):		
Inside finish:		
furred gypsum wallboard (½ in.); no insulation	.24	.24
furred, foil-backed gypsum wallboard (½ in.); no insulation	.16	.17
1 in. polystyrene insulation board (*R*-5), and ½ in. gypsum wallboard	.12	.12
Masonry (4 in. face brick, 4 in. common brick):		
Inside finish:		
furred gypsum wallboard (½ in.); no insulation	.28	.28
furred, foil-backed gypsum wallboard (½ in.); no insulation	.18	.19
1 in. polystyrene insulation board (*R*-5), and ½ in. gypsum wallboard	.13	.13
Masonry (8 in. concrete or 8 in. stone):		
Inside finish:		
furred gypsum wallboard (½ in.); no insulation	.33	.34
furred, foil-backed gypsum wallboard (½ in.); no insulation	.21	.21
1 in. polystyrene insulation board (*R*-5), and ½ in. gypsum wallboard	.14	.14
Metal with vinyl inside finish, *R*-7 (3 in. glass fiber batt)	.14	.14
PARTITIONS		
Frame (½ in. gypsum wallboard one side only): No insulation	.55	.55
Frame (½ in. gypsum wallboard each side):		
No insulation	.31	.31
R-11 insulation	.08	.08
Masonry (4 in. cinder block):		
No insulation, no finish	.40	.40
No insulation, one side furred gypsum wallboard (½ in.)	.26	.26
No insulation, both sides furred gypsum wallboard (½ in.)	.19	.19
One side 1 in. polystyrene insulation board (*R*-5), and ½ in. gypsum wallboard	.13	.13

TABLE A.7 (Continued)

Construction	U-Value in BTU/hr-ft²-F	
	Summer	Winter
CEILING–FLOOR		
Frame (asphalt tile floor, ⅝ in. plywood, 25⁄32 in. wood subfloor, finished ceiling):		
Heat flow up	.23	.23
Heat flow down	.20	.19
Concrete (asphalt tile floor, 4 in. concrete deck, air space, finished ceiling):		
Heat flow up	.34	.33
Heat flow down	.26	.25
ROOF (flat roof, no finished ceiling)		
Steel deck:		
No insulation	.64	.86
1 in. insulation (R-2.78)	.23	.25
2 in. insulation (R-5.56)	.15	.16
1 in. Wood deck:		
No insulation	.40	.48
1 in. insulation (R-2.78)	.19	.21
2 in. insulation (R-5.56)	.12	.13
2.5 in. Wood deck:		
No insulation	.25	.26
1 in. insulation (R-2.78)	.15	.16
2 in. insulation (R-5.56)	.10	.11
4 in. Wood deck:		
No insulation	.17	.18
1 in. insulation (R-2.78)	.12	.12
2 in. insulation (R-5.56)	.09	.09
ROOF–CEILING (flat roof, finished ceiling)		
Steel deck:		
No insulation	.33	.40
1 in. insulation (R-2.78)	.17	.19
2 in. insulation (R-5.56)	.12	.13
1 in. Wood deck:		
No insulation	.26	.29
1 in. insulation (R-2.78)	.15	.16
2 in. insulation (R-5.56)	.11	.11
2.5 in. Wood deck:		
No insulation	.18	.20
1 in. insulation (R-2.78)	.12	.13
2 in. insulation (R-5.56)	.09	.10
4 in. Wood deck:		
No insulation	.14	.15
1 in. insulation (R-2.78)	.10	.10
2 in. insulation (R-5.56)	.08	.08
4 in. Lightweight concrete deck:		
No insulation	.14	.15
6 in. Lightweight concrete deck:		
No insulation	.10	.11
8 in. Lightweight concrete deck:		
No insulation	.08	.09
2 in. Heavyweight concrete deck:		
No insulation	.32	.38
1 in. insulation (R-2.78)	.17	.19
2 in. insulation (R-5.56)	.11	.12
4 in. Heavyweight concrete deck:		
No insulation	.30	.36
1 in. insulation (R-2.78)	.16	.18
2 in. insulation (R-5.56)	.11	.12
6 in. Heavyweight concrete deck:		
No insulation	.28	.33
1 in. insulation (R-2.78)	.16	.17
2 in. insulation (R-5.56)	.11	.12

TABLE A.7   (Continued)

Construction	U-Value in BTU/hr-ft²-F	
	Summer	Winter
ROOF–CEILING (wood frame pitched roof, finished ceiling on rafters)		
No insulation	.28	.29
R-19 insulation (5½ in.–6½ in.)	.05	.05
ROOF–ATTIC–CEILING (attic with natural ventilation)		
No insulation	.15	.29
R-19 insulation (5½ in.–6½ in.)	.04	.05
FLOORS		
Floor over unconditioned space, no ceiling:		
Wood frame:		
No insulation	.33	.27
R-7 insulation (2 in.–2½ in.)	.09	.08
Concrete deck:		
No insulation	.59	.43
R-7 insulation	.10	.09
DOORS		
Solid wood:		
1 in. thick	.61	.64
1½ in. thick	.47	.49
2 in. thick	.42.	.43
Steel:		
1½ in. thick, mineral fiber core	.58	.59
1½ in. thick, polystyrene core	.46	.47
1½ in. thick, urethane foam core	.39	.40

TABLE A.8   OVERALL HEAT TRANSFER COEFFICIENT *U* FOR GLASS
(BTU/HR-FT²-F)   (For glass installed vertically)

Type of Glazing	Type of Frame (Sash)			
	Aluminum (with thermal break)		Wood or Vinyl	
	Winter	Summer	Winter	Summer
Single glass	1.10	1.01	0.98	0.90
Double glass				
⅜ in. air space	0.60	0.56	0.51	0.47
⅜ in. air space E-film	0.48	0.45	0.39	0.37
Triple Glass				
⅜ in. air space	0.46	0.43	0.38	0.36
⅜ in. argon space	0.34	0.33	0.25	0.24

Note: E-film is a reflective coating (E = 0.15).
Abridged with permission from the *1993 ASHRAE Handbook—Fundamentals.*

TABLE A.9 OUTDOOR DESIGN CONDITIONS

| Location, USA | LAT Deg. | Winter | | | Summer | | | |
		Degree Days	99% DB	97.5% DB	2.5% Coinc. DB	WB	Daily DB Range	2.5% WB
ALABAMA								
Birmingham	33	2550	17	21	94	75	21	77
Montgomery	32	2290	22	25	95	76	21	79
ALASKA								
Anchorage	61	10860	−23	−18	68	58	15	59
Fairbanks	64	14280	−51	−47	78	60	24	62
ARIZONA								
Phoenix	33	1770	31	34	107	71	27	75
Tucson	32	1800	28	32	102	66	26	71
ARKANSAS								
Little Rock	34	3220	15	20	96	77	22	79
CALIFORNIA								
Los Angeles	34	1350	37	40	89	70	20	71
Sacramento	38	2420	30	32	98	70	36	71
San Diego	32	1460	42	44	80	69	12	70
San Francisco	37	3000	38	40	71	62	14	62
COLORADO								
Denver	39	6280	−5	1	91	59	28	63
CONNECTICUT								
Hartford	41	6240	3	7	88	73	22	75
DELAWARE								
Wilmington	39	4930	10	14	89	74	20	76
D.C.								
Washington	38	4220	14	17	91	74	18	77
FLORIDA								
Miami	25	210	44	47	90	77	15	79
Tampa	28	680	36	40	91	77	17	79
GEORGIA								
Atlanta	33	2960	17	22	92	74	19	76
Savannah	32	1820	24	27	93	77	20	79
HAWAII								
Honolulu	21	0	62	63	86	73	12	75
IDAHO								
Boise	43	5810	3	10	94	64	31	66
ILLINOIS								
Chicago	41	5880	−3	2	91	74	15	77
Springfield	39	5430	−3	2	92	74	21	77
INDIANA								
Indianapolis	39	5700	−4	2	90	74	22	70
South Bend	41	6440	−3	1	89	73	22	75
IOWA								
Des Moines	41	6590	−10	−5	91	74	23	77
Dubuque	42	7380	−12	−7	88	73	22	75

TABLE A.9   (Continued)

Location, USA	LAT Deg.	Winter			Summer			
		Degree Days	99% DB	97.5% DB	2.5% Coinc. DB	WB	Daily DB Range	2.5% WB
KANSAS								
Topeka	39	5180	0	4	96	75	24	78
Wichita	37	4620	3	7	98	73	23	76
KENTUCKY								
Lexington	38	4680	3	8	91	73	22	76
Louisville	38	4660	5	10	93	74	23	77
LOUISIANA								
Baton Rouge	30	1560	25	29	93	77	19	80
New Orleans	30	1250	29	33	92	78	16	80
MAINE								
Portland	43	7510	−6	−1	84	71	22	72
MARYLAND								
Baltimore	39	4110	10	13	91	75	21	77
MASSACHUSETTS								
Boston	42	5630	6	9	88	71	16	74
Pittsfield	42	7580	−8	−3	84	70	23	72
MICHIGAN								
Detroit	42	6230	3	6	88	72	20	74
Lansing	42	6910	−3	1	87	72	24	74
MINNESOTA								
Minneapolis	44	8380	−16	−12	89	73	22	75
MISSISSIPPI								
Jackson	32	2240	21	25	95	76	21	78
MISSOURI								
Kansas City	39	4710	2	6	96	74	20	77
St. Louis	38	4480	3	8	94	75	18	77
MONTANA								
Billings	45	7050	−15	−10	91	64	31	66
NEBRASKA								
Omaha	41	6610	−8	−3	91	75	22	77
NEVADA								
Las Vegas	36	2710	25	28	106	65	30	70
Reno	39	6330	6	11	93	60	45	62
NEW HAMPSHIRE								
Concord	43	7380	−8	−3	87	70	26	73
NEW JERSEY								
Newark	40	4590	10	14	91	73	20	76
Trenton	40	4980	11	14	88	74	19	76
NEW MEXICO								
Albuquerque	35	4350	12	16	94	61	27	65
NEW YORK								
Albany	42	6880	−4	1	88	72	20	74
Buffalo	43	7060	2	6	85	70	21	73
NYC	40	4870	11	15	89	73	17	75

TABLE A.9 (Continued)

Location, USA	LAT Deg.	Winter			Summer			
		Degree Days	99% DB	97.5% DB	2.5% Coinc. DB	WB	Daily DB Range	2.5% WB
NORTH CAROLINA								
Charlotte	35	3190	18	22	93	74	20	76
Raleigh	35	3390	16	20	92	75	20	77
NORTH DAKOTA								
Bismark	46	8850	−23	−19	91	68	27	71
OHIO								
Cincinnati	39	4410	1	6	90	72	21	75
Cleveland	41	6350	1	5	88	72	22	74
OKLAHOMA								
Okla. City	35	3725	9	13	97	74	23	77
Tulsa	36	3860	8	13	98	75	22	78
OREGON								
Portland	45	4640	18	24	86	67	21	67
PENNSYLVANIA								
Philadelphia	39	4490	10	14	90	74	21	76
Pittsburgh	40	5050	3	7	88	71	19	73
PUERTO RICO								
San Juan	18		67	68	88		11	80
RHODE ISLAND								
Providence	41	5950	5	9	86	72	19	74
SOUTH CAROLINA								
Charleston	32	1790	25	28	92	78	13	80
SOUTH DAKOTA								
Sioux Falls	43	7840	−15	−11	91	72	24	75
TENNESSEE								
Memphis	35	3020	13	18	95	76	21	79
Nashville	36	3580	9	14	94	74	21	77
TEXAS								
Dallas	32	2360	18	22	100	75	20	78
Ft. Worth	32	2410	17	22	99	74	22	77
Houston	29	1280	28	33	95	77	18	79
UTAH								
Salt Lake City	40	6050	3	8	95	62	32	65
VERMONT								
Burlington	44	8270	−12	−7	85	70	23	72
VIRGINIA								
Richmond	37	3870	14	17	92	76	21	78
Roanoke	37	4150	12	16	91	72	23	74
WASHINGTON								
Seattle	47	4420	22	27	82	66	19	67
Spokane	47	6660	−6	2	90	63	28	64
WEST VIRGINIA								
Charleston	38	4480	7	11	90	73	20	75

TABLE A.9   (Continued)

Location, USA	LAT Deg.	Winter			Summer			
		Degree Days	99% DB	97.5% DB	2.5% Coinc. DB	WB	Daily DB Range	2.5% WB
WISCONSIN								
Milwaukee	43	7640	−8	−4	87	73	21	74
WYOMING								
Cheyenne	41	7380	−4	−1	86	58	30	62
Location, Canada								
ALBERTA								
Edmonton	53	10270	−29	−25	82	65	23	66
BRITISH COLUMBIA								
Vancouver	49	5520	15	19	77	66	17	67
MANITOBA								
Winnipeg	49	10680	−30	−27	86	71	22	73
NOVA SCOTIA								
Halifax	44	7360	1	5	76	65	16	67
ONTARIO								
Ottawa	45	8740	−17	−13	87	71	21	73
Toronto	43	6830	−5	−1	87	72	20	74
QUEBEC								
Montreal	45	7900	−16	−10	85	72	17	74
Location, Other Countries	2.5% DB							
ARGENTINA								
Buenos Aires	35s		32	34	89		22	76
AUSTRALIA								
Melbourne	38s		35	38	91		21	69
AUSTRIA								
Vienna	48		6	11	86		16	69
BRAZIL								
Rio de Janeiro	23s		58	60	92		11	79
CHINA								
Shanghai	31		23	26	92		16	81
COLOMBIA								
Bogotá	5		45	46	70		19	59
CUBA								
Havana	23		59	62	91		14	81
EGYPT								
Cairo	30		45	46	100		26	75
ENGLAND								
London	51		24	26	79		16	66
FRANCE								
Paris	49		22	25	86		21	68

TABLE A.9 (Continued)

Location, Other Countries	LAT Deg.	Winter			Summer		
		Degree Days	99% DB	97.5% DB	2.5% DB	Daily DB Range	2.5% WB
GERMANY							
Berlin	52		7	12	81	19	67
INDIA							
New Delhi	29		39	41	107	26	82
IRAN							
Tehran	36		20	24	100	27	74
ISRAEL							
Tel Aviv	32		39	41	93	16	73
ITALY							
Rome	42		30	33	92	24	73
JAPAN							
Toyko	36		26	28	89	14	80
MEXICO							
Mexico City	19		37	39	81	25	60
NIGERIA							
Lagos	6		70	71	91	12	82
POLAND							
Warsaw	52		3	8	81	19	70
RUSSIA							
Moscow	56		−11	−6	81	21	67
SAUDIA ARABIA							
Jedda	21		57	60	103	22	84
SOUTH AFRICA							
Capetown	34s		40	42	90	21	71
SPAIN							
Madrid	40		25	28	91	25	69

Room Heating Load Calculations

Project _____ Location _____ Indoor DB _____ F

Engrs. _____ Calc. by _____ Chk. by _____ Outdoor DB _____ F

Room			
Plan Size			
Heat Transfer	$U \times A \times TD = $ BTU/hr	$U \times A \times TD = $ BTU/hr	$U \times A \times TD = $ BTU/hr
Walls			
Windows			
Doors			
Roof/ceiling			
Floor			
Partition			
Heat Transfer Loss			
Infiltration	$1.1 \times A \times B \times TC = $ (CFM)	$1.1 \times A \times B \times TC = $ (CFM)	$1.1 \times A \times B \times TC = $ (CFM)
Window	1.1	1.1	1.1
Door	1.1	1.1	1.1
Infiltration Heat Loss			
Room Heating Load			

Room			
Plan Size			
Heat Transfer	$U \times A \times TD = $ BTU/hr	$U \times A \times TD = $ BTU/hr	$U \times A \times TD = $ BTU/hr
Walls			
Windows			
Doors			
Roof/ceiling			
Floor			
Partition			
Heat Transfer Loss			
Infiltration	$1.1 \times A \times B \times TC = $ (CFM)	$1.1 \times A \times B \times TC = $ (CFM)	$1.1 \times A \times B \times TC = $ (CFM)
Window	1.1	1.1	1.1
Door	1.1	1.1	1.1
Infiltration Heat Loss			
Room Heating Load			

Infiltration CFM	Column A	Column B
Windows	CFM per ft	Crack length, ft
Doors	CFM per ft^2	Area, ft^2

Figure A.1
Room heating load calculations form.

Building Heating Load Calculations

		DB, F	W′, gr/lb
Indoor			
Outdoor			
Diff.			

Project _____

Location _____ Calc. by _____

Engineers _____ Chk. by _____

Heat Transfer	$Q =$	U	$\times$	A	$\times$	TD	$=$	BTU/hr
Roof								
Walls								
Windows								
Doors								
Floor								

	BTU/hr
Heat Transfer Subtotal	
Infiltration $Q_s = 1.1 \times$ _____ CFM $\times$ _____ TC $=$	
Building Net Load	
Ventilation $Q_s = 1.1 \times$ _____ CFM $\times$ _____ TC $=$	
$Q_L = 0.68 \times$ _____ CFM $\times$ _____ gr/lb $=$	
Duct Heat Loss _____ %	
Duct Heat Leakage _____ %	
Piping and Pickup Allowance _____ %	
Service HW Load	
Boiler or Furnace Gross Load	

Figure A.2
Building heating load calculations form.

COMMERCIAL COOLING LOAD CALCULATIONS

Project _____ Bldg./Room _____

Location _____ Engrs. _____ Calc. by _____ Chk. by _____

		DB F	WB F	RH %	W' gr/lb	
Design Conditions	Outdoor					
	Room					

Daily Range _____ F   Ave. _____ F
Day _____   Time _____
Lat. _____

Conduction	Dir.	Color	U	A, ft^2		CLTD, F		SCL BTU/hr
				Gross	Net	Table	Corr.	
Glass								
Wall (Group)								
Roof/ceiling								
Floor								
Partition								
Door								

Solar	Dir.	Sh.	SHGF	A	SC	CLF	
Glass							

$$RSHR = \frac{RSCL}{RTCL} =$$
$$CFM_{sa} = \frac{RSCL}{1.1 \times (t_r - t_{sa})} =$$

Lights _____ W × 3.41 × _____ BF × _____ CLF
Lights _____ W × 3.41 × _____ BF × _____ CLF
People _____ SHG × _____ n × _____ CLF
_____ LHG × _____ n
Equipment _____ _____
Equipment _____ _____
Infiltration 1.1 × _____ CFM × _____ TC
0.68 × _____ CFM × _____ gr/lb

Subtotal

SA duct gain _____
SA duct leakage _____
_____
SA fan gain (draw through) _____

Room/Building Cooling Load

SA fan gain (blow through) _____
Ventilation 1.1 × _____ CFM × _____ TC
0.68 × _____ CFM × _____ gr/lb
RA duct gain _____
RA fan gain _____

Cooling Coil Load

Pump gain _____

Refrigeration Load

LCL BTU/hr

Total CL BTU/hr

Figure A.3
Commercial cooling load calculations form.

RESIDENTIAL COOLING LOAD CALCULATIONS

Project _____
Location _____

Out. DB _____ F
Out. WB _____ F

In. DB _____ F
In. RH _____ %

D.R. _____ F
ACH _____

Room Name																				
Plan Size																				
Wall	D.	U	A	CLTD	BTU/hr	D.	U	A	CLTD	BTU/hr	D.	U	A	CLTD	BTU/hr	D.	U	A	CLTD	BTU/hr
Roof/ceiling																				
Floor																				
Partition																				
Door																				
Windows	D.			CLF		D.			CLF		D.			CLF		D.			CLF	
Infiltration																				
People																				
Appliances																				
RSCL																				

Room Name																				
Plan Size																				
Wall	D.	U	A	CLTD	BTU/hr	D.	U	A	CLTD	BTU/hr	D.	U	A	CLTD	BTU/hr	D.	U	A	CLTD	BTU/hr
Roof/ceiling																				
Floor																				
Partition																				
Door																				
Windows	D.			CLF		D.			CLF		D.			CLF		D.			CLF	
Infiltration																				
People																				
Appliances																				
RSCL																				

Building Total

NOTES:

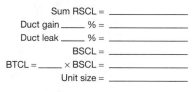

Sum RSCL = _____
Duct gain ____ % = _____
Duct leak ____ % = _____
BSCL = _____
BTCL = ____ × BSCL = _____
Unit size = _____

Figure A.4
Residential cooling load calculations form.

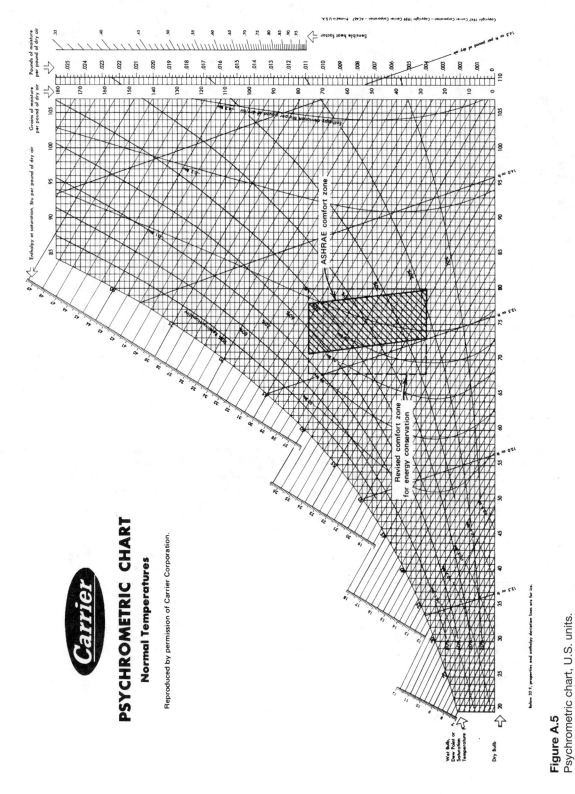

Figure A.5
Psychrometric chart, U.S. units.

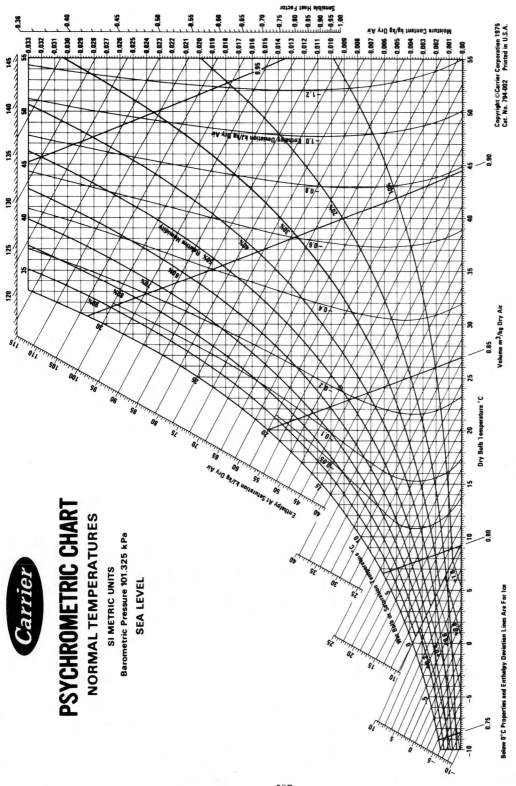

Figure A.6
Psychrometric chart, SI units.

Index

Absolute pressure, 21
Absorption refrigeration, 348
  capacity control, 351
  crystallization, 352
  cycle, 348
  equipment, 349
  installation, 352
  performance, 350
Adiabatic process, 179
Air, dry and moist, 162
Air change method, 56
Air cleaning:
  methods, 319
  purposes, 319
Air conditioning, scope, 2
Air conditioning systems:
  components, 3–6
  energy required, 324
  types, 4, 7, 304
Air-cooled condenser, 349
Air distribution, 270
  devices, 272
Air filters:
  selection, 322
  testing, 320
  types, 321
Airflow, ducts, 216
Air handling units, 316
Air mixing, 180
Air pollution, 85, 323
Air velocity:
  ducts, 233
  measurement, 420
  rooms, 270
Air-water systems, 304
All-air systems, 6, 313
All-water systems, 4, 313

Altitude, solar, 461
Anemometer, 421
Apparatus dew point, 189
Appliance load, 138
Aspect ratio, 218
Atmospheric air, 162
Azimuth, solar, 461

Balancing, 425
  air systems, 427
  hydronic systems, 428
Barometer, 24
Baseboard radiation, 112
Bernoulli equation (generalized), 200
Boilers, 73–77
  accessories, 77
  controls, 89
  energy conservation, 96
  installation, 96
  rating and selection, 92
  types, 73–75
Boiling point, 30–32
Built-up system, 7
Bypass factor, 189

Calculation form:
  commercial cooling, 149, 504
  heating, 62, 63, 502, 503
  residential cooling, 157, 505
Capillary tube, 337
Cavitation, 297
Central system, 7
Centrifugal compressor, 335
Centrifugal pump, 285
Change of state, 29
Chillers, 341
  installation, 346

selection, 340
Clearness factor, 462
Coefficient of heat transfer, 49
Coefficient of performance:
  cooling, 342, 393
  heating, 355, 395
Cogeneration, 405
Coil:
  cooling, 317
  selection, 318
Coil process line, 186
Combustion, 80
Comfort, 7
  chart, 9
Compression tank, 298
Compressors, 333
  centrifugal, 335
  helical, 334
  reciprocating, 333
  rotary, 334
  scroll, 335
Computers, 14, 410
Condensation, 30
  on surfaces, 170
Condensers, 336
Condensing units, 339
  selection, 340
Conductance, 48
Conduction, 45, 121
Conductivity, 48
Contact factor, 189
Continuity equation, 197
Controls:
  automatic, 363
  component diagram, 367
  devices, 371
  feedback, 366

functional block diagram, 365
purposes, 364
types of action, 368
valves, 374
Convection, 8, 45
Convectors, 106
Conversion factors, 17
  table, 485
Cooling coils, 317
Cooling load, 118
Cooling load factor:
  commercial, 130
  residential, 152
Cooling load temperature difference
    (CLTD), 121, 150
Cooling towers, 348
Copper tubing, 243
Crack method, 55

Dalton's Law, 163
Dampers, 279, 375
Degree day, 398
  table, 497
Dehumidification, 172, 175
Density, 19
Design:
  air system, 444–453
  hydronic system, 431–444
Design conditions, 57
  inside, 10
  outdoor:
    summer, 135
    winter, 57
  tables, 497–501
Dew point temperature, 163
Diffuse radiation, 130
Diffusers:
  air, 272
  application, 274
  location, 275
  selection, 275
  types, 272
Direct radiation, 130
Diversity, 144
Double bundle condenser, 403
Draft, boiler, 78
Dry bulb temperature, 9, 135, 162
Dual duct system, 309
Duct:
  construction, 253

design chart, 217
equivalent diameter, 218
fitting, 219
friction loss, 219
heat gain, 145
heat loss, 58
leakage, 147, 154
Duct design methods:
  equal friction, 233
  static pressure regain, 234

Effective surface temperature, 189
Efficiency:
  boiler, 83, 97
  engine, 391
  fan, 257
  pump, 287
  solar collector, 455
Electronic air cleaners, 322
Energy, definition, 25
Energy balance, 28
Energy codes and standards, 9, 386
Energy conservation, 14
  boilers, 96
  building construction, 407
  controls, 408
  cooling loads, 158
  design, 10, 417
  fans, 268
  heating loads, 64
  installation, 409
  operations, 409
  pumps, 302
  refrigeration, 361
  systems, 325
Energy efficiency ratio, 394
Energy equation, 28
Energy use, degree day method, 398
Enthalpy:
  air-water vapor mixture, 165
  definition, 27
Entropy, 41
Equal-friction method, 233
Equipment load, 138
Equivalent length, 211
Equivalent round duct, 218
Equivalent units, table, 485
Evaporative condenser, 336
Evaporative cooling, 179
Evaporator, 332

Expansion, pipe, 249
Expansion tank, 297
Expansion valve, 337

Face-and-bypass control, 193, 379
Fan:
  axial, 256
  centrifugal, 256
  characteristics, 257
  connections, 268
  construction, 267
  energy conservation, 268
  installation, 268
  laws, 266
  selection, 258
  types, 256
Fan-coil unit, 109
Filters, 319
Fin-tube, 107
First law of thermodynamics, 28
Fittings:
  duct, 219, 253
  pipe, 244
Flat-plate solar collector:
  construction, 456
  efficiency, 468
  sizing, 471
Floor slab, heat transfer, 53
Flow control devices, refrigerant, 337
Fluid flow:
  continuity equation, 197
  energy equation, 199
  measurement, 422
Fluid statics, 22
Force, 19
Forced convection, 46
Fouling factor, 341
Friction, 40
Friction loss:
  duct, 217
  pipe, 206
Fuels, 80
Furnace, 69

Gage pressure, 21
Gas constant, 39
Grilles, 274

Head, 24
  pump, 286
Heat, definition, 26

Heat gain, 120
  appliances, 138, 152
  duct, 145, 154
  fan, 146, 193
  glass solar, 128, 151
  infiltration, 138, 152
  lights, 135
  people, 137, 152
  procedure, 147, 156
  pump, 147
  ventilation, 138, 152
  walls and roofs, 121, 150
Heating load, 58, 59
Heat loss:
  basement, 51
  infiltration, 54, 59
  piping, 93
  slabs on ground, 53
Heat pipe, 402
Heat pump, 352–358
Heat recovery, 401
Heat storage, 119, 404
  solar, 473
Heat transfer, 45
  coefficient, 49
  conduction, 45
  convection, 45
  radiation, 46
Heat wheel, 401
Hermetic compressor, 333
Hot-water boiler, 73
Hot-water heating system, 100
Human comfort, 7
Humidification, 172, 175
Humidistat, 371
Humidity measurement, 424
Humidity ratio, 163
Hydronic systems, 4, 100

Ideal gas, 39, 163
Indoor air quality, 8, 323
Induction unit, 110
Infiltration, 54
  estimating:
    air change method, 56
    crack method, 55
    residential, cooling, 152
    standards, 55
Insolation, 461
  table, 463

Instrumentation, 416
  electric current, 425
  flow rate, 422
  heat flow, 424
  humidity, 424
  pressure, 419
  sound, 429
  speed, 425
  temperature, 417
  velocity, 420
Insulation:
  duct, 254
  pipe, 252
International system (SI) of units, 18

Latent heat, 34, 36, 38, 54, 120, 154
  of fusion, 39
  process, 175
  of vaporization, 35
Law of partial pressures, 163
Leakage, ducts, 147, 154
Licensing, 14
Lighting load, 135
Loss coefficient, 219

Manometer, 24, 419
Mass, 18
Mixing:
  air, 180
  energy losses, 40
Mixing box, 309
Moist air, 162
Multi-zone system, 308
Multi-zone unit, 308

Natural convection, 46
Net positive suction head, 297
Noise, 280
Noise criteria table, 277

Occupant heat gain, 137
Occupations in HVAC, 12–14
Odors, 323
Outdoor air:
  design conditions, 57, 135, 497–501
  load, 144
  requirements, 145
Overall heat transfer coefficient, 49
  tables, 491–495
Overall thermal transfer value, 386
Ozone depletion, 359

Package boiler, 75
Packaged air-conditioning unit, 315
Partial pressures, law of, 163
Perfect gas, 39
Pipe:
  expansion, 249
  friction charts, 207–209
  installation, 253
  insulation, 252
  pressure loss, 201
  specifications, 241
Pipe sizing, 213
Piping system arrangements, 100
  series loop, 100
  one-pipe main, 102
  two-pipe, 102
  three- and four-pipe, 104
Pitot tube, 203, 421
Plans and specifications, 443, 453
Power, definition, 25
Pressure:
  absolute, 21
  atmospheric, 21
  definition, 20
  gage, 21, 420
  measurement, 419
  partial, 163
  ratings, 73
  static, 202
  total, 202
  vacuum, 22
  velocity, 213
Primary air, 110, 271
Psychrometer, sling, 424
Psychrometric chart, 167, 506–507
Psychrometric processes, 171–182
Pump:
  cavitation, 297
  centrifugal, 285
  characteristics, 286
  construction, 293
  efficiency, 286
  energy conservation, 302
  net positive suction head, 297
  selection, 289
  similarity laws, 293

Radiation, 8, 46, 105
  solar, 460
Radiators, 106

Reciprocating compressors, 333
Recycling refrigerants, 360
Refrigerant, 358
Refrigerant-flow control devices, 337
Refrigeration, 35, 331
Refrigeration cycle:
    absorption, 348
    vapor compression, 331
Registers, 272
Reheat:
    control, 192
    system, 191, 307
Relative humidity, 163
Residential unit, 6
Resistance, thermal, 47
    table, 487
Return air grilles, 280
Rooftop unit, 7, 316
Room condition line, 184
Room sensible heat radio (RSHR),
    183
Room unit, 314
Rotary compressors, 334
Rounding-off data, 20
Run-around coil, 402

Saturated air, 163
Saturated state, 34
Saturated steam, 34, 486
Saturation curve, water, 33
    table, 486
Screw compressor, 334
Scroll compressor, 335
Second law of thermodynamics, 40,
    401
Secondary air, 110, 271
Self-contained unit, 315
Sensible heat, 34, 36, 120
Sensible heat ratio (factor), 183
Shade coefficient, 128
Shading:
    external, 134, 152
    internal, 130, 152
Sick building syndrome, 323
Significant figures, 20
Sling psychrometer, 424
Solar cooling, 459
Solar heat gain factor, 128
Solar heating, 455
    collector types, 455

economic analysis, 472
flat-plate collector, 456
storage, 473
system types, 458
Solar radiation, 128, 460
Solar time, 121
Sound, 280
Specific gravity, 19
Specific heat, 36
Specific humidity, 163
Specific volume, 19, 163
Split system, 315
Standard air conditions, 174
States of matter, 29
Static pressure, 202
Static regain, 204
Static regain method, 234
Steam tables, 486
Subcooled liquid, 35
Sublimation, 39
Superheat, 338
Superheated vapor, 34
Supply air conditions, 182
Supply air outlets, 272
Supports:
    duct, 254
    pipe, 249
Symbols, 483–484
System characteristics:
    duct, 265
    pipe, 291
System effect, 230

Temperature:
    apparatus dew point, 189
    definition, 26
    dew point, 163
    dry bulb, 162
    effective surface, 190
    ratings, 73
    wet bulb, 162
Terminal units, 105
Thermal conductance, 48
Thermal conductivity, 48
Thermal resistance, 47
    table, 486–490
Thermal storage, 404
Thermodynamics:
    definition, 28
    first law, 28, 390

second law, 40, 390
Thermometers, 417
Thermostatic expansion valve, 337
Thermostats, 371
Throttling range, 369
Throw, air, 271
Total energy system, 405
Total pressure, 202
Tube, copper, 242
Turning vanes, 254

Unitary equipment, 314
Unitary system, 7, 314
Unit-heaters, 108
Units, 17
    table, 485
U-tube manometer, 419

Vacuum pressure, 22
Valves, 245
    control, 374
    reversing, 354
Vapor barrier, 252, 254
Variable air-volume (VAV) system,
    311
Velocity:
    air:
        in ducts, 233
        measurement, 420
    water:
        measurement, 424
Velocity pressure, 202
Ventilation, 57, 144
    table, 145
Vibration, pipe, 250
Warm air furnace, 69
Water:
    thermodynamic properties, 486
    treatment, 361
Water chiller, 341
Weather data, 57, 135, 144
    table, 497–501
Weight, 19
Wet bulb temperature, 135, 162, 179
Window air conditioner, 314
Work, 25

Zoning, 305